Lecture Notes in Artificial Intelligence　　9925

Subseries of Lecture Notes in Computer Science

More information about this series at http://www.springer.com/series/1244

Ronald Ortner · Hans Ulrich Simon
Sandra Zilles (Eds.)

Algorithmic Learning Theory

27th International Conference, ALT 2016
Bari, Italy, October 19–21, 2016
Proceedings

 Springer

Editors
Ronald Ortner
Montanuniversität Leoben
Leoben
Austria

Hans Ulrich Simon
Ruhr-Universität Bochum
Bochum
Germany

Sandra Zilles
University of Regina
Regina, SK
Canada

ISSN 0302-9743   ISSN 1611-3349 (electronic)
Lecture Notes in Artificial Intelligence
ISBN 978-3-319-46378-0   ISBN 978-3-319-46379-7 (eBook)
DOI 10.1007/978-3-319-46379-7

Library of Congress Control Number: 2016950899

LNCS Sublibrary: SL7 – Artificial Intelligence

Printed on acid-free paper

This Springer imprint is published by Springer Nature
The registered company is Springer International Publishing AG Switzerland

Preface

This volume contains the papers presented at the 27th International Conference on Algorithmic Learning Theory (ALT 2016). ALT 2016 was co-located with the 19th International Conference on Discovery Science (DS 2016). Both conferences were held during October 19–21 in the beautiful city of Bari, Italy.

The technical program of ALT 2016 had five invited talks (presented jointly to both ALT 2016 and DS 2016) and 24 papers selected from 45 submissions by the ALT Program Committee. ALT is dedicated to the theoretical foundations of machine learning and provides a forum for high-quality talks and scientific interaction in areas such as statistical learning theory, online learning, inductive inference, query models, unsupervised learning, clustering, semi-supervised and active learning, stochastic optimization, high-dimensional and non-parametric inference, exploration–exploitation trade-off, bandit theory, reinforcement learning, planning, control, and learning with additional constraints. ALT is furthermore concerned with the analysis of the theoretical properties of existing algorithms such as boosting, kernel-based methods, SVM, Bayesian methods, graph- or manifold-based methods, methods for latent-variable estimation or clustering, decision tree methods, and information-based methods.

The present volume of LNAI contains the text of the 24 papers presented at ALT 2016 as well as the abstracts of the invited talks:

- Avrim Blum (Carnegie Mellon University, Pittsburgh):
  "Learning About Agents and Mechanisms from Opaque Transactions"
  (Invited talk for ALT 2016)
- Gianluca Bontempi (Interuniversity Institute of Bioinformatics, Brussels):
  "Perspectives of Feature Selection in Bioinformatics: From Relevance to Causal Inference"
  (Invited tutorial for DS 2016)
- Kristian Kersting (Technische Universität Dortmund):
  "Collective Attention on the Web"
  (Invited talk for DS 2016)
- Gábor Lugosi (Pompeu Fabra University, Barcelona):
  "How to Estimate the Mean of a Random Variable"
  (Invited tutorial for ALT 2016)
- John Shawe-Taylor (University College London):
  "Margin-Based Structured Output Learning"
  (Invited talk for ALT 2016 and DS 2016)

Since 1999, ALT has been awarding the E.M. Gold Award for the most outstanding student contribution. This year, the award was given to Areej Costa for her paper "Exact Learning of Juntas from Membership Queries" co-authored with Nader Bshouty.

ALT 2016 was the 27th meeting in the ALT conference series, established in Japan in 1990. The ALT series is supervised by its Steering Committee: Shai Ben-David (University of Waterloo, Canada), Marcus Hutter (Australian National University, Canberra, Australia), Sanjay Jain (National University of Singapore, Republic of Singapore), Ronald Ortner (Montanuniversität Leoben, Austria), Hans U. Simon (Ruhr-Universität Bochum, Germany), Frank Stephan (National University of Singapore, Republic of Singapore), Csaba Szepesvári (University of Alberta, Edmonton, Canada), Eiji Takimoto (Kyushu University, Fukuoka, Japan), Akihiro Yamamoto (Kyoto University, Japan), and Sandra Zilles (Chair, University of Regina, Canada).

We thank the following people and institutions who contributed to the success of the conference. Most importantly, we would like to thank the authors for contributing and presenting their work at the conference. Without their contribution this conference would not have been possible. We are very grateful to the Fondazione Puglia and to the Consorzio Interuniversitario Nazionale per l'Informatica (National Interuniversity Consortium for Informatics, CINI) for their financial support. We would also like to acknowledge the support of the European Commission through the project MAESTRA — Learning from Massive, Incompletely Annotated, and Structured Data (grant number ICT-2013-612944).

ALT 2016 and DS 2016 were organized by the University of Bari A. Moro. We thank the local arrangements chairs, Annalisa Appice, Corrado Loglisci, Gianvito Pio, Roberto Corizzo, and their team for their efforts in organizing the two conferences.

We are grateful for the collaboration with the conference series Discovery Science. In particular, we would like to thank the general chair of DS 2016 and ALT 2016, Donato Malerba, and the DS 2016 Program Committee chairs, Toon Calders and Michelangelo Ceci.

We are also grateful to EasyChair, the excellent conference management system, which was used for putting together the program for ALT 2016. EasyChair was developed mainly by Andrei Voronkov and is hosted at the University of Manchester. The system is free of charge.

We are grateful to the members of the Program Committee for ALT 2016 and the additional reviewers for their hard work in selecting a good program for ALT 2016. Special thanks go to Frank Stephan from the National University of Singapore for maintaining the ALT website. Last but not the least, we thank Springer for their support in preparing and publishing this volume in the *Lecture Notes in Artificial Intelligence* series.

July 2016   Ronald Ortner
                                                                          Hans U. Simon
                                                                          Sandra Zilles

Organization

Conference Chair

Donato Malerbo   University of Bari A. Moro, Italy

Program Committee

Nir Ailon	Israel Institute of Technology, Haifa, Israel
Dana Angluin	Yale University, USA
Peter Bartlett	UC Berkeley, USA and Queensland University of Technology, Brisbane, Australia
Shai Ben-David	University of Waterloo, Canada
Alina Beygelzimer	Yahoo! Labs, New York, USA
Corinna Cortes	Google Research, New York, USA
Malte Darnstädt	Continentale Insurance Group, Dortmund, Germany
Sanjoy Dasgupta	UC San Diego, USA
Rong Ge	Duke University, Durham, USA
Steve Hanneke	Princeton, USA
Kohei Hatano	Kyushu University, Fukuoka, Japan
Jyrki Kivinen	University of Helsinki, Finland
Wouter M. Koolen	Centrum Wiskunde & Informatica, Amsterdam, The Netherlands
Mehrdad Mahdavi	Toyota Technical Institute, Chicago, USA
Eric Martin	University of New South Wales, Sydney, Australia
Hanna Mazzawi	Google Research, Mountain View, California, USA
Mehryar Mohri	Courant Institute of Mathematical Sciences, New York, USA
Ronald Ortner (Chair)	Montanuniversität Leoben, Austria
Hans U. Simon (Chair)	Ruhr-Universität Bochum, Germany
Frank Stephan	National University of Singapore, Republic of Singapore
Gilles Stoltz	GREGHEC: HEC Paris - CNRS, France
Csaba Szepesvári	University of Alberta, Edmonton, Canada
Balázs Szörényi	Israel Institute of Technology, Haifa, Israel
György Turán	University of Illinois at Chicago, USA
Liwei Wang	Peking University, China
Akihiro Yamamoto	Kyoto University, Japan
Sandra Zilles	University of Regina, Canada

Local Arrangements

Annalisa Appice   University of Bari A. Moro, Italy
Roberto Corizzo   University of Bari A. Moro, Italy
Corrado Loglisci   University of Bari A. Moro, Italy
Gianvito Pio   University of Bari A. Moro, Italy

Additional Reviewers

Giulia Desalvo   David Pal
Christos Dimitrakakis   Bernardo Ávila Pires
András György   Lev Reyzin
Elad Hoffer   Afshin Rostamizadeh
Prateek Jain   Vikas Sindhwani
Pooria Joulani   Marta Soare
Sumeet Katariya   Daiki Suehiro
Vitaly Kuznetsov   Ambuj Tewari
Tor Lattimore   Scott Yang
Guillaume Lecué   Felix Yu
Alan Malek   Huizhen Yu

Sponsoring Institutions

Fondazione Puglia
Consorzio Interuniversitario Nazionale per l'Informatica, CINI
European Commission through the project MAESTRA (*Learning from Massive, Incompletely Annotated, and Structured Data*), Grant number ICT-2013-612944

Abstract of Invited Talks

Learning about Agents and Mechanisms from Opaque Transactions

Avrim Blum

School of Computer Science, Carnegie Mellon University,
Pittsburgh, PA 15213–3891, USA
avrim@cs.cmu.edu

In this talk I will discuss some learning problems coming from the area of algorithmic economics. I will focus in particular on settings known as combinatorial auctions in which agents have preferences over items or sets of items, and interact with an auction or allocation mechanism that determines what items are given to which agents. We consider the perspective of an outside observer who each day can only see which agents show up and what they get, or perhaps just which agents' needs are satisfied and which are not. Our goal will be from observing a series of such interactions to try to learn the agent preferences and perhaps also the rules of the allocation mechanism.

As an example, consider observing web pages where the agents are advertisers and the winners are those whose ads show up on the given page. Or consider observing the input-output behavior of a cloud computing service, where the input consists of a set of agents requesting service, and the output is a partition of them into some whose requests are actually fulfilled and the rest that are not—due to overlap of their resource needs with higher-priority requests. From such input-output behavior, we would like to learn the underlying structure. We also consider a classic Myerson single-item auction setting, where from observing who wins and also being able to participate ourselves we would like to learn the agents' valuation distributions.

In examining these problems we will see connections to decision-list learning and to Kaplan-Meier estimators from medical statistics.

This talk is based on work joint with Yishay Mansour and Jamie Morgenstern.

Perspectives of Feature Selection in Bioinformatics: From Relevance to Causal Inference

Gianluca Bontempi

Machine Learning Group, Interuniversity
Institute of Bioinformatics in Brussels (IB)^2
Université libre de Bruxelles, Bld de Triomphe, 1050 Brussels, Belgium
mlg.ulb.ac.be

A major goal of the scientific activity is to model real phenomena by studying the dependency between entities, objects or more in general variables. Sometimes the goal of the modeling activity is simply predicting future behaviors. Sometimes the goal is to understand the causes of a phenomenon (e.g. a disease). Finding causes from data is particular challenging in bioinformatics where often the number of features (e.g. number of probes) is huge with respect to the number of samples [5]. In this context, even when experimental interventions are possible, performing thousands of experiments to discover causal relationships between thousands of variables is not practical. Dimensionality reduction techniques have been largely discussed and used in bioinformatics to deal with the curse of dimensionality. However, most of the time these techniques focus on improving prediction accuracy, neglecting causal aspects. This tutorial will introduce some basics of causal inference and will discuss some open issues: may feature selection techniques be useful also for causal feature selection? Is prediction accuracy compatible with causal discovery [2]? How to deal with Markov indistinguishable settings [1]? Recent results based on information theory [3], and some learned lessons from a recent Kaggle competition [4] will be used to illustrate the issue.

References

1. Bontempi, G., Flauder, M.: From dependency to causality: a machine learning approach. JMLR **15**(16), 2437–2457 (2015)
2. Bontempi, G., Haibe-Kains, B., Desmedt, C., Sotiriou, C., Quackenbush, J.: Multiple-input multiple-output causal strategies for gene selection. BMC Bioinformatics **12**(1), 458 (2011)
3. Bontempi, G., Meyer, P.E.: Causal filter selection in microarray data. In: Proceedings of ICML (2010)
4. Bontempi, G., Olsen, C., Flauder, M.: D2C: predicting causal direction from dependency features (2014). R package version 1.1
5. Meyer, P.E., Bontempi, G.: Information-theoretic gene selection in expression data. In: Biological Knowledge Discovery Handbook. IEEE Computer Society (2014)

Margin Based Structured Output Learning

John Shawe-Taylor

Department of Computer Science, CSML,
University College London, London WC1E 6EA, UK
j.shawe-taylor@cs.ucl.ac.uk

Structured output learning has been developed to borrow strength across multi-dimensional classifications. There have been approaches to bounding the performance of these classifiers based on different measures such as microlabel errors with a fixed simple output structure. We present a different approach and analysis starting from the assumption that there is a margin attainable in some unknown or fully connected output structure. The analysis and algorithms flow from this assumption but in a way that the associated inference becomes tractable while the bounds match those attained were we to use the full structure. There are two variants depending on how the margin is estimated. Experimental results show the relative strengths of these variants, both algorithmically and statistically.

Collective Attention on the Web

Kristian Kersting

Computer Science Department, TU Dortmund University,
44221 Dortmund, Germany
Kristian.Kersting@cs.tu-dortmund.de

It's one of the most popular YouTube videos ever produced, having been viewed more than 840 million times. Its hard to understand why this clip is so famous and actually went viral, since nothing much happens. Two little boys, Charlie and Harry, are sitting in a chair when Charlie, the younger brother, mischievously bites Harrys finger. There's a shriek and then a laugh. The clip is called "Charlie Bit My Finger–Again!"

Generally, understanding the dynamics of collective attention is central to an information age where millions of people leave digital footprints everyday. So, can we capture the dynamics of collective attention mathematically? Can we even gain insights into the underlying physical resp. social processes? Is it for instance fair to call the video "viral" in an epidemiological sense?

In this talk I shall argue that computational methods of collective attention are not insurmountable. I shall review the methods we have developed to characterize, analyze, and even predict the dynamics of collective attention among millions of users to and within social media services. For instance, we found that collective attention to memes and social media grows and subsides in a highly regular manner, well explained by economic diffusion models [2, 4]. Using mathematical epidemiology, we find that so-called viral videos show very high infection rates and, hence, should indeed be called viral [1]. Moreover, the spreading processes may also be related to the underlying network structures, suggesting for instance a physically plausible model of the distance distributions of undirected networks [3]. All this favors machine learning and discovery science approaches that produce physically plausible models.

This work was partly supported by the Fraunhofer ICON project SoFWIReD and by the DFG Collaborative Research Center SFB 876 project A6.

References

1. Bauckhage, C., Hadiji, F., Kersting, K.: How viral are viral videos? In: Proceedings of the Ninth International Conference on Web and Social Media (ICWSM 2015), pp. 22–30 (2015)
2. Bauckhage, C., Kersting, K., Hadiji, F.: Mathematical models of fads explain the temporal dynamics of internet memes. In: Proceedings of the Seventh International Conference on Weblogs and Social Media (ICWSM 2013) (2013)
3. Bauckhage, C., Kersting, K., Hadiji, F.: Parameterizing the distance distribution of undirected networks. In: Proceedings of the Thirty-First Conference on Uncertainty in Artificial Intelligence (UAI 2015), pp. 121–130 (2015)

4. Bauckhage, C., Kersting, K., Rastegarpanah, B.: Collective attention to social media evolves according to diffusion models. In: 23rd International World Wide Web Conference (WWW 2014), pp. 223–224 (2014)

How to Estimate the Mean of a Random Variable?

Gabor Lugosi

Department of Economics, Pompeu Fabra University,
Ramon Trias Fargas 25-27, 08005, Barcelona, Spain
gabor.lugosi@gmail.com

Given n independent, identically distributed copies of a random variable, one is interested in estimating the expected value. Perhaps surprisingly, there are still open questions concerning this very basic problem in statistics. In this talk we are primarily interested in non-asymptotic sub-Gaussian estimates for potentially heavy-tailed random variables. We discuss various estimates and extensions to high dimensions. We apply the estimates for statistical learning and regression function estimation problems. The methods improve on classical empirical minimization techniques.

This talk is based on joint work with Emilien Joly, Luc Devroye, Matthieu Lerasle, Roberto Imbuzeiro Oliveira, and Shahar Mendelson.

Contents

Online Learning

Bandits and Reinforcement Learning

Clustering

Error Bounds, Sample Compression Schemes

A Vector-Contraction Inequality for Rademacher Complexities

Andreas Maurer[(✉)]

Adalbertstr. 55, 80799 Munich, Germany
am@andreas-maurer.eu

Abstract. The contraction inequality for Rademacher averages is extended to Lipschitz functions with vector-valued domains, and it is also shown that in the bounding expression the Rademacher variables can be replaced by arbitrary iid symmetric and sub-gaussian variables. Example applications are given for multi-category learning, K-means clustering and learning-to-learn.

Keywords: Rademacher complexities · Contraction inequality

1 Introduction

The method of Rademacher complexities has become a popular tool to prove generalization in learning theory. One has the following result [1], which gives a bound on the estimation error, uniform over a loss class $\mathcal{F}$.

Theorem 1. *Let $\mathcal{X}$ be any set, $\mathcal{F}$ a class of functions $f : \mathcal{X} \to [0,1]$ and let $X, X_1, \ldots, X_n$ be iid random variables with values in $\mathcal{X}$. Then for $\delta > 0$ with probability at least $1 - \delta$ in $\mathbf{X} = (X_1, \ldots, X_n)$ we have for every $f \in \mathcal{F}$ that*

$$\mathbb{E}f(X) \leq \frac{1}{n} \sum f(X_i) + \frac{2}{n} \mathbb{E}\left[\sup_{f \in \mathcal{F}} \sum_{i=1}^{n} \epsilon_i f(X_i) \,|\, \mathbf{X}\right] + \sqrt{\frac{9 \ln 2/\delta}{2n}}.$$

Here the $\epsilon_1, \ldots, \epsilon_n$ are (and will be throughout this paper) independent Rademacher variables, uniformly distributed on $\{-1, 1\}$. For any class $\mathcal{F}$ of real, not necessarily $[0,1]$-valued, functions defined on $\mathcal{X}$, and any vector $\mathbf{x} = (x_1, \ldots, x_n) \in \mathcal{X}^n$, the quantity

$$\mathbb{E} \sup_{f \in \mathcal{F}} \sum_{i=1}^{n} \epsilon_i f(x_i)$$

is the Rademacher complexity of the class $\mathcal{F}$ on the sample $x = (x_1, \ldots, x_n) \in \mathcal{X}^n$. Here we omit the customary factor $2/n$, as this will simplify most of our statements below.

Most applications of the method at some point or another use the so-called contraction inequality. For functions $h_i : \mathbb{R} \to \mathbb{R}$ with Lipschitz constant L, the scalar contraction inequality states that

© Springer International Publishing Switzerland 2016
R. Ortner et al. (Eds.): ALT 2016, LNAI 9925, pp. 3–17, 2016.
DOI: 10.1007/978-3-319-46379-7_1

$$\mathbb{E} \sup_{f \in \mathcal{F}} \sum_{i=1}^{n} \epsilon_i h_i \left(f \left(x_i \right) \right) \leq L \mathbb{E} \sup_{f \in \mathcal{F}} \sum_{i=1}^{n} \epsilon_i f \left(x_i \right).$$

There are situations when it is desirable to extend this result to the case when the class $\mathcal{F}$ consists of vector-valued functions and the loss functions are Lipschitz functions defined on a more than one-dimensional space. Such occurs for example in the analysis of multi-class learning, K-means clustering or learning-to-learn. At present one has dealt with these problems by passing to Gaussian averages and using Slepian's inequality (see e.g. Theorem 14 in [1]). This is sufficient for many applications, but there are two drawbacks: 1. the proof relies on a highly nontrivial result (Slepian's inequality) and 2. while Rademacher complexities are tightly bounded in terms of Gaussian complexities, it is well known ([4,12]) that bounding the latter in terms of the former incurs a factor logarithmic in the number of variables, potentially resulting in an unnecessary weakening of the results (see e.g. [13]).

In this paper we will prove the vector contraction inequality

$$\mathbb{E} \sup_{f \in \mathcal{F}} \sum_{i=1}^{n} \epsilon_i h_i \left(f \left(x_i \right) \right) \leq \sqrt{2} L \mathbb{E} \sup_{f \in \mathcal{F}} \sum_{i=1}^{n} \sum_{k=1}^{K} \epsilon_{ik} f_k \left(x_i \right), \tag{1}$$

where the members of $\mathcal{F}$ take values in $\mathbb{R}^K$ with component functions $f_k \left(\cdot \right)$, the h_i are L-Lipschitz functions from $\mathbb{R}^K$ (with Euclidean norm) to $\mathbb{R}$, and the ϵ_{ik} are an $n \times K$ matrix of independent Rademacher variables. It is also shown that the ϵ_{ik} on the right hand side of (1) can be replaced by arbitrary iid random variables as long as they are symmetric and sub-gaussian, and $\sqrt{2}$ is replaced by a suitably chosen constant. Furthermore the result extends to infinite dimensions in the sense that $\mathbb{R}^K$ can be replaced by the Hilbert space ℓ_2. The proof given is self-contained and independent of Slepian's inequality.

We illustrate applications of this inequality by showing that it applies to loss functions in a variety of relevant cases. In Sect. 3 we discuss multi-class learning, K-means clustering and learning-to-learn. We also give some indications of how the vector-valued complexity on the right hand side of (1) may be bounded. An example pertaining to the truly infinite dimensional case is given, generalizing some bounds for least-squares regression with operator valued kernels ([5,19]) to more general loss-functions.

Inequality (1) is perhaps not the most natural form of a vector-contraction inequality, and, since the right hand side is sometimes difficult to bound, one is led to look for alternatives. An attractive conjecture might be the following.

Conjecture 1. Let $\mathcal{X}$ be any set, $n \in \mathbb{N}$, $(x_1, ..., x_n) \in X^n$, let $\mathcal{F}$ be a class of functions $f : \mathcal{X} \to \ell_2$ and let $h : \ell_2 \to \mathbb{R}$ have Lipschitz norm L. Then

$$\mathbb{E} \sup_{f \in \mathcal{F}} \sum_i \epsilon_i h \left(f \left(x_i \right) \right) \leq c L \mathbb{E} \sup_{f \in \mathcal{F}} \left\| \sum_i \epsilon_i f \left(x_i \right) \right\|,$$

where c is some universal constant.

This conjecture is false and will be disproved in the sequel.

A version of the scalar contraction inequality occurs in [12], Theorem 4.12. There the absolute value of the Rademacher sum is used, and a necessary factor of two appears on right hand side. With the work of Koltchinskii and Panchenko [11] and Bartlett and Mendelson [1] Rademacher averages became attractive to the machine learning community, there was an increased interest in contraction inequalities and it was realized that the absolute value was unnecessary for most of the new applications. Meir and Zhang [18] gave a nice and simple proof of the scalar contraction inequality as stated above. Our proof of (1) is an extension of their method.

2   The Vector-Contraction Inequality

All random variables are assumed to be defined on some probability space (Ω, Σ). The space $L_p(\Omega, \Sigma)$ is abbreviated L_p. We use ℓ_2 to denote the Hilbert space of square summable sequences of real numbers. The norm on ℓ_2 and the Euclidean norm on $\mathbb{R}^K$ are denoted with $\|.\|$.

A real random variable X is called *symmetric* if $-X$ and X are identically distributed. It is called *sub-gaussian* if there exists a constant $b = b(X)$ such that for every $\lambda \in \mathbb{R}$

$$\mathbb{E}e^{\lambda X} \leq e^{\frac{\lambda^2 b^2}{2}}.$$

We call b the *sub-gaussian parameter* of X. Rademacher and standard normal variables are symmetric and sub-gaussian.

The following is the main result of this paper.

Theorem 2. *Let X be not a.s. equal to zero, symmetric and subgaussian. Then there exists a constant $C < \infty$, depending only on the distribution of X, such that for any countable set S and functions $\psi_i : S \to \mathbb{R}$, $\phi_i : S \to \ell_2$, $1 \leq i \leq n$ satisfying*

$$\forall s, s' \in S, \ \psi_i(s) - \psi_i(s') \leq \|\phi_i(s) - \phi_i(s')\|$$

we have

$$\mathbb{E} \sup_{s \in S} \sum_i \epsilon_i \psi_i(s) \leq C \ \mathbb{E} \sup_{s \in S} \sum_{i,k} X_{ik} \phi_i(s)_k,$$

where the X_{ik} are independent copies of X for $1 \leq i \leq n$ and $1 \leq k \leq \infty$, and $\phi_i(s)_k$ is the k-th coordinate of $\phi_i(s)$.

If X is a Rademacher variable we may choose $C = \sqrt{2}$, if X is standard normal $C = \sqrt{\pi/2}$.

For applications in learning theory we can at once substitute a Rademacher variable for X and $\sqrt{2}$ for C. For S we take a class $\mathcal{F}$ of vector-valued functions $f : \mathcal{X} \to \ell_2$, for the ϕ_i the evaluation functionals on a sample $(x_1, ..., x_n)$, so that $\phi_i(f) = f(x_i)$ and for ψ_i we take the evaluation functionals composed with Lipschitz loss function $h_i : \ell_2 \to \mathbb{R}$ of Lipschitz norm L, scaled by $1/L$. We obtain the following corollary.

Corollary 1. *Let $\mathcal{X}$ be any set, $(x_1, ..., x_n) \in \mathcal{X}^n$, let $\mathcal{F}$ be a class of functions $f : \mathcal{X} \to \ell_2$ and let $h_i : \ell_2 \to \mathbb{R}$ have Lipschitz norm L. Then*

$$\mathbb{E} \sup_{f \in \mathcal{F}} \sum_i \epsilon_i h_i \left(f\left(x_i \right) \right) \leq \sqrt{2} L \mathbb{E} \sup_{f \in \mathcal{F}} \sum_{i,k} \epsilon_{ik} f_k \left(x_i \right),$$

where ϵ_{ik} is an independent doubly indexed Rademacher sequence and $f_k \left(x_i \right)$ is the k-th component of $f \left(x_i \right)$.

Clearly finite dimensional versions are obtained by restricting to the subspace spanned by the first K coordinate functions in ℓ_2.

3   Examples of Loss Functions

We give some examples of seemingly complicated loss functions to which Theorem 2 and Corollary 1 can be applied. These examples are not exhaustive, in fact it seems that many applications of Slepian's inequality in the machine learning literature can be circumvented by Theorem 2 (see also [6,7,17] for applications to information retrieval and generalization of autoregressive models).

3.1   Multi-class Classification

Consider the problem of assigning to inputs taken from a space $\mathcal{X}$ a label corresponding to one of K classes. We are given a labelled iid sample $\mathbf{z} = \left(\left(x_1, y_1 \right), ..., \left(x_n, y_n \right) \right)$ drawn from some unknown distribution on $\mathcal{X} \times \{1, ..., K\}$, where the points x_i are inputs $x_i \in \mathcal{X}$ and the y_i are corresponding labels, $y_i \in \{1, ..., K\}$. Many approaches assume that there is class $\mathcal{F}$ of vector valued functions $f : \mathcal{X} \to \mathbb{R}^{K'}$, where $K' = O(K)$ (typically $K' = K$, for 1-versus-all classification, or $K' = K - 1$ for simplex coding [20]), a classification rule $c : \mathbb{R}^{K'} \to \{1, ..., K\}$, and for each label $k \in \{1, ..., K\}$ a loss function $\ell_k : \mathbb{R}^K \to \mathbb{R}_+$. The loss function ℓ_k is designed so as to upper bound, or approximate the indicator function of the set $\left\{ z \in \mathbb{R}^{K'} : c(z) \neq k \right\}$ (see [9,13]). In most cases the loss functions are Lipschitz on $\mathbb{R}^K$ relative to the euclidean norm, with some Lipschitz constant L. The empirical error incurred by a function $f \in \mathcal{F}$ is

$$\frac{1}{n} \sum_i \ell_{y_i} \left(f \left(x_i \right) \right).$$

The Rademacher complexity, which would lead to the uniform bound on the estimation error, is

$$\mathbb{E} \sup_{f \in \mathcal{F}} \sum_i \epsilon_i \ell_{y_i} \left(f \left(x_i \right) \right).$$

Using Corollary 1 with $h_i = \ell_{y_i}$ we can immediately eliminate the loss functions ℓ_{y_i}

$$\mathbb{E} \sup_{f \in \mathcal{F}} \sum_i \epsilon_i \ell_{y_i} \left(f \left(x_i \right) \right) \leq \sqrt{2} L \mathbb{E} \sup_{f \in \mathcal{F}} \sum_{i,k} \epsilon_{ik} f_k \left(x_i \right).$$

How we proceed to further bound this now depends on the nature of the vector-valued class $\mathcal{F}$. Some techniques to bound the Rademacher complexity of vector valued classes are sketched in Sect. 4 below.

3.2 K-Means Clustering

Let H be a Hilbert space and $\mathbf{x} = (x_1, ..., x_n)$ a sample of points in the unit ball B_1 of H. The algorithm seeks centers $c = (c_1, ..., c_K) \in \mathcal{S} = B_1^K$ to represent the sample.

$$c^* = \arg \min_{(c_1,...,c_K) \in \mathcal{S}} \frac{1}{n} \sum_{i=1}^n \min_{k=1}^K \|x_i - c_k\|^2 .$$

The corresponding Rademacher average to bound the estimation error is

$$R(\mathcal{S}, \mathbf{x}) = \mathbb{E} \sup_{c \in \mathcal{S}} \sum_i \epsilon_i \min_{k=1}^K \|x_i - c_k\|^2 = \mathbb{E} \sup_{c \in \mathcal{S}} \sum_i \epsilon_i \psi_i(c),$$

where we define $\psi_i(c) = \min_k \|x_i - c_k\|^2$ in preparation of an application of Theorem 2. The next step is to search for an appropriate Lipschitz constant of the ψ_i. We have, for $c, c' \in \mathcal{S}$,

$$\psi_i(c) - \psi_i(c') = \min_k \|x_i - c_k\|^2 - \min_k \|x_i - c'_k\|^2$$

$$\leq \max_k \|x_i - c_k\|^2 - \|x_i - c'_k\|^2$$

$$\leq \left(\sum_k \left(\|x_i - c_k\|^2 - \|x_i - c'_k\|^2 \right)^2 \right)^{1/2}$$

$$= \|\phi_i(c) - \phi_i(c')\|.$$

Where we defined $\phi_i : \mathcal{S} \to \mathbb{R}^K$ by $\phi_i(c) = \left(\|x_i - c_1\|^2, ..., \|x_i - c_K\|^2 \right)$. We can now apply Theorem 2 with $L = 1$ and obtain

$$2^{-1/2} R(\mathcal{S}, \mathbf{x}) \leq \mathbb{E} \sup_{c \in \mathcal{S}} \sum_{i,k} \epsilon_{ik} \|x_i - c_k\|^2$$

$$\leq 2 \mathbb{E} \sup_{c \in \mathcal{S}} \sum_{i,k} \epsilon_{ik} \langle x_i, c_k \rangle + \mathbb{E} \sup_{c \in \mathcal{S}} \sum_{i,k} \epsilon_{ik} \|c_k\|^2$$

$$\leq K \left(2 \mathbb{E} \left\| \sum_i \epsilon_i x_i \right\| + \mathbb{E} \left| \sum_i \epsilon_i \right| \right)$$

$$\leq 3K\sqrt{n}.$$

Dividing by n we obtain generalization bounds as in [3] or [15]. In this simple case it was very easy to explicitly bound the complexity of the vector-valued class.

3.3 Learning to Learn or Meta-Learning

With input space $\mathcal{X}$ suppose we have a class $\mathcal{H}$ of feature maps $h : \mathcal{X} \to \mathcal{Y} \subseteq \mathbb{R}^K$ and a loss class $\mathcal{F}$ of functions $f : \mathcal{Y} \to [0, 1]$. The loss class could be used for classification or function estimation or also in some unsupervised setting. We assume that every function $f \in \mathcal{F}$ is Lipschitz with Lipschitz constant L and that $\mathcal{F}$ is small enough for good generalization in the sense that for some $B < \infty$

$$\mathbb{E}_{Y_1,\ldots,Y_n} \left[\sup_{f \in \mathcal{F}} \mathbb{E}_Y f(Y) - \frac{1}{n} \sum_{i=1}^n f(Y_i) \right] \leq \frac{B}{\sqrt{n}} \qquad (2)$$

for any $\mathcal{Y}$-valued random variable Y and iid copies $Y_1, \ldots, Y_n$. Such conditions might be established using standard techniques, for example also Rademacher complexities.

We now want to learn a feature map $h \in \mathcal{H}$, such that empirical risk minimization (ERM) with the function class $\mathcal{F} \circ h = \{x \mapsto f(h(x)) : f \in \mathcal{F}\}$ gives good results on future, yet unseen, tasks. Of course this depends on the tasks in question, and a good feature map h can only be chosen on the basis of some kind of experience made with these tasks.

To formalize this Baxter [2] has introduced the notion of an *environment* η, which is a distribution on the set of tasks, where each task t is characterized by some distribution μ_t (e.g. on inputs and outputs). For each task $t \sim \eta$ we can then also draw an iid training sample $\mathbf{x}^t = (x_1^t, \ldots, x_n^t) \sim \mu_t^n$. In this way the environment also induces a distribution on the set of training samples. Now we can make our problem more precise.

Suppose we have T tasks and corresponding training samples $\bar{\mathbf{x}} = (\mathbf{x}^1, \ldots, \mathbf{x}^T)$ drawn iid from the environment η. For $h \in \mathcal{H}$ let

$$\psi_t(h) = \min_{f \in \mathcal{F}} \frac{1}{n} \sum_{i=1}^n f\left(h\left(x_i^t\right)\right)$$

be the training error obtained by the use of the feature map h. We propose to use the feature map

$$\hat{h} = \arg\min_{h \in \mathcal{H}} \frac{1}{T} \sum_{t=1}^T \psi_t(h).$$

To give a performance guarantee for ERM using $\mathcal{F} \circ \hat{h}$, we now seek to bound the expected *training error* $\mathbb{E}_{t \sim \eta}[\psi_t(h)]$ for a new task drawn from the environment (with corresponding training sample), in terms of the average of the observed $\psi_t(h)$, uniformly over the set of feature maps $h \in \mathcal{H}$. Observe that, given the bound on (2) such a bound will also give a bound on the expected true error when using $\hat{h}$ on new tasks in the environment η, a meta-generalization bound, so to speak (for more details on this type of argument see [14] or [16]).

The Rademacher average in question is

$$R(\mathcal{H}, \bar{\mathbf{x}}) = \mathbb{E} \sup_{h \in \mathcal{H}} \sum_{t=1}^T \epsilon_t \psi_t(h).$$

To apply Theorem 2 we look for a Lipschitz property of the ψ_t. For $h \in \mathcal{H}$ define $\phi_t(h) \in \mathbb{R}^{K \times n}$ by $[\phi_t(h)]_{k,i} = h_k(x_i^t)$. Then for $h, h' \in \mathcal{H}$

$$\psi_t(h) - \psi_t(h') \le \max_{f \in \mathcal{F}} \frac{1}{n} \sum_{i=1}^n f(h(x_i)) - f(h'(x_i))$$

$$\le \frac{L}{n} \sum_{i=1}^n \|h(x_i) - h'(x_i)\| \le \frac{L}{\sqrt{n}} \|\phi_t(h) - \phi_t(h')\|,$$

where the first inequality comes from the Lipschitz property of the functions in $\mathcal{F}$ and the second from Jensen's inequality. From Theorem 2 we conclude that

$$R(\mathcal{H}, \bar{\mathbf{x}}) \le \frac{L}{\sqrt{n}} \mathbb{E} \sup_{h \in \mathcal{H}} \sum_{tki} \epsilon_{tki} h_k(x_i^t).$$

How to proceed depends on the nature of the feature maps in $\mathcal{H}$. Examples are given in [14] or [16], but see also the next section.

4 Bounding the Rademacher Complexity of Vector-Valued Classes

At first glance the expression

$$\mathbb{E} \sup_{f \in \mathcal{F}} \sum_{i,k} \epsilon_{ik} f_k(x_i)$$

appears difficult to bound. Nevertheless there are some general techniques which can be used, such as the reduction to scalar classes, or the use of duality for linear classes. We also give an example in a truly infinite dimensional setting.

4.1 Reduction to Component Classes

Suppose $\mathcal{F}_1, ..., \mathcal{F}_K$ are classes of scalar valued functions and define a vector-valued class $\prod_k \mathcal{F}_k$ with values in $\mathbb{R}^K$ by

$$\prod_k \mathcal{F}_k = \{x \mapsto (f_1(x), ..., f_K(x)) : f_k \in \mathcal{F}_k\}.$$

Then, since the constraints are independent, the Rademacher average of the product class

$$\mathbb{E} \sup_{f \in \prod_k \mathcal{F}_k} \sum_{i,k} \epsilon_{ik} f_k(x_i) = \sum_k \mathbb{E} \sup_{f \in \mathcal{F}_k} \sum_i \epsilon_i f(x_i) \qquad (3)$$

is just the sum of the Rademacher averages of the scalar valued component classes. Now let $\mathcal{F}$ be any function class with values in $\mathbb{R}^K$ and for $k \in \{1, ..., K\}$ define a scalar-valued class $\mathcal{F}_k$ by

$$\mathcal{F}_k = \{x \mapsto f_k(x) : f = (f_1, ..., f_k, ..., f_K) \in \mathcal{F}\}.$$

Then $\mathcal{F} \subseteq \prod_k \mathcal{F}_k$, so by the identity (3)

$$\mathbb{E} \sup_{f \in \mathcal{F}} \sum_{i,k} \epsilon_{ik} f_k (x_i) \leq \sum_k \mathbb{E} \sup_{f \in \mathcal{F}_k} \sum_i \epsilon_i f (x_i).$$

This is loose in many interesting cases, but for product classes it is unimprovable.

4.2   Linear Classes Defined by Norms

Let H be a separable real Hilbert-space and let $\mathcal{B}\left(H, \mathbb{R}^K\right)$ be the set of bounded linear transformations from H to $\mathbb{R}^K$. Then every member of $\mathcal{B}\left(H, \mathbb{R}^K\right)$ is characterized by a sequence of weight vectors $(w_1, ..., w_K)$ with $w_k \in H$. Let $||| \cdot |||$ be a norm on $\mathcal{B}\left(H, \mathbb{R}^K\right)$ with dual norm $||| \cdot |||_*$. Fix some real number B, and define a class $\mathcal{F}$ of functions from H to $\mathbb{R}^K$ by

$$\mathcal{F} = \left\{ x \mapsto Wx : W \in \mathcal{B}\left(H, \mathbb{R}^K\right), |||W||| \leq B \right\}.$$

Then

$$\mathbb{E} \sup_{f \in \mathcal{F}} \sum_{i,k} \epsilon_{ik} f_k (x_i) = \mathbb{E} \sup_{|||(w_1,...,w_K)||| \leq B} \sum_k \left\langle w_k, \sum_i \epsilon_{ik} x_i \right\rangle$$

$$= \mathbb{E} \sup_{|||W||| \leq B} \operatorname{tr}(D^*W) \leq B \, \mathbb{E} |||D^*|||_*,$$

where $D \in \mathcal{B}\left(H, \mathbb{R}^K\right)$ is the random transformation

$$v \mapsto \left(\left\langle v, \sum_i \epsilon_{i1} x_i \right\rangle, ..., \left\langle v, \sum_i \epsilon_{iK} x_i \right\rangle \right).$$

The details of bounding $\mathbb{E}|||D^*|||_*$ then depend on the nature of the norm $||| \cdot |||$. The simplest case is the Hilbert-Schmidt or Frobenius norm, where

$$\mathbb{E}|||D^*|||_* = \mathbb{E}\sqrt{\sum_k \left\| \sum_i \epsilon_{ik} x_i \right\|^2} \leq \sqrt{K \sum_i \|x_i\|^2}.$$

More interesting are mixed norms or the trace norm. A valuable reference for this approach is [10].

4.3   Operator Valued Kernels

We give an example in a truly infinite dimensional setting and refer to the mechanism of learning vector valued functions as exposed in [19]. There is a generic separable Hilbert space H and a kernel $\kappa : \mathcal{X} \times \mathcal{X} \to \mathcal{L}(H)$ satisfying certain positivity and regularity properties as described in [19], where $\mathcal{X}$ is some arbitrary input space. Then there exists an induced feature-map $\Phi : \mathcal{X} \to \mathcal{L}(\ell_2, H)$ such that the kernel is given by

$$\kappa(x, y) = \Phi(x) \Phi^*(y)$$

and the class of H-valued functions to be learned is

$$\{x \mapsto f_w(x) = \Phi(x)w : \|w\| \le B\},$$

where also $\|f_w(x)\|_H = \|w\|\sqrt{\kappa(x,x)}$. Then for any sample $\mathbf{x} = (x_1, ..., x_n) \in \mathcal{X}^n$ and L-Lipschitz loss functions $h_i : H \to \mathbb{R}$ we have

$$\mathbb{E} \sup_{\|w\| \le B} \sum_i \epsilon_i h_i(\Phi(x_i)w) \le \sqrt{2}L\mathbb{E} \sup_{\|w\| \le B} \sum_{i,k} \epsilon_{ik} \langle \Phi(x_i)w, e_k \rangle$$

$$= \sqrt{2}L\mathbb{E} \sup_{\|w\| \le B} \left\langle w, \sum_{i,k} \epsilon_{ik}\Phi(x_i)^* e_k \right\rangle$$

$$\le \sqrt{2}LB\mathbb{E} \left\| \sum_{i,k} \epsilon_{ik}\Phi(x_i)^* e_k \right\|$$

$$\le \sqrt{2}LB \left(\sum_{i,k} \|\Phi(x_i)^* e_k\|^2 \right)^{1/2}$$

$$= \sqrt{2}LB \left(\sum_i tr\, \kappa(x_i, x_i) \right)^{1/2}.$$

Here we used Corollary 1 in the first and Cauchy-Schwarz in the second inequality. Then we use Jensen's inequality combined with orthonormality of the Rademacher sequence. For the result to make sense we need the $\kappa(x_i, x_i)$ to be trace class. In the case $H = \mathbb{R}$ we obtain the standard result for the scalar case, as in [1]. The bound above can be used to prove a non-asymptotic upper bound for the algorithm described in [5], where vector-valued regression with square loss and Tychonov regularization in $\|f_w\| = \|w\|$ is considered.

5   Proof of the Contraction Inequality

We start with some simple observations on subgaussian random variables.

Lemma 1. *If X is subgaussian with subgaussian-constant b and v is a unit vector in $\mathbb{R}^K$ then*

$$\Pr \left\{ \left| \sum_{k=1}^K v_k X_k \right| > t \right\} \le 2e^{-t^2/(2b^2)},$$

where $X_1, ..., X_K$ are independent copies of X.

Proof. For any $\lambda \in \mathbb{R}$

$$\mathbb{E} \exp \left(\lambda \sum_k v_k X_k \right) = \prod_k \mathbb{E} \exp \left(\lambda v_k X_k \right)$$

$$\leq \prod_k \exp \left(\lambda^2 \frac{b^2}{2} v_k^2 \right)$$

$$= \exp \left(\frac{\lambda^2 b^2}{2} \right).$$

The first line follows from independence of the X_i, the next because X is subgaussian, and the last because v is a unit vector. It then follows from Markov's inequality that

$$\Pr \left\{ \sum_k v_k X_k > t \right\} \leq \mathbb{E} \exp \left(\lambda \left(\sum_k v_k X_k - t \right) \right)$$

$$\leq \exp \left(\frac{\lambda^2 b^2}{2} - \lambda t \right)$$

$$= e^{-t^2/(2b^2)},$$

where the last identity is obtained by optimizing in λ. The conclusion follows from a union bound.

For the purpose of vector-contraction inequalities the crucial property of subgaussian random variables is the following.

Proposition 1. *Let X be nontrivial and subgaussian with subgaussian parameter b and let $\mathbf{X} = (X_1, ..., X_K, ...)$ be an infinite sequence of independent copies of X. Then*

(i) *For every $v \in \ell_2$ the sequence of random variables $Y_K = \sum_{i=1}^{K} X_k v_k$ converges in L_p for $1 \leq p < \infty$ to a random variable denoted by $\sum_{k=1}^{\infty} X_k v_k$. The map $v \mapsto \sum_{k=1}^{\infty} X_k v_k$ is a bounded linear transformation from ℓ_2 to L_p.*
(ii) *There exists a constant $C < \infty$ such that for every $v \in \ell_2$*

$$\|v\| \leq C\mathbb{E} \left| \sum_{k=1}^{\infty} X_k v_k \right|.$$

The proof, given below, is easy and modeled after the proof of the Khintchine inequalities in [12].

For Rademacher variables the best constant is $C = \sqrt{2}$ ([22], see also inequality (4.3) in [12] or Theorem 5.20 in [4]). In the standard normal case the inequality in (ii) becomes equality with $C = \sqrt{\pi/2}$. This is an easy consequence of the rotation invariance of isonormal processes.

Proof (Proof of Proposition 1). Let X have subgaussian-constant b.

(i) Assume first that $\|v\| = 1$. For $1 \le p < \infty$ it follows from integration by parts that for any $v \in \ell_2$

$$\mathbb{E} \left| \sum_{k=1}^{K} v_k X_k \right|^p = p \int_0^\infty t^{p-1} \Pr \left\{ \left| \sum_{k=1}^{K} v_k X_k \right| > t \right\} dt$$

$$\le 2p \int_0^\infty t^{p-1} e^{-t^2/(2b^2)} dt,$$

where the last inequality follows from Lemma 1. The last integral is finite and depends only on p and b. By homogeneity it follows that for some constant B and any $v \in \ell_2$

$$\left(\mathbb{E} \left| \sum_{k=1}^{K} v_k X_k \right|^p \right)^{1/p} \le B \left(\sum_{k=1}^{K} v_k^2 \right)^{1/2}$$

which implies convergence in L_p. This proves existence and boundedness of the map $v \mapsto \sum_{k=1}^{\infty} v_k X_k$. Linearity is established with standard arguments.

(ii) Let C be the finite constant

$$C := \frac{\left(8 \int_0^\infty t^3 e^{-t^2/(2b^2)} dt \right)^{1/2}}{\mathbb{E}[X^2]^{3/2}}.$$

It suffices to prove the conclusion for unit vectors $v \in \ell_2$. From the first part we obtain

$$\mathbb{E} \left| \sum_k v_k X_k \right|^4 \le 8 \int_0^\infty t^3 e^{-t^2/(2b^2)} dt$$

Combined with Hölder's inequality this implies

$$\mathbb{E}[X^2] = \mathbb{E} \left(\left| \sum_k v_k X_k \right|^2 \right) = \mathbb{E} \left(\left| \sum_k v_k X_k \right|^{4/3} \left| \sum_k v_k X_k \right|^{2/3} \right)$$

$$\le \left(\mathbb{E} \left| \sum_k v_k X_k \right|^4 \right)^{1/3} \left(\mathbb{E} \left| \sum_k v_k X_k \right| \right)^{2/3}$$

$$\le \left(8 \int_0^\infty t^3 e^{-t^2/(2b^2)} dt \right)^{1/3} \left(\mathbb{E} \left| \sum_k v_k X_k \right| \right)^{2/3}.$$

Dividing by $\mathbb{E}[X^2]$ and taking the power of $3/2$ gives

$$1 \le C\mathbb{E} \left| \sum_k v_k X_k \right|.$$

To prove the main vector contraction result we first consider only a single Rademacher variable ϵ and then complete the proof by induction.

Lemma 2. *Let X be nontrivial, symmetric and subgaussian. Then there exists a constant $C < \infty$ such that for any countable set S and functions $\psi : S \to \mathbb{R}$, $\phi : S \to \ell_2$ and $f : S \to \mathbb{R}$ satisfying*

$$\forall s, s' \in S, \; \psi(s) - \psi(s') \leq \|\phi(s) - \phi(s')\|$$

we have

$$\mathbb{E} \sup_{s \in S} \epsilon\psi(s) + f(s) \leq C\mathbb{E} \sup_{s \in S} \sum_k X_k \phi(s)_k + f(s),$$

where the X_k are independent copies of X for $1 \leq k \leq \infty$, and $\phi(s)_k$ is the k-th coordinate of $\phi(s)$.

Proof. For C we take the constant of Proposition 1 and we let $Y = CX$ and $Y_k = CX_k$ so that for every $v \in \ell_2$

$$\|v\| \leq \mathbb{E} \left| \sum_k v_k Y_k \right|. \tag{4}$$

Let $\delta > 0$ be arbitrary. Then, by definition of the Rademacher variable,

$$2\mathbb{E} \sup_{s \in S} (\epsilon\psi(s) + f(s)) - \delta$$

$$= \sup_{s_1, s_2 \in S} \psi(s_1) + f(s_1) - \psi(s_2) + f(s_2) - \delta$$

$$\leq \psi(s_1^*) - \psi(s_2^*) + f(s_1^*) + f(s_2^*) \tag{5}$$

$$\leq \|\phi(s_1^*) - \phi(s_2^*)\| + f(s_1^*) + f(s_2^*) \tag{6}$$

$$\leq \mathbb{E} \left| \sum_k Y_k (\phi(s_1^*)_k - \phi(s_2^*)_k) \right| + f(s_1^*) + f(s_2^*) \tag{7}$$

$$\leq \mathbb{E} \sup_{s_1, s_2 \in S} \left| \sum_k Y_k \phi(s_1)_k - \sum_k Y_k \phi(s_2)_k \right| + f(s_1) + f(s_2) \tag{8}$$

$$= \mathbb{E} \sup_{s_1 \in S} \sum_k Y_k \phi(s_1)_k + f(s_1) + \mathbb{E} \sup_{s_2 \in S} -\sum_k Y_k \phi(s_2)_k + f(s_2) \tag{9}$$

$$= 2 \left(\mathbb{E} \sup_{s \in S} \sum_k Y_k \phi(s)_k + f(s) \right). \tag{10}$$

In (5) we pass to approximate maximizers $s_1^*, s_2^* \in S$, in (6) we use the assumed Lipschitz property relating ψ and ϕ, and in (7) we apply inequality (4). In (8) we use linearity and bound by a supremum in s_1 and s_2. In this expression we can simply drop the absolute value, because for any fixed configuration of the Y_k the maximum will be attained when the difference is positive, since the remaining expression $f(s_1) + f(s_2)$ is invariant under the exchange of s_1 and s_2. This gives (9). The identity (10) then follows from the symmetry of the variables Y_k. Since $\delta > 0$ was arbitrary, the result follows.

Proof (Proof of Theorem 2). The constant C and the Y_k are chosen as in the previous Lemma. We prove by induction that $\forall m \in \{0, ..., n\}$

$$\mathbb{E} \sup_{s \in \mathcal{S}} \sum_i \epsilon_i \psi_i(s) \leq \mathbb{E} \left[\sup_{s \in \mathcal{S}} \sum_{i:1 \leq i \leq m} \sum_k Y_{ik} \phi_i(s)_k + \sum_{i:m < i \leq n}^n \epsilon_i \psi_i(s) \right].$$

For $m = n$ this is the desired inequality. The case $m = 0$ is an obvious identity. Assume the claim to hold for fixed $m - 1$, with $m \leq n$. We denote $\mathbb{E}_m = \mathbb{E}[. | \{\epsilon_i, Y_{ik} : i \neq m\}]$ and define $f : \mathcal{S} \to \mathbb{R}$ by

$$f(s) = \sum_{i:1 \leq i < m} \sum_k Y_{ik} \phi_i(s)_k + \sum_{i:m < i \leq n}^n \epsilon_i \psi_i(s).$$

Then

$$\mathbb{E} \sup_{s \in \mathcal{S}} \sum_i \sigma_i \psi_i(s) \leq \mathbb{E} \left[\sup_{s \in \mathcal{S}} \sum_{i:1 \leq i < m} \sum_k Y_{ik} \phi_i(s)_k + \sum_{i:m \leq i \leq n}^n \epsilon_i \psi_i(s) \right]$$

$$= \mathbb{E} \, \mathbb{E}_m \sup_{s \in \mathcal{S}} (\epsilon_m \psi_m(s) + f(s))$$

$$\leq \mathbb{E} \, \mathbb{E}_m \sup_{s \in \mathcal{S}} \sum_k Y_{mk} \phi_m(s)_k + f(s)$$

$$= \mathbb{E} \sup_{s \in \mathcal{S}} \sum_{i:1 \leq i \leq m} \sum_k Y_{ik} \phi_i(s)_k + \sum_{i:m < i \leq n} \epsilon_i \psi_i(s).$$

The first inequality is the induction hypothesis, the second is Lemma 2.

6   A Negative Result

Conjecture 1 can be refuted by a simple counterexample. Let $\mathcal{X} = \ell_2$ with canonical basis (e_i) and set $x_i = e_i$ for $1 \leq i \leq n$. Let $\mathcal{F}$ be the unit ball in the set of bounded operators $\mathcal{B}(\ell_2)$, and for h we take the function $h : x \in \ell_2 \mapsto \|x\|$, which has Lipschitz constant equal to one.

If the conjecture was true then there is a universal constant c such that

$$\mathbb{E} \sup_{T \in \mathcal{B}(H) : \|T\|_\infty \leq 1} \sum_i \epsilon_i \|T x_i\| \leq c \mathbb{E} \sup_{T \in \mathcal{B}(H) : \|T\|_\infty \leq 1} \left\| \sum_i \epsilon_i T x_i \right\|. \quad (11)$$

For any Rademacher sequence $\epsilon = (\epsilon_i)$ we let T_ϵ be the operator defined by $T_\epsilon e_i = e_i$ if $i \leq n$ and $\epsilon_i = 1$, and $T_\epsilon = 0$ in all other cases. Clearly T_ϵ has norm $\|T_\epsilon\|_\infty \leq 1$ (it is the orthogonal projection to the subspace spanned by the basis vectors e_i such that $\epsilon_i = 1$). Then

$$\frac{n}{2} = \mathbb{E} |\{i : \epsilon_i = 1\}| = \mathbb{E} \sum_i \epsilon_i \|T_\epsilon x_i\| \leq \mathbb{E} \sup_{T \in \mathcal{B}(H) : \|T\|_\infty \leq 1} \sum_i \epsilon_i \|T x_i\|.$$

But on the other hand, the orthonormality of the Rademacher sequence implies that

$$\mathbb{E} \sup_{T \in \mathcal{B}(H): \|T\|_{\infty} \leq 1} \left\| \sum_i \epsilon_i T x_i \right\| \leq \mathbb{E} \left\| \sum_i \epsilon_i e_i \right\| \leq \sqrt{n}.$$

With (11) we obtain $n/2 \leq c\sqrt{n}$ for some universal constant c, which is absurd.

References

1. Bartlett, P.L., Mendelson, S.: Rademacher and Gaussian complexities: risk bounds and structural results. J. Mach. Learn. Res. **3**, 463–482 (2002)
2. Baxter, J.: A model of inductive bias learning. J. Artif. Intell. Res. **12**, 149–198 (2000)
3. Biau, G., Devroye, L., Lugosi, G.: On the performance of clustering in Hilbert spaces. IEEE Trans. Inf. Theory **54**(2), 781–790 (2008)
4. Boucheron, S., Lugosi, G., Massart, P.: Concentration Inequalities. Oxford University Press, Oxford (2013)
5. Caponnetto, A., De Vito, E.: Optimal rates for regularized least-squares algorithm. Found. Comput. Math. **7**, 331–368 (2007)
6. Chapelle, O., Wu, M.: Gradient descent optimization of smoothed information retrieval metrics. Inf. Retr. **13**(3), 216–235 (2010)
7. Chaudhuri, S., Tewari, A.: Generalization bounds for learning to rank: does the length of document lists matter? In: ICML 2015 (2015)
8. Ciliberto, C., Poggio, T., Rosasco, L.: Convex learning of multiple tasks and their structure (2015). arXiv preprint: arXiv:1504.03101
9. Crammer, K., Singer, Y.: On the algorithmic implementation of multiclass kernel-based vector machines. J. Mach. Learn. Res. **2**, 265–292 (2002)
10. Kakade, S.M., Shalev-Shwartz, S., Tewari, A.: Regularization techniques for learning with matrices. J. Mach. Learn. Res. **13**, 1865–1890 (2012)
11. Koltchinskii, V., Panchenko, D.: Empirical margin distributions and bounding the generalization error of combined classifiers. Ann. Stat. **30**(1), 1–50 (2002)
12. Ledoux, M., Talagrand, M.: Probability in Banach Spaces: Isoperimetry and Processes. Springer, Berlin (1991)
13. Lei, Y., Dogan, U., Binder, A., Kloft, M.: Multi-class SVMs: from tighter data-dependent generalization bounds to novel algorithms. In: Advances in Neural Information Processing Systems, pp. 2026–2034 (2015)
14. Maurer, A.: Transfer bounds for linear feature learning. Mach. Learn. **75**(3), 327–350 (2009)
15. Maurer, A., Pontil, M.: K-dimensional coding schemes in Hilbert spaces. IEEE Trans. Inf. Theory **56**(11), 5839–5846 (2010)
16. Maurer, A., Pontil, M., Romera-Paredes, B.: The benefit of multitask representation learning. J. Mach. Learn. Res. **17**(81), 1–32 (2016)
17. McDonald, D.J., Shalizi, C.R., Schervish, M.: Generalization error bounds for stationary autoregressive models (2011). arXiv preprint: arXiv:1103.0942
18. Meir, R., Zhang, T.: Generalization error bounds for Bayesian mixture algorithms. J. Mach. Learn. Res. **4**, 839–860 (2003)
19. Michelli, C.A., Pontil, M.: On learning vector-valued functions. J. Mach. Learn. Res. **6**, 615–637 (2005)

20. Mroueh, Y., Poggio, T., Rosasco, L., Slotine, J.J.: Multiclass learning with simplex coding. In: Advances in Neural Information Processing Systems, pp. 2789–2797 (2012)
21. Slepian, D.: The one-sided barrier problem for Gaussian noise. Bell Syst. Tech. J. **41**, 463–501 (1962)
22. Szarek, S.: On the best constants in the Khintchine inequality. Stud. Math. **58**, 197–208 (1976)

Localization of VC Classes: Beyond Local Rademacher Complexities

Nikita Zhivotovskiy[1,2]([✉]) and Steve Hanneke[3]

[1] Moscow Institute of Physics and Technology, Moscow, Russia
nikita.zhivotovskiy@phystech.edu
[2] Institute for Information Transmission Problems, Moscow, Russia
[3] Princeton, NJ 08542, USA
steve.hanneke@gmail.com

Abstract. In statistical learning the excess risk of empirical risk minimization (ERM) is controlled by $\left(\frac{\text{COMP}_n(\mathcal{F})}{n}\right)^\alpha$, where n is a size of a learning sample, $\text{COMP}_n(\mathcal{F})$ is a complexity term associated with a given class $\mathcal{F}$ and $\alpha \in [\frac{1}{2}, 1]$ interpolates between slow and fast learning rates. In this paper we introduce an alternative localization approach for binary classification that leads to a novel complexity measure: fixed points of the local empirical entropy. We show that this complexity measure gives a tight control over $\text{COMP}_n(\mathcal{F})$ in the upper bounds under bounded noise. Our results are accompanied by a novel minimax lower bound that involves the same quantity. In particular, we practically answer the question of optimality of ERM under bounded noise for general VC classes.

Keywords: PAC learning · Local metric entropy · Local Rademacher process · Shifted empirical process · Offset Rademacher process · Empirical risk minimization · VC dimension · Star number · Alexander's capacity · Disagreement coefficient · Massart's noise condition

1 Introduction

Since the early days of statistical learning theory understanding of the generalization abilities of empirical risk minimization has been a central question. In 1968, Vapnik and Chervonenkis [23] introduced the combinatorial property of classes of classifiers which we now call the *VC dimension*, which plays a crucial role not only in statistics but in many other areas of mathematics. By now it is strongly believed that the VC dimension fully characterizes the properties of the empirical risk minimization algorithm. But this appears to be true only in the agnostic case, when no assumptions are made on the labelling mechanism. It was noticed several times in the literature, that when considering bounded noise VC dimension alone is not a right complexity measure of ERM [18,20]. Until now this phenomenon was discussed only for several specific classes. The main aim of this paper is to present this yet unknown combinatorial complexity measure.

© Springer International Publishing Switzerland 2016
R. Ortner et al. (Eds.): ALT 2016, LNAI 9925, pp. 18–33, 2016.
DOI: 10.1007/978-3-319-46379-7_2

In the last twenty years many efforts were made to understand the conditions that imply fast $\frac{1}{n}$ convergence rates, instead of slow $\frac{1}{\sqrt{n}}$ rates. At the beginning of the 2000s, so-called *localized* complexities (Bartlett et al. [3], Koltchinskii [12]) were introduced to statistical learning and became popular techniques for proving $\frac{1}{n}$ rates in different scenarios. But in addition to better rates, localization means that *only a small vicinity of the best classifier* really affects the learning complexity. We still lack tight error bounds based on localization and expressed in terms of intuitively-simple and calculable combinatorial properties of the class. Existing approaches based on localization (mainly, via *local Rademacher complexities*) are typically difficult to calculate directly, and the simpler relaxations of these bounds in the literature use localization largely to gain improvements due to the *noise conditions*, but fail to maintain the important improvements due to the local *structure of the function class* (i.e., localization of the complexity term in the bound). The present work explores this aspect of localization, resulting in a complexity measure, which correctly captures the optimal rates under bounded noise.

2   Notation and Previous Results

We define the *instance space* $\mathcal{X}$ and the *label space* $\mathcal{Y} = \{1, -1\}$, and denote $\mathcal{Z} = \mathcal{X} \times \mathcal{Y}$. We assume that the set $\mathcal{Z}$ is equipped with some σ-algebra and a probability measure P on measurable subsets is defined. We also assume that we are given a set of classifiers $\mathcal{F}$. The risk of a classifier f is its probability of error, denoted $R(f) = P(f(X) \neq Y)$. We denote the *Bayes classifier* by $f^*(x) = \text{sign}(\eta(x))$, where $\eta(x) = \mathbb{E}[Y|X = x]$. Symbol $\wedge$ will denote minimum of two real numbers, $\vee$ will denote maximum of two real numbers and $\mathbb{1}[A]$ will denote an indicator of the event A. For any subset $B \subseteq \mathcal{F}$ define the *region of disagreement* as $\text{DIS}(B) = \{x \in \mathcal{X} | \exists f, g \in B \text{ s. t. } f(x) \neq g(x)\}$. We will also consider abstract real-valued function classes, which will usually be denoted by $\mathcal{G}$. We will slightly abuse the notation and by $\log(x)$ always mean truncated logarithm: $\ln(\max(x, e))$. The notation $f(n) \lesssim g(n)$ or $g(n) \gtrsim f(n)$ will mean that for some universal constant $c > 0$ it holds that $f(n) \leq cg(n)$ for all $n \in \mathbb{N}$. Similarly, we introduce $f(n) \simeq g(n)$ to be equivalent to $g(n) \lesssim f(n) \lesssim g(n)$.

A *learner* observes $((X_1, Y_1), \ldots, (X_n, Y_n))$, an i.i.d. training sample from an unknown distribution P. Also denote $Z_i = (X_i, Y_i)$. By P_n we will denote an empirical mean. *Empirical risk minimization* (ERM) refers to any learning algorithm with the following property: given a training sample, it outputs a classifier $\hat{f}$ that minimizes $R_n(f) = P_n \mathbb{1}[f(X) \neq Y]$ among all $f \in \mathcal{F}$. At times we also refer to a *ghost sample*, which is another n i.i.d. P-distributed samples, independent of the training sample, and we denote by P'_n the empirical mean with respect to the ghost sample. We say a set $\{x_1, \ldots, x_k\} \in \mathcal{X}^k$ is shattered by $\mathcal{F}$ if there are 2^k distinct classifications of $\{x_1, \ldots, x_k\}$ realized by classifiers in $\mathcal{F}$. The *VC dimension* of $\mathcal{F}$ is the largest integer d such that there exists a set $\{x_1, \ldots, x_d\}$ shattered by $\mathcal{F}$ [23]. We define the *growth function* $\mathcal{S}_{\mathcal{F}}(n)$ as the maximum possible number of different classifications of a set of n points realized by classifiers in $\mathcal{F}$.

Definition 1 (Massart and Nédélec [18]). $(P, \mathcal{F})$ *is said to satisfy Massart's bounded noise condition if* $f^* \in \mathcal{F}$ *and for some* $h \in [0, 1]$ *it holds* $|\eta(X)| \geq h$ *with probability 1. This constant* h *is referred to as the* margin parameter.

For any $\mathcal{F}$, the set of all corresponding distributions satisfying Massart's bounded noise condition will be denoted by $\mathcal{P}(h, \mathcal{F})$. The case $h = 1$ corresponds to the so-called *realizable case*, where $Y = f^*(X)$ almost surely, and $h = 0$ corresponds to a well-specified *agnostic* case. The following result is classic [4]. Let $\mathcal{F}$ be a class with VC-dimension d. For any empirical risk minimizer $\hat{f}$ over n samples, for any $P \in \mathcal{P}(0, \mathcal{F})$, we have $\mathbb{E}(R(\hat{f}) - R(f^*)) \lesssim \sqrt{\frac{d}{n}}$. Moreover, the following lower bound exists for an output $\tilde{f}$ of *any* algorithm based on n samples: there exists $P \in \mathcal{P}(0, \mathcal{F})$ such that $\mathbb{E}(R(\tilde{f}) - R(f^*)) \gtrsim \sqrt{\frac{d}{n}} \wedge 1$. Thus we know that the VC dimension is the right complexity measure for empirical risk minimization, and indeed for optimal learning, when no restrictions are made on the probability distribution. Interestingly, this is not generally the case when $h > 0$. In this paper, we find this yet unknown essentially correct complexity measure, when h is bounded away from 0 and 1. But first, we review a refinement to the above bound for the case $h > 0$, due to Giné and Koltchinskii [6]. Specifically, consider the following definition.

Definition 2. *For* $\varepsilon_0 > 0$ *fix a set* $\mathcal{F}_{\varepsilon_0} = \{f \in \mathcal{F} : P_X(f(X) \neq f^*(X)) \leq \varepsilon_0\}$. *For* $\varepsilon \in (0, 1]$ *define* $\tau(\varepsilon) = \sup\limits_{\varepsilon_0 \geq \varepsilon} \left(\varepsilon_0^{-1} P_X\{x \in \mathcal{X} : \exists f \in \mathcal{F}_{\varepsilon_0} \text{ s.t. } f(x) \neq f^*(x)\} \right)$.

This quantity was introduced to the empirical processes literature by Alexander [1], and is referred to as *Alexander's capacity* by Giné and Koltchinskii [6]. The same quantity appeared independently in the literature on active learning, where it is referred to as the *disagreement coefficient* [7]. $\tau(\varepsilon)$ is a distribution-dependent measure of the diversity of ways in which classifiers in a relatively small vicinity of f^* can disagree with f^*. Giné and Koltchinskii [6] gave the following upper bound. Let $\mathcal{F}$ be a class of VC dimension d, and $\hat{f}$ the classifier produced by an ERM based on n training samples. For any probability measure $P \in \mathcal{P}(h, \mathcal{F})$,

$$\mathbb{E}(R(\hat{f}) - R(f^*)) \lesssim \frac{d}{nh} \log\left(\tau\left(\frac{d}{nh^2}\right)\right). \tag{1}$$

This bound is the best simple, easily calculable upper bound known so far for ERM in the case of binary classification under Massart's bounded noise condition. The proof of this bound is based on the analysis of the localized Rademacher processes. Thus we may consider this result as the best known relaxation of the local Rademacher analysis. Very recently, Hanneke and Yang [8] introduced a distribution-free complexity measure, called the *star number*. It is defined as follows.

Definition 3. *The star number* $\mathbf{s}$ *is the largest integer such that there exist distinct* $x_1, \ldots, x_{\mathbf{s}} \in \mathcal{X}$ *and* $f_0, f_1, \ldots, f_{\mathbf{s}} \in \mathcal{F}$ *such that, for all* $i \in \{1, \ldots, \mathbf{s}\}$, $DIS(\{f_0, f_i\}) \cap \{x_1, \ldots, x_{\mathbf{s}}\} = \{x_i\}$.

Similar to Alexander's capacity, the star number describes how diverse the small-size disagreements with a fixed classifier f_0 can be. One of the most interesting results about this value is its connection with the worst case of Alexander's capacity. The paper of Hanneke and Yang contains the following equality:

$$\sup_{f^* \in \mathcal{F}} \sup_{P_X} \tau(\varepsilon) = \mathbf{s} \wedge \tfrac{1}{\varepsilon}.$$

An immediate corollary of this and (1) is that, for any $P \in \mathcal{P}(h, \mathcal{F})$, $\mathbb{E}(R(\hat{f}) - R(f^*)) \lesssim \frac{d}{nh} \log\left(\frac{nh^2}{d} \wedge \mathbf{s}\right)$. Since $\mathbf{s}$ controls Alexander's capacity with equality, there is no room for any kind of improvement using the bound of Giné and Koltchinskii if we consider distribution-free upper bounds.

3   Preliminaries from Empirical Processes

Given a function class $\mathcal{G}$ mapping $\mathcal{Z}$ to $\mathbb{R}$, one may consider the following quantity: $\sup_{g \in \mathcal{G}} (P - P_n) g$. This random value plays in important role in statistical learning theory. Since the pioneering paper of Vapnik and Chervonenkis [23], the analysis of learning algorithms is usually performed by the tight uniform control over the process $(P - P_n) g$ for a special class of functions. The behaviour of the supremum of this empirical process is controlled by a supremum of the so-called *Rademacher process*: $\frac{1}{n} \mathbb{E}_\varepsilon \max_{g \in \mathcal{G}} \left(\sum_{i=1}^{n} \varepsilon_i g_i \right)$, where g_i denotes $g(Z_i)$, ε_i are independent Rademacher variables taking values ± 1 with equal probabilities, and $\mathbb{E}_\varepsilon$ denoted the expectation over the ε_i random variables (conditioning on the Z_i variables). We will instead consider different quantities, so-called *shifted empirical* processes, introduced by Lecué and Mitchell [14]. Given $c > 0$, we consider $\sup_{g \in \mathcal{G}} (P - (1 + c)P_n) g$. The second important quantity is an expected supremum of the *offset Rademacher process*, introduced recently by Liang, Rakhlin, and Sridharan [16]: $\frac{1}{n} \mathbb{E}_\varepsilon \max_{g \in \mathcal{G}} \left(\sum_{i=1}^{n} \varepsilon_i g_i - c' g_i^2 \right)$. This quantity was introduced for the analysis of a specific aggregation procedure under the squared loss and so far has not been related to a shifted process. In this paper, we will investigate some new properties of these processes and will show how they may be used in the classification framework. The following short lemma appears in a more general form in [16] (Lemma 5).

Lemma 1. *Let $V \subset \{0, 1\}^n$ be a finite set of binary vectors of cardinality N. Then for any $c > 0$,*

$$\frac{1}{n} \mathbb{E}_\varepsilon \max_{v \in V} \left(\sum_{i=1}^{n} \varepsilon_i v_i - c v_i \right) \leq \frac{1}{2c} \frac{\log(N)}{n}.$$

Lemma 2 (Shifted Symmetrization in Expectation). *Let $\mathcal{G}$ be a function class and $c \geq 0$ an absolute constant. Then*

$$\mathbb{E}\sup_{g \in \mathcal{G}}((P - (1+c)P_n)g) \leq \frac{c+2}{n}\mathbb{E}\mathbb{E}_\varepsilon \sup_{g \in \mathcal{G}}\left(\sum_{i=1}^{n}\varepsilon_i g(Z_i) - \frac{c}{c+2}g(Z_i)\right).$$

Proof. Denote $g(Z_i)$ by g_i. Using the symmetrization trick and Jensen's inequality,

$$\mathbb{E}\sup_{g \in \mathcal{G}}((P - (1+c)P_n)g) \leq \mathbb{E}\sup_{g \in \mathcal{G}}(P_n'g - (1+c)P_n g)$$

$$= \mathbb{E}\sup_{g \in \mathcal{G}}((1+c/2)(P_n'g - P_n g) - cP_n'g/2 - cP_n g/2)$$

$$\leq \mathbb{E}\mathbb{E}_\varepsilon \sup_{g \in \mathcal{G}}\left(\frac{c+2}{n}\sum_{i=1}^{n}\varepsilon_i g_i - cP_n g\right) = (c+2)\mathbb{E}\mathbb{E}_\varepsilon \sup_{g \in \mathcal{G}}\left(\frac{1}{n}\sum_{i=1}^{n}\varepsilon_i g_i - \frac{c}{c+2}P_n g\right).$$

$\square$

Let $\mathbf{s}$ be the star number of a class of binary classifiers $\mathcal{F}$. Hanneke [9] recently proved that $\mathbb{E}P(\mathrm{DIS}(\mathcal{V}_n)) \leq \frac{\mathbf{s}}{n+1}$, where $\mathcal{V}_n = \{f \in \mathcal{F}|P_n\mathbb{1}[f(X) \neq f^*(X)] = 0\}$ is a *version space*, and used this fact to bound the risk of ERM. In this same spirit, this inequality will be important in our next theorem, one of the novel contributions of the present work. Its proof is in the appendix.

Theorem 1. *Let $\mathbf{s}$ be the star number of a class of binary classifiers $\mathcal{F}$. In the realizable case, for any ERM $\hat{f}$,*

$$\mathbb{E}R(\hat{f}) \lesssim \frac{\log\left(\mathcal{S}_\mathcal{F}\left(\mathbf{s} \wedge n\right)\right)}{n}.$$

Example 1. Theorem 1 yields examples showing the gaps in the distribution-free bound (1) in the realizable case. Specifically, suppose $\mathcal{X} = \{x_1, \ldots, x_\mathbf{s}\}$, define class $\mathcal{F}_1$ as the classifiers on this $\mathcal{X}$ with at most d points classified 1, and class $\mathcal{F}_2$ as the classifiers having at most $d - 1$ points classified 1 among $\{x_1, \ldots, x_{d-1}\}$ and at most one point classified 1 among $\{x_d, \ldots, x_\mathbf{s}\}$. For both $\mathcal{F}_1$ and $\mathcal{F}_2$, the VC dimension is d and the star number is $\mathbf{s}$. However, for $\mathcal{F}_1$ Theorem 1 gives a bound of order $\frac{d\log\left(\frac{\mathbf{s}}{d}\right)}{n}$, but for $\mathcal{F}_2$ it gives a smaller bound of order $\frac{d+\log(\mathbf{s}\wedge n)}{n}$. In both cases, these are known to be tight characterizations of ERM in the realizable case [9,10]. It should be noted, however, that one can also construct examples where Theorem 1 is itself not tight.

4   Local Metric Entropy

This section presents our main result. Toward this end, we introduce a new complexity measure: the *worst-case local empirical packing numbers*. Given a set of n points we fix some $f \in \mathcal{F}$ and construct a Hamming ball of the radius γ:

$$\mathcal{B}_H(f, \gamma, \{x_1, \ldots, x_n\}) = \{g \in \mathcal{F}|\rho_H(f, g) \leq \gamma\},$$

where $\rho_H(f,g) = |\{i \in \{1,\ldots,n\} : f(x_i) \neq g(x_i)\}|$. When $x_1,\ldots,x_n$ are clear from the context, we sometimes simply write $\mathcal{B}_H(f,\gamma)$. We further introduce

$$\mathcal{M}_1^{loc}(\mathcal{F},\gamma,n,h) = \max_{x_1,\ldots,x_n} \max_{f \in \mathcal{F}} \max_{\varepsilon \geq \gamma} \mathcal{M}_1(\mathcal{B}_H(f,\varepsilon/h,\{x_1,\ldots,x_n\}),\varepsilon/2),$$

where $\mathcal{M}_1(\mathcal{H},\varepsilon)$ denotes the size of a maximal ε-packing of $\mathcal{H}$ under ρ_H distance (for the given $x_1,\ldots,x_n$ points). This quantity measures how one can pack a ball in $\mathcal{F}$ by balls of smaller radius. For any $h,h' \in (0,1]$, define

$$\gamma_{h,h'}^{loc}(n,\mathcal{F}) = \max\{\gamma \in \mathbb{N} : h\gamma \leq \log(\mathcal{M}_1^{loc}(\mathcal{F},\gamma,n,h'))\}.$$

When $\mathcal{F}$ is clear from the context, we simply write $\gamma_{h,h'}^{loc}(n)$ instead of $\gamma_{h,h'}^{loc}(n,\mathcal{F})$. The quantity $\gamma_{h,h'}^{loc}(n)$ defines the *fixed point of a local empirical entropy*.

We note that, because $1 \leq d < \infty$ in this work, when $h,h' > 0$ the set on the right in this definition is finite and nonempty, so that $\gamma_{h,h'}^{loc}(n)$ is a well-defined strictly-positive integer. Indeed, for any $h,h' \in (0,1]$, the value $\gamma = \lfloor \frac{1}{h} \rfloor$ satisfies $h\gamma \leq 1$, so that (because $\log(\cdot)$ is the truncated logarithm) this γ is contained in the set; in particular, this implies $h\gamma_{h,h'}^{loc}(n,\mathcal{F}) \geq h\lfloor \frac{1}{h} \rfloor \geq \frac{1}{2}$ always. The next theorem is the main upper bound of this paper. The rest of this section is devoted to its proof.

Theorem 2. *Fix any function class $\mathcal{F}$; denote its VC dimension d and star number $\mathbf{s}$. Fix any $h \in \left(\sqrt{\frac{d}{n}},1\right]$ and suppose $\gamma_{h,h}^{loc}(n) > 0$. If $P \in \mathcal{P}(h,\mathcal{F})$, then for any ERM $\hat{f}$,*

$$\mathbb{E}(R(\hat{f}) - R(f^*)) \lesssim \frac{\gamma_{h,h}^{loc}(n)}{n}. \tag{2}$$

Moreover,

$$\frac{d + \log(nh^2 \wedge \mathbf{s})}{h} \lesssim \gamma_{h,h}^{loc}(n) \lesssim \frac{d\log\left(\frac{nh^2}{d} \wedge \mathbf{s}\right)}{h} + \frac{d\log\left(\frac{1}{h}\right)}{h}. \tag{3}$$

Our complexity term (3) is not worse than the upper bound of Giné and Koltchinskii (1) when h is bounded from 0 by a constant. Another interesting property is that the bound (2) involves neither the VC dimension nor the star number explicitly. At the same time one can control the complexity term with both of them from below and above. We should mention that the connection between global covering numbers and VC dimension is well known [11].

Consider the *excess loss class* $\mathcal{G}_\mathcal{Y} = \{(x,y) \to \mathbb{1}[f(x) \neq y] - \mathbb{1}[f^*(x) \neq y]$ for $f \in \mathcal{F}\}$ and the *class* $\mathcal{G}_{f^*} = \{x \to \mathbb{1}[f(x) \neq f^*(x)]$ for $f \in \mathcal{F}\}$, which may be interpreted as an excess loss class in the realizable case. For any $g \in \mathcal{G}_\mathcal{Y}$ it holds $g^2(x,y) = \mathbb{1}[f(x) \neq f^*(x)] = \frac{1}{2}|f(x) - f^*(x)| = \frac{1}{4}(f(x) - f^*(x))^2$. And also for any $g \in \mathcal{G}_\mathcal{Y}$ it holds $g(x,y) = \frac{y(f^*(x) - f(x))}{2}$ and $R(f^*) \leq \frac{1}{2}(1-h)$ [5].

Lemma 3 (Contraction). *Let $\mathcal{G}_y$ be an excess loss class associated with a given class $\mathcal{F}$, and fix any $h \in [0,1]$. For any $c \in [0,1]$ and any $P \in \mathcal{P}(h, \mathcal{F})$,*

$$\mathbb{E}\mathbb{E}_\varepsilon \sup_{g \in \mathcal{G}_y} \left(\sum_{i=1}^n \varepsilon_i g(X_i, Y_i) - cg(X_i, Y_i) \right) \leq \frac{5}{4} \mathbb{E}\mathbb{E}_\xi \sup_{g' \in \mathcal{G}_{f^*}} \left(\sum_{i=1}^n \xi_i g'(X_i) - \frac{4}{5} hcg'(X_i) \right),$$

where $\xi_1, \ldots, \xi_n$ are r. v. conditionally independent given $X_1, \ldots, X_n$, with $\mathbb{E}[\xi_i | X_1, \ldots, X_n] = 0$ and $\mathbb{E}[\exp(\lambda \xi_i) | X_1, \ldots, X_n] \leq \exp(\frac{\lambda^2}{2})$ for all λ.

Proof. We will denote $g(X_i, Y_i)$ by g_i. First we notice that any $g \in \mathcal{G}_y$ may be defined by some $f \in \mathcal{F}$. Then note that

$$\mathbb{E}\mathbb{E}_\varepsilon \sup_{g \in \mathcal{G}_y} \left(\sum_{i=1}^n \varepsilon_i g_i - cg_i \right) = \mathbb{E}\mathbb{E}_\varepsilon \sup_{f \in \mathcal{F}} \left(\sum_{i=1}^n \frac{1}{2} \varepsilon_i Y_i (f(X_i) - f^*(X_i)) - cg_i \right)$$

$$= \mathbb{E}\mathbb{E}_\varepsilon \sup_{f \in \mathcal{F}} \left(\sum_{i=1}^n \frac{1}{2} \varepsilon_i (f(X_i) - f^*(X_i)) - cg_i \right) = \frac{1}{4} \mathbb{E}\mathbb{E}_\varepsilon \sup_{g \in \mathcal{G}_y} \left(\sum_{i=1}^n \varepsilon_i g_i^2 - 4cg_i \right).$$

Now consider the term $- \sum_{i=1}^n g(X_i, Y_i)$. Denoting $h_i' = 1 - 2P(f^*(X_i) \neq Y_i | X_i)$ (an X_i-dependent random variable), we know that $1 \geq h_i' \geq h$ almost surely. Furthermore, the event that $f^*(X_i) \neq Y_i$ has conditional probability (given X_i) equal $\frac{1}{2}(1 - h_i')$, and on this event we have $g^2(X_i, Y_i) = -g(X_i, Y_i)$. Similarly, the event that $f^*(X_i) = Y_i$ occurs with conditional probability (given X_i) equal $\frac{1}{2}(1 + h_i')$, and on this event we have $g^2(X_i, Y_i) = g(X_i, Y_i)$. Thus, defining $\xi_i^{(h')} = h_i' + \mathbb{1}[f^*(X_i) \neq Y_i] - \mathbb{1}[f^*(X_i) = Y_i]$, these $\xi_1^{(h')}, \ldots, \xi_n^{(h')}$ random variables are conditionally independent given $X_1, \ldots, X_n$, with $\mathbb{E}[\xi_i^{(h')} | X_1, \ldots, X_n] = 0$. In particular, if $h_i' = 0$ for all i, these are Rademacher random variables, while if $h_i' = 1$ these random variables are equal to 0 with probability 1. Now note that, by the above reasoning about these events $- \sum_{i=1}^n g_i = - \sum_{i=1}^n h_i' g_i^2 + \sum_{i=1}^n \xi_i^{(h')} g_i^2 \leq -(\min_i h_i') \sum_{i=1}^n g_i^2 + \sum_{i=1}^n \xi_i^{(h')} g_i^2$.
Therefore, denoting $\xi_i' = \varepsilon_i + 4c\xi_i^{(h')}$ (which are also conditionally independent over i given $X_1, \ldots, X_n$) and using the fact that $h \leq h_i'$ almost surely, we have $\frac{1}{4} \mathbb{E}\mathbb{E}_\varepsilon \sup_{g \in \mathcal{G}_y} \left(\sum_{i=1}^n \varepsilon_i g_i^2 - 4cg_i \right) \leq \frac{1}{4} \mathbb{E}\mathbb{E}_\varepsilon \sup_{g \in \mathcal{G}_y} \left(\sum_{i=1}^n \xi_i' g_i^2 - 4hcg_i^2 \right) = \frac{1}{4} \mathbb{E}_X \mathbb{E}_{\xi'} \sup_{g' \in \mathcal{G}_{f^*}} \left(\sum_{i=1}^n \xi_i' g'(X_i) - 4hcg'(X_i) \right)$. Finally, because ε_i and $\xi_i^{(h')}$ both have zero conditional mean, so does ξ_i', and since we also have $-5 + 4ch_i' \leq \xi_i' \leq 5 + 4ch_i'$, Hoeffding's lemma ([5] Lemma 8.1) implies $\mathbb{E}[\exp(\lambda \xi_i') | X_1, \ldots, X_n] \leq \exp(25\lambda^2/2)$. The lemma easily follows, taking $\xi_i = \xi_i'/5$. $\quad\square$

Lemma 4 (Localization). *Let $\mathcal{G}$ be a set of functions taking binary values, containing the zero function, and let $c \in [0, \frac{1}{4}]$ be a constant. Let $\xi_1, \ldots, \xi_n$ be any random variables conditionally independent given $X_1, \ldots, X_n$ with*

$\mathbb{E}[\xi_i|X_1,\ldots,X_n] = 0$ and $\mathbb{E}[\exp(\lambda\xi_i)|X_1,\ldots,X_n] \leq \exp(\frac{\lambda^2}{2})$ for all λ. Then if $c\gamma_{c,c}^{loc}(n,\mathcal{G}) \gtrsim 1$,

$$\frac{1}{n}\mathbb{E}\max_{g\in\mathcal{G}}\left(\sum_{i=1}^{n}\xi_i g(X_i) - 4cg(X_i)\right) \lesssim \frac{\gamma_{c,c}^{loc}(n,\mathcal{G})}{n}.$$

The proof of this lemma is deferred to the appendix.

Proof (Theorem 2). Let $\hat{f}$ be an ERM and $\hat{g}$ be a corresponding function in the excess loss class $\mathcal{G}_y$. We obviously have $\mathbb{E}(R(\hat{f}) - R(f^*)) = \mathbb{E}P\hat{g}$ and $P_n\hat{g} \leq 0$. Then $\forall c > 0$, $\mathbb{E}(R(\hat{f}) - R(f^*)) \leq \mathbb{E}(P\hat{g} - (1+c)P_n\hat{g}) \leq \mathbb{E}\sup_{g\in\mathcal{G}_y}(Pg - (1+c)P_ng)$.

Now using the symmetrization lemma (Lemma 2) we have

$$\mathbb{E}\sup_{g\in\mathcal{G}_y}(Pg - (1+c)P_ng) \leq \frac{c+2}{n}\mathbb{E}\mathbb{E}_\varepsilon\sup_{g\in\mathcal{G}_y}\left(\sum_{i=1}^{n}\varepsilon_i g(X_i,Y_i) - \frac{c}{c+2}g(X_i,Y_i)\right).$$

Applying Lemma 3, we have $\frac{c+2}{n}\mathbb{E}\mathbb{E}_\varepsilon\sup_{g\in\mathcal{G}_y}\left(\sum_{i=1}^{n}\varepsilon_i g(X_i,Y_i) - \frac{c}{c+2}g(X_i,Y_i)\right) \leq$

$\frac{5(c+2)}{4n}\mathbb{E}\mathbb{E}_\varepsilon\sup_{g'\in\mathcal{G}_{f^*}}\left(\sum_{i=1}^{n}\xi_i g'(X_i) - \frac{4ch}{5(c+2)}g'(X_i)\right)$. Now we are ready to apply the localization lemma (Lemma 4). The conditions on the ξ_i variables required for Lemma 4 are supplied by Lemma 3, and all functions in $\mathcal{G}_{f^*}$ take only binary values. Thus, for a fixed c,

$$\frac{5(c+2)}{4n}\mathbb{E}\mathbb{E}_\varepsilon\sup_{g\in\mathcal{G}_{f^*}}\left(\sum_{i=1}^{n}\xi_i g'(X_i) - \frac{4ch}{5(c+2)}g'(X_i)\right) \lesssim \frac{\gamma_{h,h}^{loc}(n)}{n}.$$

The following proposition finishes the proof of Theorem 2. Its proof is in the appendix.

Proposition 1. *Let d be the VC-dimension and $\mathbf{s}$ be the star number of $\mathcal{F}$. For any $h \in (0,1]$, it holds*

$$\frac{d + \log\left(nh^2 \wedge \mathbf{s}\right)}{h} \wedge \sqrt{dn} \lesssim \gamma_{h,h}^{loc}(n) \lesssim \frac{d\log\left(\frac{nh^2}{d} \wedge \mathbf{s}\right)}{h} + \frac{d\log(\frac{1}{h})}{h}.$$

5   Minimax Lower Bound

In this section we prove that under Massart's bounded noise condition, fixed points of the local empirical entropy appear in minimax lower bounds. Results are based on classic lower bound techniques from the literature [18,20,25], previously used only for specific classes.

Definition 4. *Fix a class of classifiers $\mathcal{F}$. Assume that there exists a positive constant $c \geq 1$ such that for any N the supremum with respect to the radius in $\mathcal{M}_1^{loc}(\mathcal{F}, \gamma_{h,1}^{loc}(N), N, 1)$ is achieved at some $\varepsilon_h(N) \leq c\gamma_{h,1}^{loc}(N)$. This class will be referred to as c-pseudoconvex.*

Theorem 3. *Let $\tilde{f}$ be the output of any learning algorithm. Fix any $c_{\mathcal{F}}$- pseudo-convex class $\mathcal{F}$ and any h satisfying $\sqrt{\frac{d}{n}} \leq h \leq 1$. Then there exists a $P \in \mathcal{P}(h, \mathcal{F})$ such that*

$$\mathbb{E}(R(\tilde{f}) - R(f^*)) \gtrsim \frac{d}{nh} + \frac{1}{c_{\mathcal{F}}} \frac{(1-h)\gamma_{h,1}^{\text{loc}}\left(\lceil \frac{nc_{\mathcal{F}}h}{(1-h)} \rceil\right)}{n}. \tag{4}$$

Conditions involving the constant $c_{\mathcal{F}}$ can be relaxed in different ways. We may remove the pseudoconvexity assumptions by redefining the local empirical entropy (4) by removing the maximum with respect to the radius. Alternatively one can remove the maximum by introducing certain monotonicity assumptions, which were used implicitly in previous papers [6,20]. In both cases our lower bound holds with $c_{\mathcal{F}} = 1$. Finally, we note that these monotonicity problems do not appear for convex classes, as noted by Mendelson in [19]. The next lemma is given in [17] (Corollary 2.18).

Lemma 5 (Birgé). *Let $\{P_i\}_{i=0}^N$ be a finite family of distributions defined on the same measurable space and $\{A_i\}_{i=0}^N$ be a family of disjoint events. Then*

$$\min_{0 \leq i \leq N} P_i(A_i) \leq 0.71 \vee \frac{\sum_{i=1}^{N} KL(P_i \| P_0)}{N \log(N+1)}.$$

Proof (Theorem 3). First we consider the value $\mathcal{M}_1^{\text{loc}}(\mathcal{F}, \gamma_{h,1}^{\text{loc}}(N), N, 1)$. Recall that the definition of this value considers suprema over $f \in \mathcal{F}$ and over N-element subsets of $\mathcal{X}^n$. Without loss of generality we assume that these suprema are achieved at some classifier $g \in \mathcal{F}$, some $\varepsilon_h(N) \in [\gamma_{h,1}^{\text{loc}}(N), N]$ and at some particular set $\mathcal{X}_N = \{x_1, \ldots, x_N\}$. Let k_i define the number of copies of x_i in $\mathcal{X}_N$. We define $P_{\mathcal{X}_N}(\{x_i\}) = \frac{k_i}{N}$. If all elements are distinct this measure is just a uniform measure on $\mathcal{X}_N$. We introduce a natural para-metrization: any classifier is represented by an N-dimensional binary vector and two vectors (for classifiers g, f) disagree only on a set corresponding to $\text{DIS}(\{g, f\}) \cap \mathcal{X}_N$. The set of binary vectors corresponding to classifiers in $\mathcal{F}$ will be denoted by $\mathcal{B}$. For a given binary vector b define $P_b = P_{\mathcal{X}_N} \times P_{Y|X}^b$, where $P_{Y=1|X_i}^b = \frac{1+(2b_i-1)h}{2}$. Let $\tilde{f}_b$ denote the classifier $\tilde{f}$ produced by the learning algorithm when P_b is the data distribution, and let $\tilde{b}$ denote the binary vector corresponding to $\tilde{f}_b$; thus, $\tilde{b}$ is a random vector, which depends on the parameter b only through the n data points having distribution P_b. It is known [5] that $R(\tilde{f}) - R(f^*) = \mathbb{E}(|\eta(X)|\mathbb{1}[\tilde{f}(X) \neq f^*(X)]|\tilde{f}) \geq hP((x,y) : \tilde{f}(x) \neq f^*(x))$, when $P \in \mathcal{P}(h, \mathcal{F})$. Furthermore, when P_b is the data distribution, we have $P_b((x,y) : \tilde{f}_b(x) \neq f^*(x)) = \frac{\rho_H(\tilde{b}, b)}{N}$. Thus, we have $\sup_{P \in \mathcal{P}(h,\mathcal{F})} \mathbb{E}(R(\tilde{f}) - R(f^*)) \geq$ $\max_{b \in \mathcal{B}} \mathbb{E}\left(hP_b((x,y) : \tilde{f}_b(x) \neq f^*(x))\right) \geq \frac{h}{N} \max_{b \in \mathcal{B}} \mathbb{E}(\rho_H(\tilde{b}, b))$. Let b^* be the binary vector in $\mathcal{B}$ corresponding to the classifier g defined above, and fix a maximal sub-set $\mathcal{B}^{\text{loc}} \subset \mathcal{B}$ satisfying the properties that for any $b' \in \mathcal{B}^{\text{loc}}$ we have $\rho_H(b', b^*) \leq$

$\varepsilon_h(N)$ and for any two $b', b'' \in \mathcal{B}^{\mathrm{loc}}$ we have $\rho_H(b', b'') > \varepsilon_h(N)/2$. Next, define $\check{b}$ as the minimizer of $\rho_H(\check{b}, \tilde{b})$ among $\mathcal{B}^{\mathrm{loc}}$. In particular, if $b \in \mathcal{B}^{\mathrm{loc}}$, we have $\rho_H(\check{b}, \tilde{b}) \leq \rho_H(b, \tilde{b})$, so that $\rho_H(\check{b}, b) \leq \rho_H(\check{b}, \tilde{b}) + \rho_H(\tilde{b}, b) \leq 2\rho_H(\tilde{b}, b)$. Therefore, $\frac{h}{N} \max_{b \in \mathcal{B}} \mathbb{E}(\rho_H(\tilde{b}, b)) \geq \frac{h}{N} \max_{b \in \mathcal{B}^{\mathrm{loc}}} \mathbb{E}(\rho_H(\tilde{b}, b)) \geq \frac{h}{2N} \max_{b \in \mathcal{B}^{\mathrm{loc}}} \mathbb{E}(\rho_H(\check{b}, b))$. Recalling that $\check{b}$ is a deterministic function of $\tilde{f}$, which itself is a function of the n data points, we may define disjoint subsets A_b of $(\mathcal{X} \times \mathcal{Y})^n$, for $b \in \mathcal{B}^{\mathrm{loc}}$, where A_b corresponds to the collection of data sets that would yield $\check{b} = b$. Now, from Markov's inequality and the fact that the vectors in $\mathcal{B}^{\mathrm{loc}}$ are $\frac{\varepsilon_h(N)}{2}$-separated, we have $\mathbb{E}(\rho_H(\check{b}, b)) \geq \frac{\varepsilon_h(N)}{2} P(\check{b} \neq b) = \frac{\varepsilon_h(N)}{2}(1 - P_b^n(A_b))$. Thus we have that $\frac{h}{2N} \max_{b \in \mathcal{B}^{\mathrm{loc}}} \mathbb{E}(\rho_H(\check{b}, b)) \geq \frac{h\varepsilon_h(N)}{4N} \left(1 - \min_{b \in \mathcal{B}^{\mathrm{loc}}} P_b^n(A_b) \right)$. We are interested in using Lemma 5 to upperbound $\min_{b \in \mathcal{B}^{\mathrm{loc}}} P_b^n(A_b)$. Toward this end, note that for any $b', b'' \in \mathcal{B}^{\mathrm{loc}}$, simple calculations show that $\mathrm{KL}(P_{b'}^n \| P_{b''}^n) = \frac{n}{N} h \ln \left(\frac{1+h}{1-h} \right) \rho_h(b', b'')$. Because for $x > 0$ we have $\ln(x+1) \leq x$, it holds that $h \ln \left(\frac{1+h}{1-h} \right) \leq \frac{2h^2}{1-h}$. Furthermore, for any $b', b'' \in \mathcal{B}^{\mathrm{loc}}$ we have $\rho_H(b', b'') \leq 2\varepsilon_h(N)$. Therefore, $\mathrm{KL}(P_{b'}^n \| P_{b''}^n) \leq \frac{4nh^2\varepsilon_h(N)}{N(1-h)}$. Thus, by Lemma 5, $\min_{b \in \mathcal{B}^{\mathrm{loc}}} P_b^n(A_b) \leq 0.71 \vee \frac{\frac{4nh^2\varepsilon_h(N)}{N(1-h)}}{\log(|\mathcal{B}^{\mathrm{loc}}|)}$. Noting that $\log(|\mathcal{B}^{\mathrm{loc}}|) = \log(\mathcal{M}_1^{\mathrm{loc}}(\mathcal{F}, \varepsilon_h(N), N, 1)) \geq h\gamma_{h,1}^{\mathrm{loc}}(N) \geq h\varepsilon_h(N)/c_{\mathcal{F}}$, choosing $N = \left\lceil \frac{6nc_{\mathcal{F}}h}{(1-h)} \right\rceil$ yields $\frac{4nh^2\varepsilon_h(N)}{N(1-h)} \leq \frac{2h\varepsilon_h(N)}{3c_{\mathcal{F}}} \leq \frac{2}{3} \log(|\mathcal{B}^{\mathrm{loc}}|)$, so that $\min_{b \in \mathcal{B}^{\mathrm{loc}}} P_b^n(A_b) \leq 0.71$. Finally, we have that for $h < 1$, $\sup_{P \in \mathcal{P}(h, \mathcal{F})} \mathbb{E}(R(\tilde{f}) - R(f^*)) \geq 0.29 \frac{h\varepsilon_h(N)}{4N} \geq \frac{0.29}{48c_{\mathcal{F}}} \frac{(1-h)\varepsilon_h(N)}{n} \geq \frac{0.29}{48c_{\mathcal{F}}} \frac{(1-h)\gamma_{h,1}^{\mathrm{loc}}(N)}{n}$. The term $\frac{d}{nh}$ for $h > \sqrt{\frac{d}{n}}$ is a part of the classic lower bound of [18]. $\square$

6 Discussion and Open Problems

Local entropies are well known in statistics since the early work of Le Cam [13]. Since then local metric entropies appear in minimax lower bounds. Simultaneously, the upper bounds are usually given in terms of global empirical entropies. Interestingly, it is sometimes possible to recover optimal rates by considering only global packings [21, 25]. Generally, empirical covering numbers of classes in statistics have two types of behaviour. There are *parametric* and *VC-type* classes where the logarithm of covering numbers scales as $\log(\frac{1}{\varepsilon})$ and expressive *nonparametric classes* where it scales as ε^{-p} for some $p > 0$. It was proved in [25] that for nonparametric classes local and global entropies are of the same order. Thus for such classes localization of the class does not give any significant improvement. We also note that questions similar to ours have been considered recently by Mendelson [19] and by Lecué and Mendelson [15]. Both papers show that in the convex regression setup for subgaussian classes distribution dependent fixed points of particular local entropies give optimal upper and lower bounds. However, the direct comparison with their results is problematic due to the fact that in the VC case we do not have convexity assumptions: they are replaced by noise assumptions and specifically used by our approach.

We have compared our bound with some of the best known relaxations of the bounds based on local Rademacher processes (1). However, the title of our paper demands also a direct comparison with the bounds based *solely* on local Rademacher complexities. For this, we need the following result.

Theorem 4 (Sudakov Minoration for Bernoulli Process [22]). *Let $V \subset \mathbb{R}^n$ be a finite set such that for any $v_1, v_2 \in V$ if $v_1 \neq v_2$ then $\|v_1 - v_2\|_2 \geq a$ for some $a > 0$ and for any $v \in V$ it holds $\|v\|_\infty \leq b$ for some $b > 0$. Then*

$$\mathbb{E}_\varepsilon \sup_{v \in V} \sum_{i=1}^n \varepsilon_i v_i \gtrsim a\sqrt{\log |V|} \wedge \frac{a^2}{b}. \tag{5}$$

For simplicity, we will consider only the realizable case. However we note that similar arguments will also work under bounded noise and general distributions P_X. Fix a sample $x_1, \ldots, x_n$. Applying Corollary 5.1 from [3] we have $\mathbb{E}R(\hat{f}) \lesssim \sup_{x_1, \ldots, x_n} r^*$, where r^* is a fixed point of the local empirical Rademacher complexity, that is a solution of the following equality $\frac{1}{n}\mathbb{E}_\varepsilon \sup_{g \in \text{star}(\mathcal{G}_{f*}), P_n g \leq 2r} \sum_{i=1}^n \varepsilon_i g(x_i) = r$, where star$(\mathcal{G})$ denotes the *star-hull* of a class $\mathcal{G}$: that is, the class of functions αg, where $g \in \mathcal{G}$ and $\alpha \in [0, 1]$. Since star$(\mathcal{G}_{f*})$ is star-shaped, it can be simply proven (see appropriate discussions in [19]) that local empirical entropies are not increasing in its radius. Using this fact together with (5) it can be shown

$$\mathbb{E}_\varepsilon \sup_{g \in \text{star}(\mathcal{G}_{f*}), P_n g \leq \frac{2\gamma}{n}} \sum_{i=1}^n \varepsilon_i g(x_i) \gtrsim \sqrt{\gamma}\sqrt{\log(\mathcal{M}_1^{\text{loc}}(\mathcal{F}, \gamma, n, 1))} \wedge \gamma.$$ From this it

easily follows that $\frac{\gamma_{1,1}^{\text{loc}}(n)}{n} \lesssim r^*$. Thus our bounds are not generally worse than the bounds based solely on the local Rademacher complexities.

There are still interesting questions and possible directions that are out of the scope of this paper. At first, we are focusing on a distribution free analysis. At the same time one may obtain a distribution dependent version of Theorem 2. Recently, Balcan and Long [2] have proved that for some special distributions and classes of homogenous linear separators rates of convergence of ERM may be faster than if we consider worst-case distributions. It will be interesting to generalize our results using distribution dependent fixed points of the local empirical entropy and also to miss-specified models, when $f^* \notin \mathcal{F}$.

Acknowledgments. The authors would like to thank Sasha Rakhlin for his suggestion to use offset Rademacher processes to analyze binary classification under Tsybakov noise conditions and anonymous reviewers for their helpful comments. NZ was supported solely by the Russian Science Foundation grant (project 14-50-00150).

Appendix

Proof (Theorem 1). Let DIS_0 be a disagreement set of the version space of first $\lfloor n/2 \rfloor$ instances of the learning sample. The random error set will be denoted by $E_1 = \{x \in \mathcal{X} | \hat{f}(x) \neq f^*(x)\}$. Using symmetrization Lemmas 2 and 1 we have

$$\mathbb{E}P(E_1) = \mathbb{E}R(\hat{f}) \leq \mathbb{E} \sup_{g \in \mathcal{G}_{f^*}} (Pg - (1+c)P_n g) \leq \frac{2\left(1+\frac{c}{2}\right)^2}{c} \frac{\log(S_{\mathcal{F}}(n))}{n} \text{ for } c > 0.$$

We fix $c = 2$ and prove that for any distribution $\mathbb{E}P(E_1) \leq \frac{4\log(S_{\mathcal{F}}(n))}{n}$. Now we use $R(\hat{f}) = P(E_1|\mathrm{DIS}_0)P(\mathrm{DIS}_0)$. Let $\xi = |\mathrm{DIS}_0 \cap \{X_{\lfloor n/2 \rfloor+1}, \ldots, X_n\}|$. Conditionally on the first $\lfloor n/2 \rfloor$ instances ξ has binomial distribution. Expectations with respect to the first and the last parts of the sample will be denoted respectfully by $\mathbb{E}$ and $\mathbb{E}'$. Conditionally on $\{x_1, \ldots, x_{\lfloor n/2 \rfloor}\}$ we introduce two events: $A_1 : \xi < \frac{nP(\mathrm{DIS}_0)}{4}$ and $A_2 : \xi > \frac{3nP(\mathrm{DIS}_0)}{4}$. Using Chernoff bounds we have $P(A_1) \leq \exp\left(-\frac{nP(\mathrm{DIS}_0)}{16}\right)$ and $P(A_2) \leq \exp\left(-\frac{nP(\mathrm{DIS}_0)}{16}\right)$. Denote $A = A_1 \cup A_2$. Then $\mathbb{E}'P(E_1|\mathrm{DIS}_0) = \mathbb{E}'\left[P(E_1|\mathrm{DIS}_0)\big|\overline{A}\right]P(\overline{A}) + \mathbb{E}'\left[P(E_1|\mathrm{DIS}_0)\big|A\right]P(A)$. For the first term we have $\mathbb{E}'\left[P(E_1|\mathrm{DIS}_0)\big|\overline{A}\right]P(\overline{A}) \leq \frac{16\log\left(S_{\mathcal{F}}\left(\frac{3nP(\mathrm{DIS}_0)}{4}\right)\right)}{nP(\mathrm{DIS}_0)}$ We can directly prove for the second term that $\mathbb{E}'\left[P(E_1|\mathrm{DIS}_0)\big|A\right]P(\mathrm{DIS}_0)P(A) \leq \frac{12}{n}$. It easy to see, that for all natural k, r we have $(S_{\mathcal{F}}(kr))^{\frac{1}{r}} \leq S_{\mathcal{F}}(k)$. Finally,

$$\mathbb{E}R(\hat{f}) \leq \mathbb{E}\frac{16\log\left(S_{\mathcal{F}}\left(\frac{3nP(\mathrm{DIS}_0)}{4}\right)\right)}{n} + \frac{12}{n} \leq \frac{40\log(S_{\mathcal{F}}(s))}{n} + \frac{12}{n}. \qquad \square$$

Proof (Lemma 4). Once again, given $X_1, \ldots, X_n$, let $V = \{(g(X_1), \ldots, g(X_n)) : g \in \mathcal{G}\}$ denote the set of binary vectors corresponding to the values of functions in $\mathcal{G}$. As above, for a fixed γ and fixed minimal γ-covering subset $\mathcal{N}_\gamma \subseteq V$, for each $v \in V$, $p(v)$ will denote the closest vector to v in $\mathcal{N}_\gamma$. We will denote by $\mathbb{E}_\xi$ the conditional expectation over the ξ_i variables, given $X_1, \ldots, X_n$. We follow the decomposition proposed by Liang, Rakhlin, and Sridharan [16]:

$$\frac{1}{n}\mathbb{E}_\xi \max_{v \in V} \left(\sum_{i=1}^n \xi_i v_i - cv_i\right) \leq \frac{1}{n}\mathbb{E}_\xi \max_{v \in V} \left(\sum_{i=1}^n \xi_i(v_i - p(v)_i)\right)$$

$$+ \frac{1}{n}\mathbb{E}_\xi \max_{v \in V} \left(\sum_{i=1}^n \frac{c}{4}p(v)_i - cv_i\right) + \frac{1}{n}\mathbb{E}_\xi \max_{v \in V} \left(\sum_{i=1}^n \xi_i p(v)_i - \frac{c}{4}p(v)_i\right).$$

The first term is $\lesssim \frac{\gamma}{n}$ by the γ-cover property and the fact that $|\xi_i| \lesssim 1$. Furthermore it is easy to show that the second term is at most $\frac{c}{4}\frac{\gamma}{n}$. Now we analyze the last term carefully. First we use the standard peeling argument. Given a set W of binary vectors we define $W[a, b] = \{w \in W | a \leq \rho_H(w, 0) < b\}$.

$$\mathbb{E}_\xi \max_{v \in V} \left(\sum_{i=1}^n \xi_i p(v)_i - \frac{c}{4} p(v)_i \right) = \mathbb{E}_\xi \max_{v \in \mathcal{N}_\gamma} \left(\sum_{i=1}^n \xi_i v_i - \frac{c}{4} v_i \right)$$

$$\leq \mathbb{E}_\xi \max_{v \in \mathcal{N}_\gamma[0,2\gamma/c]} \left(\xi_i v_i - \frac{c}{4} v_i \right) + \sum_{k=1}^\infty \mathbb{E}_\xi \max_{\mathcal{N}_\gamma[2^k\gamma/c, 2^{k+1}\gamma/c]} \left(\sum_{i=1}^n \xi_i v_i - \frac{c}{4} v_i \right)_+.$$

The first term is upper bounded by $\frac{2 \log(\mathcal{M}_1^{\mathrm{loc}}(V,\gamma,n,c))}{cn}$ by Lemma 1 and by noting that $|\mathcal{N}_\gamma[0, 2\gamma/c]| \leq \mathcal{M}_1(\mathcal{B}_H(0, (2\gamma)/c, \{X_1, \ldots, X_n\}), (2\gamma)/2) \leq \mathcal{M}_1^{\mathrm{loc}}(V, \gamma, n, c)$. Now we upper-bound the second term. We start with an arbitrary summand. For any $\lambda > 0$, we have

$$\mathbb{E}_\xi \max_{v \in \{0\} \cup \mathcal{N}_\gamma[2^k\gamma/c, 2^{k+1}\gamma/c]} \left(\sum_{i=1}^n \xi_i v_i - \frac{c}{4} v_i \right)$$

$$\leq \frac{1}{\lambda} \ln \left(\sum_{v \in \mathcal{N}_\gamma[2^k\gamma/c, 2^{k+1}\gamma/c]} \mathbb{E}_\xi \exp \left\{ \sum_{i=1}^n \lambda \xi_i v_i - \frac{\lambda c}{4} v_i \right\} + 1 \right)$$

$$\leq \frac{1}{\lambda} \ln \left(\left| \mathcal{N}_\gamma \left[2^k\gamma/c, 2^{k+1}\gamma/c \right] \right| \exp \left\{ 2^{k-2}\gamma(4\lambda^2 - \lambda c)/c \right\} + 1 \right)$$

$$\leq \frac{1}{\lambda} \ln \left(\left(\mathcal{M}_1^{\mathrm{loc}}(\mathcal{G}, 2\gamma, n, c) \right)^{2^{k+1}} \exp \left\{ 2^{k-2}\gamma(4\lambda^2 - \lambda c)/c \right\} + 1 \right).$$

Here we used that $\left| \mathcal{M}_\gamma \left[0, 2^{k+1}\gamma/c \right] \right| \leq \left| \mathcal{M}_1^{\mathrm{loc}}(\mathcal{G}, 2\gamma, n, c) \right|^{2^{k+1}}$ and that any minimal covering is also a packing. We fix $\gamma = K\gamma_{c,c}^{\mathrm{loc}}(n)$ for some $K > 2$. Observe that local entropy is nonincreasing and $K\gamma_{c,c}^{\mathrm{loc}}(n) > 2\gamma_{c,c}^{\mathrm{loc}}(n) \geq \gamma_{c,c}^{\mathrm{loc}}(n) + 1$. Thus,

$$\ln \left(\exp \left\{ 2^{k+1} \log \left(\mathcal{M}_1^{\mathrm{loc}}(V, 2K\gamma_{c,c}^{\mathrm{loc}}(n), n, c) \right) + 2^{k-2} K\gamma_{c,c}^{\mathrm{loc}}(n)(4\lambda^2 - \lambda c)/c \right\} + 1 \right)$$

$$\leq \ln \left(\exp \left\{ 2^{k+1} c(\gamma_{c,c}^{\mathrm{loc}}(n) + 1) + 2^{k-2} K\gamma_{c,c}^{\mathrm{loc}}(n)(4\lambda^2 - \lambda c)/c \right\} + 1 \right).$$

Then we have for $\lambda = \frac{c}{8}$,

$$\sum_{k=1}^\infty \frac{8}{c} \ln \left(\exp \left(2^{k+1} \log \left(\mathcal{M}_1^{\mathrm{loc}}(\mathcal{G}, 2K\gamma_{c,c}^{\mathrm{loc}}(n), n) \right) \right) \exp \left(-2^{k-6} Kc\gamma_{c,c}^{\mathrm{loc}}(n) \right) + 1 \right)$$

$$\leq \sum_{k=1}^\infty \frac{8}{c} \ln \left(\exp \left(2^{k+2} c\gamma_{c,c}^{\mathrm{loc}}(n) - 2^{k-6} Kc\gamma_{c,c}^{\mathrm{loc}}(n) \right) + 1 \right).$$

We set $K = 2^9$ and have $\sum_{k=1}^\infty \ln \left(\exp \left(2^{k+2} c\gamma_{c,c}^{\mathrm{loc}}(n) - 2^{k-6} Kc\gamma_{c,c}^{\mathrm{loc}}(n) \right) + 1 \right) \leq C$, where $C > 0$ is an absolute constant. Here we used that $\ln(x + 1) \leq x$ for $x > 0$ and $c\gamma_{c,c}^{\mathrm{loc}} \gtrsim 1$. Combining with the first two terms we finish the proof. $\qquad\square$

Proof (Proposition 1). The first part of the proof closely follows the proof of Theorem 17 in [8], with slight modifications, to arrive at an upper bound on $\mathcal{M}_1^{\mathrm{loc}}(\mathcal{F}, \gamma, n, h)$. The suprema in the definition of local empirical entropy are achieved at some set $\{x_1, \ldots, x_n\}$, some function $f \in \mathcal{F}$, and some $\varepsilon \in [\gamma, n]$. Letting $r = \varepsilon/n$, denote by $\mathcal{M}_r$ the maximal $(rn/2)$-packing (under ρ_H) of $\mathcal{B}_H(f, rn/h, \{x_1, \ldots, x_n\})$, so that $|\mathcal{M}_r| = \mathcal{M}_1^{\mathrm{loc}}(\mathcal{F}, \gamma, n, h)$. Also introduce a uniform probability measure P_X on $\{x_1, \ldots, x_n\}$ and fix $m = \lceil \frac{4}{r} \log(|\mathcal{M}_r|) \rceil$. Let $X_1, \ldots, X_m$ be m independent P_X-distributed random variables, and let A denote the event that, for all $g, g' \in \mathcal{M}_r$ with $g \neq g'$, there exists an $i \in \{1, \ldots, n\}$ such that $g(X_i) \neq g'(X_i)$. For a given pair of distinct functions $g, g' \in \mathcal{M}_r$, they disagree on some X_i with probability $1 - (1 - P_X(g(X) \neq g'(X)))^m > 1 - \exp(-rm/2) \geq 1 - \frac{1}{|\mathcal{M}_r|^2}$. Using a union bound and summing over all possible unordered pairs $g, g' \in \mathcal{M}_r$ will give us that $\mathbb{P}(A) > \frac{1}{2}$. On the event A, functions in $\mathcal{M}_r$ realize distinct classifications of $X_1, \ldots, X_m$. For any $X_i \notin \mathrm{DIS}(\mathcal{B}_H(f, rn/h, \{x_1, \ldots, x_n\}))$, all classifiers in $\mathcal{M}_r$ agree. Thus, $|\mathcal{M}_r|$ is bounded by the number of classifications $\{X_1, \ldots, X_m\} \cap \mathrm{DIS}(\mathcal{B}_H(f, rn/h))$ realized by classifiers in $\mathcal{F}$. By the Chernoff bound, on an event B with $\mathbb{P}(B) \geq \frac{1}{2}$ we have $|\{X_1, \ldots, X_m\} \cap \mathrm{DIS}(\mathcal{B}_H(f, rn/h))| \leq 1 + 2e P_X(\mathrm{DIS}(\mathcal{B}_H(f, rn/h))m$. Using the definition of $\tau(\cdot)$ (Definition 2) we have $1 + 2e P_X(\mathrm{DIS}(\mathcal{B}_H(f, rn/h)))m \leq 1 + 2e\tau\left(\frac{r}{h}\right) \frac{r}{h} m \leq 11 e \tau\left(\frac{r}{h}\right) \frac{\log(|\mathcal{M}_r|)}{h}$. With probability at least $\frac{1}{2}$, $|\{X_1, \ldots, X_m\} \cap \mathrm{DIS}(\mathcal{B}_H(f, rn/h))| \leq 11 e \tau\left(\frac{r}{h}\right) \frac{\log(|\mathcal{M}_r|)}{h}$. Using the union bound, we have that with positive probability there exists a sequence of at most $11 e \tau\left(\frac{r}{h}\right) \frac{\log(|\mathcal{M}_r|)}{h}$ elements, such that all functions in $\mathcal{M}_r$ classify this sequence distinctly. By the VC lemma [23], we therefore have that $|\mathcal{M}_r| \leq \left(\frac{11 e^2 \tau\left(\frac{r}{h}\right) \frac{\log(|\mathcal{M}_r|)}{h}}{d} \right)^d$.

Using Corollary 4.1 from [24] we have $\log(|\mathcal{M}_r|) \leq 2d \log\left(11 e^2 \tau\left(\frac{r}{h}\right) \frac{1}{h}\right)$. Using $\tau\left(\frac{r}{h}\right) \leq \mathbf{s} \wedge \frac{h}{r} \leq \mathbf{s} \wedge \frac{nh}{\gamma}$ (Theorem 10 in [8]) we finally have $\log(\mathcal{M}_1^{\mathrm{loc}}(\mathcal{F}, \gamma, n, h)) \leq 2d \log\left(11 e^2 \left(\frac{n}{\gamma} \wedge \frac{\mathbf{s}}{h}\right)\right)$. Observe that $h \gamma_{h,h}^{\mathrm{loc}}(n) \leq 2d \log\left(11 e^2 \left(\frac{n}{\gamma_{h,h}^{\mathrm{loc}}(n)} \wedge \frac{\mathbf{s}}{h}\right)\right)$. We have $\gamma_{h,h}^{\mathrm{loc}}(n) \leq \frac{2d \log\left(11 e^2 \frac{\mathbf{s}}{h}\right)}{h}$. If $\gamma = \frac{2d \log\left(11 e^2 \frac{nh}{d}\right)}{h}$, then $h\gamma = 2d \log\left(11 e^2 \frac{nh}{d}\right)$, but $2d \log\left(11 e^2 \frac{n}{\gamma}\right) \leq 2d \log\left(11 e^2 \frac{nh}{d}\right)$ if $h > \frac{d}{11 en}$. Finally, we have $\gamma_{h,h}^{\mathrm{loc}}(n) \leq \frac{2d \log\left(11 e^2 \left(\frac{nh}{d} \wedge \frac{\mathbf{s}}{h}\right)\right)}{h}$. Now we prove the lower bound. From (2) established above, we know that $\frac{\gamma_{h,h}^{\mathrm{loc}}(n)}{n}$ is, up to an absolute constant, a distribution-free upper bound for $\mathbb{E}(R(\hat{f}) - R(f^*))$, holding for all ERM learners $\hat{f}$. Then a lower bound on $\sup_{P \in \mathcal{P}(h, \mathcal{F})} \mathbb{E}(R(\hat{f}) - R(f^*))$ holding for any ERM learner is also a lower bound for $\frac{\gamma_{h,h}^{\mathrm{loc}}(n)}{n}$. In particular, it is known [9,18] that for any learning procedure $\tilde{f}$, if $h \geq \sqrt{\frac{d}{n}}$, then $\sup_{P \in \mathcal{P}(h, \mathcal{F})} \mathbb{E}(R(\tilde{f}) - R(f^*)) \gtrsim \frac{d + (1-h) \log(nh^2 \wedge \mathbf{s})}{nh}$, while if $h < \sqrt{\frac{d}{n}}$ then $\sup_{P \in \mathcal{P}(h, \mathcal{F})} \mathbb{E}(R(\tilde{f}) - R(f^*)) \gtrsim \sqrt{\frac{d}{n}}$. Furthermore, in the particular case of ERM, [9] proves that any upper bound on $\sup_{P \in \mathcal{P}(1, \mathcal{F})} \mathbb{E}(R(\hat{f}) - R(f^*))$ holding for all ERM learners $\hat{f}$ must have size, up to an absolute constant, at least $\frac{\log(n \wedge \mathbf{s})}{n}$. Together, these lower bounds imply $\gamma_{h,h}^{\mathrm{loc}}(n) \gtrsim \frac{d + \log(nh^2 \wedge \mathbf{s})}{h} \wedge \sqrt{dn}$. $\qquad\square$

References

1. Alexander, K.S.: Rates of growth and sample moduli for weighted empirical processes indexed by sets. Probab. Theory Relat. Fields **75**, 379–423 (1987)
2. Balcan, M.F., Long, P.M.: Active and passive learning of linear separators under log-concave distributions. In: 26th Conference on Learning Theory (2013)
3. Bartlett, P.L., Bousquet, O., Mendelson, S.: Local Rademacher complexities. Ann. Stat. **33**(4), 1497–1537 (2005)
4. Boucheron, S., Bousquet, O., Lugosi, G.: Theory of classification: a survey of recent advances. ESAIM: Probab. Stat. **9**, 323–375 (2005)
5. Devroye, L., Györfi, L., Lugosi, G.: A Probabilistic Theory of Pattern Recognition. Applications of Mathematics, vol. 31. Springer, New York (1996)
6. Giné, E., Koltchinskii, V.: Concentration inequalities and asymptotic results for ratio type empirical processes. Ann. Probab. **34**(3), 1143–1216 (2006)
7. Hanneke, S.: Theory of disagreement-based active learning. Found. Trends Mach. Learn. **7**(2–3), 131–309 (2014)
8. Hanneke, S., Yang, L.: Minimax analysis of active learning. J. Mach. Learn. Res. **16**(12), 3487–3602 (2015)
9. Hanneke, S.: Refined error bounds for several learning algorithms (2015). http://arXiv.org/abs/1512.07146
10. Haussler, D., Littlestone, N., Warmuth, M.: Predicting $\{0, 1\}$-functions on randomly drawn points. Inf. Comput. **115**, 248–292 (1994)
11. Haussler, D.: Sphere packing numbers for subsets of the Boolean n-cube with bounded Vapnik–Chervonenkis dimension. J. Combin. Theory Ser. A **69**, 217–232 (1995)
12. Koltchinskii, V.: Local Rademacher complexities and oracle inequalities in risk minimization. Ann. Stat. **34**(6), 2593–2656 (2006)
13. Le Cam, L.M.: Convergence of estimates under dimensionality restrictions. Ann. Statist. **1**, 38–53 (1973)
14. Lecué, G., Mitchell, C.: Oracle inequalities for cross-validation type procedures. Electron. J. Stat. **6**, 1803–1837 (2012)
15. Lecué, G., Mendelson, S.: Learning subgaussian classes: upper and minimax bounds (2013). http://arXiv.org/abs/1305.4825
16. Liang, T., Rakhlin, A., Sridharan, K.: Learning with square loss: localization through offset Rademacher complexity. In: Proceedings of The 28th Conference on Learning Theory (2015)
17. Massart, P.: Concentration Inequalties and Model Selection. Ecole dEtè de Probabilités, Saint Flour. Springer, New York (2003)
18. Massart, P., Nédélec, E.: Risk bounds for statistical learning. Ann. Stat. **34**(5), 2326–2366 (2006)
19. Mendelson, S.: 'Local' vs. 'global' parameters – breaking the Gaussian complexity barrier (2015). http://arXiv.org/abs/1504.02191
20. Raginsky, M., Rakhlin, A.: Lower bounds for passive and active learning. In: Advances in Neural Information Processing Systems 24, NIPS (2011)
21. Rakhlin, A., Sridharan, K., Tsybakov, A.B.: Empirical entropy, minimax regret and minimax risk. Bernoulli (2015, forthcoming)
22. Talagrand, M.: Upper and Lower Bounds for Stochastic Processes. Springer, Heidelberg (2014)
23. Vapnik, V., Chervonenkis, A.: On the uniform convergence of relative frequencies of events to their probabilities. Proc. USSR Acad. Sci. **181**(4), 781–783 (1968). English tranlation: Soviet Math. Dokl. **9**, 915–918

24. Vidyasagar, M.: Learning and Generalization with Applications to Neural Networks, 2nd edn. Springer, Heidelberg (2003)
25. Yang, Y., Barron, A.: Information-theoretic determination of minimax rates of convergence. Ann. Stat. **27**, 1564–1599 (1999)

Labeled Compression Schemes for Extremal Classes

Shay Moran[1,2,3]($\boxtimes$) and Manfred K. Warmuth[4]

[1] Technion, Israel Institute of Technology, 32000 Haifa, Israel
shaymoran1@gmail.com
[2] Microsoft Research, Herzliya, Israel
[3] Max Planck Institute for Informatics, Saarbrücken, Germany
[4] Computer Science Department, University of California,
Santa Cruz, USA
manfred@ucsc.edu

Abstract. It is a long-standing open problem whether there exists
a compression scheme whose size is of the order of the Vapnik-
Chervonienkis (VC) dimension d. Recently compression schemes of size
exponential in d have been found for any concept class of VC dimension d. Previously, compression schemes of size d have been given for
maximum classes, which are special concept classes whose size equals
an upper bound due to Sauer-Shelah. We consider a generalization of
maximum classes called extremal classes. Their definition is based on a
powerful generalization of the Sauer-Shelah bound called the Sandwich
Theorem, which has been studied in several areas of combinatorics and
computer science. The key result of the paper is a construction of a sample compression scheme for extremal classes of size equal to their VC
dimension. We also give a number of open problems concerning the combinatorial structure of extremal classes and the existence of unlabeled
compression schemes for them.

1 Introduction

Generalization and compression/simplification are two basic facets of "learning".
Generalization concerns the expansion of existing knowledge and compression
concerns simplifying our explanations of it. In machine learning, compression
and generalization are deeply related: learning algorithms perform compression
and the ability to compress guarantees good generalization.

A simple form of this connection is how Occam's Razor [5] is manifested
in Machine Learning: if the input sample can be compressed to a small number of bits which encodes a hypothesis consistent with the input sample, then
good generalization is guaranteed. A more sophisticated notion of compression is
given by "sample compression schemes" [20]. In these schemes the input sample
is compressed to a carefully chosen small subsample that encodes a hypothesis
consistent with the input sample. For example support vector machine can be

---

Supported by NSF grant IIS-1118028. See [25] for a slightly more detailed version.

R. Ortner et al. (Eds.): ALT 2016, LNAI 9925, pp. 34–49, 2016.
DOI: 10.1007/978-3-319-46379-7_3

seen as compressing the original sample to the subset of support vectors which represent a maximum margin hyperplane that is consistent with the entire original sample.

What is the connection to generalization? In the Occam's razor setting, the generalization error decreases with the number of bits that are used to encode the output hypothesis. Similarly for compression schemes, the generalization error decreases with the sample size.

A core question is what parameter of the concept class characterizes the sample size required for good generalization? The Vapnik-Chervonenkis (VC) dimension serves as such a parameter [4], where the exact definition of generalization underlying our discussion is specified by the Probably Approximately Correct (PAC) model of learning [32]. The size of the best compression scheme is an alternate parameter and has several additional advantages: (i) Compression schemes frame many natural algorithms (e.g. support vector machines). (ii) Unlike the VC dimension, the definition of sample compression schemes as well as the fact that they yield low generalization error extends naturally to multi label concept classes [29]. This is particularly interesting when the number of labels is very large (or possibly infinite), because for that case there is no known combinatorial parameter that characterizes the sample complexity in the PAC model (See [9]).

Previous Work. In 1986, [20] defined *sample compression schemes* and showed that the sample size required for learning grows linearly with the size of the subsamples the scheme compresses to. They have also posed the other direction as an open question: Does every concept class have a compression scheme of size depending only on its VC dimension? Later [12,33], refined this question: Does every class of VC dimension d have a sample compression scheme of size $O(d)$.

[3] proved a compactness theorem for sample compression schemes. It essentially says that existence of compression schemes for infinite classes follows[1] from the existence of such schemes for finite classes. Thus, it suffices to consider only finite concept classes. [12] constructed sample compression schemes of size $\log |C|$ for every concept class C. More recently [24] have constructed sample compression schemes of size $\exp(d) \log \log |C|$ where $d = VCdim(C)$. Finally, [26] have constructed sample compression scheme of size $\exp(d)$, resolving Littlestone and Warmuth's question. Their compression scheme is based on an earlier compression scheme which was defined in the context of boosting (see [13]). This sample compression scheme is of variable size: It compresses samples of size m to subsamples of size $O(d \log m)$.

For many natural and important families of concept classes, sample compression schemes of size equal the VC dimension were constructed (e.g. [3,8,21,28]). However, the question whether there exists a compression scheme whose size is equal or linear in the VC dimension remains open.

[12] observed that in order to prove the conjecture it suffices to consider only maximal classes (A class C is maximal if no concept can be added without

---

[1] The proof of that theorem is however non-constructive.

increasing the VC dimension). Furthermore, they constructed sample compression schemes of size d for every *maximum class* of VC dimension d. These classes are maximum in the sense that their size equals an upper bound (due to Sauer-Shelah) on the size of any concept class of VC dimension d. Later, [17,28] provided even more efficient sample compression schemes for maximum classes that are called unlabeled compression schemes because the labels of the subsample are not needed to encode the output hypothesis.

One possibility of making a progress on Floyd and Warmuth's question is by extending the optimal compression schemes for maximum classes to a more general family. In this paper we consider a natural and rich generalization of maximum classes which are known as extremal classes (or shattering extremal classes). Similar to maximum classes, these classes are defined when a certain inequality, which generalizes the Sauer-Shelah bound, and known as The Sandwich Theorem is tight. The Sandwich Theorem as well as extremal classes were discovered several times and independently by several groups of researchers and in several contexts such as Functional analysis [27], Discrete-geometry [18], Phylogenetic Combinatorics [2,11] and Extremal Combinatorics [6,7]. Even though a lot of knowledge regarding the structure of extremal classes has been accumulated, the understanding of these classes is still considered incomplete by several authors [6,15].

Our Results. Our main result is a construction of sample compression scheme of size d for every extremal class of VC dimension d. When the concept class is maximum, then our scheme specializes to the compression scheme for maximum classes given in [12]. Our generalized sample compression scheme for extremal classes is still easy to describe. However its analysis requires more combinatorics and heavily exploits the rich structure of extremal classes. Despite being more general, the construction is simple. We also give explicit examples of maximal classes that are extremal but not maximum (see Example 5).

We also discuss a certain greedy peeling method for producing an unlabeled compressions scheme. Such schemes were first conjectured in [17] and later proven to exist for maximum classes [28]. However the existence of such schemes for extremal classes remains open. We relate the existence of such schemes to basic open questions concerning the combinatorial structure of extremal classes.

Organization. In Sect. 2 we give some preliminary definitions and define extremal classes. We also discuss some basic properties and give some examples of extremal classes which demonstrate their generality over maximum classes. In Sect. 3 we give a labeled compression scheme for any extremal class of VC dimension d. Finally, in Sect. 4 we relate unlabeled compression schemes for extremal classes with basic open questions concerning extremal classes.

2   Extremal Classes

2.1   Preliminaries

Concepts, Concept Classes, and the One-Inclusion Graph. A concept c is a mapping from some domain to $\{0,1\}$. We assume that the domain of c (denoted by $dom(c)$) is finite and allow the case that $dom(c) = \emptyset$. A concept c can also be viewed as a characteristic function of a subset of $dom(c)$, i.e. for any domain point $x \in dom(c)$, $c(x) = 1$ iff $x \in c$. A concept class C is a set of concepts with the same domain (denoted by $dom(C)$). A concept class can be represented by a binary table (see Fig. 1), where the rows correspond to concepts and the columns to the elements of $dom(C)$. Whenever the elements in $dom(C)$ are clear from the context, then we represent concepts as bit strings of length $|dom(C)|$ (See Fig. 1).

The concept class C can also be represented as a subgraph of the Boolean hypercube with $|dom(C)|$ dimensions. Each dimension corresponds to a particular domain element, the vertices are the concepts in C and two concepts are connected with an edge if they disagree on the label of a single element (Hamming distance 1). This graph is called the *one-inclusion graph* of C. Note that each edge is naturally labeled by the single dimension/element on which the incident concepts disagree (See Fig. 1).

	$x_1\, x_2\, x_3\, x_4\, x_5\, x_6$
c_1	0 0 0 0 0 0
c_2	0 0 1 0 0 0
c_3	0 1 0 0 0 0
c_4	1 0 0 0 0 0
c_5	0 0 1 0 1 0
c_6	0 0 1 1 0 0
c_7	1 0 1 0 0 0
c_8	1 1 0 0 0 0
c_9	0 0 1 0 1 1
c_{10}	0 0 1 1 1 0
c_{11}	0 0 1 1 0 1
c_{12}	1 0 1 1 0 0
c_{13}	1 1 1 0 0 0
c_{14}	1 1 0 1 0 0
c_{15}	0 0 1 1 1 1
c_{16}	1 0 1 1 0 1
c_{17}	1 1 1 1 0 0
c_{18}	1 0 1 1 1 1

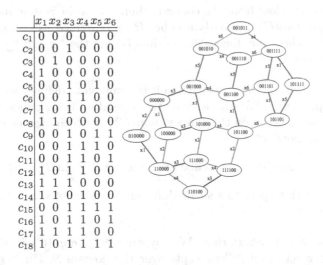

Fig. 1. The table and the one-inclusion graph of an extremal class C of VC dimension 2. The reduction $C^{x_2} = \{00000, 10000, 11000, 11100\}$ has the domain $\{x_1, x_3, x_4,\ x_5, x_6\}$. Notice that each concept in C^{x_2} corresponds to an edge labelled with x_2. Similarly $C^{\{x_3, x_4\}}$ consists of the single concept $\{1100\}$ over the reduced domain $\{x_1, x_2, x_5, x_6\}$. Notice that this concept corresponds to the single cube of C with dimension set $\{x_3, x_4\}$.

Restrictions and Samples. We denote the *restriction/sample* of a concept c onto $S \subseteq dom(c)$ as $c|S$. This concept has the restricted domain S and labels this domain consistently with c. Essentially concept $c|S$ is obtained by removing from row c in the table all columns not in S. The *restriction/set of samples* of an entire class C onto $S \subseteq dom(C)$ is denoted as $C|S$. A table for $C|S$ is produced by simply removing all columns not in S from the table for C and collapsing identical rows.[2] Also the one-inclusion graph for the restriction $C|S$ is now a subgraph of the boolean hypercube with $|S|$ dimensions instead of the full dimension $|dom(C)|$. We also use $C - S$ as shorthand for $C|(dom(C) \setminus S)$ (since the columns labeled with S are removed from the table). Note that the sub domain $S \subseteq dom(C)$ induces an equivalence class on C: Two concepts $c, c' \in C$ are equivalent iff $c|S = c'|S$. Thus there is one equivalence class per concept of $C|S$.

Cubes. A concept class B is called a cube if for some subset S of the domain $dom(B)$, the restriction $B|S$ is the set of all $2^{|S|}$ concepts over the domain S and the class $B - S$ contains a single concept. We denote this single concept by $tag(B)$. In this case, we say that S is the dimension set of B (denoted as $\dim(B)$). For example, if B contains two concepts that are incident to an edge labeled x then B is a cube with $\dim(B) = \{x\}$. We say that B *is a cube of* concept class C if B is a cube that is a subset of C. We say that B is a maximal cube of C if there exists no other cube of C which strictly contains B. When the dimensions are clear from the context, then a concept is described as a bit string of length $dom(C)$. Similarly a cube, B, is described as an expression in $\{0, 1, *\}^{|dom(C)|}$, where the dimensions of $\dim(B)$ are the $*$'s and the remaining bits is the concept $tag(B)$.

Reductions. In addition to the restriction it is common to define a second operation on concept classes. We will describe this operation using cubes. The *reduction* C^S is a concept class on the domain $dom(C) \setminus S$ which has one concept per cube with dimensions set S

$$C^S := \{tag(B) : B \text{ is a cube of } C \text{ such that } \dim(B) = S\}.$$

The reduction with respect to a single dimension x is denoted as C^x. See Fig. 1 for some examples.

Shattering and Strong Shattering. We say that $S \subseteq dom(C)$ is *shattered* by C, if $C|S$ is the set of all $2^{|S|}$ concepts over the domain S. Furthermore, S is *strongly shattered* by C, if C has a cube with dimensions set S. We use $s(C)$ to denote all shattered sets of C and $st(C)$ to denote all strongly shattered sets, respectively. Clearly, both $s(C)$ and $st(C)$ are closed under the subset relation, and $st(C) \subseteq s(C)$.

The following theorem is the result of accumulated work by different authors, and parts of it were rediscovered independently several times [1,6,11,27].

---

[2] We define $c|\emptyset = \emptyset$. Note that $C|\emptyset = \{\emptyset\}$ if $C \neq \emptyset$ and $\emptyset$ otherwise.

Theorem 1 (Sandwich Theorem). *For any concept class C,* $|st(C)| \leq |C| \leq |s(C)|$.

This theorem has been discovered independently several times and has several proofs (see [23] for more details).

The inequalities in this theorem can be strict: Let $C \subseteq \{0,1\}^n$ be such that C contains all boolean vectors with an even number of $1's$. Then $st(C)$ contains only the empty set and $s(C)$ contains all subsets of $\{1, \ldots, n\}$ of size at most $n - 1$. Thus in this example, $|st(C)| = 1$, $|C| = 2^{n-1}$, and $|s(C)| = 2^n - 1$.

The *VC dimension* [4] is defined as: $VCdim(C) = \max\{|S| : S \in s(C)\}$. Clearly, $s(C) \subseteq \{S \subseteq dom(C) : |S| \leq VCdim(C)\}$ and hence the cardinality $|C| \leq \sum_{i=0}^{VCdim(C)} \binom{|dom(C)|}{i}$. Thus the Sandwich theorem implies the well-known Sauer-Shelah Lemma [30, 31].

2.2   Definition of Extremal Classes and Examples

Maximum classes are defined as concept classes which satisfy the Sauer-Shelah inequality with equality. Analogously, *extremal classes* are defined as concept classes which satisfy the inequalities[3] in the Sandwich Theorem with equality: A concept class C is *extremal* if for every shattered set S of C there is a cube of C with dimension set S, i.e. $s(C) = st(C)$.

Every maximum class is an extremal class. Moreover, maximum classes of VC dimension d are precisely the extremal classes for which the shattered sets consist of all subsets of the domain of size up to d. The other direction does not hold - there are extremal classes that are not maximum. All the following examples are extremal but not maximum.

Example 1. Consider the concept class C over the domain $\{x_1, \ldots, x_6\}$ given in Fig. 1. In this example $st(C) = s(C) = \{\emptyset, \{x_1\}, \{x_2\}, \{x_3\}, \{x_4\}, \{x_5\}, \{x_6\}, \{x_1, x_2\}, \{x_1, x_3\}, \{x_1, x_4\}, \{x_1, x_5\}, \{x_1, x_6\}, \{x_2, x_3\}, \{x_2, x_4\}, \{x_3, x_4\}, \{x_4, x_5\}, \{x_4, x_6\}, \{x_5, x_6\}\}$. This example also demonstrates the cubical structure of extremal classes.

Example 2 **(Downward-Closed Classes).** A standard example of a maximum class of VC dimension d is

$$C = \{c \in \{0,1\}^n : \text{the number of 1's in } c \text{ is at most } d\}.$$

This is simply the hamming ball of radius d around the all 0's concept. A natural generalization of such classes are downward closed classes. We say that C is downward closed if for all $c \in C$ and for all $c' \leq c$, also $c' \in C$. Here $c' \leq c$ means that for every $x \in dom(C)$, $c'(x) \leq c(x)$. It is not hard to verify that every downward closed class is extremal.

---

[3] There are two inequalities in the Sandwich Theorem, but every class which satisfies one of them with equality also satisfies the other with equality (See Theorem 2).

Example 3 **(Hyper-Planes Arrangements in a Convex Domain).** Another standard set of examples for maximum classes comes from geometry (see e.g. [14]). Let H be an arrangement of hyperplanes in $\mathbb{R}^d$. For each hyperplane $p_i \in H$, pick one of half-planes determined by p_i to be its *positive side* and the other its *negative side*. The hyperplanes of H cut $\mathbb{R}^d$ into open regions (*cells*). Each cell defines a binary mapping with domain H:

$$c(p_i) = \begin{cases} 1 & \text{if } c \text{ is in the positive side of } p_i \\ 0 & \text{if } c \text{ is in the negative side of } p_i. \end{cases}$$

It is known that if the hyperplanes are in general position, then the set C of all cells is a maximum class of VC dimension d.

Consider the following generalization of these classes: Let $K \subseteq \mathbb{R}^d$ be a convex set. Instead of taking the vectors corresponding to all of the cells, take only those that correspond to cells that intersect K:

$$C_K = \{c : c \text{ corresponds to a cell that intersects } K\}.$$

C_K is extremal. In fact, for C_K to be extremal it is not even required that the hyperplanes are in general position. It suffices to require that no $d+1$ hyperplanes have a non-empty intersection (e.g. parallel hyperplanes are allowed). Figure 2 illustrates such a class C_K in the plane. These classes were studied in [23].

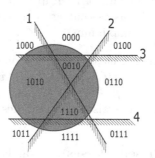

Fig. 2. An extremal class that correspond to the cells of a hyperplane arrangement of a convex set. An arrangement of 4 lines is given which partitions the plane to 10 cells. Each cell corresponds to a binary vector which specifies its location relative to the lines. For example the cell corresponding to 1010 is on the positive sides of lines 1 and 3 and on the negative side of lines 3 and 4. Here the convex set K is an ellipse and the extremal concept class consisting of the cells the ellipse intersects is $C_K = \{1000, 1010, 1011, 1111, 1110, 0010, 0000, 0110\}$ (the cells $0100, 0111$ are not intersected by the ellipse). The class C_K here has VC dimension 2. Note that it's shattered sets of size 2 are exactly the pairs of lines whose intersection point lies in the ellipse K.

Interestingly, extremal classes also arise in the context of graph theory:

Example 4 (**Edge-Orientations Which Preserve Connectivity** [16]). Let $G = (V, E)$ be an undirected simple graph and let $\overrightarrow{E}$ be a fixed reference orientation. Now an arbitrary orientation of E is a function $d : E \to \{0, 1\}$: If $d(e) = 0$ then e is oriented as in $\overrightarrow{E}$ and if $d(e) = 1$ then e is oriented opposite to $\overrightarrow{E}$. Now let $s, t \in V$ be two fixed vertices, and consider all orientations of E for which there exists a directed path from s to t. The corresponding class of orientations $E \to \{0, 1\}$ is an extremal concept class over the domain E.

Moreover, the extremality of this class yields the following result in graph theory: The number of orientations for which there exists a directed path from s to t equals the number of subgraphs for which there exists an undirected path from s to t. For a more thorough discussion and other examples of extremal classes related to graph orientations see [16].

Example 5 (**General Construction of a Maximal Class that is Extremal but not Maximum**). Take a k-dimensional cube and glue to each of its vertices an edge of a new distinct dimension. The resulting class has 2^{k+1} concepts and $n = 2^k + k$ dimensions. Let C be the complement of that class.

Claim. C is an extremal maximal class of VC dimension $n - 2$ which is not maximum.

A proof of this claim is given in the full version of this paper [25]. Note that $|C| = 2^n - 2^{k+1} = 2^{2^k+k} - 2^{k+1}$ and maximum classes of VCdim $d = n - 2$ over n dimensions have size $2^n - n - 1 = 2^{2^k+k} - 2^k - k - 1$. So the maximum classes of VCdim $n - 2$ are by $2^k - k - 1$ larger than the constructed extremal maximal class of VCdim $n - 2$.

2.3 Basic Properties of Extremal Classes

Extremal classes have a rich combinatorial structure (See [23] and references within for more details). We discuss some of parts which are relevant to compression schemes.

The following theorem provides alternative characterizations of extremal classes:

Theorem 2 ([2,6]). *The following statements are equivalent:*

1. C *is extremal, i.e.* $s(C) = \text{st}(C)$.
2. $|s(C)| = |\text{st}(C)|$.
3. $|\text{st}(C)| = |C|$.
4. $|C| = |s(C)|$.
5. $\{0, 1\}^n \setminus C$ *is extremal.*

The following theorem shows that the property of "being an extremal class" is preserved under standard operations. It was also proven independently by several authors (e.g. [2,6]).

Theorem 3. *Let C be any extremal class, $S \subseteq dom(C)$, and B be any cube such that $dom(B) = dom(C)$. Then $C - S$ and C^S are extremal concept classes over the domain $dom(C) - S$ and $B \cap C$ is an extremal concept class over the domain $dom(C)$.*

Note that if C is maximum then $C - S$ and C^S are also maximum, but $B \cap C$ is not necessarily maximum. This is an example of the advantage extremal classes have over the more restricted notion of maximum classes.

Interestingly, the fact that extremal classes are preserved under intersecting with cubes yields a rather simple proof (communicated to us by Ami Litman) of the fact that every extremal class is "distance preserving". This property also holds for maximum classes [14], however the proof for extremal classes is much simpler than the previous proof for maximum classes (given in [14]):

Theorem 4 [15]. *Let C be any extremal class. Then for every $c_0, c_1 \in C$, the distance between c_0 and c_1 in the one-inclusion graph of C equals the hamming distance between c_1 and c_2.*

The proof is given in the full version of this paper [25].

The following lemma brings out the special cubical structure of extremal classes. We will use it to prove the correctness of the compression scheme given in the following section. It shows that if B_1 and B_2 are two maximal cubes of an extremal class C then their dimensions sets $\dim(B_1)$ and $\dim(B_2)$ are incomparable.

Lemma 1. *Given B_1 and B_2 are two cubes of an extremal class C. If B_1 is maximal, then*

$$\dim(B_1) \subseteq \dim(B_2) \implies B_1 = B_2.$$

A proof is given in the full version of this paper [25].

3   A Labeled Compression Scheme for Extremal Classes

Let C be a concept class. On a high level, a sample compression scheme for C compresses every sample of C to a subsample of size at most k and this subsample represents a hypothesis on the entire domain of C that must be consistent with the original sample. More formally, a labeled compression scheme of size k for C consists of a compression map κ and a reconstruction map ρ. The domain of the compression map consists of all samples from concepts in C: For each sample s, κ compresses it to a subsample s' of size at most k. The domain of the reconstruction function ρ is the set of all samples of C of size at most k. Each such sample is used by ρ to reconstruct a concept h with $dom(h) = dom(C)$. The sample compression scheme must satisfy that for all samples s of C, $\rho(\kappa(s)) \, |dom(s) = s$. The sample compression scheme is said to be *proper* if the reconstructed hypothesis h always belongs to the original concept class C.

A proper labeled compression scheme for extremal classes of size at most the VC dimension is given in Algorithm 1. Let C be an extremal concept class and s be a sample of C. In the compression phase the algorithm finds any *maximal cube* B of $C|dom(s)$ that contains the sample s and compresses s to the subsample determined by the dimensions set of that maximal cube. Note that the size of the dimension set (and the compression scheme) is bounded by the VC dimension.

How should we reconstruct? Consider all concepts of C that are consistent with the sample s:

$$H_s = \{h \in C : h|dom(s) = s\}.$$

Correctness means that we need to reconstruct to one of those concepts. Let s' be the input for the reconstruction function and let $D := dom(s')$. During the reconstruction, the domain $dom(s)$ of the original sample s is not known. All that is known at this point is that D is the dimensions set of a maximal cube B of $C|dom(s)$ that contained the sample s. The reconstruction map of the algorithm outputs a concept in the following set H_B (Fig. 3):

$$\{h \in C : h \text{ in cube } B' \text{ of } C \text{ s.t. } \dim(B') = \dim(B) \text{ and } h|\dim(B) = s|\dim(B)\}.$$

For the correctness of the compression scheme it suffices to show that for all choices of the maximal cube B of $C|dom(s)$, H_B is non-empty and a subset of H_s. The following Lemma guarantees the non-emptiness.

Lemma 2. *Let C be an extremal class and let $D \subseteq dom(C)$ be the dimensions set of some cube of $C|dom(s)$. Then D is also the dimensions set of some cube of C.*

Proof. Clearly the dimension set D is shattered by $C|dom(s)$ and therefore it is also shattered by C. By the extremality of C, D is also strongly shattered by it, and thus there exists a cube B of C with dimensions set D.

The second lemma show that for each choice of the maximal cube B, $H_B \subseteq H_s$.

Algorithm 1. (Labeled compression scheme for any extremal classes C)

The compression map.

- Input: A sample s of C.
- Output: A subsample $s' = s|\dim(B)$, where B is any maximal cube of $C|dom(s)$ that contains the sample s.

The reconstruction map.

- Input: A sample s' of size at most $VCdim(C)$.
- Output: Any concept h which is consistent with s' on $dom(s')$ and belongs to a cube B of C with dimensions set $dom(s')$.

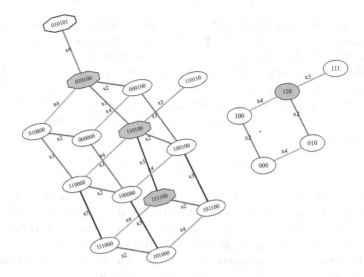

Fig. 3. The one-inclusion graph of an extremal concept class C is given on the left. Consider the sample $s = \overset{x_2 x_4 x_5}{\mathbf{1\,1\,0}}$. There are 4 concepts $c \in C$ consistent with this sample (the octagonal vertices), i.e. $H_s = \{111100, 110100, 010100, 010101\}$. There are 2 maximal cubes of $C|dom(s)$ (graph on right) that contain the sample s (in grey) with dimension sets $\{x_5\}$ and $\{x_2, x_4\}$, respectively. Let B be the maximal cube with dimension set $D = \{x_2, x_4\}$. There are 3 cubes of C (on left) with the same dimension set D. Each contains a concept h (shaded grey) that is consistent with the original sample on D, i.e. $h|D = s|D = \overset{x_2 x_4}{\mathbf{1\,1}}$ and therefore $H_B = \{111100, 110100, 010100\}$. For the correctness we need that H_B (grey nodes on left) is non-empty and a subset of H_s (octagon nodes on left). Note that in this case H_B is a strict subset.

Lemma 3. *Let s be a sample of an extremal class C, let B be any maximal cube of $C|dom(s)$ that contains s, and let D denote the dimensions set of B. Then for any cube B' of C with $\dim(B') = D$, the concept $h \in B'$ that is consistent with s on D is also consistent with s on $dom(s) \setminus D$.*

Proof. Since B is a cube with dimensions set D, $B|(dom(s) \setminus D)$ contains the single concept $tag(B)$.

Let B' be any cube of C with $\dim(B') = D$, and let h be the concept in B' which is consistent with s on D. Now consider the cube $B'|dom(s)$. We will show that $B'|dom(s) = B$. This will finish the proof as it shows that both $h|dom(s)$ and s belong to $B'|dom(s) = B$ which means that $tag(B) = h|(dom(s) \setminus D) = s|(dom(s) \setminus D)$. Moreover, by the definition of h, $h|D = s|D$, and therefore $h|dom(s) = s$ as required.

We now show that $B'|dom(s) = B$. Indeed, since B' is a cube of C with dimension set $D \subseteq dom(s)$, the cube $B'|dom(s)$ is a cube of $C|dom(s)$ with the same dimension set D. Thus the dimension set of $B'|dom(s)$ contains the dimension set of the maximal cube B of $C|dom(s)$. Therefore, since $C|dom(s)$ is extremal (Theorem 3) it follows by Lemma 1 that $B'|dom(s) = B$.

4 Unlabeled Sample Compression Schemes and Combinatorial Conjectures

Alternate "unlabeled" compression schemes have also been found for maximum classes and a natural question is whether these schemes again generalize to extremal classes. As we shall see there is an excellent match between the combinatorics of unlabeled compression schemes and extremal classes. The existence of such schemes remains open at this point. We can however relate their existence to some natural conjectures about extremal classes.

An unlabeled compression schemes compresses a sample s of the concept class C to an (unlabeled) subset of the domain of the sample s. In other words, in an unlabeled compression scheme the labels of the original sample are not used by the reconstruction map. The size of the compression scheme is now the maximum size of the subset that the sample is compressed to. Consider an unlabeled compression scheme for C of size $VCdim(C)$. For a moment restrict your attention to samples of C over some fixed domain $S \subseteq dom(C)$. Each such sample is a concept in the restriction $C|S$. Note that two different concepts in $C|S$ must be compressed to different subsets of S, otherwise if they were compressed to the same subset, the reconstruction of it would not be consistent with one of them. For maximum classes, the number of concepts in $C|S$ is exactly the number of subsets of S of size up to the VC dimension. Intuitively, this "tightness" makes unlabeled compression schemes combinatorially rich and interesting.

Previous unlabeled compression schemes for maximum classes were based on "representation maps"; these are one-to-one mappings between C and subsets of $dom(C)$ of size at most $VCdim(C)$. Representation maps were used in the following way: each sample s is compressed to a subset of $dom(s)$ which represents a consistent hypothesis with s, and each subset of size at most $VCdim(C)$ of $dom(C)$ is reconstructed to the hypothesis it represents. The key combinatorial property that enabled finding representation maps for maximum classes was a "non clashing" condition [17]. This property was used to show that for any sample s of C there is exactly one concept c that is consistent with s and $r(c) \subseteq dom(s)$. This immediately implies an unlabeled compression scheme based on non clashing representation maps: Compress to the unique subset of the domain of the sample that represents a concept consistent with the given sample.

The first representation maps for maximum classes were derived via a recursive construction [17]. Alternate representation maps were also proposed in [17] based on a certain greedy "peeling" algorithm that iteratively assigns a representation to a concept and removes this concept from the class. The correctness of the representation maps based on peeling was finally established in [28].

Representation Maps. For any concept class C a *representation map* is any one-to-one mapping from concepts to subsets of the domain, i.e. $r : C \to \mathcal{P}(dom(C))$. We say that $c \in C$ *is represented* by the *representation set* $r(c)$. Furthermore we say that two different concepts c, c' *clash* with respect to r if they are consistent with each other on the union of their representation sets, i.e. $c|(r(c) \cup r(c')) = c'|(r(c) \cup r(c'))$. If no two concepts clash then we say that r is *non clashing*.

Example 6 (**Non Clashing Representation Map for Distance Preserving Classes**). Let C be a distance preserving class, that is for every $u, v \in C$, the distance between u, v in the one-inclusion graph of C equals to their hamming distance. For every $c \in C$, define $deg_C(c) = \{x \in dom(C) : c$ is incident to an x-edge of $C\}$. The representation map $r(c) := deg_C(c)$ has the property that for every $c \neq c' \in C$, c and c' disagree on $r(c)$. To see this, note that any shortest path from c to c' in C traverses exactly the dimensions on which c and c' disagrees. In particular, the first edge leaving c in this path traverses a dimension x for which $c(x) \neq c'(x)$. By the definition of $deg_C(c)$ we have that $x \in deg_C(c)$ and indeed c and c' disagree on $deg_C(c)$.

In fact, this gives a stronger property for distance preserving classes, which is summarized in the following lemma. This lemma will be useful in our analysis.

Lemma 4. *Let C be a distance preserving class and let $c \in C$. Then $deg_C(c)$ is a teaching set for c with respect to C. That is, for all $c' \in C$, such that $c' \neq c$ there is $x \in deg_C(c)$ such that $c(x) \neq c'(x)$.*

Clearly the representation map $r(c) = deg_C(c)$ is non clashing. The following lemma establishes that certain non clashing representation maps immediately give unlabeled compression schemes:

Algorithm 2. (Unlabeled compression scheme from a representation map)

The compression map.

- Input: A sample s of C.
  Let $c \in C$ be the unique concept which satisfies $c|dom(s) = s$, and $r(c) \subseteq dom(s)$.
- Output $r(c)$.

The reconstruction map.

- Input: a set $S' \in st(C)$.
  Since r is a bijection between C and $st(C)$, there is a unique c such that $r(c) = S'$.
- Output c.

Lemma 5. *Let r be any representation map that is a bijection between an extremal class C and $st(C)$. Then the following two statements are equivalent: (i) r is non clashing. (ii) For every sample s of C, there is exactly one concept $c \in C$ that is consistent with s and $r(c) \subseteq dom(s)$.*

A proof is given in the full version of this paper [25].

Based on this lemma it is easy to see that a representation mapping r for an extremal concept class C defines a compression scheme as follows (See Algorithm 2). For any sample s of C we *compress* s to the unique representative $r(c)$ such that c is consistent with s and $r(c) \subseteq dom(s)$. Reconstruction is even simpler, since r is bijective: If s is compressed to the set $r(c)$, then we reconstruct $r(c)$ to the concept c.

Corner Peeling Yields Good Representation Maps. We now present a natural conjecture concerning extremal classes and relate it to the construction of non clashing representation maps. A concept c of an extremal class C is a *corner* of C if $C \setminus \{c\}$ is extremal. By Lemma 1 we have that for each $S \subseteq dom(C)$ there is at most one maximal cube with dimension set S and if S. Therefore $st(C \setminus \{c\}) = st(C) \setminus \{\dim(B) : B$ is maximal cube of C containing $c\}$.

For $C \setminus \{c\}$ to be extremal, $|st(C \setminus \{c\})|$ must be $|C| - 1$ (by Theorem 2) and therefore c is a corner of an extremal class C iff c lies in exactly one maximal cube of C.

Conjecture 1. Every non empty extremal class C has at least one corner.

A related conjecture for maximum classes was presented in [17]. For these latter classes, the conjecture was finally proved in [28]. This conjecture also has been proven for other special cases such as extremal classes of VC dimension at most 2 [19,22].

In fact [19] proved a stronger statement: For every two extremal classes $C_1 \subseteq C_2$ such that $VCdim(C_2) \leq 2$ and $|C_2 \setminus C_1| \geq 2$, there exists an extremal class C such that $C_1 \subset C \subset C_2$ (i.e. C is a strict subset of C_2 and a strict superset of C_1). Indeed, this statement is stronger as by repeatedly picking a larger extremal class $C_1 \subseteq C_2$ eventually a $c \in C_2$ is obtained such that $C_2 - \{c\}$ is extremal. For general extremal classes this stronger statement also remains open.

Conjecture 2. For every two extremal classes $C_1 \subseteq C_2$ with $|C_2 \setminus C_1| \geq 2$ there exists an extremal class C such that $C_1 \subset C \subset C_2$.

How does Conjecture 1 yield a representation map? Define an order[4] $c_1 \ldots c_{|C|}$ on the concept clase C such that for every i, c_i is a corner of $C_i = \{c_j : j \geq i\}$, and define a map $r : C \to st(C)$ such that $r(c_i) = \dim(B_i)$ where B_i is the unique maximal cube of C_i that c_i belongs to. We claim that r is a representation map. Indeed, r is a one-to-one mapping from C to $st(C)$ (and since C is extremal r is a bijection). To see that r is non clashing, note that $r(c_i) = \dim(B_i) = \deg_{C_i}(c_i)$. C_i is extremal and therefore distance preserving (Theorem 4). Thus, Lemma 4 implies that $r(c_i)$ is a teaching set of c_i with respect to C_i. This implies that r is indeed non clashing.

---

[4] Such orderings are related to the recursive teaching dimension which was studied by [10].

5   Discussion

We studied the conjecture of [12] which asserts that every concept classes has a sample compression scheme of size linear in its VC dimension. We extended the family of concept classes for which the conjecture is known to hold by showing that every extremal class has a sample compression scheme of size equal to its VC dimension. We demonstrated that extremal classes form a natural and rich generalization of maximum classes for which the conjecture had been proved before [12], and further related basic conjectures concerning the combinatorial structure of extremal classes with the existence of optimal unlabeled compression schemes. These connections may also be used in the future to provide a better understanding on the combinatorial structure of extremal classes, which is considered to be incomplete by several authors [6,15].

Our compression schemes for extremal classes yield another direction of attacking the general conjecture of Floyd and Warmuth: it is enough to show that an arbitrary maximal concept class of VC dimension d can be covered by $\exp(d)$ extremal classes of VC dimension $O(d)$. Note it takes additional $O(d)$ bits to specify which of the $\exp(d)$ extremal classes is used in the compression.

Acknowledgements. We thank Michał Dereziński for a good feedback on the writing of the paper and Ami Litman for helpful combinatorial insights.

References

1. Anstee, R., Rónyai, L., Sali, A.: Shattering news. Graphs Comb. **18**(1), 59–73 (2002)
2. Bandelt, H., Chepoi, V., Dress, A., Koolen, J.: Combinatorics of lopsided sets. Eur. J. Comb. **27**(5), 669–689 (2006)
3. Ben-David, S., Litman, A.: Combinatorial variability of Vapnik-Chervonenkis classes with applications to sample compression schemes. Discret. Appl. Math. **86**(1), 3–25 (1998)
4. Blumer, A., Ehrenfeucht, A., Haussler, D., Warmuth, M.K.: Learnability and the Vapnik-Chervonenkis dimension. J. Assoc. Comput. Mach. **36**(4), 929–965 (1989)
5. Blumer, A., Ehrenfeucht, A., Haussler, D., Warmuth, M.K.: Occam's razor. Inf. Process. Lett. **24**(6), 377–380 (1987)
6. Bollobás, B., Radcliffe, A.J.: Defect Sauer results. J. Comb. Theory Ser. A **72**(2), 189–208 (1995)
7. Bollobás, B., Radcliffe, A.J., Leader, I.: Reverse Kleitman inequalities. Proc. Lond. Math. Soc. Ser. A (3) **58**, 153–168 (1989)
8. Chernikov, A., Simon, P.: Externally definable sets and dependent pairs. Isr. J. Math. **194**(1), 409–425 (2013)
9. Daniely, A., Shalev-Shwartz, S.: Optimal learners for multiclass problems. In: COLT, pp. 287–316 (2014)
10. Doliwa, T., Simon, H.U., Zilles, S.: Recursive teaching dimension, learning complexity, and maximum classes. In: Hutter, M., Stephan, F., Vovk, V., Zeugmann, T. (eds.) ALT 2015. Lecture Notes in Artificial Intelligence (LNAI), vol. 6331, pp. 209–223. Springer, Heidelberg (2010). doi:10.1007/978-3-642-16108-7_19

11. Dress, A.: Towards a theory of holistic clustering. DIMACS Ser. Discret. Math. Theoret. Comput. Sci. **37**, 271–289 (1997). (Amer. Math. Soc.)
12. Floyd, S., Warmuth, M.K.: Sample compression, learnability, and the Vapnik-Chervonenkis dimension. Mach. Learn. **21**(3), 269–304 (1995)
13. Freund, Y., Schapire, R.E.: Boosting: Foundations and Algorithms. Adaptive Computation and Machine Learning. MIT Press, Cambridge (2012)
14. Gartner, B., Welzl, E.: Vapnik-Chervonenkis dimension and (pseudo-)hyperplane arrangements. Discret. Comput. Geom. (DCG) **12**, 399–432 (1994)
15. Greco, G.: Embeddings and the trace of finite sets. Inf. Process. Lett. **67**(4), 199–203 (1998)
16. Kozma, L., Moran, S.: Shattering, graph orientations, and connectivity. Electron. J. Comb. **20**(3), P44 (2013)
17. Kuzmin, D., Warmuth, M.K.: Unlabeled compression schemes for maximum classes. J. Mach. Learn. Res. **8**, 2047–2081 (2007)
18. Lawrence, J.: Lopsided sets and orthant-intersection by convex sets. Pac. J. Math. **104**(1), 155–173 (1983)
19. Litman, A., Moran, S.: Unpublished results (2012)
20. Littlestone, N., Warmuth, M.: Relating data compression and learnability (1986, Unpublished)
21. Livni, R., Simon, P.: Honest compressions and their application to compression schemes. In: COLT, pp. 77–92 (2013)
22. Mészáros, T., Rónyai, L.: Shattering-extremal set systems of VC dimension at most 2. Electron. J. Comb. **21**(4), P4.30 (2014)
23. Moran, S.: Shattering-extremal systems (2012). CoRR abs/1211.2980
24. Moran, S., Shpilka, A., Wigderson, A., Yehudayoff, A.: Teaching and compressing for low VC-dimension. In: ECCC TR15-025 (2015)
25. Moran, S., Warmuth, M.K.: Labeled compression schemes for extremal classes (2015). CoRR abs/1506.00165. http://arXiv.org/abs/1506.00165
26. Moran, S., Yehudayoff, A.: Sample compression schemes for VC classes. J. ACM **63**(3), 21:1–21:10 (2016)
27. Pajor, A.: Sous-espaces l_1^n des espaces de banach. Travaux en Cours. Hermann, Paris (1985)
28. Rubinstein, B.I.P., Rubinstein, J.H.: A geometric approach to sample compression. J. Mach. Learn. Res. **13**, 1221–1261 (2012)
29. Samei, R., Yang, B., Zilles, S.: Generalizing labeled and unlabeled sample compression to multi-label concept classes. In: Auer, P., Clark, A., Zeugmann, T., Zilles, S. (eds.) ALT 2015. Lecture Notes in Artificial Intelligence (LNAI), vol. 8776, pp. 275–290. Springer, Heidelberg (2014). doi:10.1007/978-3-319-11662-4_20
30. Sauer, N.: On the density of families of sets. J. Comb. Theory Ser. A **13**, 145–147 (1972)
31. Shelah, S.: A combinatorial problem; stability and order for models and theories in infinitary languages. Pac. J. Math. **41**, 247–261 (1972)
32. Valiant, L.: A theory of the learnable. Commun. ACM **27**, 1134–1142 (1984)
33. Warmuth, M.K.: Compressing to VC dimension many points. In: Schölkopf, B., Warmuth, M.K. (eds.) COLT/Kernel 2003. LNCS (LNAI), vol. 2777, pp. 743–744. Springer, Heidelberg (2003)

On Version Space Compression

Shai Ben-David[1] and Ruth Urner[2(✉)]

[1] University of Waterloo, Waterloo, Canada
[2] Max Planck Institute for Intelligent Systems, Stuttgart, Germany
ruth.urner@tuebingen.mpg.de

Abstract. We study compressing labeled data samples so as to maintain version space information. While classic compression schemes [11] only ask for recovery of a samples' labels, many applications, such as distributed learning, require compact representations of more diverse information which is contained in a given data sample. In this work, we propose and analyze various frameworks for compression schemes designed to allow for recovery of version spaces. We consider exact versus approximate recovery as well as compression to subsamples versus compression to subsets of the version space. For all frameworks, we provide some positive examples and sufficient conditions for compressibility while also pointing out limitations by formally establishing impossibility of compression for certain classes.

1 Introduction

Sample compression schemes, introduced in [11], have received considerable attention by the machine learning theory community throughout the past three decades [3,6,9,13,14,16]. These "classic" compression schemes for fixed hypothesis classes ask for a small size representation of a training sample that allows recovery of all the labels of points in that sample. However, training samples carry more information than that. Given a labeled sample S, one may want to infer more generally which domain points' labels are determined and which points' labels are *not* determined by S (given that the labeling belongs to some known concept class H). Such extra information may well get lost in the classic definition of compression.

In this work, we initiate the study of compressing such information of determined and undetermined labels. In other words, we wish to develop compact representations of a sample S from which the induced *version space* [12] can be recovered. The version space of a sample with respect to a class H is the set of all hypotheses in H that are consistent with the labels in the sample. As a motivating example, consider the class of initial segments on the real line. It is easy to see that the full version space can be recovered from at most 2 sample points (the rightmost positively and leftmost negatively labeled points). We are interested in understanding when this type (or alternative types) of compression can be extended to other concept classes.

An obvious application of such version space compression schemes is distributed parallel learning [1]. Version space compression provides a solution to

© Springer International Publishing Switzerland 2016
R. Ortner et al. (Eds.): ALT 2016, LNAI 9925, pp. 50–64, 2016.
DOI: 10.1007/978-3-319-46379-7_4

learning when the training sample is distributed between several agents, and one wishes to use that full training set for learning while controlling the volume of between-agents communication. Other applications include various non-standard learning settings such as active [8, 20] and transfer learning.

Here, we propose and explore various frameworks for compressing version space information, such as exact versus approximate compression, and compression to subsamples versus compression to subsets of the version space. Our conclusions are mixed. While we provide sufficient conditions and examples of compressibility for all frameworks, we also formally establish some strong impossibility results.

Outline of Results. We start by providing two natural definitions of exact version space compression schemes. In Sect. 3, we introduce our notions of compressing to subsamples and to subsets of the version space that allow for exact recovery of version space information. We provide various examples and sufficient conditions for when such exact compression and recovery is possible. In the last part of that section however, we provide impossibility results for these notions based on a new complexity measure of hypothesis classes. We also show that our impossibility results apply to many natural classes.

In light of these impossibility results, we turn to investigate notions of approximate version space compression in Sect. 4. That is, we consider notions, where the recovery function is allowed a certain amount of error, or where the requirement for compressibility is relaxed to only hold with high probability over samples that are generated by some distribution. Here we provide some positive approximate compression results employing schemes that incorporate additional unlabeled data samples. Again, we also provide some impossibility results characterizing situations of a "bad match" between a hypothesis class and the data generating distribution.

We discuss related work in Sect. 5 and conclude with Sect. 6.

2 Definitions and Notation

We let X denote some domain set. A *hypothesis* is a binary function $h : X \rightarrow \{0, 1\}$ over X and *hypothesis class* is a set $H \subseteq \{0, 1\}^X$ of hypotheses. We also consider hypotheses as subsets of the domain, that is, we identify binary functions $h : X \rightarrow \{0, 1\}$ with $h^{-1}(1)$. We use $S \subseteq X \times \{0, 1\}$ to denote a finite labeled *sample*. Abusing notation, we also use the notation S for the samples' projection to the domain X, that is the set of elements without labels. A hypothesis h is *consistent* with a sample S if $h(x) = y$ for all $(x, y) \in S$.

The *version space* of a sample S with respect to a hypothesis class H is the set of all hypotheses in H that are consistent with S, formally

$$V_H(S) = \{h \in H : \forall (x, y) \in S, \ h(x) = y\}.$$

For a set H' of classifiers, we define their *consensus recovery function* as

$$\operatorname{consrec}(H')(x) = \begin{cases} 0 & \text{if } \forall h \in H', \ h(x) = 0 \\ 1 & \text{if } \forall h \in H', \ h(x) = 1 \\ \star & \text{otherwise} \end{cases}$$

The version space of a sample S can now equivalently be viewed as its consensus recovery function $F_H(S)$. That is, for all x, we set

$$F_H(S)(x) = \operatorname{consrec}(V_H(S))(x)$$

Note that one can easily recover the set $V_H(S)$ and the function $F_H(S)$ from each other and we use them interchangeably to refer to the version space of S.

In the statistical setup, we consider distributions D over $X \times \{0, 1\}$. The error of a hypothesis with respect to a distribution is $\operatorname{err}_D(h) = \mathbb{E}_{(x,y)\sim D}\mathbb{I}[h(x) \neq y]$. We focus on the realizable case, that is we assume that there is a function $h \in H$ with $\operatorname{err}_D(h) = 0$. We also assume realizability of samples S throughout.

2.1 Hypothesis Classes

For ease of reference, we here provide notation for some specific hypothesis classes that we frequently use as illustrating examples. We expect that the reader is familiar with most of these classes, and do not formally define most them. Whenever we consider classes defined by thresholds over a euclidean space, we allow both open and closed upper and lower thresholds.

- H^1_{intrv} intervals in $\mathbb{R}^1$
- H^1_{init} initial segments in $\mathbb{R}^1$
- H^n_{rec} axis aligned rectangles in $\mathbb{R}^n$
- H^n_{lin} linear halfspaces in $\mathbb{R}^n$
- H_{sing} singletons over some infinite set
- H^n_{dest} decision stumps in $\mathbb{R}^n$
- H^n_{conj} boolean conjunctions over $\{0, 1\}^n$
- $H^{\prec}_{\text{init}}$ initial segments with respect to a partial order $\prec$
- H^n_{star} the class of stars with n arms

Recall that a binary relation over X is a *partial order* if it is reflexive, antisymmetric and transitive. An *initial segment* with respect to a partial order $\prec$ is defined as all elements of X that are "smaller" than some element x, formally $I_x := \{z \in X \ : \ z \prec x\}$. Thus, we get $H^{\prec}_{\text{init}} := \{I_x \ : \ x \in X\}$.

To define the class H^n_{star}, consider the domain X that is a disjoint union of n copies of $\mathbb{R}^+$, all sharing the same 0-element ("glued together at 0"). We call the copies the *arms* of the domain X. Now a star is a union of initial segments in each of the arms. Note that this class has VC-dimension n.

3 Exact Version Space Compression (VSC)

In this section we discuss the most demanding framework. Given a hypothesis class H and a sample S, we seek to compress the sample to a compact representation (either in form of a small subsample or in form of a small subset of the version space) from which the full version space can be exactly recovered. We propose and analyze two notions of such exact version space compression schemes. For both notions we provide some examples and sufficient conditions for classes to admit such compression. However, we then proceed to show that for many classes exact version space compression is too much to hope for by providing formal impossibility results.

3.1 Definitions of Exact VSC

We start with what may be considered the most natural notion of version space compression. For *instance based version space compression* we require compressing a sample to a small subset of sample points, which contain the full information about the samples' version space. This notion is closest to the "classic" sample compression schemes.

Definition 1 (Instance Based Exact VSC). *We say that H admits* instance based (exact) version space compression (IBE-VSC) *to size d if there exists a function ρ from finite labeled samples to the set of subsets of H such that for every finite labeled sample $S \subseteq X \times \{0,1\}$, there is $C \subseteq S$ of size $\leq d$ such that $\rho(C) = V_H(S)$.*

Remark 1.

1. In some parts of the paper we use κ to denote the mapping of samples to their compressed images. Thus, a *version space compression scheme* can be defined as a pair of functions κ, ρ with $\kappa(S) \subseteq S$ and $\rho(\kappa(S)) = F_H(S)$ for all S.
2. It is natural to also allow the compressed image of a sample to contain some *side information* rather than only sample points. That is, for a sample S, $\kappa(S)$ is a pair (C, b), where $C \subseteq S$ and b is a binary string. The size of such a compressed image is then defined as $|C| + |b|$.
3. Another natural variant are *ordered* version space compression schemes. Here the compressed set is an ordered sequence of elements in S. Note that one can turn an (unordered) compressed set of size d into an ordered one by adding $d \log d$ bits of side information.

Alternatively to compressing a sample S to some subsample, one may consider version space compression to a subset of the version space that encapsulates the full version space. To rule out trivial encodings (via some bijection from the class to version spaces), we enforce such a compressed set to encode the version space in a specific manner.

Definition 2 (Span of a Subclass). *Let $H' \subseteq H$ be a subclass of some hypothesis class H. Then the* span *of H' in H is defined as the maximal set H'' with*

$H' \subseteq H'' \subseteq H$ such that $\mathrm{consrec}(H') = \mathrm{consrec}(H'')$. That is, the span of H' contains all functions in H that agree with H' on the agreement region of H'. Note that for all H, H', H'', if $H'' \subseteq H'$, then $\mathrm{span}_H(H') \subseteq \mathrm{span}_H(H'')$.

We now define concept based version space compression as follows:

Definition 3 (Concept Based Exact VSC). *We say that H admits concept based (exact) version space compression (CBE-VSC) to size d if for every finite labeled sample $S \subseteq X \times \{0,1\}$, there exists $H_S \subseteq H$ such that $|H_S| \leq d$ and $V_H(S) = \mathrm{span}_H(H_S)$.*

Example 1. It is easy to see that the class $\mathrm{H}^1_{\mathrm{init}}$ of initial segments on the real line admits both IBE-VSC and CBE-VSC to size 2. Similarly, the class $\mathrm{H}^1_{\mathrm{intrv}}$ of intervals on the real line admits IBE-VSC to size 4 and CBE-VSC to size 2, for samples S that contain at least one positive example (we will see later that for $\mathrm{H}^1_{\mathrm{intrv}}$, samples containing only negative examples have no finite size VSC, neither IBE-VSC nor CBE-VSC). The class $\mathrm{H}^\infty_{\mathrm{star}}$ of stars with infinitely many arms, admits CBE-VSC but not IBE-VSC to any finite size (as we will see later).

3.2   Existence of Exact VSC

In this subsection we provide some sufficient conditions for the existence of VSC for hypothesis classes. The following condition allows a particularly simple case of CBE-VSC.

Definition 4. *We call a class H linearly consistent if for every $h \in H$ and every finite $S \subseteq X$, there exist functions $\overline{h}_S, \underline{h}_S \in H$ such that both functions agree with h on S and for every $g \in H$ that agrees with h on S, we have $g^{-1}(1) \subseteq \overline{h}_S^{-1}(1)$ and $g^{-1}(0) \subseteq \underline{h}_S^{-1}(0)$.*

Example 2. The class $\mathrm{H}^1_{\mathrm{init}}$ of initial segments on the real line is linearly consistent. Further, the classes $\mathrm{H}^n_{\mathrm{star}}$ of stars with n arms are linearly consistent. The class $\mathrm{H}^1_{\mathrm{intrv}}$ is "almost" linear consistent - it satisfies the above requirement for h on a set S, once for some $x \in S$, $h(x) = 1$.

Claim. If a class H is linearly consistent, then H admits CBE-VSC to size 2.

Next, we present a general condition that implies IBE-VSC.

Definition 5.   *We define the width $w(H)$ of a class H as the maximum size of a domain subset $T \subseteq X$ such that $\{\{t\} : t \in T\} \subseteq \{h \cap T : h \in H\}$. The definition can be generalized by replacing the all zero base function by any function $f \in H$. Namely $w_f(H)$ is the maximum size of a domain subset T such that $\forall t \in T \exists h \in H$ such that $\{x \in T : h(x) \neq f(x)\} = \{t\}$.*

Example 3. The class $\mathrm{H}^1_{\mathrm{init}}$ of initial segments has width $w(\mathrm{H}^1_{\mathrm{init}}) = 1$. The classes $\mathrm{H}^n_{\mathrm{star}}$ of stars with n arms, have width $w(\mathrm{H}^n_{\mathrm{star}}) = n$. For the class $\mathrm{H}^n_{\mathrm{lin}}$ of halfspaces, the class of singletons $\mathrm{H}_{\mathrm{sing}}$, and the class of axis aligned rectangles $\mathrm{H}^n_{\mathrm{rec}}$, we have $w(\mathrm{H}^n_{\mathrm{lin}}) = w(\mathrm{H}_{\mathrm{sing}}) = w(\mathrm{H}^n_{\mathrm{rec}}) = \infty$.

With this, we can provide a sufficient condition for the existence of VSCs.

Theorem 1. *Classes H with* $\mathrm{VCdim}(H) = 1$ *admit IBE-VSC to size* $2w(H)$.

For the proof we need the following structural result characterizing classes of VC-dimension 1. This result has appeared in [2]. We prove it for completeness.

A partial ordering $\prec$ over X is called a *forest* if for every $x \in X$ the initial segment $I_x = \{y : y \prec x\}$ that x induces, is linearly ordered by $\prec$.

Theorem 2 (Theorem 4 in [2]). *A class H over any domain set X has VC dimension at most 1 if and only if there exist a partial ordering of the domain set X, which is a forest, and such that every set $h \in H$ is a linearly ordered initial segment under that ordering.*

Proof (Proof of Theorem 2). First, it is not hard to verify that a class of linearly ordered initial segments over some partial order has VC dimension 1.

For the other implication, let H be a class of binary functions, over some domain set X, and fix some $h_0 \in H$. Assume, w.l.o.g., that for every $x \neq y \in X$, there exists some $h \in H$ so that $h(x) \neq h(y)$. We define a binary relation $\preccurlyeq_H$ over $X \times X$ by $\preccurlyeq_H = \{(x, y) : \forall h \in H, \, h(y) \neq h_0(y) \Rightarrow h(x) \neq h_0(x)\}$.

We now argue that $\preccurlyeq_H$ is a partial ordering. Namely, it is reflexive, transitive and anti symmetric: Being reflexive and transitive follows trivially from the definition. For anti-symmetry, let x, y be such that both $x \preccurlyeq_H y$ and $y \preccurlyeq_H x$ hold. It is easy to see that this implies that for all $h \in H$, $h(x) = h(y)$.

Now, if $\mathrm{VCdim}(H) < 2$ then, for every $h \in H$, $A_h \overset{def}{=} \{x : h(x) \neq h_0(x)\}$ is a linearly ordered initial segment under $\preccurlyeq_H$. That is,

1. For every $x, y \in A_h$, either $x \preccurlyeq_H y$ or $y \preccurlyeq_H x$.
2. If $x \in A_h$ and $y \preccurlyeq_H x$ then $y \in A_h$.

Note that the second point follows immediately from the definition of the relation $\preccurlyeq_H$. As for the first, assume both $x \preccurlyeq_H y$ and $y \preccurlyeq_H x$ fail. Then there exist $h_1, h_2 \in H$ s.t. $h_1(x) \neq h_0(x)$ and $h_1(y) = h_0(y)$ and $h_2(x) = h_0(x)$ and $h_2(y) \neq h_0(y)$. Since $x, y \in A_h$, we also have $h(x) \neq h_0(x)$ and $h(y) \neq h_0(y)$. It follows that the set $\{h_0, h, h_1, h_2\}$ shatters $\{x, y\}$, contradicting the assumption that $\mathrm{VCdim}(H) < 2$. This completes the proof the other implication.

With this, we can proceed to prove our existence theorem for IBE-VSC.

Proof (of Theorem 1). To simplify the notation let us assume w.l.o.g. that the all zero function (or the empty set) is a member of H and that for every $x \in X$ there is some $h \in H$ such that $h(x) = 1$. We set this all-zero hypothesis to be h_0 in the definition of the partial order $\preccurlyeq_H$ in the proof of Theorem 2. Let T be a maximal size domain subset such that $\{\{t\} : t \in T\} \subseteq \{h \cap T : h \in H\}$. For each $t \in T$, pick $h_t \in H$ such that $h_t \cap T = \{t\}$. For $h \in H$ let $A_h := \{x : h(x) \neq h_0(x)\}$. Given a finite labeled sample S, for every $t \in T$, let x_t^+ be the maximal (under $\preccurlyeq_H$) 1- labeled point in $S \cap A_{h_t}$ (if there exists a 1- labeled point in $S \cap A_{h_t}$), and let x_t^- be the minimal (under $\preccurlyeq_H$) 0-labeled point in $S \cap A_{h_t}$ (if there exists a 0- labeled point in $S \cap A_{h_t}$).

Finally, let $\kappa(S) = \{x_t^+ : t \in T\} \cup \{x_t^- : t \in T\}$. For the decompression function ρ, note that every point that is below some x_t^+ (w.r.t $\preccurlyeq_H$) should be labeled 1 by $F_H(S)$, every point that is above some x_t^- (w.r.t $\preccurlyeq_H$) should be labeled 0 by $F_H(S)$, and any other point should be labeled $\star$ by $F_H(S)$. Therefore the set $\kappa(S)$ that we defined allows full reconstruction of $F_H(S)$ (and, by construction, has size at most $2w(H)$).

3.3   Closure Properties

In this section, we show that the existence of (instance and concept based) exact version space compression is closed under certain set operations on the hypothesis classes. Together with the existence results of the previous section, these yield a significant family of classes that admit exact VSCs.

Unions and Intersections. Finite size exact VSC is closed under union and intersections. Recall the notion of ordered VSC from Remark 1.

Lemma 1. *Let H and H' be hypothesis classes over some domain X that admit IBE-VSC (CBE-VSC) to size d and d' respectively. Then the hypothesis class $H \cap H'$ admits ordered IBE-VSC (CBE-VSC) to size $d + d'$, and the hypothesis class $H \cup H'$ admits ordered IBE-VSC (CBE-VSC) to size $d + d'$.*

Proof. Let κ_H and $\kappa_{H'}$ denote the compression functions of H and H' respectively. Note that for every sample S, we have $V_{H \cup H'}(S) = V_H(S) \cup V'_H(S)$ and $V_{H \cap H'}(S) = V_H(S) \cap V'_H(S)$. Thus, given S, use $\kappa_H(S)$ and $\kappa_{H'}(S)$ to recover the two version spaces and then do union (or intersection) on those version spaces.

We may alternatively encode the order with $(d + d') \log(d + d')$ bits (Remark 1).

Corollary 1. *For every n, the class of decision stumps H_{dest}^n over $\mathbb{R}^n$ admits IBE-VSC to size $\leq 2n$ samples plus $2n \log 2n$ bits.*

Fixed Intersection c-Unions. We define the *hypothesis class union* (c-union in short) of two classes H and H' as $HcH' := \{h \cup h' : h \in H \text{ and } h' \in H'\}$. We say that this is a *fixed intersection c-union* if for some $A \subseteq X$, $h \cap h' = A$ for all $h \in H, h' \in H'$. The family of classes that admit IBE-VSC (CBE-VSC) is closed under fixed intersection c-unions.

Lemma 2. *Let H and H' be hypothesis classes over some domain X that admit IBE-VSC (CBE-VSC) to size d and d' respectively. Then the hypothesis class HcH' admits IBE-VSC (CBE-VSC) to size $d + d'$ samples with $d + d'$ bits if the c-union has a fixed intersection.*

Proof. First, note that if A is the fixed intersection of any pair of hypotheses belonging to different classes, then $A = \bigcup\{h \in H\} \cap \bigcup\{h \in H'\}$. Let S be a labeled sample consistent with HcH'. Let $h \in H$ and $h' \in H'$ be such that S is consistent with $h \cup h'$. Note that any negative example in S would be labeled negative by both h and h'. Furthermore, every positive example in S, is either in the common intersection A of the two classes, in which case it is labeled positively by both h and h' or, if it is outside A is either in $\bigcup\{h \in H\}$ or in $\bigcup\{h \in H'\}$ but not in both. Therefore, we can determine samples S_1, $S_2 \subseteq S$ such that S_1 is consistent with H and S_2 is consistent with H' and $V_{HcH'}(S) = V_H(S_1)cV_{H'}(S_2)$. By the assumption that both classes have compression schemes of sizes d and d' respectively, we can now form $\kappa(S_1)$ and $\kappa(S_2)$ (of sizes d and d' respectively), and recover $V_{HcH'}(S)$ from their union, again adding $d + d'$ bits of side information to encode membership to the $\kappa(S_i)$'s.

Projections. Let H be a hypothesis class over domain X, that is $H \subseteq 2^X$. For some subset S of X, we let H_S denote the *projection* of H on S, that is

$$H_S = \{h_S \subseteq S \ : \ \exists h \in H \text{ with } h_S = h \cap S\}.$$

The existence of both instance and concept based exact compression schemes is closed under projections.

Lemma 3. *Let H be a hypothesis class that admits IBE-VSC (CBE-VSC) of size d and let $S \subseteq X$. Then H_S admits IBE-VSC (CBE-VSC) of size d as well.*

Proof. The claim of the lemma is obvious for IBE-VSC. We now argue that it also holds for CBE-VSC. Fix a labeled sample $S \subseteq X \times \{0,1\}$ and consider some subset $A \subseteq S$. Let $(h^1, \ldots h^d)$ be the compression of $V_H(A)$, the version space of A with respect to the original class H. Then $V_H(A) = \mathrm{span}(h^1, \ldots h^d)$. Let h_S^i denote the projections of the functions in the compressed set to S. It suffices to argue that $V_{H_S}(A) = \mathrm{span}(h_S^1, \ldots h_S^d)$.

Let $h \in H_S$ be consistent with A, that is $h \in V_{H_S}(A)$. Let h^X denote a preimage of h in H (that is $h = h^X \cap S$). Then we have $h^X \in V_H(A)$, hence $h^X \in \mathrm{span}(h^1, \ldots h^d)$. To see that $h \in \mathrm{span}(h_S^1, \ldots h_S^d)$, let $x \in S$ with $h_S^1(x) = \ldots = h_S^d(x) =: y$. This implies $h^1(x) = \ldots = h^d(x) = y$, hence $h^X(x) = y$ (since $h^X \in \mathrm{span}(h^1, \ldots h^d)$) and hence $h(x) = y$. This completes the proof.

3.4   Impossibility Results for Exact VSC

In this section we prove that many natural simple classes fail to admit IBE-VSC or CBE-VSC. Our impossibility results are based on the following complexity measure for hypothesis classes.

Definition 6. *We say that a labeled sample S is* independent *with respect to a class H if for every subsample $A \subseteq S$ and for every $x \in S \setminus A$ there exist $h_0, h_1 \in V_H(A)$ such that $h_0(x) \neq h_1(x)$.*

Example 4.

- Let $\mathrm{H}_{\mathrm{rec}}^2$ be the class of axis aligned rectangles in $\mathbb{R}^2$. Let S be a finite (or discrete) set of points on a line with slope -1, for example $S = \{(x, -x) : x \in \mathbb{R}\}$, all labeled 0. Then S is independent with respect to the class $\mathrm{H}_{\mathrm{rec}}^2$.

- Let S be a finite (or discrete) set of points all on the same half of a sphere in $\mathbb{R}^d$ and all having the same label (say, $\{(x_1, \ldots, x_n) : x_1 \geq 0 \text{ and } \Sigma_{i=1}^n x_i^2 = 1\}$ all with the label 1). Then S is independent with respect to the class $\mathrm{H}_{\mathrm{lin}}^n$ of halfspaces in $\mathbb{R}^d$.
- If $T \subseteq X$ is a witness of the width of H w.r.t. some function, f (see Definition 5), than $\{(x, f(x)) : x \in T\}$ is an independent set w.r.t. H.

Theorem 3. *Let H be a hypothesis class and $m \in \mathbb{N}$. If there exists a sample of size at least m that is independent with respect to H, then H does not admit an IBE-VSC of size smaller than $m/\log(m)$.*

Proof. Let H satisfy the above assumption and assume that (κ, ρ) is a instance based exact VSC scheme of size d for H. Let S be a labeled sample of size m that is independent with respect to H.

We now argue that the subsets of S have pairwise different version spaces: Let $A, A' \subseteq S$ with $A \neq A'$. Let x be in the symmetric difference of A and A', without loss of generality we assume $x \in A \setminus A'$. Since $x \notin A'$, and since S is independent with respect to H, there exist functions h and h' in $V_H(A')$ with $h(x) \neq h'(x)$. Since $x \in A$, at least one of h and h' is not consistent with A, thus not a member of the version space of A. That is, for all $A, A' \subseteq S$, we have shown that $A \neq A'$ implies $V_H(A) \neq V_H(A')$.

Note that a d-size compressing function κ for H can take at most $\sum_{i=1}^d \binom{m}{i}$ many values over the union of all subsamples $\{A : A \subseteq S\}$. However, there are 2^m sets of corresponding version spaces (since, as argued above every subset A of S induces a unique version space). We therefore get $\Sigma_{i=0}^d \binom{m}{i} \geq 2^m$, which implies, by a simple calculation, that $d \geq \frac{m}{\log(m)}$. The claim of the theorem now follows by invoking Lemma 3.

Corollary 2.

1. The classes $\mathrm{H}_{\mathrm{rec}}^n$ of axis aligned rectangles in $\mathbb{R}^d$, do not admit finite size IBE-VSC.
2. The classes $\mathrm{H}_{\mathrm{lin}}^n$ of linear half spaces in $\mathbb{R}^d$ do not admit finite size IBE-VSC.
3. The class $\mathrm{H}_{\mathrm{conj}}^n$ of boolean conjunctions over the propositional (binary) variables $p_1 \ldots, p_n$ does not have an IBE-VSC of any size $< \frac{2^n}{n}$. (Just note that the domain set X has size 2^n and, since for every instance $x \in \{0,1\}^{\{1,\ldots,n\}}$, there is a conjunction c_x that only the assignment x satisfies, therefore the set X is a witness that $w(\mathrm{H}_{\mathrm{conj}}^n) = 2^n$. Now apply the third point in Example 4 above).

We now turn to proving a similar impossibility result for CBE-VSC. The result applies to classes that have a finite VC-dimension. Recall that, by Sauers' lemma (e.g. Theorem 6.10 in [18]) for classes of finite VC-dimension, the number of behaviors of a class on some finite subset $S \subseteq X$ is bounded as follows:

$$|H_S| \leq \Sigma_{i=0}^{\mathrm{VCdim}(H)} \binom{|S|}{i}$$

Theorem 4.   *Let H be a hypothesis class with a finite VC-dimension, v, and $m \in \mathbb{N}$. If there exists a sample of size at least m that is independent with respect to H, then H does not admit an CBE-VSC of size smaller than $\frac{m}{v \log(m)}$.*

Proof. By Lemma 3, it suffices to argue that the projection of H on some domain subset $S \subseteq X$ does not admit finite size CBE-VSC.

Let S be an m-size independent sample with respect to H, project the class to the domain of S and consider subsamples of S. Let H_S denote the projected class. As in the proof of Theorem 3, we consider subsamples A of S. We again observe that there are 2^m many such subsamples A, each with a unique version space. However, since we consider only the domain of S, if two hypotheses in H_S agree on the domain of S, using one of them for the compression is equivalent to using the other.

The claim follows now by applying Sauer's lemma to bound the number of possible fixed size sets of hypotheses (up to their behavior over the domain of S). There are at most $\Sigma_{i=0}^{v} \binom{m}{i} < m^v$ functions in H_S and thus, given a compression size d, there are only m^{dv} possible compressed sets (namely, subsets of cardinality at most d of H_S). We therefore get that $m^{vd} \geq 2^m$, implying $d \geq \frac{m}{v \log(m)}$, and then the claim of the theorem follows by invoking Lemma 3.

The next example shows that the assumption of bounded VC-dimension is needed for the impossibility result for CBE-VSC.

Example 5. Let H_{star}^{∞} be the class of stars with infinitely many arms. It has infinite VC-dim and infinite width. However, it does admit CBE-VSC to size 2.

4   Approximate VSC

In view of the above negative results, we now investigate some relaxations of the definition of exact VSC. The following subsection proposes two degrees such a relaxation. First, we relax the requirement on the output on the decompression to only be *approximately correct*, i.e. correct on most of the instance space (with respect to the data generating distribution). We then further relax the requirement on compression to only hold for *most data sets S*, i.e. with high probability over the data generation. The notions we propose are still strong enough to allow, for example, for distributed parallelization of learning. The relaxed notions we introduce here are for both instance based and concept based compression and we phrase the definitions in general terms to apply to both settings.

4.1   Definitions of Approximate VSC

Recall that we use κ to denote the compression function (either to a subset of S or a subset of H), and use ρ to denote the recovery function, from $\kappa(S)$ to a $\{0, 1, *\}$-valued predictor. Given a loss function for such three valued predictors (see Definition 8 below), we can define approximate VSC as follows.

Definition 7 (Approximate VSC). *A pair of functions, (κ, ρ), as described above, is an ϵ-approximate compression scheme for a class H with respect and a distribution D, if for every sample S that is realizable by H*

$$L_D^H(S, \rho(\kappa(S))) \leq \epsilon.$$

The loss L_D^H can be specified as follows. Recall that $F_H(S)$ is the $\{0, 1, *\}$-valued function reflecting the consensus of the version space of S.

Definition 8 (Version Space Loss). *Given a class of binary valued functions H, a binary labeled sample S realizable by H, a three valued $h : X \to \{0, 1, \star\}$, and non-negative parameters $\{a, b\}$. Define a loss pointwise for all $x \in X$:*

$$\ell^H(S, h, x) = \begin{cases} 0 & \text{if } F_H(S)(x) = h(x) \\ a & \text{if } F_H(S)(x) = \star \text{ and } h(x) \neq \star \\ b & \text{if } F_H(S)(x) \neq \star \text{ and } h(x) \neq F_H(S)(x) \end{cases}$$

This yields a loss with respect to a data distribution D over the domain set X:

$$L_D^H(S, h) = \mathbb{E}_{x \sim D} \ell^H(S, h, x),$$

where $\mathbb{E}$ denotes the expectation. For a finite subset $U \subseteq X$, we let $L_U^H(S, h)$ denote the loss with respect to the uniform distribution over U. $L_U^H(S, h)$ can be viewed as the empirical loss of h with respect to an unlabeled sample U.

Remark 2.

1. By setting $a = 0$ and a very high value (say, ∞) for b, one can guarantee that any approximate version space compression scheme is a compression scheme in the "classic" sense, namely, it provides exact recovery of all labels of the input sample S, but there is no loss penalty on instances whose labels are not determined by the sample S. With $a = b = \infty$, we recover the exact version space compression of Sect. 3.

2. The parameters $\{a, b\}$ could, be extended to a full 3×3 confusion matrix, with different values for any possible type of misclassification of h with respect to the behavior of $F_H(S)$. From now, we will focus on the case $a = b = 1$.

We now propose a natural statistical extension of the above definition, where we require the approximate recovery of Definition 7 to only hold with high probability over input samples.

Definition 9 (Statistical Approximate VSC). *Let H, κ, ρ be as above and D a probability distribution over the domain set X. A pair of functions, (κ, ρ) is an (m, ϵ, δ)-approximate version space compression scheme for a class H with respect to the distribution D, if for every labeling function $h \in H$, with probability at least $(1 - \delta)$ over samples S of size m generated i.i.d. by D and labeled by h we have*

$$L_D(S, \rho(\kappa(S))) \leq \epsilon.$$

4.2   A Semi-supervised ERM Paradigm for Approximate VSC

We now present a compression paradigm for approximate, concept based version space compression in the statistical setup.

Recall that for a set of classifiers $H' = \{h_1, \ldots, h_T\}$ their consensus recovery function is denoted by $\mathrm{consrec}(\{h_1, \ldots, h_T\})$. Note that as long as $h_i \in V_H(S)$ for all i, the only type of error that $\mathrm{consrec}(\{h_1, \ldots, h_T\})$ may make is assigning a label in $\{0, 1\}$ to a point x for which $F_H(S)(x) = \star$.

We consider the case where we also have an unlabeled sample U, generated i.i.d. by the underlying marginal distribution. We search for a set of hypotheses $H' = \{h_1, \ldots h_T\} \subseteq V_H(S)$, that minimize $L_D^H(S, \mathrm{consrec}(\{h_1, \ldots, h_T\}))$. In other words, minimize the weight (with respect to D) of the set of instances on which some function in the version space of S disagree but all members of H' agree. Namely, minimize the weight of

$$B(S, H') = \{x : \exists h, h' \in V_H(S), \ h(x) \neq h'(x), \ \text{but } \forall h, h' \in H', \ h(x) = h'(x)\}.$$

Since a learner does not have access to D, we will instead search for functions $\{h_1, \ldots, h_T\}$ that minimizes the empirical loss $L_U^H(S, \mathrm{consrec}(\{h_1, \ldots, h_T\}))$ with respect to an unlabeled sample U. The following result provides a finite sample size guarantee for this ERM approach:

Theorem 5. *There is a constant C such that for every ϵ, $\delta > 0$, every $T \in \mathbb{N}$, every class H, every probability distribution D over X, and every labeled S, if U is an i.i.d. unlabeled sample generated by D of size*

$$|U| \geq m_H(\epsilon, \delta, T, \mathrm{VCdim}(H)) = C \frac{T^4 \mathrm{VCdim}(H) + \log(1/\delta)}{\epsilon^2},$$

then with probability at least $(1 - \delta)$ over sampling U, for every $\{h_1, \ldots h_T\} \subseteq H$,

$$|L_U^H(S, \mathrm{consrec}(\{h_1, \ldots, h_T\})) - L_D^H(S, \mathrm{consrec}(\{h_1, \ldots, h_T\}))| \leq \epsilon.$$

Proof. First, note that

$$L_U^H(S, \mathrm{consrec}(\{h_1, \ldots, h_T\})) = |U \cap (\{x : F_H(S)(x) = \star\} \setminus \bigcup_{i,j \leq T} h_i \Delta h_j)| \ / \ |U|,$$

and similarly,

$$L_D^H(S, \mathrm{consrec}(\{h_1, \ldots, h_T\})) = D(\{x : F_H(S)(x) = \star\} \setminus \bigcup_{i,j \leq T} h_i \Delta h_j).$$

Consequently, it suffices to show that the U empirical estimates of each of the sets $h_i \Delta h_j$ and the set $\{x : F_H(S) = \star\}$ are within $\frac{\epsilon}{\binom{T}{2}+1}$ of their true D probabilities.

The statement of the theorem now follows by invoking Vapnik-Chervonenkis' ϵ-approximation theorem [19]. That is, with probability larger $(1 - \delta)$, an i.i.d. D-sample of that size is an $\frac{\epsilon}{T^2}$-approximation of $H \Delta H$ with respect to D.

4.3   Impossibility Results for Approximate VSC

The previous subsection offers a paradigm for obtaining approximate, concept based version space compression for marginal distributions that allow such compression. However, we will show in this subsection that, for many natural concept classes, there exist marginal distributions which are "hard for compression". We start by stating an impossibility result for the, rather artificial, class of singletons, and then show how it implies that for many natural classes there exist marginal distributions with respect to which approximate VSC is impossible.

Theorem 6. *Let X be a finite domain set and let $\mathrm{H}^X_{\mathrm{sing}}$ denote the class of singletons $\mathrm{H}^X_{\mathrm{sing}} = \{\{x\} : x \in X\}$ over X. $\mathrm{H}^X_{\mathrm{sing}}$ does not have a finite size (independent of $|X|$) $1/6$-approximate VSC with respect to the uniform distribution over X and the $0--1$ loss (neither concept based approximate VSC nor instance based approximate VSC).*

Proof. Let S, S' be two samples in which all instances are labeled 0. Note that

$$L_U(S, F_H(S')) = |S \Delta S'| \, / \, |X|,$$

where U denotes the uniform distribution over X and $F_H(S')$ is the version space function of a sample S'. Lemma 4 below shows that there exists a family of size exponential in $|X|$ of subsets of X such that each two of them have symmetric difference of size $\geq |X|/3$. For every $S \neq S'$ in such a family, an $1/6$-approximate compression (ρ, κ) will require that $\kappa(S) \neq \kappa(S')$ (regardless of whether this is a instance based compression or a concept based compression). The proof is now established by invoking an argument similar to the ones in the above impossibility results (Theorem 3 and Theorem 4).

Lemma 4 ([4]). *For every $n \in \mathbb{N}$ there exists a family $W_n \subseteq \{0,1\}^n$ such that $|W_n| = 2^{cn}$ (for some constant c) and $|\Delta(A, B)| \geq n/3$ for every $A \neq B \in W_n$.*

Corollary 3. *Let H be a concept class over some domain set X. Let $T \subseteq X$ be such that $\{h \cap T : h \in H\} \supseteq \{\{t\} : t \in T\}$. Then H does not have a $1/6$-approximate version space compression scheme of size $o(\log(|T|))$ for the uniform distribution over T. In particular, neither the class of linear half spaces in $\mathbb{R}^2$ nor the class of axis aligned rectangles in $\mathbb{R}^2$ have a $1/6$-approximate version space compression scheme of size that is independent of the distribution with respect to which the approximation is defined.*

5   Related Work

Several earlier works have considered the task of algorithmically conveying not just label predictions but also "awareness of lack of knowledge". The earliest of those is probably a study by Rivest and Sloan [15]. Later work in that direction was presented under the name of KWIK (knows what it knows) algorithms [10,17,21]. There are several significant differences between that work and ours.

First, those papers address the possibility of algorithms to be aware of their uncertainties without addressing the issue of compressing that knowledge. Secondly, that body of work is focused on online learning, while we investigate batch learning. Finally, we are considering the knowledge encapsulated in a training sample rather than the knowledge derived by an algorithm.

Another related direction is that of distributed learning [1,5]. In that case, the concern is compression size, as it effects the communication volume between agents in a distributed learning scenario. However, those works allow iterations of the communications between different parts of the data, while we are considering "one time" compression.

Another notion related to this work is Teaching Dimension (TD). It has been introduced in the context of exact learning [7]. A variant of teaching dimension has been introduced for the purpose of analyzing active learning paradigms [8,20]. This notion of TD is equivalent to some notion of version space compression [8], the consistent variant of our instance based exact VSC. More precisely, that study employs a notion of VSC where the recovery function ρ is restricted to being the consensus recovery function. It also characterizes the size of such compression by star number parameter. That parameter is closely related to our notion of width of a class.

6   Discussion

In this work, we have introduced and analyzed a variety of formal frameworks for version space compression schemes. Such schemes capture not only what a sample labels but also the instances whose labels the sample does not determine. We prove positive results, including general families of classes for which such compression schemes are possible, closure properties, as well as convergence guarantees of a natural ERM paradigm that uses additional unlabeled samples.

On the other hand, we introduce some novel parameters of concept classes that imply lower bounds on the size of such compressions as well as cases for which no such compression is possible.

In view of various potential applications of such rich compression schemes, we hope that our work may open a line of research that may be both theoretically rich and practically relevant. Possible directions for future research range from exploring further alternative variants of compression notions, in particular notions of version space compression that are more closely tied with a particular application, to investigating algorithmic and computational complexity aspects.

References

1. Balcan, M.-F., Blum, A., Fine, S., Mansour, Y.: Distributed learning, communication complexity and privacy. In: Proceedings of the 25th Annual Conference on Learning Theory (COLT), pp. 26.1–26.22 (2012)
2. Ben-David, S.: 2 notes on classes with Vapnik-Chervonenkis dimension 1 (2015). CoRR arXiv:1507.05307

3. Ben-David, S., Litman, A.: Combinatorial variability of Vapnik-Chervonenkis classes with applications to sample compression schemes. Discrete Appl. Math. **86**(1), 3–25 (1998)
4. Ben-David, S.: Low-sensitivity functions from unambiguous certificates. In: Electronic Colloquium on Computational Complexity (ECCC), vol. 23, no. 84 (2016)
5. Chen, S.-T., Balcan, M.-F., Chau, D.H.: Communication efficient distributed agnostic boosting (2015). CoRR arXiv:1506.06318
6. Floyd, S., Warmuth, M.K.: Sample compression, learnability, and the Vapnik-Chervonenkis dimension. Mach. Learn. **21**(3), 269–304 (1995)
7. Goldman, S.A., Kearns, M.J.: On the complexity of teaching. J. Comput. Syst. Sci. **50**(1), 20–31 (1995)
8. Hanneke, S., Yang, L.: Minimax analysis of active learning. J. Mach. Learn. Res. **16**, 3487–3602 (2015)
9. Kuzmin, D., Warmuth, M.K.: Unlabeled compression schemes for maximum classes. J. Mach. Learn. Res. **8**, 2047–2081 (2007)
10. Li, L., Littman, M.L., Walsh, T.J., Strehl, A.L.: Knows what it knows: a framework for self-aware learning. Mach. Learn. **82**(3), 399–443 (2011)
11. Littlestone, N., Warmuth, M.K.: Relating data compression and learnability (1986, unpublished manuscript)
12. Mitchell, T.M.: Version spaces: a candidate elimination approach to rule learning. In: Proceedings of the 5th International Joint Conference on Artificial Intelligence, pp. 305–310 (1977)
13. Moran, S., Shpilka, A., Wigderson, A., Yehudayoff, A.: Compressing and teaching for low VC-dimension. In: Proceedings of IEEE 56th Annual Symposium on Foundations of Computer Science (FOCS), pp. 40–51 (2015)
14. Moran, S., Warmuth, M.K.: Labeled compression schemes for extremal classes (2015). CoRR arXiv:1506.00165
15. Rivest, R.L., Sloan, R.H.: Learning complicated concepts reliably and usefully. In: Proceedings of the 7th National Conference on Artificial Intelligence, pp. 635–640 (1988)
16. Samei, R., Semukhin, P., Yang, B., Zilles, S.: Sample compression for multi-label concept classes. In: Proceedings of The 27th Conference on Learning Theory (COLT), pp. 371–393 (2014)
17. Sayedi, A., Zadimoghaddam, M., Blum, A.: Trading off mistakes and don't-know predictions. In: Advances in Neural Information Processing Systems 23: 24th Annual Conference on Neural Information Processing Systems (NIPS), pp. 2092–2100 (2010)
18. Shalev-Shwartz, S., Ben-David, S.: Understanding Machine Learning. Cambridge University Press, Cambridge (2014)
19. Vapnik, V.N., Chervonenkis, A.J.: On the uniform convergence of relative frequencies of events to their probabilities. Theor. Probab. Appl. **16**(2), 264–280 (1971)
20. Wiener, Y., Hanneke, S., El-Yaniv, R.: A compression technique for analyzing disagreement-based active learning. J. Mach. Learn. Res. **16**, 713–745 (2015)
21. Zhang, C., Chaudhuri, K.: The extended littlestone's dimension for learning with mistakes and abstentions (2016). CoRR arXiv:1604.06162

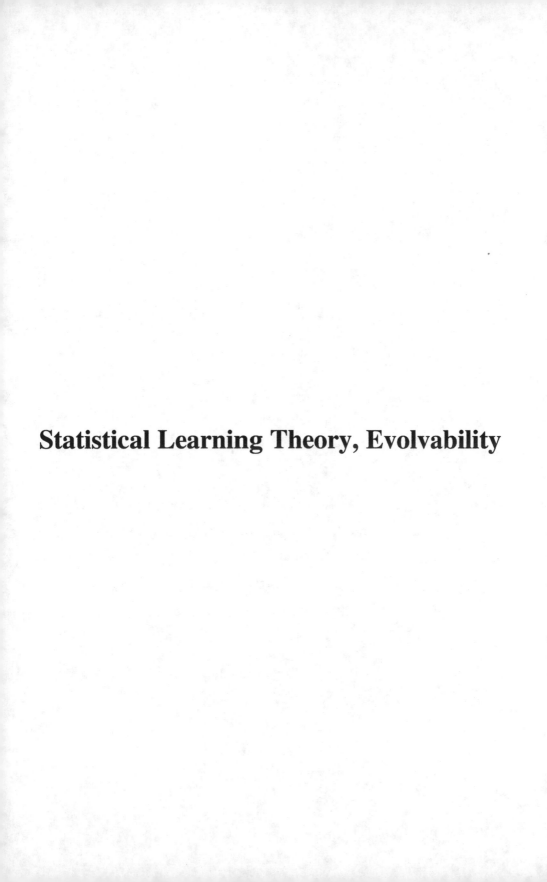

Statistical Learning Theory, Evolvability

Learning with Rejection

Corinna Cortes[1], Giulia DeSalvo[2(✉)], and Mehryar Mohri[1,2]

[1] Google Research, 111 8th Avenue, New York, NY, USA
[2] Courant Institute of Mathematical Sciences,
251 Mercer Street, New York, NY, USA
desalvo@cims.nyu.edu

Abstract. We introduce a novel framework for classification with a rejection option that consists of simultaneously learning two functions: a classifier along with a rejection function. We present a full theoretical analysis of this framework including new data-dependent learning bounds in terms of the Rademacher complexities of the classifier and rejection families as well as consistency and calibration results. These theoretical guarantees guide us in designing new algorithms that can exploit different kernel-based hypothesis sets for the classifier and rejection functions. We compare and contrast our general framework with the special case of confidence-based rejection for which we devise alternative loss functions and algorithms as well. We report the results of several experiments showing that our kernel-based algorithms can yield a notable improvement over the best existing confidence-based rejection algorithm.

1 Introduction

We consider a flexible binary classification scenario where the learner is given the option to reject an instance instead of predicting its label, thereby incurring some pre-specified cost, typically less than that of a random prediction. While classification with a rejection option has received little attention in the past, it is in fact a scenario of great significance that frequently arises in applications. Incorrect predictions can be costly, especially in applications such as medical diagnosis and bioinformatics. In comparison, the cost of abstaining from prediction, which may be that of additional medical tests, or that of routing a call to a customer representative in a spoken-dialog system, is often more acceptable. From a learning perspective, abstaining from fitting systematic outliers can also result in a more accurate predictor. Accurate algorithms for learning with rejection can further be useful to developing solutions for other learning problems such as active learning [4].

Various problems related to the scenario of learning with a rejection option have been studied in the past. The trade-off between error rate and rejection rate was first studied by Chow [5,6] who also provided an analysis of the Bayes optimal decision for this setting. Later, several publications studied an optimal rejection rule based on the ROC curve and a subset of the training data [16, 26,29], while others used rejection options or *punting* to reduce misclassification rate [2,15,20,24,27], though with no theoretical analysis or guarantee.

R. Ortner et al. (Eds.): ALT 2016, LNAI 9925, pp. 67–82, 2016.
DOI: 10.1007/978-3-319-46379-7_5

More generally, few studies have presented general error bounds in this area, but some have given risk bounds for specific scenarios. Freund et al. [14] studied an ensemble method and presented an algorithm that predicts with a weighted average of the hypotheses while abstaining on some examples without incurring a cost. Herbei and Wegkamp [18] considered classification with a rejection option that incurs a cost and provided bounds for these ternary functions.

One of the most influential works in this area has been that of Bartlett and Wegkamp [1] who studied a natural discontinuous loss function taking into account the cost of a rejection. They used consistency results to define a convex and continuous *Double Hinge Loss* (DHL) surrogate loss upper-bounding that rejection loss, which they also used to derive an algorithm. A series of follow-up articles further extended this publication, including [33] which used the same convex surrogate while focusing on the l_1 penalty. Grandvalet et al. [17] derived a convex surrogate based on [1] that aims at estimating conditional probabilities only in the vicinity of the threshold points of the optimal decision rule. They also provided some preliminary experimental results comparing the DHL algorithm and their variant with a naive rejection algorithm. Under the same rejection rule, Yuan and Wegkamp [32] studied the infinite sample consistency for classification with a reject option.

Using a different approach based on active learning, El-Yaniv and Wiener [11] studied the trade-off between the coverage and accuracy of classifiers and, in a subsequent paper [12] provided a strategy to learn a certain type of selective classification, which they define as *weakly optimal*, that has diminishing rejection rate under some Bernstein-type conditions. Finally, several papers have discussed learning with rejection in the multi-class setting [3,10,28], reinforcement learning [22], and in online learning [34].

There are also several learning scenarios tangentially related to the rejection scenario we consider, though they are distinct and hence require a very different approach. Sequential learning with budget constraints is a related framework that admits two stages: first a classifier is learned, next the classifier is fixed and a rejection function is learned [30,31]. Since it assumes a fixed predictor and only admits the rejection function as an argument, the corresponding loss function is quite different from ours. Another somewhat similar approach is that of cost-sensitive learning where a class-dependent cost can be used [13]. One could think of adopting that framework here to account for the different costs for rejection and incorrect prediction. However, the cost-sensitive framework assumes a distribution over the classes or labels, which, here, would be the set $\{-1, 1, ®\}$, with ® the rejection symbol. But, ® is not a class and there is no natural distribution over that set in our scenario.

In this paper, we introduce a novel framework for classification with a rejection option that consists of simultaneously learning a pair of functions (h, r): a predictor h along with a rejection function r, each selected from a different hypothesis set. This is a more general framework than that the special case of confidence-based rejection studied by Bartlett and Wegkamp [1] and others, where the rejection function is constrained to be a thresholded function of the predictor's scores. Our novel framework opens up a new perspective on the problem of learning with rejection for which we present a full theoretical analysis, including new data-dependent learning bounds in terms of the Rademacher complexities of the classifier and rejection families, as well as consistency and calibration results. We derive convex surrogates for this framework that are realizable $(\mathcal{H}, \mathcal{R})$-consistent. These guarantees in turn guide the design of a variety of algorithms for learning with rejection. We describe in depth two different types of algorithms: the first type uses kernel-based hypothesis classes, the second type confidence-based rejection functions. We report the results of experiments comparing the performance of these algorithms and that of the DHL algorithm.

The paper is organized as follows. Section 2 introduces our novel learning framework and contrasts it with that of Bartlett and Wegkamp [1]. Section 3 provides generalization guarantees for learning with rejection. It also analyzes two convex surrogates of the loss along with consistency results and provides margin-based learning guarantees. In Sect. 4, we present an algorithm with kernel-based hypothesis sets derived from our learning bounds. In Sect. 5, we further examine the special case of confidence-based rejection by analyzing various algorithmic alternatives. Lastly, we report the results of several experiments comparing the performance of our algorithms with that of DHL (Sect. 6).

2 Learning Problem

Let $\mathcal{X}$ denote the input space. We assume as in standard supervised learning that training and test points are drawn i.i.d. according to some fixed yet unknown distribution $\mathcal{D}$ over $\mathcal{X} \times \{-1, +1\}$. We present a new general model for learning with rejection, which includes the confidence-based models as a special case.

2.1 General Rejection Model

The learning scenario we consider is that of binary classification with rejection. Let ⓇR denote the rejection symbol. For any given instance $x \in \mathcal{X}$, the learner has the option of abstaining or *rejecting* that instance and returning the symbol Ⓡ, or assigning to it a label $\widehat{y} \in \{-1, +1\}$. If the learner rejects an instance, it incurs some loss $c(x) \in [0, 1]$; if it does not reject but assigns an incorrect label, it incurs a cost of one; otherwise, it suffers no loss. Thus, the learner's output is a pair (h, r) where $h \colon \mathcal{X} \to \mathbb{R}$ is the hypothesis used for predicting a label for points not rejected using $\mathrm{sign}\,(h)$ and where $r \colon \mathcal{X} \to \mathbb{R}$ is a function determining the points $x \in \mathcal{X}$ to be rejected according to $r(x) \leq 0$.

The problem is distinct from a standard multi-class classification problem since no point is inherently labeled with Ⓡ. Its natural loss function L is defined by

$$L(h, r, x, y) = 1_{yh(x) \leq 0} 1_{r(x) > 0} + c(x) 1_{r(x) \leq 0}, \tag{1}$$

for any pair of functions (h, r) and labeled sample $(x, y) \in \mathcal{X} \times \{-1, +1\}$, thus extending the loss function considered by [1]. In what follows, we assume for simplicity that c is a constant function, though part of our analysis is applicable to the general case. Observe that for $c \geq \frac{1}{2}$, on average, there is no incentive for rejection since a random guess can never incur an expected cost of more than $\frac{1}{2}$. For biased distributions, one may further limit c to the fraction of the smallest class. For $c = 0$, we obtain a trivial solution by rejecting all points, so we restrict c to the case of $c \in]0, \frac{1}{2}[$.

Let $\mathcal{H}$ and $\mathcal{R}$ denote two families of functions mapping $\mathcal{X}$ to $\mathbb{R}$. The learning problem consists of using a labeled sample $S = ((x_1, y_1), \ldots, (x_m, y_m))$ drawn i.i.d. from $\mathcal{D}^m$ to determine a pair $(h, r) \in \mathcal{H} \times \mathcal{R}$ with a small expected rejection loss $R(h, r)$

$$R(h, r) = \underset{(x,y) \sim \mathcal{D}}{\mathbb{E}} \left[1_{yh(x) \leq 0} 1_{r(x) > 0} + c 1_{r(x) \leq 0} \right]. \tag{2}$$

We denote by $\widehat{R}_S(h, r)$ the empirical loss of a pair $(h, r) \in \mathcal{H} \times \mathcal{R}$ over the sample S and use $(x, y) \sim S$ to denote the draw of (x, y) according to the empirical distribution defined by S: $\widehat{R}_S(h, r) = \mathbb{E}_{(x,y) \sim S} \left[1_{yh(x) \leq 0} 1_{r(x) > 0} + c 1_{r(x) \leq 0} \right]$.

2.2   Confidence-Based Rejection Model

Learning with rejection based on two independent yet jointly learned functions h and r introduces a completely novel approach to this subject. However, our new framework encompasses much of the previous work on this problem, e.g. [1], is a special case where rejection is based on the magnitude of the value of the predictor h, that is $x \in \mathcal{X}$ is rejected if $|h(x)| \leq \gamma$ for some $\gamma \geq 0$. Thus, r is implicitly defined in the terms of the predictor h by $r(x) = |h(x)| - \gamma$.

This specific choice of the rejection function r is natural when considering the Bayes solution (h^*, r^*) of the learning problem, that is the one where the distribution $\mathcal{D}$ is known. Indeed, for any $x \in \mathcal{X}$, let $\eta(x)$ be defined by $\eta(x) = \mathbb{P}[Y = +1|x]$. For a standard binary classification problem, it is known that the predictor h^* defined for any $x \in \mathcal{X}$ by $h^*(x) = \eta(x) - \frac{1}{2}$ is optimal since $\text{sign}(h^*(x)) = \max\{\eta(x), 1 - \eta(x)\}$. For any $x \in \mathcal{X}$, the misclassification loss of h^* is $\mathbb{E}[1_{yh(x) \leq 0}|x] = \min\{\eta(x), 1 - \eta(x)\}$. The optimal rejection r^* should therefore be defined such that $r^*(x) \leq 0$, meaning x is rejected, if and only if

$$\min\{\eta(x), 1 - \eta(x)\} \geq c \Leftrightarrow 1 - \max\{\eta(x), 1 - \eta(x)\} \geq c$$
$$\Leftrightarrow \max\{\eta(x), 1 - \eta(x)\} \leq 1 - c$$
$$\Leftrightarrow \max\{\eta(x) - \tfrac{1}{2}, \tfrac{1}{2} - \eta(x)\} \leq \tfrac{1}{2} - c \Leftrightarrow |h^*(x)| \leq \tfrac{1}{2} - c.$$

Thus, we can choose h^* and r^* as in Fig. 1, which also provides an illustration of confidence-based rejection. However, when predictors are selected out of a limited

$$h^*(x) = \eta(x) - \tfrac{1}{2} \quad \text{and}$$
$$r^*(x) = |h^*(x)| - (\tfrac{1}{2} - c).$$

Fig. 1. Mathematical expression and illustration of the optimal classification and rejection function for the Bayes solution. Note, as c increases, the rejection region shrinks.

Fig. 2. The best predictor h is defined by the threshold θ: $h(x) = x - \theta$. For $c < \frac{1}{2}$, the region defined by $X \leq \eta$ should be rejected. Note that the corresponding rejection function r defined by $r(x) = x - \eta$ cannot be defined as $|h(x)| \leq \gamma$ for some $\gamma > 0$.

subset $\mathcal{H}$ of all measurable functions over $\mathcal{X}$, requiring the rejection function r to be defined as $r(x) = |h(x)| - \gamma$, for some $h \in \mathcal{H}$, can be too restrictive. Consider, for example, the case where $\mathcal{H}$ is a family of linear functions. Figure 2 shows a simple case in dimension one where the optimal rejection region cannot be defined simply as a function of the best predictor h. The model for learning with rejection that we describe where a pair (h, r) is selected is more general. In the next section, we study the problem of learning such a pair.

3   Theoretical Analysis

We first give a generalization bound for the problem of learning with our rejection loss function as well as consistency results. Next, to devise efficient learning algorithms, we give general convex upper bounds on the rejection loss. For several of these convex surrogate losses, we prove margin-based guarantees that we subsequently use to define our learning algorithms (Sect. 4).

3.1   Generalization Bound

Theorem 1. *Let $\mathcal{H}$ and $\mathcal{R}$ be families of functions taking values in $\{-1, +1\}$. Then, for any $\delta > 0$, with probability at least $1 - \delta$ over the draw of a sample S of size m from $\mathcal{D}$, the following holds for all $(h, r) \in \mathcal{H} \times \mathcal{R}$:*

$$R(h, r) \leq \widehat{R}_S(h, r) + \mathfrak{R}_m(\mathcal{H}) + (1 + c)\mathfrak{R}_m(\mathcal{R}) + \sqrt{\frac{\log \frac{1}{\delta}}{2m}}.$$

Proof. Let $\mathcal{L}_{\mathcal{H}, \mathcal{R}}$ be the family of functions $\mathcal{L}_{\mathcal{H}, \mathcal{R}} = \{(x, y) \mapsto L(h, r, x, y), h \in \mathcal{H}, r \in \mathcal{R}\}$. Since the loss function L takes values in $[0, 1]$, by the general Rademacher complexity bound [19], with probability at least $1 - \delta$, the following holds for all $(h, r) \in \mathcal{H} \times \mathcal{R}$: $R(h, r) \leq \widehat{R}_S(h, r) + 2\mathfrak{R}_m(\mathcal{L}_{\mathcal{H}, \mathcal{R}}) + \sqrt{\frac{\log 1/\delta}{2m}}$. Now, the Rademacher complexity can be bounded as follows:

$$\mathfrak{R}_m(\mathcal{L}_{\mathcal{H}, \mathcal{R}}) = \mathbb{E}_\sigma \left[\sup_{(h,r) \in \mathcal{H} \times \mathcal{R}} \frac{1}{m} \sum_{i=1}^m \sigma_i 1_{y_i h(x_i) \leq 0} 1_{r(x_i) > 0} + \sigma_i c 1_{r(x_i) \leq 0} \right]$$

$$\leq \mathbb{E}_\sigma \left[\sup_{(h,r) \in \mathcal{H} \times \mathcal{R}} \frac{1}{m} \sum_{i=1}^m \sigma_i 1_{h(x_i) \neq y_i} 1_{r(x_i) = +1} \right] + c \mathbb{E}_\sigma \left[\sup_{r \in \mathcal{R}} \frac{1}{m} \sum_{i=1}^m \sigma_i 1_{r(x_i) = -1} \right].$$

By Lemma 1 (below), the Rademacher complexity of products of indicator functions can be bounded by the sum of the Rademacher complexities of each indicator function class, thus, $\mathbb{E}_{\sigma}\left[\sup_{(h,r)\in\mathcal{H}\times\mathcal{R}}\frac{1}{m}\sum_{i=1}^{m}\sigma_i 1_{h(x_i)\neq y_i}1_{r(x_i)=+1}\right] \leq$ $\mathbb{E}_{\sigma}\left[\sup_{h\in\mathcal{H}}\frac{1}{m}\sum_{i=1}^{m}\sigma_i 1_{h(x_i)\neq y_i}\right] + \mathbb{E}_{\sigma}\left[\sup_{r\in\mathcal{R}}\frac{1}{m}\sum_{i=1}^{m}\sigma_i 1_{r(x_i)=+1}\right]$. The proof can be completed by using the known fact that the Rademacher complexity of indicator functions based on a family of functions taking values in $\{-1,+1\}$ is equal to one half the Rademacher complexity of that family.   □

To derive an explicit bound in terms of $\mathcal{H}$ and $\mathcal{R}$ in Theorem 1, we make use of the following lemma relating the Rademacher complexity of a product of two (or more) families of functions to the sum of the Rademacher complexity of each family, whose proof can be found in [9].

Lemma 1. *Let $\mathcal{F}_1$ and $\mathcal{F}_2$ be two families of functions mapping $\mathcal{X}$ to $[-1,+1]$. Let $\mathcal{F} = \{f_1 f_2 \colon f_1 \in \mathcal{F}_1, f_2 \in \mathcal{F}_2\}$. Then, the empirical Rademacher complexities of $\mathcal{F}$ for any sample S of size m are bounded: $\widehat{\mathfrak{R}}_S(\mathcal{F}) \leq 2(\widehat{\mathfrak{R}}_S(\mathcal{F}_1) + \widehat{\mathfrak{R}}_S(\mathcal{F}_2))$.*

The theorem gives generalization guarantees for learning with a family of predictors $\mathcal{H}$ and rejection function $\mathcal{R}$ mapping to $\{-1,+1\}$ that admit Rademacher complexities in $O(1/\sqrt{m})$. For such families, it suggests to select the pair (h,r) to minimize the right-hand side. As with the zero-one loss, minimizing $\widehat{R}_S(h,r)$ is computationally hard for most families of functions. Thus, in the next section, we study convex upper bounds that lead to more efficient optimization problems, while admitting favorable learning guarantees as well as consistency results.

3.2   Convex Surrogate Losses

We first present general convex upper bounds on the rejection loss. Let $u \mapsto \Phi(-u)$ and $u \mapsto \Psi(-u)$ be convex functions upper-bounding $1_{u\leq 0}$. Since for any $a,b \in \mathbb{R}$, $\max(a,b) = \frac{a+b+|b-a|}{2} \geq \frac{a+b}{2}$, the following inequalities hold with $\alpha > 0$ and $\beta > 0$:

$$L(h,r,x,y) = 1_{yh(x)\leq 0}1_{r(x)>0} + c\,1_{r(x)\leq 0} = \max\left(1_{yh(x)\leq 0}1_{-r(x)<0}, c\,1_{r(x)\leq 0}\right)$$

$$\leq \max\left(1_{\max(yh(x),-r(x))\leq 0}, c\,1_{r(x)\leq 0}\right) \leq \max\left(1_{\frac{yh(x)-r(x)}{2}\leq 0}, c\,1_{r(x)\leq 0}\right)$$

$$\leq \max\left(1_{\alpha\frac{yh(x)-r(x)}{2}\leq 0}, c\,1_{\beta r(x)\leq 0}\right)$$

$$\leq \max\left(\Phi\left(\frac{\alpha}{2}(r(x)-yh(x))\right), c\,\Psi(-\beta r(x))\right) \tag{3}$$

$$\leq \Phi\left(\frac{\alpha}{2}(r(x)-yh(x))\right) + c\,\Psi(-\beta r(x)). \tag{4}$$

Since Φ and Ψ are convex, their composition with an affine function of h and r is also a convex function of h and r. Since the maximum of two convex functions is convex, the right-hand side of (3) is a convex function of h and r. Similarly, the right-hand side of (4) is a convex function of h and r. In the specific case where the Hinge loss is used for both $u \mapsto \Phi(-u)$ and $u \mapsto \Psi(-u)$, we obtain the following two convex upper bounds, Max Hinge (MH) and Plus Hinge (PH), also illustrated in Fig. 3:

$$L_{\mathrm{MH}}(h, r, x, y) = \max\left(1 + \tfrac{\alpha}{2}(r(x) - yh(x)), c\,(1 - \beta r(x)), 0\right)$$

$$L_{\mathrm{PH}}(h, r, x, y) = \max\left(1 + \tfrac{\alpha}{2}(r(x) - yh(x)), 0\right) + \max\left(c\,(1 - \beta r(x)), 0\right).$$

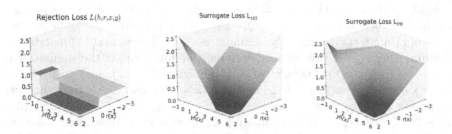

Fig. 3. From the left, the figures show the rejection loss L, the convex surrogate loss L_{MH}, and the convex surrogate loss L_{PH} as a function of $yh(x)$ and $r(x)$, for the cost value $c = 0.4$. The convex surrogates have a steeper left surface reflecting the rejection loss's penalty of incorrectly classifying a point while their gentler right surface of the surrogates reflects the lower cost c of abstaining. Also, the figures clearly show that the surrogate loss L_{PH} is an upper bound on L_{MH}.

3.3   Consistency Results

In this section, we present a series of theoretical results related to the consistency of the convex surrogate losses introduced. We first prove the calibration and consistency for specific choices of the parameters α and β. Next, we show that the excess risk with respect to the rejection loss can be bounded by its counterpart defined via our surrogate loss. We further prove a general realizable $(\mathcal{H}, \mathcal{R})$-consistency for our surrogate losses.

Calibration. The constants $\alpha > 0$ and $\beta > 0$ are introduced in order to calibrate the surrogate loss with respect to the Bayes solution. Let $(h_{\mathrm{M}}^{*}, r_{\mathrm{M}}^{*})$ be a pair attaining the infimum of the expected surrogate loss $\mathbb{E}_{(x,y)}(L_{\mathrm{MH}}(h, r, x, y))$ over all measurable functions. Recall from Sect. 2, the Bayes classifier is denoted by (h^{*}, r^{*}). The following lemma shows that for $\alpha = 1$ and $\beta = \frac{1}{1-2c}$, the loss L_{MH} is calibrated, that is the sign of $(h_{\mathrm{M}}^{*}, r_{\mathrm{M}}^{*})$ matches the sign of (h^{*}, r^{*}).

Theorem 2. *Let (h_{M}^{*}, r_{M}^{*}) denote a pair attaining the infimum of the expected surrogate loss, $\mathbb{E}_{(x,y)}[L_{\mathrm{MH}}(h_{M}^{*}, r_{M}^{*}, x, y)] = \inf_{(h,r)\in\mathrm{meas}} \mathbb{E}_{(x,y)}[L_{\mathrm{MH}}(h, r, x, y)]$. Then, for $\beta = \frac{1}{1-2c}$ and $\alpha = 1$,*

1. the surrogate loss L_{MH} is calibrated with respect to the Bayes classifier: $\mathrm{sign}\,(h^{*}) = \mathrm{sign}\,(h_{M}^{*})$ and $\mathrm{sign}\,(r^{*}) = \mathrm{sign}\,(r_{M}^{*})$;

2. *furthermore, the following equality holds for the infima over pairs of measurable functions:*

$$\inf_{(h,r)} \mathbb{E}_{(x,y)\sim\mathcal{D}}[L_{\mathrm{MH}}(h,r,x,y)] = (3-2c)\inf_{(h,r)} \mathbb{E}_{(x,y)\sim\mathcal{D}}[L(h,r,x,y)].$$

Proof Sketch. The expected surrogate loss can be written in terms of $\eta(x)$: $\mathbb{E}_{(x,y)\sim\mathcal{D}}[L_{\mathrm{MH}}(h,r,x,y)] = \mathbb{E}_x[\eta(x)\phi(-h(x),r(x))+(1-\eta(x))\phi(h(x),r(x))]$, with $\phi(-h(x),r(x)) = \max(1 + \frac{1}{2}(r(x) - h(x)), c(1 - \frac{1}{1-2c}r(x)), 0)$. Let the argument of the expectation, $\eta(x)\phi(-h(x),r(x))+(1-\eta(x))\phi(h(x),r(x))$, be denoted by $\mathcal{L}_\phi(\eta(x),h(x),r(x))$. Since the infimum is over all measurable functions, to determine $(h_{\mathrm{M}}^*,r_{\mathrm{M}}^*)$ it suffices to determine, for any fixed x the minimizer of $(u,v) \mapsto \mathcal{L}_\phi(\eta(x),u,v)$. For a fixed x, minimizing $\mathcal{L}_\phi(\eta(x),u,v)$ with respect to (u,v) is equivalent to minimizing seven LPs. One can check that the optimal points of these LPs are in the set $(u,v) \in \{(0,(2c-2)(1-2c)),(3-2c,1-2c),(-3+2c,1-2c)\}$. Evaluating $\mathcal{L}_\phi(\eta(x),u,v)$ at these points, we find that $\mathcal{L}_\phi(\eta(x),3-2c,1-2c) = (3-2c)(1-\eta(x))$, $\mathcal{L}_\phi(\eta(x),-3+2c,1-2c) = (3-2c)(\eta(x))$, and $\mathcal{L}_\phi(\eta(x),0,(2c-2)(1-2c)) = (3-2c)c$. Thus, we can conclude that the minimum of $\mathcal{L}_\phi(\eta(x),u,v)$ is attained at $(3-2c)\big[\eta(x)\mathbf{1}_{\eta(x)<c} + c\mathbf{1}_{c\le\eta(x)\le1-c} + (1-\eta(x))\mathbf{1}_{\eta(x)>1-c}\big]$, which completes the proof. □

Excess Risk Bound. Here, we show upper bounds on the excess risk in terms of the surrogate loss excess risk. Let R^* denote the Bayes rejection loss, that is $R^* = \inf_{(h,r)} \mathbb{E}_{(x,y)\sim\mathcal{D}}[L(h,r,x,y)]$, where the infimum is taken over all measurable functions and similarly let R_L^* denote $\inf_{(h,r)} \mathbb{E}_{(x,y)\sim\mathcal{D}}[L_{\mathrm{MH}}(h,r,x,y)]$.

Theorem 3. *Let $R_L(h,r) = \mathbb{E}_{(x,y)\sim\mathcal{D}}[L_{\mathrm{MH}}(h,r,x,y)]$ denote the expected surrogate loss of a pair (h,r). Then, the surrogate excess of (h,r) is upper bounded by its surrogate excess error as follows:*

$$R(h,r) - R^* \le \tfrac{1}{(1-c)(1-2c)}\big(R_L(h,r) - R_L^*\big).$$

Proof Sketch. Let $\mathcal{L}^*(\eta(x))$ denote the expected loss of the Bayes solution conditioned on x, $\mathcal{L}^*(\eta(x)) = \eta(x)\mathbf{1}_{\eta(x)<c} + c\mathbf{1}_{c\le\eta(x)\le1-c} + (1-\eta(x))\mathbf{1}_{\eta(x)>1-c}$. Then

$$R(h,r) - R(h^*,r^*) = \mathbb{E}_x\big[(\eta(x) - \mathcal{L}^*(\eta(x)))\mathbf{1}_{h(x)<0,r>(x)0} \tag{5}$$

$$+ (1-\eta(x) - \mathcal{L}^*(\eta(x)))\mathbf{1}_{h(x)\ge0,r(x)>0} + (c - \mathcal{L}^*(\eta(x)))\mathbf{1}_{r(x)\le0}].$$

Since $\mathcal{L}^*(\eta(x))$ admits three values, we can distinguish three cases and give a proof for each. When $c \le \eta(x) \le 1 - c$, $\mathcal{L}^*(\eta(x)) = c$, that is $r^* \le 0$ and $r_L^* \le 0$, by calibration. In that case, Eq. 5 can be written as $R(h,r)-R(h^*,r^*) = \mathbb{E}_x\big((\eta(x)-c)\mathbf{1}_{h(x)<0,r(x)>0}+(1-\eta(x)-c)\mathbf{1}_{h(x)\ge0,r(x)>0}\big)$. Note that the indicator functions on the right-hand side are mutually exclusive, thus, it suffices to show that each component is bounded. ⊔

$(\mathcal{H}, \mathcal{R})$-Consistency. The standard notion of loss consistency does not take into account the hypothesis set H used since it assumes an optimization carried out over the set of all measurable functions. Long and Servedio [23] proposed instead a notion of H-*consistency* precisely meant to take the hypothesis set used into consideration. They showed empirically that using loss functions that are H-consistent can lead to significantly better performances than using a loss function known to be consistent. Here, we prove that our surrogate losses are *realizable* $(\mathcal{H}, \mathcal{R})$-*consistent*, a hypothesis-set-specific notion of consistency under our framework. The realizable setting in learning with rejection means that there exists a function that never rejects and correctly classifies all points. A loss l is *realizable* $(\mathcal{H}, \mathcal{R})$-*consistent* if for any distribution $\mathcal{D}$ over $\mathcal{X} \times \mathcal{Y}$ and any $\epsilon > 0$, there exists $\delta > 0$ such that if $\big| \mathbb{E}_{(x,y)\sim\mathcal{D}}\left[l(h, r, x, y)\right] - \inf_{(h,r)\in(\mathcal{H},\mathcal{R})} \mathbb{E}_{(x,y)\sim\mathcal{D}}[l(h, r, x, y)] \big| \leq \delta$, then $\mathbb{E}_{(x,y)\sim\mathcal{D}}[L(h, r, x, y)] \leq \epsilon$.

Theorem 4. *Let* $(u, v) \mapsto \Phi(-u, -v)$ *be a non-increasing function upperbounding* $(u, v) \mapsto 1_{u\leq0}1_{v>0} + c1_{v\leq0}$ *such that for any fixed* v, $\lim_{u\to+\infty} \Phi(-u, -v) = 0$ *and for any fixed* v, $u \mapsto \Phi(-u, -v)$ *is bounded over* $\mathbb{R}_{+}$. *Let* $(\mathcal{H}, \mathcal{R})$ *be pair of families of functions mapping* $\mathcal{X}$ *to* $\mathbb{R}$ *where* $\mathcal{H}$ *is closed under multiplication by a positive scalar (*$\mathcal{H}$ *is a cone). Then, the loss function* $(h, r, x, y) \mapsto \Phi(-yh(x), -r(x))$ *is realizable* $(\mathcal{H}, \mathcal{R})$-*consistent.*

Proof. Let $\mathcal{D}$ be a distribution for which $(h^*, r^*) \in (\mathcal{H}, \mathcal{R})$ achieves zero error, thus $yh^*(x) > 0$ and $r^*(x) > 0$ for all x in the support of $\mathcal{D}$. Fix $\epsilon > 0$ and assume that $\big| \mathbb{E}\left[\Phi\left(- yh(x), -r(x)\right)\right] - \inf_{(h,r)\in(\mathcal{H},\mathcal{R})} \mathbb{E}\left[\Phi\left(- yh(x), -r(x)\right)\right] \big| \leq \epsilon$ for some $(h, r) \in (\mathcal{H}, \mathcal{R})$. Then, since $1_{u\leq0}1_{v>0} + c1_{v\leq0} \leq \Phi(-u, -v)$ and since μh^* is in $\mathcal{H}$ for any $\mu > 0$, the following holds for any $\mu > 0$:

$$\mathbb{E}\left[L(h, r, x, y)\right] \leq \mathbb{E}\left[\Phi\left(- yh(x), -r(x)\right)\right] \leq \mathbb{E}\left[\Phi\left(- \mu yh^*(x), -r^*(x)\right)\right] + \epsilon$$
$$\leq \mathbb{E}\left[\Phi\left(- \mu yh^*(x), -r^*(x)\right)|r^*(x) > 0\right] \mathbb{P}[r^*(x) > 0] + \epsilon.$$

Now, $u \mapsto \Phi(-\mu yh^*(x), -r^*(x))$ is bounded for $yh^*(x) > 0$ and $r^*(x) > 0$; since $\lim_{\mu\to+\infty} \Phi(-\mu yh^*(x), -r^*(x)) = 0$, by Lebesgue's dominated convergence theorem $\lim_{\mu\to+\infty} \mathbb{E}[\Phi(-\mu yh^*(x), -r^*(x))|r^*(x) > 0] = 0$. Thus, $\mathbb{E}[L(h, r, x, y)] \leq \epsilon$ for all $\epsilon > 0$, which concludes the proof. $\qquad\square$

The conditions of the theorem hold in particular for the exponential and the logistic functions as well as hinge-type losses. Thus, the theorem shows that the general convex surrogate losses we defined are realizable $(\mathcal{H}, \mathcal{R})$-consistent when the functions Φ or Ψ are exponential or logistic functions.

3.4   Margin Bounds

In this section, we give margin-based learning guarantees for the loss function L_{MH}. Since L_{PH} is a simple upper bound on L_{MH}, its margin-based learning bound can be derived similarly. In fact, the same technique can be used to derive margin-based guarantees for the subsequent convex surrogate loss functions we present.

For any $\rho, \rho' > 0$, the margin-loss associated to L_{MH} is given by
$$L_{\mathrm{MH}}^{\rho,\rho'}(h, r, x, y) = \max\left(\max\left(1 + \tfrac{\alpha}{2}\left(\tfrac{r(x)}{\rho'} - \tfrac{yh(x)}{\rho}\right), 0\right), \max\left(c\left(1 - \beta\tfrac{r(x)}{\rho'}\right), 0\right)\right).$$

The theorem enables us to derive margin-based learning guarantees. The proof requires dealing with this max-based surrogate loss, which is a non-standard derivation.

Theorem 5. *Let $\mathcal{H}$ and $\mathcal{R}$ be families of functions mapping $\mathcal{X}$ to $\mathbb{R}$. Then, for any $\delta > 0$, with probability at least $1 - \delta$ over the draw of a sample S of size m from $\mathcal{D}$, the following holds for all $(h, r) \in \mathcal{H} \times \mathcal{R}$:*

$$R(h,r) \leq \mathop{\mathbb{E}}_{(x,y)\sim S}[L_{\text{MH}}(h,r,x,y)] + \alpha \mathfrak{R}_m(\mathcal{H}) + (2\beta c + \alpha)\mathfrak{R}_m(\mathcal{R}) + \sqrt{\frac{\log \frac{1}{\delta}}{2m}}.$$

Proof. Let $\mathcal{L}_{\text{MH},\mathcal{H},\mathcal{R}}$ be the family of functions defined by $\mathcal{L}_{\mathcal{H},\mathcal{R}} = \{(x,y) \mapsto \min\left(L_{\text{MH}}(h,r,x,y),1\right), h \in \mathcal{H}, r \in \mathcal{R}\}$. Since $\min(L_{\text{MH}},1)$ is bounded by one, by the general Rademacher complexity generalization bound [19], with probability at least $1 - \delta$ over the draw of a sample S, the following holds:

$$R(h,r) \leq \mathop{\mathbb{E}}_{(x,y)\sim\mathcal{D}}[\min(L_{\text{MH}}(h,r,x,y),1)] \leq \mathop{\mathbb{E}}_{(x,y)\sim S}[\min(L_{\text{MH}}(h,r,x,y),1)] +$$

$$2\mathfrak{R}_m(\mathcal{L}_{\text{MH},\mathcal{H},\mathcal{R}}) + \sqrt{\frac{\log 1/\delta}{2m}} \leq \mathop{\mathbb{E}}_{(x,y)\sim S}[L_{\text{MH}}(h,r,x,y)] + 2\mathfrak{R}_m(\mathcal{L}_{\text{MH},\mathcal{H},\mathcal{R}}) + \sqrt{\frac{\log 1/\delta}{2m}}.$$

Observe that we can express L_{MH} as follows: $\max\left(\max\left(1 + \frac{\alpha}{2}(r(x) - yh(x)), 0\right), \max\left(c\left(1 - \beta r(x)\right), 0\right)\right)$. Therefore, since for any $a, b \in \mathbb{R}$, $\min\left(\max(a,b), 1\right) = \max\left(\min(a,1), \min(b,1)\right)$, we can re-write $\min(L_{\text{MH}}, 1)$ as:

$$\max\left(\min\left(\max(1 + \tfrac{\alpha}{2}(r(x) - yh(x)), 0), 1\right), \min\left(\max(c\left(1 - \beta r(x)), 0\right), 1\right)\right)$$

$$\leq \min\left(\max(1 + \tfrac{\alpha}{2}(r(x) - yh(x)), 0), 1\right) + \min\left(\max(c\left(1 - \beta r(x)), 0\right), 1\right).$$

Since $u \mapsto \min\left(\max(1 + \frac{\alpha u}{2}, 0), 1\right)$ is $\frac{\alpha}{2}$-Lipschitz and $u \mapsto \min\left(\max(c(1 - \beta u), 0), 1\right)$ is $c\beta$-Lipschitz, by Talagrand's contraction lemma [21],

$$\mathfrak{R}_m(\mathcal{L}_{\text{MH},\mathcal{H},\mathcal{R}}) \leq \tfrac{\alpha}{2}\mathfrak{R}_m\left(\{(x,y) \mapsto r(x) - yh(x)\}\right) + \beta c \mathfrak{R}_m\left(\{(x,y) \mapsto r(x)\}\right)$$

$$\leq \tfrac{\alpha}{2}\left(\mathfrak{R}_m(\mathcal{R}) + \mathfrak{R}_m(\mathcal{H})\right) + \beta c \mathfrak{R}_m(\mathcal{R}) = \frac{\alpha}{2}\mathfrak{R}_m(\mathcal{H}) + \left(\beta c + \tfrac{\alpha}{2}\right)\mathfrak{R}_m(\mathcal{R}),$$

which completes the proof. □

The following corollary is then a direct consequence of the theorem above.

Corollary 1. *Let $\mathcal{H}$ and $\mathcal{R}$ be families of functions mapping $\mathcal{X}$ to $\mathbb{R}$. Fix $\rho, \rho' > 0$. Then, for any $\delta > 0$, with probability at least $1 - \delta$ over the draw of an i.i.d. sample S of size m from $\mathcal{D}$, the following holds for all $(h, r) \in \mathcal{H} \times \mathcal{R}$:*

$$R(h,r) \leq \mathop{\mathbb{E}}_{(x,y)\sim S}[L_{\text{MH}}^{\rho,\rho'}(h,r,x,y)] + \frac{\alpha}{\rho}\mathfrak{R}_m(\mathcal{H}) + \frac{2\beta c + \alpha}{\rho'}\mathfrak{R}_m(\mathcal{R}) + \sqrt{\frac{\log \frac{1}{\delta}}{2m}}.$$

Then, via [19], the bound of Corollary 1 can be shown to hold uniformly for all $\rho, \rho' \in (0,1)$, at the price of a term in $O\left(\sqrt{\frac{\log\log 1/\rho}{m}} + \sqrt{\frac{\log\log 1/\rho'}{m}}\right)$.

4   Algorithms for Kernel-Based Hypotheses

In this section, we devise new algorithms for learning with a rejection option when $\mathcal{H}$ and $\mathcal{R}$ are kernel-based hypotheses. We use Corollary 1 to guide the optimization problems for our algorithms.

Let $\mathcal{H}$ and $\mathcal{R}$ be hypotheses sets defined in terms of PSD kernels K and K' over $\mathcal{X}$:

$$\mathcal{H} = \{x \to \boldsymbol{w} \cdot \boldsymbol{\Phi}(x) \colon \|\boldsymbol{w}\| \leq \Lambda\} \text{ and } \mathcal{R} = \{x \to \boldsymbol{u} \cdot \boldsymbol{\Phi}'(x) \colon \|\boldsymbol{u}\| \leq \Lambda'\},$$

where $\boldsymbol{\Phi}$ is the feature mapping associated to K and $\boldsymbol{\Phi}'$ the feature mapping associated to K' and where $\Lambda, \Lambda' \geq 0$ are hyperparameters. One key advantage of this formulation is that different kernels can be used to define $\mathcal{H}$ and $\mathcal{R}$, thereby providing a greater flexibility for the learning algorithm. In particular, when using a second-degree polynomial for the feature vector $\boldsymbol{\Phi}'$, the rejection function corresponds to abstaining on an ellipsoidal region, which covers confidence-based rejection. For example, the Bartlett and Wegkamp [1] solution consists of choosing $\boldsymbol{\Phi}'(x) = \boldsymbol{\Phi}(x)$, $\boldsymbol{u} = \boldsymbol{w}$, and the rejection function, $r(x) = |h(x)| - \gamma$.

Corollary 2. *Let $\mathcal{H}$ and $\mathcal{R}$ be the hypothesis spaces as defined above. Then, for any $\delta > 0$, with probability at least $1 - \delta$ over the draw of a sample S of size m from $\mathcal{D}$, the following holds for all $(h, r) \in \mathcal{H} \times \mathcal{R}$:*

$$R(h,r) \leq \mathop{\mathbb{E}}_{(x,y)\sim S}[L_{\mathrm{MH}}^{\rho,\rho'}(h,r,x,y)] + \alpha\sqrt{\frac{(\kappa\Lambda/\rho)^2}{m}} + (2\beta c + \alpha)\sqrt{\frac{(\kappa'\Lambda'/\rho')^2}{m}} + \sqrt{\frac{\log\frac{1}{\delta}}{2m}}$$

where $\kappa^2 = \sup_{x\in\mathcal{X}} K(x,x)$ and $\kappa'^2 = \sup_{x\in\mathcal{X}} K'(x,x)$.

Proof. By standard kernel-based bounds on Rademacher complexity [25], we have that $\mathfrak{R}_m(\mathcal{H}) \leq \Lambda\sqrt{\frac{Tr[\mathbf{K}]}{m}} \leq \sqrt{\frac{(\kappa\Lambda)^2}{m}}$ and similarly $\mathfrak{R}_m(\mathcal{R}) \leq \Lambda'\sqrt{\frac{Tr[\mathbf{K}']}{m}} \leq \sqrt{\frac{(\kappa'\Lambda')^2}{m}}$. Applying this bounds to Corollary 1 completes the proof. □

This learning bound guides directly the definition of our first algorithm based on the L_{MH} (see full version [7] for details) resulting in the following optimization:

$$\min_{\boldsymbol{w},\boldsymbol{u},\boldsymbol{\xi}} \quad \frac{\lambda}{2}\|\boldsymbol{w}\|^2 + \frac{\lambda'}{2}\|\boldsymbol{u}\|^2 + \sum_{i=1}^{m}\xi_i \quad \text{subject to: } \xi_i \geq c(1 - \beta(\boldsymbol{u}\cdot\boldsymbol{\Phi}'(x_i) + b')),$$

$$\text{and} \quad \xi_i \geq 1 + \frac{\alpha}{2}\left(\boldsymbol{u}\cdot\boldsymbol{\Phi}'(x_i) + b' - y_i\boldsymbol{w}\cdot\boldsymbol{\Phi}(x_i) - b\right), \xi_i \geq 0,$$

where $\lambda, \lambda' \geq 0$ are parameters and b and b' are explicit offsets for the linear functions h and r. Similarly, we use the learning bound to derive a second algorithm based on the loss L_{PH} (see full paper [7]). We have implemented and tested the dual of both algorithms, which we will refer to as CHR algorithms (short for convex algorithms using $\mathcal{H}$ and $\mathcal{R}$ families). Both the primal and dual optimization are standard QP problems whose solution can be readily found via both general-purpose and specialized QP solvers. The flexibility of the kernel choice and the QP formulation for both primal and dual are key advantages of the CHR algorithms. In Sect. 6 we report experimental results with these algorithms as well as the details of our implementation.

5   Confidence-Based Rejection Algorithms

In this section, we explore different algorithms for the confidence-based rejection model (Sect. 2.2). We thus consider a rejection function $r(x) = |h(x)| - \gamma$ that abstains on points classified with confidence less than a given threshold γ.

The most standard algorithm in this setting is the DHL algorithm, which is based on a double hinge loss, a hinge-type convex surrogate that has favorable consistency properties. The double hinge loss, L_{DHinge}, is an upper bound of the rejection loss only when $0 \le \gamma \le 1 - c$, making DHL algorithm only valid for these restricted γ values. Moreover, it is important to note that the hinge loss is in fact a tighter convex upper bound than the double hinge loss for these possible values of γ. We have $L_\gamma(h) \le L_{\text{Hinge}}(h) \le L_{\text{DHinge}}(h)$ where $L_\gamma(h) = 1_{yh(x) \le 0} 1_{|h(x)| > \gamma} + c(x) 1_{|h(x)| \le \gamma}$ is the rejection loss in this setting. Thus, a natural alternative to the DHL algorithm is simply minimizing the hinge loss. The DHL solves a QCQP optimization problem while the natural alternative solve a standard SVM-type dual.

The aforementioned confidence based algorithms only apply for $\gamma \in [0, 1 - c]$ but a robust surrogate should majorate the rejection loss L_γ for all possible values. In [7] we present an algorithm that upper-bounds the rejection error for all values of $\gamma \in [0, 1]$. We provide further details of all these confidence-based algorithm as well as report several experimental results in [7]. While the alternative algorithms we described are based on tighter surrogate losses for the rejection loss than that of DHL, empirical evidence suggests that DHL outperforms these alternatives. Thus, in the experiments with our CHR algorithm, we will use DHL as the baseline for comparison (Sect. 6).

6   Experiments

In this section, we present the results of several experiments comparing our CHR algorithms with the DHL algorithm. All algorithms were implemented using CVX [8]. We tested the algorithms on seven data sets from the UCI data repository, specifically `australian`, `cod`, `skin`, `liver`, `banknote`, `haberman`, and `pima`. For each data set, we performed standard 5-fold cross-validation. We randomly divided the data into training, validation and test set in the ratio 3:1:1. We then repeated the experiments five times where each time we used a different random partition.

The cost values ranged over $c \in \{0.05, 0.1, \ldots, 0.5\}$ and the kernels for both algorithms were polynomial kernels of degree $d \in \{1, 2, 3\}$ and Gaussian kernels with widths in the set $\{1, 10, 100\}$. The regularization parameters λ, λ' for the CHR algorithms varied over $\lambda, \lambda' \in \{10^i : i = -5, \ldots, 5\}$ and the threshold γ for DHL ranged over $\gamma \in \{0.08, 0.16, \ldots, 0.96\}$.

For each fixed value of c, we chose the parameters with the smallest average rejection loss on the validation set. For these parameter values, Table 1 shows the corresponding rejection loss on the test set for the CHR algorithm based on L_{MH} and the DHL algorithm both with cost $c = 0.25$. The table also shows

Table 1. For the DHL algorithm and the CHR algorithm of L_{MH} with cost values $c = 0.25$, we report the mean and standard deviations on the test set of the following quantities: the left two columns contain the rejection loss, the next two columns the fraction of points rejected, followed by two columns with the classification error on the non-rejected points. The rightmost column provides the error on the non-rejected points of the DHL algorithm if its rejection threshold is changed so it rejects the same fraction of points as the CHR algorithm.

Data sets	Rejection loss DHL	Rejection loss CHR	Fraction rejected DHL	Fraction rejected CHR	Non-rejected error DHL	Non-rejected error CHR	Non-rejected err (incr. thrh.) DHL
cod	$0.176 \pm .030$	$0.098 \pm .037$	$0.186 \pm .055$	$0.024 \pm .028$	$0.130 \pm .043$	$0.092 \pm .039$	$0.186 \pm .033$
skin	$0.158 \pm .041$	$0.043 \pm .020$	$0.093 \pm .033$	$0.052 \pm .027$	$0.135 \pm .037$	$0.030 \pm .024$	$0.135 \pm .041$
bank	$0.061 \pm .022$	$0.030 \pm .006$	$0.066 \pm .016$	$0.036 \pm .022$	$0.045 \pm .018$	$0.021 \pm .008$	$0.044 \pm .016$
haber	$0.261 \pm .033$	$0.211 \pm .037$	$0.875 \pm .132$	$0.439 \pm .148$	$0.043 \pm .027$	$0.102 \pm .048$	$0.252 \pm .110$
pima	$0.241 \pm .025$	$0.171 \pm .017$	$0.055 \pm .007$	$0.700 \pm .055$	$0.227 \pm .025$	$0.043 \pm .023$	$0.112 \pm .060$
australian	$0.115 \pm .026$	$0.111 \pm .021$	$0.136 \pm .008$	$0.172 \pm .024$	$0.081 \pm .025$	$0.068 \pm .023$	$0.349 \pm .100$
liver	$0.236 \pm .040$	$0.248 \pm .005$	$0.397 \pm .047$	$0.980 \pm .019$	$0.136 \pm .044$	$0.003 \pm .006$	$0.292 \pm .120$

the fraction of points rejected by each algorithm and the classification error on non-rejected points (see full paper version [7] for similar tables for all cost values). The rejection loss results of Table 1 show that the CHR algorithm yields an improvement in the rejection loss over the DHL algorithm. These findings are statistically significant at the 1 % level or higher with one-sided paired t–test for all data sets except for the liver and australian data sets. Table 1 also reveals that the DHL algorithm rejects at a different rate than the CHR algorithm and often predicts the wrong label on the non-rejected points at a much higher rate. In order to level the playing field for the two algorithms, for the optimal settings of the DHL algorithm, we changed the rejection threshold till the fraction rejected by the DHL algorithm matched the fraction rejected by the CHR algorithm and recorded the error on the remaining non-rejected points. These results are included in the right-most column of Table 1 and demonstrate that the CHR algorithm rejects the hard cases and obtains a significantly better error rate on the remaining ones. In Fig. 4, we show the rejection loss as a function of the cost for six of our data sets. These plots demonstrate that the difference in accuracy between the two algorithms holds consistently for almost all values of c across all the data sets.

We also analyzed the rejection regions of the two algorithms. Unlike the DHL algorithm, we found that the CHR algorithms do not restrict their rejection regions to only areas of low confidence. On the other hand, the DHL algorithm only rejects around the boundary of the classification surface, see Fig. 5. In [7], we further analyze the difference between the rejection functions found by the two algorithms. We also provide more results for the CHR algorithm including results for the CHR algorithm based on L_{PH}. We find that on average the CHR with L_{MH} performs slightly better than the CHR with L_{PH} as is expected since the loss L_{PH} is an upper bound of the loss L_{MH}.

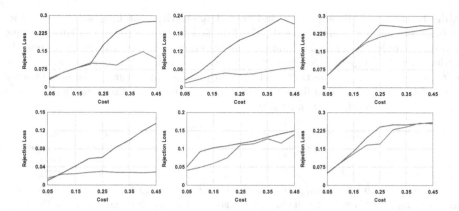

Fig. 4. Average rejection loss on the test set as a function of cost c for the DHL algorithm and the CHR algorithm for six datasets and polynomial kernels. The blue line is the DHL algorithm while the red line is the CHR algorithm based on L_{MH}. The figures on the top starting from the left are for the `cod`, `skin`, and `haberman` data set while the figures on the bottom are for `banknote`, `australian` and `pima` data sets. These figures show that the CHR algorithm outperforms the DHL algorithm for most values of cost, c, across all data sets.

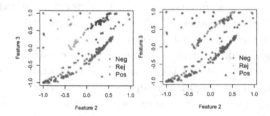

Fig. 5. The left figure shows CHR's classification of sample test points from the `skin` dataset with respect to different feature vectors. The right figure shows their classification by DHL and demonstrates how DHL rejects in areas of low confidence.

7   Conclusion

We presented a detailed study of the problem of learning with rejection, which is a key question in a number of applications. We gave a general formulation of the problem for which we provided a theoretical analysis, including generalization guarantees, the derivation of different convex surrogates that are calibrated and consistent, and margin bounds that helped us devise new algorithms. The empirical results we reported demonstrate the effectiveness of our algorithms in several datasets. Our general formulation can further inspire the design of other algorithms as well as new theoretical insights and studies, one such a potential area being active learning. Furthermore, a natural extension of our framework is to include a constraint on the maximum fraction of points that can be rejected. Such an additional constraint will require new algorithms and generalization bounds.

Acknowledgments. This work was partly funded by NSF IIS-1117591, CCF-1535987, and DGE-1342536.

References

1. Bartlett, P., Wegkamp, M.: Classification with a reject option using a hinge loss. JMLR **9**, 1823–1840 (2008)
2. Bounsiar, A., Grall, E., Beauseroy, P.: Kernel based rejection method for supervised classification. WASET **3**, 312–321 (2007)
3. Capitaine, H.L., Frelicot, C.: An optimum class-rejective decision rule and its evaluation. In: ICPR (2010)
4. Chaudhuri, K., Zhang, C.: Beyond disagreement-based agnostic active learning. In: NIPS (2014)
5. Chow, C.: An optimum character recognition system using decision function. IEEE T. C. (1957)
6. Chow, C.: On optimum recognition error and reject trade-off. IEEE T. C. (1970)
7. Cortes, C., DeSalvo, G., Mohri, M.: Learning with rejection. In: arXiv (2016)
8. I. Cvx Research. CVX: Matlab software for disciplined convex programming, version 2.0, August 2012
9. DeSalvo, G., Mohri, M., Syed, U.: Learning with deep cascades. In: Chaudhuri, K., Gentile, C., Zilles, S. (eds.) ALT 2015. LNCS, vol. 9355, pp. 254–269. Springer, Heidelberg (2015). doi:10.1007/978-3-319-24486-0_17
10. Dubuisson, B., Masson, M.: Statistical decision rule with incomplete knowledge about classes. Pattern Recognit. **26**, 155–165 (1993)
11. El-Yaniv, R., Wiener, Y.: On the foundations of noise-free selective classification. JMLR **11**, 1605–1641 (2010)
12. El-Yaniv, R., Wiener, Y.: Agnostic selective classification. In: NIPS (2011)
13. Elkan, C.: The foundations of cost-sensitive learning. In: IJCAI (2001)
14. Freund, Y., Mansour, Y., Schapire, R.: Generalization bounds for averaged classifiers. Ann. Stat. (2004)
15. Fumera, G., Roli, F.: Support vector machines with embedded reject option. In: ICPR (2002)
16. Fumera, G., Roli, F., Giacinto, G.: Multiple reject thresholds for improving classification reliability. In: ICAPR (2000)
17. Grandvalet, Y., Keshet, J., Rakotomamonjy, A., Canu, S.: Suppport vector machines with a reject option. In: NIPS (2008)
18. Herbei, R., Wegkamp, M.: Classification with reject option. Can. J. Stat. (2005)
19. Koltchinskii, V., Panchenko, D.: Empirical margin distributions and bounding the generalization error of combined classifiers. Ann. Stat. **30**, 1–50 (2002)
20. Landgrebe, T., Tax, D., Paclik, P., Duin, R.: Interaction between classification and reject performance for distance-based reject-option classifiers. PRL **27**, 908–917 (2005)
21. Ledoux, M., Talagrand, M.: Probability in Banach Spaces: Isoperimetry and Processes. Springer, New York (1991)
22. Littman, M., Li, L., Walsh, T.: Knows what it knows: a framework for self-aware learning. In: ICML (2008)
23. Long, P.M., Servedio, R.A.: Consistency versus realizable H-consistency for multiclass classification. In: ICML, vol. 3, pp. 801–809 (2013)
24. Melvin, I., Weston, J., Leslie, C.S., Noble, W.S.: Combining classifiers for improved classification of proteins from sequence or structure. BMCB **9**, 1 (2008)

25. Mohri, M., Rostamizadeh, A., Talwalkar, A.: Foundations of Machine Learning. The MIT Press, Cambridge (2012)
26. Pereira, C.S., Pires, A.: On optimal reject rules and ROC curves. PRL **26**, 943–952 (2005)
27. Pietraszek, T.: Optimizing abstaining classifiers using ROC. In: ICML (2005)
28. Tax, D., Duin, R.: Growing a multi-class classifier with a reject option. Pattern Recognit. Lett. **29**, 1565–1570 (2008)
29. Tortorella, F.: An optimal reject rule for binary classifiers. In: ICAPR (2001)
30. Trapeznikov, K., Saligrama, V.: Supervised sequential classification under budget constraints. In: AISTATS (2013)
31. Wang, J., Trapeznikov, K., Saligrama, V.: An LP for sequential learning under budgets. In: JMLR (2014)
32. Yuan, M., Wegkamp, M.: Classification methods with reject option based on convex risk minimizations. In: JMLR (2010)
33. Yuan, M., Wegkamp, M.: SVMs with a reject option. In: Bernoulli (2011)
34. Zhang, C., Chaudhuri, K.: The extended Littlestone's dimension for learning with mistakes and abstentions. In: COLT (2016)

Sparse Learning for Large-Scale and High-Dimensional Data: A Randomized Convex-Concave Optimization Approach

Lijun Zhang[1(✉)], Tianbao Yang[2], Rong Jin[3], and Zhi-Hua Zhou[1]

[1] National Key Laboratory for Novel Software Technology,
Nanjing University, Nanjing 210023, China
{zhanglj,zhouzh}@lamda.nju.edu.cn
[2] Department of Computer Science, The University of Iowa, Iowa City 52242, USA
tianbao-yang@uiowa.edu
[3] Alibaba Group, Seattle, USA
jinrong.jr@alibaba-inc.com

Abstract. In this paper, we develop a randomized algorithm and theory for learning a sparse model from large-scale and high-dimensional data, which is usually formulated as an empirical risk minimization problem with a sparsity-inducing regularizer. Under the assumption that there exists a (approximately) sparse solution with high classification accuracy, we argue that the dual solution is also sparse or approximately sparse. The fact that both primal and dual solutions are sparse motivates us to develop a randomized approach for a general convex-concave optimization problem. Specifically, the proposed approach combines the strength of random projection with that of sparse learning: it utilizes random projection to reduce the dimensionality, and introduces ℓ_1-norm regularization to alleviate the approximation error caused by random projection. Theoretical analysis shows that under favored conditions, the randomized algorithm can accurately recover the optimal solutions to the convex-concave optimization problem (i.e., recover both the primal and dual solutions).

Keywords: Random projection · Sparse learning · Convex-concave optimization · Primal solution · Dual solution

1 Introduction

Learning the sparse representation of a predictive model has received considerable attention in recent years [4]. Given a set of training examples $\{(\mathbf{x}_i, \mathbf{y}_i)\}_{i=1}^n$ with $\mathbf{x}_i \in \mathbb{R}^d$ and $y_i \in \mathbb{R}$, the optimization problem is generally formulated as

$$\min_{\mathbf{w} \in \Omega} \frac{1}{n} \sum_{i=1}^{n} \ell(y_i \mathbf{x}_i^\top \mathbf{w}) + \gamma \psi(\mathbf{w}) \tag{1}$$

where $\ell(\cdot)$ is a convex function such as the logistic loss to measure the empirical error, and $\psi(\cdot)$ is a sparsity-inducing regularizer such as the elastic net [38]

© Springer International Publishing Switzerland 2016
R. Ortner et al. (Eds.): ALT 2016, LNAI 9925, pp. 83–97, 2016.
DOI: 10.1007/978-3-319-46379-7_6

to avoid overfitting [13]. When both d and n are very large, directly solving (1) could be computationally expensive. A straightforward way to address this challenge is first reducing the dimensionality of the data, then solving a low-dimensional problem, and finally mapping the solution back to the original space. The limitation of this approach is that the final solution, after mapping from the low-dimensional space to the original high-dimensional space, may not be sparse.

The goal of this paper is to develop an efficient algorithm for solving the problem in (1), and at the same time preserve the (approximate) sparsity of the solution. Our approach is motivated by the following simple observation:

If there exists a sparse model with high prediction accuracy, the dual solution to (1) is also sparse or approximately sparse.

To see this, let us formulate (1) as a convex-concave optimization problem. By writing $\ell(z)$ in its convex conjugate form, i.e.,

$$\ell(z) = \max_{\lambda \in \Gamma} \lambda z - \ell_*(\lambda),$$

where $\ell_*(\cdot)$ is the Fenchel conjugate of $\ell(\cdot)$ [27] and Γ is the domain of the dual variable, we get the following convex-concave formulation:

$$\max_{\lambda \in \Gamma^n} \min_{\mathbf{w} \in \Omega} \gamma n \psi(\mathbf{w}) - \sum_{i=1}^{n} \ell_*(\lambda_i) + \sum_{i=1}^{n} \lambda_i y_i \mathbf{x}_i^\top \mathbf{w}. \tag{2}$$

Denote the optimal solutions to (2) by $(\mathbf{w}_*, \boldsymbol{\lambda}_*)$. By the Fenchel conjugate theory [9, Lemma 11.4], we have

$$[\boldsymbol{\lambda}_*]_i = \ell'(y_i \mathbf{x}_i^\top \mathbf{w}_*).$$

Let us consider the squared hinge loss for classification [31], where $\ell(z) = \max(0, 1 - z)^2$. Therefore, $y_i \mathbf{x}_i^\top \mathbf{w}_* \geq 1$ indicates that $[\boldsymbol{\lambda}_*]_i = 0$. As a result, when most of the examples can be classified by a large margin (which is likely to occur in large-scale and high-dimensional setting), it is reasonable to assume that the dual solution is sparse. Similarly, for logistic regression, we can argue the dual solution is approximately sparse.

Abstracting (2) slightly, in the following, we will study a general convex-concave optimization problem:

$$\max_{\boldsymbol{\lambda} \in \Delta} \min_{\mathbf{w} \in \Omega} g(\mathbf{w}) - h(\boldsymbol{\lambda}) - \mathbf{w}^\top A \boldsymbol{\lambda} \tag{3}$$

where $\Delta \subseteq \mathbb{R}^n$ and $\Omega \subseteq \mathbb{R}^d$ are the domains for $\boldsymbol{\lambda}$ and $\mathbf{w}$, respectively, $g(\cdot)$ and $h(\cdot)$ are two convex functions, and $A \in \mathbb{R}^{d \times n}$ is a matrix. The benefit of analyzing (3) instead of (1) is that the convex-concave formulation allows us to exploit the prior knowledge that *both* $\mathbf{w}_*$ and $\boldsymbol{\lambda}_*$ are sparse or approximately sparse. The problem in (3) has been widely studied in the optimization community, and when n and d are medium size, it can be solved iteratively by gradient based methods [21,22].

We assume the two convex functions $g(\cdot)$ and $h(\cdot)$ are relatively simple such that evaluating their values or gradients takes $O(d)$ and $O(n)$ complexities, respectively. The bottleneck is the computations involving the bilinear term $\mathbf{w}^\top A\boldsymbol{\lambda}$, which have $O(nd)$ complexity in both time and space. To overcome this difficulty, we develop a randomized algorithm that solves (3) approximately but at a significantly lower cost. The proposed algorithm combines two well-known techniques—*random projection* and ℓ_1-*norm regularization* in a principled way. Specifically, random projection is used to find a low-rank approximation of A, which not only reduces the storage requirement but also accelerates the computations. The role of ℓ_1-norm regularization is twofold. One one hand, it is introduced to compensate for the distortion caused by randomization, and on the other hand it enforces the sparsity of the final solutions. Under mild assumptions about the optimization problem in (3), the proposed algorithm has a small recovery error provided the optimal solutions to (3) are sparse or approximately sparse.

2 Related Work

Random projection has been widely used as an efficient algorithm for dimensionality reduction [6,16]. In the case of unsupervised learning, it has been proved that random projection is able to preserve the distance [11], inner product [3], volumes and distance to affine spaces [18]. In the case of supervised learning, random projection is generally used as a preprocessing step to find a low-dimensional representation of the data, and thus reduces the computational cost of training. For classification, theoretical studies mainly focus on examining the generalization error or the preservation of classification margin in the low-dimensional space [5,24,28]. For regression, there do exist theoretical guarantees for the recovery error, but they only hold for the least squares problem [19].

Our work is closely related to Dual Random Projection (DRP) [35,36] and Dual-sparse Regularized Randomized Reduction (DSRR) [34], which also investigate random projection from the perspective of optimization. However, both DRP and DSRR are limited to the special case that $\psi(\mathbf{w}) = \|\mathbf{w}\|_2^2$, which leads to a simple dual problem. In contrast, our algorithm is designed for the case that $\psi(\cdot)$ is a sparsity-inducing regularizer, and built upon the convex-concave formulation. Similar to DSRR, our algorithm makes use of the sparsity of the dual solution, but we further exploit the sparsity of the primal solution. A noticeable advantage of our analysis is the mild assumption about the data matrix A. To recover the primal solution, DRP assumes the data matrix is low-rank and DSRR assumes it satisfies the restricted eigenvalue condition, in contrast, our algorithm only requires columns or rows of A are bounded.

There are many literatures that study the statistical property of the sparse learning problem in (1) [2,23,33,37]. For example, in the context of compressive sensing [12], it has been established that a sparse signal can be recovered up to an $O(\sqrt{s \log d/n})$ error, where s is the sparsity of the unknown signal. We note that the statistical error is not directly comparable to the optimization error derived in this paper. That is because the analysis of statistical error relies on heavy assumptions about the data, e.g., the RIP condition [8]. On the other hand, the optimization error is derived under very weak conditions.

3 Algorithm

To reduce the computational cost of (3), we first generate a random matrix $R \in \mathbb{R}^{n \times m}$, where $m \ll \min(d, n)$. Define $\widehat{A} = AR \in \mathbb{R}^{d \times m}$, we propose to solve the following problem

$$\max_{\boldsymbol{\lambda} \in \Delta} \min_{\mathbf{w} \in \Omega} g(\mathbf{w}) - h(\boldsymbol{\lambda}) - \mathbf{w}^\top \widehat{A} R^\top \boldsymbol{\lambda} + \gamma_w \|\mathbf{w}\|_1 - \gamma_\lambda \|\boldsymbol{\lambda}\|_1 \qquad (4)$$

where γ_w and γ_λ are two regularization parameters. The construction of the random matrix R, as well as the values of the two regularization parameters γ_w and γ_λ will be discussed later. The optimization problem in (4) can be solved by algorithms designed for composite convex-concave problems [10,14].

Compared to (3), the main advantage of (4) is that it only needs to load $\widehat{A}$ and R into the memory, making it convenient to deal with large-scale problems. With the help of random projection, the computational complexity for evaluating the value and gradient is reduced from $O(dn)$ to $O(dm+nm)$. Compared to previous randomized algorithms [5,34,35], (4) has two new features: (i) the optimization is still performed in the original space; and (ii) the ℓ_1-norm is introduced to regularize both primal and dual solutions. As we will prove later, the combination of these two features will ensure the solutions to (4) are approximately sparse. Finally, note that in (4) $RR^\top$ is inserted at the right side of A, it can also be put at the left side of A. In this case, we have the following optimization problem

$$\max_{\boldsymbol{\lambda} \in \Delta} \min_{\mathbf{w} \in \Omega} g(\mathbf{w}) - h(\boldsymbol{\lambda}) - \mathbf{w}^\top R \widehat{A} \boldsymbol{\lambda} + \gamma_w \|\mathbf{w}\|_1 - \gamma_\lambda \|\boldsymbol{\lambda}\|_1 \qquad (5)$$

where $R \in \mathbb{R}^{d \times m}$ is a random matrix, and $\widehat{A} = R^\top A \in \mathbb{R}^{m \times n}$.

Let $(\mathbf{w}_*, \boldsymbol{\lambda}_*)$ and $(\widehat{\mathbf{w}}, \widehat{\boldsymbol{\lambda}})$ be the optimal solution to the convex-concave optimization problem in (3) and (4)/(5), respectively. Under suitable conditions, we will show that

$$\|\widehat{\mathbf{w}} - \mathbf{w}_*\|_2 \leq O\left(\sqrt{\frac{\|\mathbf{w}_*\|_0 \|\boldsymbol{\lambda}_*\|_0 \log n}{m}}\right) \text{ and}$$

$$\|\widehat{\boldsymbol{\lambda}} - \boldsymbol{\lambda}_*\|_2 \leq O\left(\sqrt{\frac{\|\mathbf{w}_*\|_0 \|\boldsymbol{\lambda}_*\|_0 \log d}{m}}\right)$$

implying a small recovery error when $\mathbf{w}_*$ and $\boldsymbol{\lambda}_*$ are sparse. A similar recovery guarantee also holds when the optimal solutions to (3) are approximately sparse, i.e., when they can be well-approximated by sparse vectors.

4 Main Results

We first introduce common assumptions that we make, and then present theoretical guarantees.

4.1 Assumptions

Assumptions About (3). We make the following assumptions about (3).

- $g(\mathbf{w})$ is α-strongly convex with respect to the Euclidean norm. Let's take the optimization problem in (2) as an example. (2) will satisfy this assumption if some strongly convex function (e.g., $\|\mathbf{w}\|_2^2$) is a part of the regularizer $\psi(\mathbf{w})$.
- $h(\boldsymbol{\lambda})$ is β-strongly convex with respect to the Euclidean norm. For the problem in (2), if $\ell(\cdot)$ is a smooth function (e.g., the logistic loss), then its convex conjugate $\ell_*(\cdot)$ will be strongly convex [15,27].
- Either columns or rows of A have bounded ℓ_2-norm. Without loss of generality, we assume

$$\|A_{i*}\|_2 \leq 1, \ \forall i \in [d], \tag{6}$$

$$\|A_{*j}\|_2 \leq 1, \ \forall j \in [n]. \tag{7}$$

The above assumption can be satisfied by normalizing rows or columns of A.

Assumptions About R. We assume the random matrix $R \in \mathbb{R}^{n \times m}$ has the following property.

- With a high probability, the linear operator $R^\top : \mathbb{R}^n \mapsto \mathbb{R}^m$ is able to preserve the ℓ_2-norm of its input. In mathematical terms, we need the following property.

Property 1. There exists a constant $c > 0$, such that

$$\Pr\left\{(1-\varepsilon)\|\mathbf{x}\|_2^2 \leq \|R^\top \mathbf{x}\|_2^2 \leq (1+\varepsilon)\|\mathbf{x}\|_2^2\right\} \geq 1 - 2\exp(-m\varepsilon^2/c)$$

for any fixed $\mathbf{x} \in \mathbb{R}^d$ and $0 < \epsilon \leq 1/2$.

The above property is widely used to prove the famous Johnson–Lindenstrauss lemma [11]. Let $R = \frac{1}{\sqrt{m}}S$. Previous studies [1,3] have proved that Property 1 is true if $\{S_{ij}\}$ are independent random variables sampled from the Gaussian distribution $\mathcal{N}(0,1)$, uniform distribution over $\{\pm 1\}$, or the following database-friendly distribution

$$X = \begin{cases} \sqrt{3}, & \text{with probability } 1/6; \\ 0, & \text{with probability } 2/3; \\ -\sqrt{3}, & \text{with probability } 1/6. \end{cases}$$

More generally, a sufficient condition for Property 1 is that columns of R are independent, isotropic, and subgaussian vectors [20].

4.2 Theoretical Guarantees

Sparse Solutions. We first consider the case that both $\mathbf{w}_*$ and $\boldsymbol{\lambda}_*$ are sparse. Define

$$s_w = \|\mathbf{w}_*\|_0, \text{ and } s_\lambda = \|\boldsymbol{\lambda}_*\|_0.$$

We have the following theorem for the optimization problem in (4).

Theorem 1. *Let* $(\widehat{\mathbf{w}}, \widehat{\boldsymbol{\lambda}})$ *be the optimal solution to the problem in* (4). *Set*

$$\gamma_\lambda \geq 2\|A^\top \mathbf{w}_*\|_2 \sqrt{\frac{c}{m} \log \frac{4n}{\delta}}, \tag{8}$$

$$\gamma_w \geq 2\|\boldsymbol{\lambda}_*\|_2 \sqrt{\frac{c}{m} \log \frac{4d}{\delta}} + \frac{6\gamma_\lambda\sqrt{s_\lambda}}{\beta} \left(1 + 7\sqrt{\frac{c}{m}\left(\log \frac{4d}{\delta} + 16s_\lambda \log \frac{9n}{8s_\lambda}\right)}\right). \tag{9}$$

With a probability at least $1 - 3\delta$, *we have*

$$\|\widehat{\mathbf{w}} - \mathbf{w}_*\|_2 \leq \frac{3\gamma_w\sqrt{s_w}}{\alpha}, \quad \|\widehat{\mathbf{w}} - \mathbf{w}_*\|_1 \leq \frac{12\gamma_w s_w}{\alpha}, \quad and \quad \frac{\|\widehat{\mathbf{w}} - \mathbf{w}_*\|_1}{\|\widehat{\mathbf{w}} - \mathbf{w}_*\|_2} \leq 4\sqrt{s_w}$$

provided

$$m \geq 4c \log \frac{4}{\delta} \tag{10}$$

where c *is the constant in Property* 1.

Notice that $\|\widehat{\mathbf{w}} - \mathbf{w}_*\|_1 / \|\widehat{\mathbf{w}} - \mathbf{w}_*\|_2 \leq 4\sqrt{s_w}$ indicates that $\widehat{\mathbf{w}} - \mathbf{w}_*$ is approximately sparse [25,26]. Combining with the fact $\mathbf{w}_*$ is sparse, we conclude that $\widehat{\mathbf{w}}$ is also approximately sparse.

Then, we discuss the recovery guarantee for the sparse learning problem in (1) or (2). Since $A^\top \mathbf{w}_* \in \mathbb{R}^n$, we can take $\|A^\top \mathbf{w}_*\|_2 = O(\sqrt{n})$. Since $\|\boldsymbol{\lambda}_*\|_0 = s_\lambda$, we can assume $\|\boldsymbol{\lambda}_*\|_2 = O(\sqrt{s_\lambda})$. According to the theoretical analysis of regularized empirical risk minimization [17,29,32], the optimal γ, that minimizes the generalization error, can be chosen as $\gamma = O(1/\sqrt{n})$, and thus $\alpha = O(\gamma n) = O(\sqrt{n})$. When the loss $\ell(\cdot)$ is smooth, we have $\beta = O(1)$. The following corollary provides a simplified result based on the above discussions.

Corollary 1. *Assume* $\|A^\top \mathbf{w}_*\|_2 = O(\sqrt{n})$, $\|\boldsymbol{\lambda}_*\|_2 = O(\sqrt{s_\lambda})$, $\alpha = O(\sqrt{n})$, *and* $\beta = O(1)$. *When* $m \geq O(s_\lambda \log n)$, *we can choose*

$$\gamma_\lambda = O\left(\sqrt{\frac{n \log n}{m}}\right) \quad and \quad \gamma_w = O\left(\sqrt{\frac{s_\lambda \log d}{m}} + \gamma_\lambda\sqrt{s_\lambda}\right) = O\left(\sqrt{\frac{n s_\lambda \log n}{m}}\right)$$

such that with a high probability

$$\|\widehat{\mathbf{w}} - \mathbf{w}_*\|_2 \leq O\left(\frac{\gamma_w\sqrt{s_w}}{\sqrt{n}}\right) = O\left(\sqrt{\frac{s_w s_\lambda \log n}{m}}\right) \quad and \quad \frac{\|\widehat{\mathbf{w}} - \mathbf{w}_*\|_1}{\|\widehat{\mathbf{w}} - \mathbf{w}_*\|_2} \leq 4\sqrt{s_w}.$$

A natural question to ask is whether similar recovery guarantees for $\widehat{\boldsymbol{\lambda}}$ can be proved under the conditions in Theorem 1. Unfortunately, we are not able to give a positive answer, and only have the following theorem.

Theorem 2. *Assume* γ_λ *satisfies the condition in* (8). *With a probability at least* $1 - \delta$, *we have*

$$\|\widehat{\boldsymbol{\lambda}} - \boldsymbol{\lambda}_*\|_2 \leq \frac{3\gamma_\lambda\sqrt{s_\lambda}}{\beta} + \frac{2}{\beta}\left(1 + \|RR^\top - I\|_2\right)\|A^\top(\widehat{\mathbf{w}} - \mathbf{w}_*)\|_2$$

provided (10) *holds.*

The upper bound in the above theorem is quite loose, because $\|RR^\top - I\|_2$ is roughly on the order of $n\log n/m$ [30].

Due to the symmetry between $\boldsymbol{\lambda}$ and $\mathbf{w}$, we can recover $\boldsymbol{\lambda}_*$ via (5) instead of (4). Then, by replacing $\mathbf{w}_*$ in Theorem 1 with $\boldsymbol{\lambda}_*$, $\widehat{\mathbf{w}}$ with $\widehat{\boldsymbol{\lambda}}$, n with d, and so on, we obtain the following theoretical guarantee.

Theorem 3. *Let $(\widehat{\mathbf{w}}, \widehat{\boldsymbol{\lambda}})$ be the optimal solution to the problem in* (5). *Set*

$$\gamma_w \geq 2\|A\boldsymbol{\lambda}_*\|_2\sqrt{\frac{c}{m}\log\frac{4d}{\delta}},$$

$$\gamma_\lambda \geq 2\|\mathbf{w}_*\|_2\sqrt{\frac{c}{m}\log\frac{4n}{\delta}} + \frac{6\gamma_w\sqrt{s_w}}{\alpha}\left(1 + 7\sqrt{\frac{c}{m}\left(\log\frac{4n}{\delta} + 16s_w\log\frac{9d}{8s_w}\right)}\right).$$

With a probability at least $1 - 3\delta$, we have

$$\|\widehat{\boldsymbol{\lambda}} - \boldsymbol{\lambda}_*\|_2 \leq \frac{3\gamma_\lambda\sqrt{s_\lambda}}{\beta}, \quad \|\widehat{\boldsymbol{\lambda}} - \boldsymbol{\lambda}_*\|_1 \leq \frac{12\gamma_\lambda s_\lambda}{\beta}, \quad and \quad \frac{\|\widehat{\boldsymbol{\lambda}} - \boldsymbol{\lambda}_*\|_1}{\|\widehat{\boldsymbol{\lambda}} - \boldsymbol{\lambda}_*\|_2} \leq 4\sqrt{s_\lambda}$$

provided (10) *holds.*

To simplify the above theorem, we can take $\|A\boldsymbol{\lambda}_*\|_2 = O(\sqrt{d})$ since $A\boldsymbol{\lambda}_* \in \mathbb{R}^d$. Because (1) has both a constraint and a regularizer, we can assume the optimal primal solution is well-bounded, that is, $\|\mathbf{w}_*\|_2 = O(1)$. Finally, we assume $d \leq O(n)$, and have the following corollary.

Corollary 2. *Assume $\|A\boldsymbol{\lambda}_*\|_2 = O(\sqrt{d})$, $\|\mathbf{w}_*\|_2 = O(1)$, $\alpha = O(\sqrt{n})$, $\beta = O(1)$, and $d \leq O(n)$. When $m \geq O(s_w\log d)$, we can choose*

$$\gamma_w = O\left(\sqrt{\frac{d\log d}{m}}\right) \quad and \quad \gamma_\lambda = O\left(\sqrt{\frac{\log n}{m}} + \gamma_w\sqrt{\frac{s_w}{n}}\right) \leq O\left(\sqrt{\frac{s_w\log d}{m}}\right)$$

such that with a high probability

$$\|\widehat{\boldsymbol{\lambda}} - \boldsymbol{\lambda}_*\|_2 \leq O\left(\gamma_\lambda\sqrt{s_\lambda}\right) = O\left(\sqrt{\frac{s_w s_\lambda\log d}{m}}\right) \quad and \quad \frac{\|\widehat{\boldsymbol{\lambda}} - \boldsymbol{\lambda}_*\|_1}{\|\widehat{\boldsymbol{\lambda}} - \boldsymbol{\lambda}_*\|_2} \leq 4\sqrt{s_\lambda}.$$

Approximately Sparse Solutions. We now proceed to study the case that the optimal solutions to (3) are only approximately sparse.

With a slight abuse of notation, we assume $\mathbf{w}_*$ and $\boldsymbol{\lambda}_*$ are two sparse vectors, with $\|\mathbf{w}_*\|_0 = s_w$ and $\|\boldsymbol{\lambda}_*\|_0 = s_\lambda$, that solve (3) approximately in the sense that

$$\|\nabla g(\mathbf{w}_*) - A\boldsymbol{\lambda}_*\|_\infty \leq \varsigma, \tag{11}$$

$$\|\nabla h(\boldsymbol{\lambda}_*) + A^\top\mathbf{w}_*\|_\infty \leq \varsigma, \tag{12}$$

for some small constant $\varsigma > 0$. The above conditions can be considered as sub-optimality conditions [7] of $\mathbf{w}_*$ and $\boldsymbol{\lambda}_*$ measured in the ℓ_∞-norm. After a similar analysis, we have the following theorem.

Theorem 4. *Let $(\widehat{\mathbf{w}}, \widehat{\boldsymbol{\lambda}})$ be the optimal solution to the problem in (4). Assume (11) and (12) hold. Set*

$$\gamma_\lambda \geq 2\|A^\top \mathbf{w}_*\|_2 \sqrt{\frac{c}{m} \log \frac{4n}{\delta}} + 2\varsigma,$$

$$\gamma_w \geq 2\|\boldsymbol{\lambda}_*\|_2 \sqrt{\frac{c}{m} \log \frac{4d}{\delta}} + \frac{6\gamma_\lambda \sqrt{s_\lambda}}{\beta}\left(1 + 7\sqrt{\frac{c}{m}\left(\log \frac{4d}{\delta} + 16 s_\lambda \log \frac{9n}{8 s_\lambda}\right)}\right) + 2\varsigma.$$

With a probability at least $1 - 3\delta$, we have

$$\|\widehat{\mathbf{w}} - \mathbf{w}_*\|_2 \leq \frac{3\gamma_w \sqrt{s_w}}{\alpha}, \quad \|\widehat{\mathbf{w}} - \mathbf{w}_*\|_1 \leq \frac{12\gamma_w s_w}{\alpha}, \quad and \quad \frac{\|\widehat{\mathbf{w}} - \mathbf{w}_*\|_1}{\|\widehat{\mathbf{w}} - \mathbf{w}_*\|_2} \leq 4\sqrt{s_w}$$

provided (10) holds.

When ς is small enough, the upper bound in Theorem 4 is on the same order as that in Theorem 1. To be specific, we have the following corollary.

Corollary 3. *Assume $\|A^\top \mathbf{w}_*\|_2 = O(\sqrt{n})$, $\|\boldsymbol{\lambda}_*\|_2 = O(\sqrt{s_\lambda})$, $\alpha = O(\sqrt{n})$, $\beta = O(1)$, and $\varsigma = O(\sqrt{n \log n}/m)$. When $m \geq O(s_\lambda \log n)$, we can choose γ_λ and γ_w as in Corollary 1 such that with a high probability*

$$\|\widehat{\mathbf{w}} - \mathbf{w}_*\|_2 = O\left(\frac{\gamma_w \sqrt{s_w}}{\sqrt{n}}\right) = O\left(\sqrt{\frac{s_w s_\lambda \log n}{m}}\right) \quad and \quad \frac{\|\widehat{\mathbf{w}} - \mathbf{w}_*\|_1}{\|\widehat{\mathbf{w}} - \mathbf{w}_*\|_2} \leq 4\sqrt{s_w}.$$

5   Analysis

Due to the limitation of space, we only provide proofs of Theorem 1 and related lemmas. The omitted proofs will be included in a supplementary.

5.1   Proof of Theorem 1

To facilitate the analysis, we introduce a pseudo optimization problem

$$\max_{\boldsymbol{\lambda} \in \Delta} -h(\boldsymbol{\lambda}) - \mathbf{w}_*^\top \widehat{A} R^\top \boldsymbol{\lambda} - \gamma_\lambda \|\boldsymbol{\lambda}\|_1$$

whose optimal solution is denoted by $\widetilde{\boldsymbol{\lambda}}$. In the following, we will first discuss how to bound the difference between $\widetilde{\boldsymbol{\lambda}}$ and $\boldsymbol{\lambda}_*$, and then bound the difference between $\widehat{\mathbf{w}}$ and $\mathbf{w}_*$ in a similar way.

From the optimality of $\widetilde{\boldsymbol{\lambda}}$ and $\boldsymbol{\lambda}_*$, we derive the following lemma to bound their difference.

Lemma 1. *Denote*

$$\rho_\lambda = \left\|(RR^\top - I)A^\top \mathbf{w}_*\right\|_\infty. \tag{13}$$

By choosing $\gamma_\lambda \geq 2\rho_\lambda$, we have

$$\|\widetilde{\boldsymbol{\lambda}} - \boldsymbol{\lambda}_*\|_2 \leq \frac{3\gamma_\lambda \sqrt{s_\lambda}}{\beta}, \quad \|\widetilde{\boldsymbol{\lambda}} - \boldsymbol{\lambda}_*\|_1 \leq \frac{12\gamma_\lambda s_\lambda}{\beta}, \quad and \quad \frac{\|\widetilde{\boldsymbol{\lambda}} - \boldsymbol{\lambda}_*\|_1}{\|\widetilde{\boldsymbol{\lambda}} - \boldsymbol{\lambda}_*\|_2} \leq 4\sqrt{s_\lambda}.$$

Based on the property of the random matrix R described in Property 1, we have the following lemma to bound ρ_λ in (13).

Lemma 2. *With a probability at least* $1 - \delta$, *we have*

$$\rho_\lambda = \left\| (RR^\top - I)A^\top \mathbf{w}_* \right\|_\infty \leq \|A^\top \mathbf{w}_*\|_2 \sqrt{\frac{c}{m} \log \frac{4n}{\delta}}$$

provided (10) *holds.*

Combining Lemma 1 with Lemma 2, we immediately obtain the following lemma.

Lemma 3. *Set*

$$\gamma_\lambda \geq 2\|A^\top \mathbf{w}_*\|_2 \sqrt{\frac{c}{m} \log \frac{4n}{\delta}}.$$

With a probability at least $1 - \delta$, *we have*

$$\|\widetilde{\boldsymbol{\lambda}} - \boldsymbol{\lambda}_*\|_2 \leq \frac{3\gamma_\lambda \sqrt{s_\lambda}}{\beta}, \quad \|\widetilde{\boldsymbol{\lambda}} - \boldsymbol{\lambda}_*\|_1 \leq \frac{12\gamma_\lambda s_\lambda}{\beta}, \quad and \quad \frac{\|\widetilde{\boldsymbol{\lambda}} - \boldsymbol{\lambda}_*\|_1}{\|\widetilde{\boldsymbol{\lambda}} - \boldsymbol{\lambda}_*\|_2} \leq 4\sqrt{s_\lambda}$$

provided (10) *holds.*

We are now in a position to formulate the key lemmas that lead to Theorem 1. Similar to Lemma 1, we introduce the following lemma to characterize the relation between $\widehat{\mathbf{w}}$ and $\mathbf{w}_*$.

Lemma 4. *Denote*

$$\rho_w = \left\| A(I - RR^\top)\boldsymbol{\lambda}_* \right\|_\infty + \left\| ARR^\top(\boldsymbol{\lambda}_* - \widetilde{\boldsymbol{\lambda}}) \right\|_\infty. \tag{14}$$

By choosing $\gamma_w \geq 2\rho_w$, *we have*

$$\|\widehat{\mathbf{w}} - \mathbf{w}_*\|_2 \leq \frac{3\gamma_w \sqrt{s_w}}{\alpha}, \quad \|\widehat{\mathbf{w}} - \mathbf{w}_*\|_1 \leq \frac{12\gamma_w s_w}{\alpha}, \quad and \quad \frac{\|\widehat{\mathbf{w}} - \mathbf{w}_*\|_1}{\|\widehat{\mathbf{w}} - \mathbf{w}_*\|_2} \leq 4\sqrt{s_w}.$$

The last step of the proof is to derive an upper bound for ρ_w based on Property 1 and Lemma 3.

Lemma 5. *Assume the conclusion in Lemma 3 happens. With a probability at least* $1 - 2\delta$, *we have*

$$\rho_w \leq \|\boldsymbol{\lambda}_*\|_2 \sqrt{\frac{c}{m} \log \frac{4d}{\delta}} + \frac{3\gamma_\lambda \sqrt{s_\lambda}}{\beta} \left(1 + 7\sqrt{\frac{c}{m} \left(\log \frac{4d}{\delta} + 16 s_\lambda \log \frac{9n}{8 s_\lambda} \right)} \right)$$

provided (10) *holds.*

5.2 Proof of Lemma 1

Notations. For a vector $\mathbf{x} \in \mathbb{R}^d$ and a set $\mathcal{D} \subseteq [d]$, we denote by $\mathbf{x}_\mathcal{D}$ the vector which coincides with $\mathbf{x}$ on $\mathcal{D}$ and has zero coordinates outside $\mathcal{D}$.

Let Ω_λ include the subset of non-zeros entries in $\boldsymbol{\lambda}_*$ and $\bar{\Omega}_\lambda = [n] \setminus \Omega_\lambda$. Define

$$\mathcal{L}(\boldsymbol{\lambda}) = -h(\boldsymbol{\lambda}) + \min_{\mathbf{w} \in \Omega} g(\mathbf{w}) - \mathbf{w}^\top A \boldsymbol{\lambda},$$

$$\widetilde{\boldsymbol{\lambda}}(\boldsymbol{\lambda}) = -h(\boldsymbol{\lambda}) - \mathbf{w}_*^\top \widehat{A} R^\top \boldsymbol{\lambda} - \gamma_\lambda \|\boldsymbol{\lambda}\|_1.$$

Let $\mathbf{v} \in \partial \|\boldsymbol{\lambda}_*\|_1$ be any subgradient of $\|\cdot\|_1$ at $\boldsymbol{\lambda}_*$. Then, we have[1]

$$\mathbf{u} = -\nabla h(\boldsymbol{\lambda}_*) - RR^\top A^\top \mathbf{w}_* - \gamma_\lambda \mathbf{v} \in \partial \widetilde{\boldsymbol{\lambda}}(\boldsymbol{\lambda}_*).$$

Using the fact that $\widetilde{\boldsymbol{\lambda}}$ maximizes $\widetilde{\boldsymbol{\lambda}}(\cdot)$ over the domain Δ and $h(\cdot)$ is β-strongly convex, we have

$$
\begin{aligned}
0 \geq \widetilde{\boldsymbol{\lambda}}(\boldsymbol{\lambda}_*) - \widetilde{\boldsymbol{\lambda}}(\widetilde{\boldsymbol{\lambda}}) &\geq \langle -(\widetilde{\boldsymbol{\lambda}} - \boldsymbol{\lambda}_*), \mathbf{u} \rangle + \frac{\beta}{2} \|\boldsymbol{\lambda}_* - \widetilde{\boldsymbol{\lambda}}\|_2^2 \\
&= \left\langle \widetilde{\boldsymbol{\lambda}} - \boldsymbol{\lambda}_*, \nabla h(\boldsymbol{\lambda}_*) + RR^\top A^\top \mathbf{w}_* + \gamma_\lambda \mathbf{v} \right\rangle + \frac{\beta}{2} \|\boldsymbol{\lambda}_* - \widetilde{\boldsymbol{\lambda}}\|_2^2.
\end{aligned}
\tag{15}
$$

By setting $v_i = \text{sign}(\widetilde{\lambda}_i)$, $\forall i \in \bar{\Omega}_\lambda$, we have $\langle \widetilde{\boldsymbol{\lambda}}_{\bar{\Omega}_\lambda}, \mathbf{v}_{\bar{\Omega}_\lambda} \rangle = \|\widetilde{\boldsymbol{\lambda}}_{\bar{\Omega}_\lambda}\|_1$. As a result,

$$\langle \widetilde{\boldsymbol{\lambda}} - \boldsymbol{\lambda}_*, \mathbf{v} \rangle = \langle \widetilde{\boldsymbol{\lambda}}_{\bar{\Omega}_\lambda}, \mathbf{v}_{\bar{\Omega}_\lambda} \rangle + \langle \widetilde{\boldsymbol{\lambda}}_{\Omega_\lambda} - \boldsymbol{\lambda}_*, \mathbf{v}_{\Omega_\lambda} \rangle \geq \|\widetilde{\boldsymbol{\lambda}}_{\bar{\Omega}_\lambda}\|_1 - \|\widetilde{\boldsymbol{\lambda}}_{\Omega_\lambda} - \boldsymbol{\lambda}_*\|_1. \tag{16}$$

Combining (15) with (16), we have

$$\left\langle \widetilde{\boldsymbol{\lambda}} - \boldsymbol{\lambda}_*, \nabla h(\boldsymbol{\lambda}_*) + RR^\top A^\top \mathbf{w}_* \right\rangle + \frac{\beta}{2} \|\boldsymbol{\lambda}_* - \widetilde{\boldsymbol{\lambda}}\|_2^2 + \gamma_\lambda \|\widetilde{\boldsymbol{\lambda}}_{\bar{\Omega}_\lambda}\|_1 \leq \gamma_\lambda \|\widetilde{\boldsymbol{\lambda}}_{\Omega_\lambda} - \boldsymbol{\lambda}_*\|_1. \tag{17}$$

From the fact that $\boldsymbol{\lambda}_*$ maximizes $\mathcal{L}(\cdot)$ over the domain Δ, we have

$$\langle \nabla \mathcal{L}(\boldsymbol{\lambda}_*), \boldsymbol{\lambda} - \boldsymbol{\lambda}_* \rangle = \langle -\nabla h(\boldsymbol{\lambda}_*) - A^\top \mathbf{w}_*, \boldsymbol{\lambda} - \boldsymbol{\lambda}_* \rangle \leq 0, \quad \forall \boldsymbol{\lambda} \in \Delta. \tag{18}$$

Then,

$$
\begin{aligned}
&\left\langle \widetilde{\boldsymbol{\lambda}} - \boldsymbol{\lambda}_*, \nabla h(\boldsymbol{\lambda}_*) + RR^\top A^\top \mathbf{w}_* \right\rangle \\
&= \left\langle \widetilde{\boldsymbol{\lambda}} - \boldsymbol{\lambda}_*, \nabla h(\boldsymbol{\lambda}_*) + A^\top \mathbf{w}_* \right\rangle + \left\langle \widetilde{\boldsymbol{\lambda}} - \boldsymbol{\lambda}_*, (RR^\top - I)A^\top \mathbf{w}_* \right\rangle \\
&\overset{(18)}{\geq} -\|\widetilde{\boldsymbol{\lambda}} - \boldsymbol{\lambda}_*\|_1 \|(RR^\top - I)A^\top \mathbf{w}_*\|_\infty \\
&\overset{(13)}{=} -\rho_\lambda \|\widetilde{\boldsymbol{\lambda}} - \boldsymbol{\lambda}_*\|_1 = -\rho_\lambda \left(\|\widetilde{\boldsymbol{\lambda}}_{\bar{\Omega}_\lambda}\|_1 + \|\widetilde{\boldsymbol{\lambda}}_{\Omega_\lambda} - \boldsymbol{\lambda}_*\|_1 \right).
\end{aligned}
\tag{19}
$$

From (17) and (19), we have

$$\frac{\beta}{2} \|\widetilde{\boldsymbol{\lambda}} - \boldsymbol{\lambda}_*\|_2^2 + (\gamma_\lambda - \rho_\lambda)\|\widetilde{\boldsymbol{\lambda}}_{\bar{\Omega}_\lambda}\|_1 \leq (\gamma_\lambda + \rho_\lambda)\|\widetilde{\boldsymbol{\lambda}}_{\Omega_\lambda} - \boldsymbol{\lambda}_*\|_1.$$

---

[1] In the case that $h(\cdot)$ is non-smooth, $\nabla h(\boldsymbol{\lambda}_*)$ refers to a subgradient of $h(\cdot)$ at $\boldsymbol{\lambda}_*$. In particular, we choose the subgradient that satisfies (18).

Since $\gamma_\lambda \geq 2\rho_\lambda$, we have

$$\frac{\beta}{2}\|\widetilde{\boldsymbol{\lambda}} - \boldsymbol{\lambda}_*\|_2^2 + \frac{\gamma_\lambda}{2}\|\widetilde{\boldsymbol{\lambda}}_{\bar{\Omega}_\lambda}\|_1 \leq \frac{3\gamma_\lambda}{2}\|\widetilde{\boldsymbol{\lambda}}_{\Omega_\lambda} - \boldsymbol{\lambda}_*\|_1.$$

And thus,

$$\frac{\beta}{2}\|\widetilde{\boldsymbol{\lambda}} - \boldsymbol{\lambda}_*\|_2^2 \leq \frac{3\gamma_\lambda}{2}\|\widetilde{\boldsymbol{\lambda}}_{\Omega_\lambda} - \boldsymbol{\lambda}_*\|_1 \leq \frac{3\gamma_\lambda\sqrt{s_\lambda}}{2}\|\widetilde{\boldsymbol{\lambda}}_{\Omega_\lambda} - \boldsymbol{\lambda}_*\|_2$$

$$\Rightarrow \|\widetilde{\boldsymbol{\lambda}} - \boldsymbol{\lambda}_*\|_2 \leq \frac{3\gamma_\lambda\sqrt{s_\lambda}}{\beta},$$

$$\frac{\beta}{2s_\lambda}\|\widetilde{\boldsymbol{\lambda}}_{\Omega_\lambda} - \boldsymbol{\lambda}_*\|_1^2 \leq \frac{\beta}{2}\|\widetilde{\boldsymbol{\lambda}} - \boldsymbol{\lambda}_*\|_2^2 \leq \frac{3\gamma_\lambda}{2}\|\widetilde{\boldsymbol{\lambda}}_{\Omega_\lambda} - \boldsymbol{\lambda}_*\|_1$$

$$\Rightarrow \|\widetilde{\boldsymbol{\lambda}}_{\Omega_\lambda} - \boldsymbol{\lambda}_*\|_1 \leq \frac{3\gamma_\lambda s_\lambda}{\beta},$$

$$\frac{\gamma_\lambda}{2}\|\widetilde{\boldsymbol{\lambda}}_{\bar{\Omega}_\lambda}\|_1 \leq \frac{3\gamma_\lambda}{2}\|\widetilde{\boldsymbol{\lambda}}_{\Omega_\lambda} - \boldsymbol{\lambda}_*\|_1$$

$$\Rightarrow \|\widetilde{\boldsymbol{\lambda}}_{\bar{\Omega}_\lambda}\|_1 \leq 3\|\widetilde{\boldsymbol{\lambda}}_{\Omega_\lambda} - \boldsymbol{\lambda}_*\|_1 \Rightarrow \|\widetilde{\boldsymbol{\lambda}} - \boldsymbol{\lambda}_*\|_1 \leq \frac{12\gamma_\lambda s_\lambda}{\beta},$$

$$\frac{\|\widetilde{\boldsymbol{\lambda}} - \boldsymbol{\lambda}_*\|_1}{\|\widetilde{\boldsymbol{\lambda}} - \boldsymbol{\lambda}_*\|_2} \leq \frac{4\|\widetilde{\boldsymbol{\lambda}}_{\Omega_\lambda} - \boldsymbol{\lambda}_*\|_1}{\|\widetilde{\boldsymbol{\lambda}} - \boldsymbol{\lambda}_*\|_2} \leq \frac{4\sqrt{s_\lambda}\|\widetilde{\boldsymbol{\lambda}}_{\Omega_\lambda} - \boldsymbol{\lambda}_*\|_2}{\|\widetilde{\boldsymbol{\lambda}} - \boldsymbol{\lambda}_*\|_2} \leq 4\sqrt{s_\lambda}.$$

5.3   Proof of Lemma 2

We first introduce one lemma that is central to our analysis. From the property that R preserves the ℓ_2-norm, it is easy to verify that it also preserves the inner product [3]. Specifically, we have the following lemma.

Lemma 6. *Assume R satisfies Property 1. For any two fixed vectors $\mathbf{u} \in \mathbb{R}^n$ and $\mathbf{v} \in \mathbb{R}^n$, with a probability at least $1 - \delta$, we have*

$$\left|\mathbf{u}^\top R R^\top \mathbf{v} - \mathbf{u}^\top \mathbf{v}\right| \leq \|\mathbf{u}\|_2 \|\mathbf{v}\|_2 \sqrt{\frac{c}{m}\log\frac{4}{\delta}}.$$

provided (10) *holds.*

Let $\mathbf{e}_j$ be the j-th standard basis vector of $\mathbb{R}^n$. From Lemma 6, we have with a probability at least $1 - \delta$,

$$\left|[(RR^\top - I)A^\top \mathbf{w}_*]_j\right| = |\mathbf{e}_j^\top(RR^\top - I)A^\top \mathbf{w}_*| \leq \|A^\top \mathbf{w}_*\|_2 \sqrt{\frac{c}{m}\log\frac{4}{\delta}}$$

for each $j \in [n]$. We complete the proof by taking the union bound over all $j \in [n]$.

5.4   Proof of Lemma 5

We first upper bound ρ_w as

$$
\rho_w \leq \underbrace{\left\| A(I - RR^\top)\boldsymbol{\lambda}_* \right\|_\infty}_{:=U_1} + \underbrace{\left\| A(\boldsymbol{\lambda}_* - \widetilde{\boldsymbol{\lambda}}) \right\|_\infty}_{:=U_2} + \underbrace{\left\| A(RR^\top - I)(\boldsymbol{\lambda}_* - \widetilde{\boldsymbol{\lambda}}) \right\|_\infty}_{:=U_3}.
$$

Bounding U_1. From Lemma 6, we have with a probability at least $1 - \delta$,

$$
\left| \left[A(I - RR^\top)\boldsymbol{\lambda}_* \right]_i \right| = \left| A_{i*}(I - RR^\top)\boldsymbol{\lambda}_* \right|
$$

$$
\leq \max_{i \in [d]} \| A_{i*} \|_2 \| \boldsymbol{\lambda}_* \|_2 \sqrt{\frac{c}{m} \log \frac{4}{\delta}} \stackrel{(6)}{\leq} \| \boldsymbol{\lambda}_* \|_2 \sqrt{\frac{c}{m} \log \frac{4}{\delta}}
$$

for each $i \in [d]$. Taking the union bound over all $i \in [d]$, we have with a probability at least $1 - \delta$,

$$
\left\| A(I - RR^\top)\boldsymbol{\lambda}_* \right\|_\infty \leq \| \boldsymbol{\lambda}_* \|_2 \sqrt{\frac{c}{m} \log \frac{4d}{\delta}}.
$$

Bounding U_2. From our assumption, we have

$$
\left\| A(\boldsymbol{\lambda}_* - \widetilde{\boldsymbol{\lambda}}) \right\|_\infty \leq \max_{i \in [d]} \| A_{i*} \|_2 \| \boldsymbol{\lambda}_* - \widetilde{\boldsymbol{\lambda}} \|_2 \stackrel{(6)}{\leq} \| \boldsymbol{\lambda}_* - \widetilde{\boldsymbol{\lambda}} \|_2.
$$

Bounding U_3. Notice that the arguments for bounding U_1 cannot be used to upper bound U_3, that is because $\boldsymbol{\lambda}_* - \widetilde{\boldsymbol{\lambda}}$ is a random variable that depends on R and thus we cannot apply Lemma 6 directly. To overcome this challenge, we will exploit the fact that $\boldsymbol{\lambda}_* - \widetilde{\boldsymbol{\lambda}}$ is approximately sparse to decouple the dependence. Define

$$
\mathcal{K}_{n,16s_\lambda} = \left\{ \mathbf{x} \in \mathbb{R}^n : \| \mathbf{x} \|_2 \leq 1, \| \mathbf{x} \|_1 \leq 4\sqrt{s_\lambda} \right\}.
$$

When the conclusion in Lemma 3 happens, we have

$$
\frac{\widetilde{\boldsymbol{\lambda}} - \boldsymbol{\lambda}_*}{\| \widetilde{\boldsymbol{\lambda}} - \boldsymbol{\lambda}_* \|_2} \in \mathcal{K}_{n,16s_\lambda} \tag{20}
$$

and thus

$$
U_3 = \| \boldsymbol{\lambda}_* - \widetilde{\boldsymbol{\lambda}} \|_2 \left\| A(RR^\top - I)\frac{\boldsymbol{\lambda}_* - \widetilde{\boldsymbol{\lambda}}}{\| \boldsymbol{\lambda}_* - \widetilde{\boldsymbol{\lambda}} \|_2} \right\|_\infty
$$

$$
\stackrel{(20)}{\leq} \| \boldsymbol{\lambda}_* - \widetilde{\boldsymbol{\lambda}} \|_2 \underbrace{\sup_{\mathbf{z} \in \mathcal{K}_{n,16s_\lambda}} \left\| A(RR^\top - I)\mathbf{z} \right\|_\infty}_{:=U_4}.
$$

Then, we will utilize techniques of covering number to provide an upper bound for U_4.

Lemma 7. *With a probability at least* $1 - \delta$, *we have*

$$\sup_{\mathbf{z} \in \mathcal{K}_{n,16s_\lambda}} \|A(RR^\top - I)\mathbf{z}\|_\infty \leq 2(2 + \sqrt{2})\sqrt{\frac{c}{m}\left(\log\frac{4d}{\delta} + 16s_\lambda \log\frac{9n}{8s_\lambda}\right)}.$$

Putting everything together, we have

$$\rho_w$$

$$\leq \|\boldsymbol{\lambda}_*\|_2\sqrt{\frac{c}{m}\log\frac{4d}{\delta}}$$

$$+ \|\boldsymbol{\lambda}_* - \tilde{\boldsymbol{\lambda}}\|_2\left(1 + 2(2 + \sqrt{2})\sqrt{\frac{c}{m}\left(\log\frac{4d}{\delta} + 16s_\lambda \log\frac{9n}{8s_\lambda}\right)}\right)$$

$$\leq \|\boldsymbol{\lambda}_*\|_2\sqrt{\frac{c}{m}\log\frac{4d}{\delta}} + \frac{3\gamma_\lambda\sqrt{s_\lambda}}{\beta}\left(1 + 7\sqrt{\frac{c}{m}\left(\log\frac{4d}{\delta} + 16s_\lambda \log\frac{9n}{8s_\lambda}\right)}\right).$$

6 Conclusion and Future Work

In this paper, a randomized algorithm is proposed to solve the convex-concave optimization problem in (3). Compared to previous studies, a distinctive feature of the proposed algorithm is that ℓ_1-norm regularization is introduced to control the damage cased by random projection. Under mild assumptions about the optimization problem, we demonstrate that it is able to accurately recover the optimal solutions to (3) provided they are sparse or approximately sparse.

From the current analysis, we need to solve two different problems if our goal is to recover both $\mathbf{w}_*$ and $\boldsymbol{\lambda}_*$ accurately. It is unclear whether this is an artifact of the proof technique or actually unavoidable. We will investigate this issue in the future. Since the proposed algorithm is designed for the case that the optimal solutions are (approximately) sparse, it is practically important to develop a pre-precessing procedure that can estimate the sparsity of solutions before applying our algorithm. We plan to utilize random sampling to address this problem. Last but not least, we will investigate the empirical performance of the proposed algorithm.

Acknowledgments. This work was partially supported by NSFC (61333014, 61272217), JiangsuFS (BK20160658, BK20131278), NSF (1463988, 1545995), and the Collaborative Innovation Center of Novel Software Technology and Industrialization of Nanjing University.

References

1. Achlioptas, D.: Database-friendly random projections: Johnson-lindenstrauss with binary coins. J. Comput. Syst. Sci. **66**(4), 671–687 (2003)

2. Agarwal, A., Negahban, S., Wainwright, M.J.: Fast global convergence of gradient methods for high-dimensional statistical recovery. Ann. Stat. **40**(5), 2452–2482 (2012)
3. Arriaga, R.I., Vempala, S.: An algorithmic theory of learning: robust concepts and random projection. Mach. Learn. **63**(2), 161–182 (2006)
4. Bach, F., Jenatton, R., Mairal, J., Obozinski, G.: Optimization with sparsity-inducing penalties. Found. Trends Mach. Learn. **4**(1), 1–106 (2012)
5. Balcan, M.F., Blum, A., Vempala, S.: Kernels as features: on kernels, margins, and low-dimensional mappings. Mach. Learn. **65**(1), 79–94 (2006)
6. Bingham, E., Mannila, H.: Random projection in dimensionality reduction: applications to image and text data. In: Proceedings of the 7th ACM SIGKDD International Conference on Knowledge Discovery and Data Mining, pp. 245–250 (2001)
7. Boyd, S., Vandenberghe, L.: Convex Optimization. Cambridge University Press, Cambridge (2004)
8. Candès, E.J.: The restricted isometry property and its implications for compressed sensing. C.R. Math. **346**(9–10), 589–592 (2008)
9. Cesa-Bianchi, N., Lugosi, G.: Prediction, Learning, and Games. Cambridge University Press, Cambridge (2006)
10. Chambolle, A., Pock, T.: A first-order primal-dual algorithm for convex problems with applications to imaging. J. Math. Imaging Vis. **40**(1), 120–145 (2011)
11. Dasgupta, S., Gupta, A.: An elementary proof of a theorem of Johnson and lindenstrauss. Random Struct. Algorithms **22**(1), 60–65 (2003)
12. Davenport, M.A., Duarte, M.F., Eldar, Y.C., Kutyniok, G.: Introduction to compressed sensing (chap. 1). In: Compressed Sensing, Theory and Applications, pp. 1–64. Cambridge University Press (2012)
13. Hastie, T., Tibshirani, R., Friedman, J.: The Elements of Statistical Learning. Springer Series in Statistics. Springer, New York (2009)
14. He, Y., Monteiro, R.D.: An accelerated hpe-type algorithm for a class of composite convex-concave saddle-point problems. Technical report, Georgia Institute of Technology (2014)
15. Kakade, S.M., Shalev-Shwartz, S., Tewari, A.: On the duality of strong convexity and strong smoothness: learning applications and matrix regularization. Technical report, Toyota Technological Institute at Chicago (2009)
16. Kaski, S.: Dimensionality reduction by random mapping: fast similarity computation for clustering. In: Proceedings of the 1998 IEEE International Joint Conference on Neural Networks, vol. 1, pp. 413–418 (1998)
17. Koltchinskii, V.: Oracle Inequalities in Empirical Risk Minimization and Sparse Recovery Problems. Springer, Heidelberg (2011)
18. Magen, A.: Dimensionality reductions that preserve volumes and distance to affine spaces, and their algorithmic applications. In: Rolim, J.D.P., Vadhan, S.P. (eds.) RANDOM 2002. LNCS, vol. 2483, pp. 239–253. Springer, Heidelberg (2002)
19. Mahoney, M.W.: Randomized algorithms for matrices and data. Found. Trends Mach. Learn. **3**(2), 123–224 (2011)
20. Mendelson, S., Pajor, A., Tomczak-Jaegermann, N.: Uniform uncertainty principle for Bernoulli and subgaussian ensembles. Constr. Approximation **28**(3), 277–289 (2008)
21. Nemirovski, A.: Prox-method with rate of convergence $O(1/t)$ for variational inequalities with Lipschitz continuous monotone operators and smooth convex-concave saddle point problems. SIAM J. Optim. **15**(1), 229–251 (2005)
22. Nesterov, Y.: Smooth minimization of non-smooth functions. Math. Program. **103**(1), 127–152 (2005)

23. Omidiran, D., Wainwright, M.J.: High-dimensional variable selection with sparse random projections: measurement sparsity and statistical efficiency. J. Mach. Learn. Res. **11**, 2361–2386 (2010)
24. Paul, S., Boutsidis, C., Magdon-Ismail, M., Drineas, P.: Random projections for support vector machines. In: Proceedings of the 16th International Conference on Artificial Intelligence and Statistics, pp. 498–506 (2013)
25. Plan, Y., Vershynin, R.: One-bit compressed sensing by linear programming. Commun. Pure Appl. Math. **66**(8), 1275–1297 (2013)
26. Plan, Y., Vershynin, R.: Robust 1-bit compressed sensing and sparse logistic regression: a convex programming approach. IEEE Trans. Inf. Theor. **59**(1), 482–494 (2013)
27. Rockafellar, R.T.: Convex Analysis. Princeton University Press, Princeton (1997)
28. Shi, Q., Shen, C., Hill, R., van den Hengel, A.: Is margin preserved after random projection? In: Proceedings of the 29th International Conference on Machine Learning (2012)
29. Sridharan, K., Shalev-shwartz, S., Srebro, N.: Fast rates for regularized objectives. Adv. Neural Inf. Process. Syst. **21**, 1545–1552 (2009)
30. Tropp, J.A.: User-friendly tail bounds for sums of random matrices. Found. Comput. Math. **12**, 389–434 (2012)
31. Tsochantaridis, I., Joachims, T., Hofmann, T., Altun, Y.: Large margin methods for structured and interdependent output variables. J. Mach. Learn. Res. **6**, 1453–1484 (2005)
32. Wu, Q., Zhou, D.X.: Svm soft margin classifiers: linear programming versus quadratic programming. Neural Comput. **17**(5), 1160–1187 (2005)
33. Xiao, L., Zhang, T.: A proximal-gradient homotopy method for the ℓ_1-regularized least-squares problem. In: Proceedings of the 29th International Conference on Machine Learning, pp. 839–846 (2012)
34. Yang, T., Zhang, L., Jin, R., Zhu, S.: Theory of dual-sparse regularized randomized reduction. In: Proceedings of the 32nd International Conference on Machine Learning (2015)
35. Zhang, L., Mahdavi, M., Jin, R., Yang, T., Zhu, S.: Recovering the optimal solution by dual random projection. In: Proceedings of the 26th Annual Conference on Learning Theory (COLT), pp. 135–157 (2013)
36. Zhang, L., Mahdavi, M., Jin, R., Yang, T., Zhu, S.: Random projections for classification: a recovery approach. IEEE Trans. Inf. Theor. **60**(11), 7300–7316 (2014)
37. Zhang, L., Yang, T., Jin, R., Zhou, Z.H.: A simple homotopy algorithm for compressive sensing. In: Proceedings of the 18th International Conference on Artificial Intelligence and Statistics (2015)
38. Zou, H., Hastie, T.: Regularization and variable selection via the elastic net. J. Roy. Stat. Soc. Series B (Stat. Methodol.) **67**(2), 301–320 (2005)

On the Evolution of Monotone Conjunctions: Drilling for Best Approximations

Dimitrios I. Diochnos[(✉)]

Department of Computer Science, University of Virginia, Charlottesville, VA, USA
diochnos@virginia.edu

Abstract. We study the evolution of monotone conjunctions using local search; the fitness function that guides the search is correlation with Boolean loss. Building on the work of Diochnos and Turán [6], we generalize Valiant's algorithm [19] for the evolvability of monotone conjunctions from the uniform distribution $\mathcal{U}_n$ to *binomial distributions* $\mathcal{B}_n$.

With a drilling technique, for a frontier q, we exploit a structure theorem for best q-approximations. We study the algorithm using hypotheses from their natural representation ($\mathcal{H} = \mathcal{C}$), as well as when hypotheses contain at most q variables ($\mathcal{H} = \mathcal{C}_{\leq q}$). Our analysis reveals that $\mathcal{U}_n$ is a very special case in the analysis of binomial distributions with parameter p, where $p \in \mathcal{F} = \{2^{-1/k} \mid k \in \mathbb{N}^*\}$. On instances of dimension n, we study approximate learning for $0 < p < 2^{-\frac{1}{n-1}}$ when $\mathcal{H} = \mathcal{C}$ and for $0 < p < \sqrt[n-1]{2/3}$ when $\mathcal{H} = \mathcal{C}_{\leq q}$. Thus, in either case, approximate learning can be achieved for any $0 < p < 1$, for sufficiently large n.

Keywords: Evolution · Evolvability · PAC learning · Noise · Evolutionary algorithms · Optimization · Local search · Distribution-specific learning · Binomial distributions · Correlation · Boolean loss

1 Introduction

Valiant introduced in [19] a framework for analyzing evolution, called *evolvability*. The purpose is to allow and explain the evolution of complex mechanisms in realistic population sizes within realistic time periods. Evolution is treated as a form of computational learning from examples (experiences) and is a restricted form of the *probably approximately correct* (PAC) model of learning [18].

Noise was first studied in the framework of PAC learning by Angluin and Laird [3] and many subsequent results have been obtained in the statistical queries model which is due to Kearns [15]; see also [4,17]. Apart from classification noise, noise on the attributes has also been considered [12]. Noise is natural in evolvability as the functionalities that evolve over time realize their fitness through interaction with the environment (sampling); not by interpreting tiny differences of the true fitness values given in some compact representation. In fact, Feldman showed in [8] that evolvability is equivalent to learning with *correlational statistical queries* [5]. However, as also pointed out by Feldman,

© Springer International Publishing Switzerland 2016
R. Ortner et al. (Eds.): ALT 2016, LNAI 9925, pp. 98–112, 2016.
DOI: 10.1007/978-3-319-46379-7_7

this translation is not necessarily the most efficient or intuitive method in general. Hence, it is common to discuss *distribution-specific* results on the analysis of intuitive algorithms in the framework of evolvability; e.g. [2,14,16]. Thus, the study of simple and intuitive evolvability algorithms using Valiant's original Boolean loss is of interest for specific distributions. Our aim is to understand better such algorithms in the framework of evolvability as well as in the broader framework of *optimization* and *evolutionary algorithms* (EAs) [7,22].

Previous work in evolvability includes [2,8–11,13,14,16,19–21]. In [19] Valiant introduced a swapping-type algorithm and proved the evolvability of monotone conjunctions under the uniform distribution $(\mathcal{U}_n)$. The analysis was simplified by Diochnos and Turán in [6] and in fact it was shown that monotone conjunctions are evolvable in $\mathcal{O}\left(\log(1/\varepsilon)\right)$ generations. The result was strengthened to general conjunctions under $\mathcal{U}_n$ by Kanade, Valiant and Vaughan in [14] including target drift. Further, Feldman in [8] showed that conjunctions are evolvable for any fixed distribution in $\widetilde{\mathcal{O}}(n)$ generations, where $\widetilde{\mathcal{O}}(\cdot)$ ignores polylog factors. Kanade in [13] extended Valiant's model to include genetic recombination where it follows that conjunctions are evolvable in $\mathcal{O}\left((\log(n)/\varepsilon)^2\right)$ generations. On the other hand, one open question from [13] was whether the analysis of Diochnos and Turán could be generalized to distributions beyond $\mathcal{U}_n$.

In this paper we address this last question by considering binomial distributions $\mathcal{B}_n$ with parameter p. We do so by exploiting a structure theorem for best approximations with a *drilling* technique. Drilling improves our estimates of the fitness function by increasing the sample size. In turn, we can discover any important variable for targets up to a certain size beyond the frontier q of our search. Hence, even if we have the power to form some targets precisely, the evolutionary mechanism only forms a best approximation for them. This way, targets with many variables are dealt in an easy way. Our analysis reveals the family $\mathcal{F} = \{2^{-1/k} \mid k \in \mathbb{N}^*\}$, where $\mathcal{U}_n$ is the first member and is obtained for $k = 1$; i.e. $p = \frac{1}{2}$. As we consider larger values of p in the $(0,1)$ interval, every time we encounter one more member of $\mathcal{F}$, we *drill deeper*, thus allowing evolution to identify variables from targets containing one more variable. Evolvability follows for any fixed distribution in $\mathcal{O}\left(\log_{\frac{1}{p}}(1/\varepsilon)\right)$ generations; the setup of [6] for $\mathcal{U}_n$ is recovered as a special case. Our analysis reveals an interesting non-trivial connection between the parameters, which is captured in Fig. 3.

The paper is structured as follows. Section 2 gives the definition of evolvability and Sect. 3 preliminaries specific to our setup. Section 4 lays the foundations for the evolvability of monotone conjunctions. Section 5 discusses adaptation. Section 6 discusses the convergence. Section 7 analyzes the complexity. We conclude with further remarks in Sect. 8. Due to space limitations some proofs are sketched or omitted in this version.

2   Definition of Evolvability

The truth values TRUE and FALSE are represented by 1 and −1 respectively. The fitness function that guides the search is called *performance*. For a target c and a fixed distribution $\mathcal{D}_n$ over $\{0,1\}^n$, the performance of a hypothesis h is

$$\mathrm{Perf}_{\mathcal{D}_n}(\mathrm{h,c}) = \sum_{x \in \{0,1\}^n} \mathrm{h}(x) \cdot \mathrm{c}(x) \cdot \mathbf{Pr}_{x \sim \mathcal{D}_n}(x), \qquad (1)$$

called the *correlation* of h and c. Evolution starts with an initial hypothesis h_0, and produces a sequence of hypotheses using a local-search procedure in $\mathcal{H}$. Similarity between h and c in an underlying distribution $\mathcal{D}_n$ is measured by the *empirical performance* function $\mathrm{Perf}_{\mathcal{D}_n}(\mathrm{h,c},|S|)$ which is evaluated approximately by drawing a random sample S and computing $\mathrm{Perf}_{\mathcal{D}_n}(\mathrm{h,c},|S|) = \frac{1}{|S|}\sum_{x \in S} \mathrm{h}(x) \cdot \mathrm{c}(x)$. Valiant's original definition of evolvability treated the confidence parameter δ and the error parameter ε as one. Below, even if we draw the definitions from [19], we modify them slightly to also include δ explicitly.

Definition 1 (Modified From [19]). *For a polynomial $p(\cdot,\cdot)$ and a representation class R a p-neighborhood N on R is a pair M_1, M_2 of randomized polynomial time Turing machines such that the numbers n (in unary), $\lceil 1/\varepsilon \rceil$ and a representation $r \in R_n$ act as follows: M_1 outputs all the members of a set $Neigh_N(r, \varepsilon) \subseteq R_n$, that contains r and may depend on random coin tosses of M_1, and has size at most $p(n, 1/\varepsilon)$. If M_2 is then run on this output of M_1, it in turn outputs one member of $Neigh_N(r, \varepsilon)$, with member r_1 being output with a probability $\mathbf{Pr}_N(r, r_1) \geq 1/p(n, 1/\varepsilon)$.*

Definition 2 (Modified From [19]). *For confidence parameter δ, error parameter ε, positive integers n and s, an ideal function $f \in \mathcal{C}_n$, a representation class R with $p(n, 1/\varepsilon)$-neighborhood N on R, a distribution $\mathcal{D}$, a representation $r \in R_n$ and a real number t, the mutator $Mu(f, p(n, 1/\varepsilon), R, N, \mathcal{D}, s, r, t)$ is a random variable that on input $r \in R_n$ takes a value $r_1 \in R_n$ determined as follows: For each $r_1 \in Neigh_N(r, \varepsilon)$ it first computes an empirical value of $v(r_1) = Perf_{\mathcal{D}_n}(r_1, f, s)$. Let Bene be the set $\{r_1 \mid v(r_1) > v(r) + t\}$ and Neut be the set difference $\{r_1 \mid v(r_1) \geq v(r) - t\} \setminus$ Bene. If Bene $\neq \emptyset$ then output $r_1 \in$ Bene with probability $\mathbf{Pr}_N(r, r_1) / \sum_{r_1 \in Bene} \mathbf{Pr}_N(r, r_1)$. Otherwise (Bene $= \emptyset$), output an $r_1 \in$ Neut, the probability of a specific r_1 being $\mathbf{Pr}_N(r, r_1) / \sum_{r_1 \in Neut} \mathbf{Pr}_N(r, r_1)$.*

Definition 3 (Modified From [19]). *For a mutator $Mu(f, p(n, 1/\varepsilon), R, N, \mathcal{D}, s, r, t)$ a t-evolution step on input $r_1 \in R_n$ is the random variable $r_2 = Mu(f, p(n, 1/\varepsilon), R, N, \mathcal{D}, s, r_1, t)$. We then say $r_1 \to r_2$ or $r_2 \leftarrow Evolve(f, p(n, 1/\varepsilon), R, N, \mathcal{D}_n, s, r_1, t)$.*

We say that polynomials $t_\ell(x, y)$ and $t_u(x, y)$ are *polynomially related* if for some $\eta > 1$ for all $x, y(0 < x, y < 1)(t_u(x, y))^\eta \leq t_\ell(x, y) \leq t_u(x, y)$.

Definition 4 (Modified From [19]). *For a mutator $Mu(f, p(n, 1/\varepsilon), R, N, \mathcal{D}, s, r, t)$ a (t_ℓ, t_u)-evolution sequence for $r_1 \in R_n$ is a random variable that takes as values sequences $r_1, r_2, r_3, \ldots$ such that for all i $r_i \leftarrow Evolve(f, p(n, 1/\varepsilon), R, N, \mathcal{D}, s, r_{i-1}, t_i)$, where $t_\ell(1/n, \varepsilon) \leq t_i \leq t_u(1/n, \varepsilon)$, t_ℓ and t_u are polynomially related polynomials, and t_i is the output of a TM T on input r_{i-1}, n, ε and δ.*

Definition 5 (Goal of Evolution; Modified From [19]**).** *For polynomials* $p(n, 1/\varepsilon), s(n, 1/\varepsilon, 1/\delta), t_\ell(1/n, \varepsilon)$ *and* $t_u(1/n, \varepsilon)$, *a representation class* R *and* $p(n, 1/\varepsilon)$-*neighborhood* N *on* R, *the class* $\mathcal{C}$ *is* (t_ℓ, t_u)-*evolvable by* $(p(n, 1/\varepsilon), R,$- $N, s(n, 1/\varepsilon, 1/\delta))$ *over distribution* $\mathcal{D}$ *if there is a polynomial* $g(n, 1/\varepsilon, 1/\delta)$ *and a Turing machine* T, *which computes a tolerance bounded between* t_ℓ *and* t_u, *such that for every positive integer* n, *every* $f \in \mathcal{C}_n$, *every* $\delta > 0$, *every* $\varepsilon > 0$, *and every* $r_0 \in R_n$ *it is the case that with probability at least* $1 - \delta$, *a* (t_ℓ, t_u)-*evolution sequence* $r_0, r_1, r_2, \ldots$, *where* $r_i \leftarrow Evolve(f, p(n, 1/\varepsilon), R, N, \mathcal{D}_n, s(n, 1/\varepsilon, 1/\delta),$- $r_{i-1}, T(r_{i-1}, n, \varepsilon))$, *will have* $Perf_{\mathcal{D}_n}\left(r_{g(n,1/\varepsilon,1/\delta)}, f\right) \geq 1 - \varepsilon$.

The number of generations needed for evolution is upper bounded by $g\left(n, \frac{1}{\varepsilon}, \frac{1}{\delta}\right)$.

Definition 6 (Modified From [19]**).** *A class* $\mathcal{C}$ *is evolvable by* $(p(n, 1/\varepsilon), R,$- $N, s(n, 1/\varepsilon, 1/\delta))$ *over* $\mathcal{D}$ *iff for some pair of polynomially related polynomials* t_ℓ, t_u, $\mathcal{C}$ *is* (t_ℓ, t_u)-*evolvable by* $(p(n, 1/\varepsilon), R, N, s(n, 1/\varepsilon, 1/\delta))$ *over* $\mathcal{D}$.

Definition 7 (Modified From [19]**).** *A class* $\mathcal{C}$ *is evolvable by* R *over* $\mathcal{D}$ *iff for some polynomials* $(p(n, 1/\varepsilon)$ *and* $s(n, 1/\varepsilon, 1/\delta))$, *and some* $p\left(n, \frac{1}{\varepsilon}\right)$-*neighborhood* N *on* R, $\mathcal{C}$ *is evolvable by* $(p(n, 1/\varepsilon), R, N, s(n, 1/\varepsilon, 1/\delta))$ *over* $\mathcal{D}$.

3 Preliminaries

Given a set of Boolean variables $x_1, \ldots, x_n$, we assume that there is an unknown target $c \in \mathcal{C}$, a monotone conjunction of some of these variables. Let $\mathcal{C}$ be the concept class of all possible conjunctions in their natural representation. For a threshold q, let $\mathcal{C}_{\leq q}$ be the set of monotone conjunctions from $\mathcal{C}$ that contain at most q variables. Further, let $\mathcal{C}_{>q} = \mathcal{C} \setminus \mathcal{C}_{\leq q}$ be the set of conjunctions from $\mathcal{C}$ that are not included in $\mathcal{C}_{\leq q}$.

By Definition 2, the neighborhood is split in 3 parts by the *increase* in performance that the hypotheses in the neighborhood offer. There are *beneficial*, *neutral*, and *deleterious* mutations. Thus, we need an oracle for computing

$$\Delta = Perf_{\mathcal{D}_n}\left(h', c\right) - Perf_{\mathcal{D}_n}\left(h, c\right), \tag{2}$$

and hence, for a given t, determine the set where $h' \in N$ lies. Now let

$$h = \bigwedge_{i=1}^{m} x_i \wedge \bigwedge_{\ell=1}^{r} y_\ell \quad \text{and} \quad c = \bigwedge_{i=1}^{m} x_i \wedge \bigwedge_{k=1}^{u} w_k. \tag{3}$$

The x's are *mutual* variables, the y's are called *redundant* and the w's are called *undiscovered* or *missing*. Variables in the target c are called *good*, otherwise *bad*. With $|h|$ we denote the *size* (or *length*) of a conjunction; the number of variables that it contains. A *binomial distribution* over $\{0, 1\}^n$ is specified by the probability p of setting each variable x_i to 1. A truth assignment $(a_1, \ldots, a_n) \in \{0, 1\}^n$ has probability $\prod_{i=1}^{n} p^{a_i} \cdot (1 - p)^{1-a_i}$. We write $\mathcal{B}_n$ to denote a fixed binomial distribution, omitting p for simplicity. On an instance of dimension

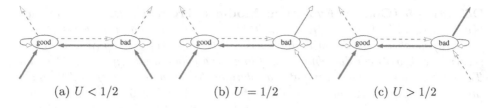

(a) $U < 1/2$ (b) $U = 1/2$ (c) $U > 1/2$

Fig. 1. Arrows pointing towards the nodes indicate addition of one variable and arrows pointing away from a node indicate removal of one variable. This is consistent with arrows indicating swapping a pair of variables. Thick solid lines indicate $\Delta > 0$. Simple lines indicate $\Delta = 0$. Dashed lines indicate $\Delta < 0$. Let U be the weight of the undiscovered variables. Figure 1(a) holds when $U < 1/2$, Fig. 1(b) holds when $U = 1/2$ and Fig. 1(c) holds when $U > 1/2$.

n we say that $\mathcal{B}_n$ has *low density* when $0 < p < \frac{1}{2}$, *medium density* when $\frac{1}{2} \leq p \leq 2^{-\frac{1}{n}}$, *high density* when $2^{-\frac{1}{n}} < p < \sqrt[n]{2/3}$, and *very high density* when $\sqrt[n]{2/3} \leq p < 1$. Consider a target c and a hypothesis h as in (3). Then (1) gives

$$\text{Perf}_{\mathcal{B}_n}(\text{h}, \text{c}) = 1 - 2p^{m+r} - 2p^{m+u} + 4p^{m+r+u}. \tag{4}$$

Figure 1 presents the sign of Δ that guides the search. Note that while the sign of an arrow may be fully determined, it is the value of the tolerance t that defines the two sets of interest (Bene and Neut) that guide the search. Figure 1(a) refers to the *expansion phase*, Fig. 1(b) to the *identification phase* and Fig. 1(c) to the *shrinking phase*.

3.1   The Swapping Algorithm

The swapping algorithm for monotone conjunctions was introduced by Valiant in [19] and was also analyzed in [6]. The neighborhood N of a conjunction h is the set of conjunctions that arise by *adding* a variable (neighborhood N^+), *removing* a variable (neighborhood N^-), or *swapping* a variable with another one (neighborhood N^{+-}), plus the conjunction itself[1]. Thus, $N = N^- \cup N^+ \cup N^{+-} \cup \{h\}$. As an example, let h $= x_1 \wedge x_2$, and $n = 3$. Then, $N^- = \{x_1, x_2\}$, $N^+ = \{x_1 \wedge x_2 \wedge x_3\}$, and $N^{+-} = \{x_3 \wedge x_2, x_1 \wedge x_3\}$. Note that $|N| = \mathcal{O}(n |\text{h}|)$ in general. Algorithm 1 presents the mutator function for the swapping algorithm.

Compute-q uses Table 2 or (11) to set q depending on the hypothesis class $\mathcal{H}$ that is used for evolution. (Table 2 used for Compute-q, already incorporates a modified ε when needed.) Line 6 computes the minimum non-zero value A of $\mathcal{A}(u) = |1 - 2p^u|$ for $u \in \{0, \ldots, n\}$ using Table 1 from Sect. 4.1. Tolerance t is normally t_ℓ; however, when $\mathcal{H} = \mathcal{C}$ and $|\text{h}| > q$ then $t = t_u$. We discuss tolerance in Sects. 4.2 and 7. Performance computes the empirical performance of h w.r.t. c over the distribution $\mathcal{B}_n$ with parameter p, within ϵ_s of its true value, with probability at least $1 - \delta_s$; see Sect. 7. SetWeight assigns the same

---

[1] As h will be clear from the context, we write N instead of $N(\text{h})$.

Algorithm 1. Mutator function for a binomial distribution

Input: dimension n, $p \in (0,1)$, $\delta \in (0,1)$, $\varepsilon \in (0,2)$, $\mathcal{H} \in \{\mathcal{C}_{\leq q}, \mathcal{C}\}$, $h \in \mathcal{H}$
Output: a new hypothesis h'

1 $q \leftarrow$ Compute-q$(p, \varepsilon, \mathcal{H})$; $\vartheta \leftarrow \left\lfloor \log_{\frac{1}{p}}(2) \right\rfloor$;

2 **if** $|h| > 0$ **then** Generate N^- **else** $N^- \leftarrow \emptyset$;

3 **if** $|h| < q$ **then** Generate N^+ **else** $N^+ \leftarrow \emptyset$;

4 **if** $|h| \leq q$ **then** Generate N^{+-} **else** $N^{+-} \leftarrow \emptyset$;

5 $Bene \leftarrow \emptyset$; $Neutral \leftarrow \{h\}$;

6 $A \leftarrow \min_{\neq 0}\{|1 - 2p^u|\}$; $\mu \leftarrow \min\left\{2p^{q+\vartheta}, A\right\}$;

7 $t_\ell \leftarrow p^{q-1}\mu(1-p)$; $t_u \leftarrow 4p^q(1-p)$;

8 **if** $(\mathcal{H} = \mathcal{C})$ and $(|h| > q)$ **then** $t \leftarrow t_u$; $\epsilon_s \leftarrow t_u/4$; $\delta_s \leftarrow \delta/4$;

9 **else** $t \leftarrow t_\ell$; $\epsilon_s \leftarrow t_\ell$; $\delta_s \leftarrow \delta/2$;

10 SetWeight(h, h, N^-, N^+, N^{+-}); $\nu_h \leftarrow$ Perf$(p, h, \epsilon_s, \delta_s)$;

11 **for** $x \in N^+, N^-, N^{+-}$ **do**

12 $\quad$ SetWeight(x, h, N^-, N^+, N^{+-}); $\nu_x \leftarrow$ Perf$(p, x, \epsilon_s, \delta_s)$;

13 $\quad$ **if** $\nu_x > \nu_h + t$ **then** $Bene \leftarrow Bene \cup \{x\}$;

14 $\quad$ **else if** $\nu_x \geq \nu_h - t$ **then** $Neutral \leftarrow Neutral \cup \{x\}$;

15 **if** $Bene \neq \emptyset$ **then return** Select$(Bene)$ **else return** Select$(Neutral)$;

---

weight to all members of $\{h\} \cup N^- \cup N^+$ so that they add up to $\frac{1}{2}$, and the same weight to all the members of N^{+-} so that they add up to $\frac{1}{2}$. Select computes the sum of weights W of the conjunctions in the set passed as argument, and returns a hypothesis h' with probability $w_{h'}/W$, where $w_{h'}$ is the weight of h'.

4   Foundations for Evolvability

Let $\log_{\frac{1}{p}}(x)$ be the logarithm of x in base $\frac{1}{p}$. Given a size q and an extension ϑ, a hypothesis h is called *short* when $|h| \leq q$, *medium* when $q < |h| \leq q + \vartheta$ and *long* when $|h| > q + \vartheta$. Given a target c and a size q, we will be interested in the best size q approximation of c. The reason is Theorem 1 below, first proved in [6] for $\mathcal{U}_n$. Note that the best approximation is not necessarily unique.

Definition 8 (Best q-Approximation). *Let h be a hypothesis such that $|h| \leq q$ and $\forall h' \neq h, |h'| \leq q : Perf_{\mathcal{D}_n}(h', c) \leq Perf_{\mathcal{D}_n}(h, c)$. We call h a best q-approximation of c.*

Theorem 1 (Best Approximations Under Binomial Distributions; [6]). *The best q-approximation of a target c is c if $|c| \leq q$, or any hypothesis formed by q good variables if $|c| > q$.*

Lemma 1 (Performance Lower Bound, Medium Target). *Let $\mathcal{B}_n$ be a binomial distribution. Let c be a medium target. A best q-approximation h has $Perf_{\mathcal{B}_n}(h, c) > 1 - 2p^q$.*

Lemma 2 (Performance Lower Bound, Long Target). *Let $\mathcal{B}_n$ be a binomial distribution. Let h be a hypothesis such that $|h| \geq q$ and consider a long target c. Then, $\mathrm{Perf}_{\mathcal{B}_n}(h, c) > 1 - 2p^q \left(1 + p^{1+\vartheta}\right)$.*

We now examine the difference Δ between the current hypothesis h and a hypothesis h' that is generated in each neighborhood.

Comparing $h' \in N^+$ with h. We introduce a variable z in the hypothesis h. If z is good, $\Delta = 2p^{|h|}(1 - p) > 0$. If z is bad, $\Delta = 2p^{|h|}(1 - 2p^u)(1 - p)$.

Comparing $h' \in N^-$ with h. We remove a variable z from the hypothesis h. If z is good, $\Delta = -2p^{|h|-1}(1 - p) < 0$. If z is bad, $\Delta = -2p^{|h|-1}(1 - 2p^u)(1 - p)$.

Comparing $h' \in N^{+-}$ with h. Replacing a good with a bad variable gives $\Delta = -4p^{|h|+u}(1 - p)$. Replacing a good with a good, or a bad with a bad variable gives $\Delta = 0$. Replacing a bad with a good variable gives $\Delta = 4p^{|h|+u-1}(1 - p)$.

Our aim for short and medium targets is to have the ability to determine the signs of the differences Δ in every case. For long targets, we want to determine the signs of the Δ's for the mutations that arise in the N^+ and N^- neighborhoods; not necessarily for those in the N^{+-} neighborhood. We denote

$$\mathcal{A}(u) = |1 - 2p^u| \,, \quad u \in \{0, \ldots, n\}. \tag{5}$$

As $\mathcal{A}(u)$ appears in the Δ's for the mutations in the N^+ and N^- neighborhoods, we need to study the minimum non-zero value that $\mathcal{A}(u)$ can attain for $u \in \{0, 1, \ldots, n\}$ under an arbitrary $\mathcal{B}_n$. The zeros of $\mathcal{A}(u)$ are found in the family

$$\mathcal{F} = \left\{ 2^{-\frac{1}{k}} \mid k \in \mathbb{N}^* \right\}. \tag{6}$$

4.1   On the Minimum Non-zero Value of $\mathcal{A}(u)$, $u \in \{0, \ldots, n\}$

Lemma 3. *Consider the polynomials $f_k(p) = p^{k+1} + p^k - 1$ defined respectively in the intervals $\mathcal{J}_k = \left[2^{-1/k}, 2^{-1/(k+1)}\right]$ with $k \in \mathbb{N}^*$. Then, each f_k is monotone increasing in $\mathcal{J}_k$ and has a (unique) root ξ_k in the open interval $\mathcal{J}_k$.*

Table 1 and Fig. 2 present $\min_{\neq 0} \{\mathcal{A}(u)\}$ as p ranges in $(0, 1)$.

4.2   On Tolerance and Design Requirements

The critical part of the evolution will be evolving short hypotheses. In this part we want to identify swaps precisely for short and medium targets and thus $|\Delta| \geq 4p^{2q+\vartheta-1}(1 - p)$. Regarding additions and removals we want to be able to identify the sign of Δ precisely, regardless of the target; thus, using (5), for the non-zero values of Δ, $|\Delta| = 2p^{|h|-1} \cdot \mathcal{A}(u) \cdot (1-p) \geq 2p^{q-1} \cdot \min_{\neq 0} \{\mathcal{A}(u)\} \cdot (1-p)$. Therefore, in order to determine the tolerance, we want to determine

$$2p^{q+\vartheta} < \min_{\neq 0} \{\mathcal{A}(u)\} = \min_{\neq 0} \{|1 - 2p^u|\} = A. \tag{7}$$

Table 1. $\min_{\neq 0}\{\mathcal{A}(u)\}$, attained for specific u by some target c, as p ranges in $(0,1)$. When $2^{-1/k} < p < 2^{-1/(k+1)}$, then ξ_k is the root from Lemma 3.

Density	p	$\min_{\neq 0}\{\mathcal{A}(u)\}$	For u	Obtained by target c
Low	$0 < p < 1/2$	$1 - 2p$	1	$1 \leq |c| \leq \min\{n, q+1\}$
Medium	$2^{-1/k}$, with $1 \leq k \leq n-1$	$1 - p$	$k+1$	$k+1 \leq |c| \leq \min\{n, q+k+1\}$
	$2^{-1/n}$	$(1-p)/p$	$n-1$	$n-1 \leq |c| \leq n$
	$2^{-1/k} < p \leq \xi_k$ with $1 \leq k = \left\lfloor \log_{\frac{1}{p}}(2) \right\rfloor \leq n-1$	$2p^k - 1$	k	$k \leq |c| \leq \min\{n, q+k\}$
	$\xi_k \leq p < 2^{-1/(k+1)}$ with $1 \leq k = \left\lfloor \log_{\frac{1}{p}}(2) \right\rfloor \leq n-1$	$1 - 2p^{k+1}$	$k+1$	$k+1 \leq |c| \leq \min\{n, q+k+1\}$
(Very) high	$2^{-1/n} < p < 1$	$2p^n - 1$	n	$|c| = n$

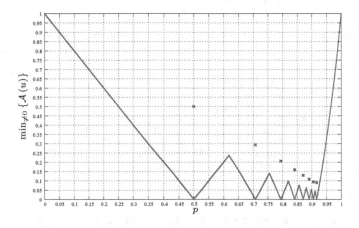

Fig. 2. $\min_{\neq 0}\{\mathcal{A}(u)\}$ for $n = 8$, as presented in Table 1.

We now let

$$\mu = \min\left\{2p^{q+\vartheta}, \min_{\neq 0}\{\mathcal{A}(u)\}\right\} \quad \text{and} \quad t_\ell = p^{q-1}\mu(1-p). \qquad (8)$$

Tolerance is set by (8) when evolution takes place in $\mathcal{C}_{\leq q}$. When $\mathcal{H} = \mathcal{C}_{\leq q}$, $t_\ell = t_u$ and this is a special case, *fixed-tolerance* evolvability. On the other hand, if $\mathcal{H} = \mathcal{C} = \mathcal{C}_{\leq q} \cup \mathcal{C}_{>q}$, the approach in $\mathcal{C}_{>q}$ relies on setting the tolerance t_u large enough so that a random walk can be performed and eventually form a hypothesis in $\mathcal{C}_{\leq q}$. The neighborhood in $\mathcal{C}_{>q}$ is $N^- \cup \{h\}$; see Algorithm 1. Thus, $|\Delta| \leq 2p^{|h|-1}(1-p) \leq 2p^q(1-p)$ and t_u is set to be $t_u = 2 \cdot \max\{|\Delta|\}$, that is,

$$t_u = 4p^q(1-p). \qquad (9)$$

Thus, requiring $(t_u)^\eta \leq t_\ell \leq t_u < 1$ for some $\eta > 1$, we get the constraints

$$
\begin{array}{c|c|c|c}
p \leq \frac{1}{4} & \frac{1}{4} < p < \frac{1}{2} & p = \frac{1}{2} & \frac{1}{2} < p \\
\hline
q \geq 1 & q \geq 2 & q > 1 & q \geq \log_{\frac{1}{p}}(2)
\end{array}
\qquad (10)
$$

5   Adaptation

Corollary 1. $q \geq \log_{\frac{1}{p}}\left(\frac{3}{\varepsilon}\right), \vartheta \geq 0, |h| = q < |c| \leq q + \vartheta \Rightarrow Perf_{\mathcal{B}_n}(h, c) > 1 - \frac{2\varepsilon}{3}$.

Corollary 2. $q \geq \log_{\frac{1}{p}}\left(\frac{3}{\varepsilon}\right), \vartheta \geq \log_{\frac{1}{p}}(2p), |h| \geq q, |c| > q + \vartheta \Rightarrow Perf_{\mathcal{B}_n}(h, c) > 1 - \varepsilon$.

5.1   Evolution When $\mathcal{H} = \mathcal{C}$

In light of Corollaries 1 and 2, setting $q = \left\lceil \log_{\frac{1}{p}}\left(\frac{8}{\varepsilon}\right) \right\rceil$ and $\vartheta = \left\lfloor \log_{\frac{1}{p}}(2) \right\rfloor$ would also satisfy the requirements in (10) for every $0 < \varepsilon < 2$. However, we can improve the frontier q. Depending on p, let q and ϑ be defined from Table 2.

Table 2. Definition of q and ϑ depending on p when evolving on $\mathcal{H} = \mathcal{C} = \mathcal{C}_{\leq q} \cup \mathcal{C}_{>q}$.

p	q	ϑ
$p \leq \frac{1}{4}$	$\left\lceil \log_{\frac{1}{p}}(3/\varepsilon) \right\rceil$	0
$\frac{1}{4} < p < \frac{1}{2}$	$\max\left\{ \left\lceil \log_{\frac{1}{p}}(3/\varepsilon) \right\rceil, 2 \right\}$	0
$p = \frac{1}{2}$	$\max\left\{ \left\lceil \log_{\frac{1}{p}}(3/\varepsilon) \right\rceil, 2 \right\}$	1
$p > \frac{1}{2}$	$\max\left\{ \left\lceil \log_{\frac{1}{p}}(3/\varepsilon) \right\rceil, \left\lfloor \log_{\frac{1}{p}}(2) \right\rfloor \right\}$	$\left\lfloor \log_{\frac{1}{p}}(2) \right\rfloor$

Learnability on a Fixed Dimension. Let $\lambda > 0$. Then, $\log_{\frac{1}{p}}\left(\frac{3}{\varepsilon}\right) \leq \lambda \Rightarrow \varepsilon \geq 3p^{\lambda}$. Approximate learning degenerates to exact, when $\log_{\frac{1}{p}}\left(\frac{3}{\varepsilon}\right) > n - 1$, as due to rounding $q \geq n$. However, we also need to be able to achieve $q = n$ for a value of ε in the $(0, 2)$ interval. Thus, on any fixed dimension n, it makes sense to discuss about approximate learning ($\lambda \leq n - 1$) when $0 < p < {}^{n-1}\!\sqrt{2/3}$ and about exact ($\lambda = n$) when $p < \sqrt[n]{2/3}$. That is, when $p \geq \sqrt[n]{2/3}$ then the dimension is *too low* to allow even exact learning with our method. Regarding $\lceil \log_{\frac{1}{p}}(2) \rceil$ from Table 2, approximate learning can be done when $0 < p < 2^{-\frac{1}{n-1}}$ and exact when $0 < p < 2^{-\frac{1}{n}}$. Hence, on an instance of dimension n, we study approximate learning when $0 < p < 2^{-\frac{1}{n-1}}$ and exact learning when $0 < p < 2^{-\frac{1}{n}}$.

Adaptation for Large Input Error. For $\frac{1}{4} < p < \frac{1}{2}$, when $\varepsilon \geq 3p$, evolution will reset ε to $\varepsilon' = 3p^2 \geq \frac{3}{16}$. In the end it will return a hypothesis that has accuracy $1 - \varepsilon' > 1 - \varepsilon$ but through a (t_ℓ, t_u)-evolutionary sequence.

Let $\mathcal{I}_k = [2^{-\frac{1}{k}}, 2^{-\frac{1}{k+1}})$. When $p \in \mathcal{I}_1$, if $\varepsilon \geq 3p$, it will be reset to $\varepsilon' = 3p^2 > \frac{3}{4}$. When $p \in \mathcal{I}_k$ with $\lfloor \log_{\frac{1}{p}}(2) \rfloor = k \in \{2, \ldots, n - 1\}$, if $\varepsilon \geq 3p^k$, then setting $\varepsilon' = 3p^{k+1} \geq 3 \cdot 2^{-\frac{k+1}{k}} > 1$ implies $q = \lceil \log_{\frac{1}{p}}(3/\varepsilon') \rceil = \lfloor \log_{\frac{1}{p}}(2) \rfloor = k + 1 \leq n$. Thus, we will treat q as if it is defined *solely* by $q = \lceil \log_{\frac{1}{p}}\left(\frac{3}{\varepsilon}\right) \rceil$ in Table 2. If the input ε is too large, evolution will adapt it to an appropriate constant.

5.2   Evolution When $\mathcal{H} = \mathcal{C}_{\leq q}$

Working strictly on $\mathcal{H} = \mathcal{C}_{\leq q}$, one need no longer respect the requirements in (10) as we have fixed-tolerance evolvability; see Sect. 4.2. Hence, we let

$$q = \left\lceil \log_{\frac{1}{p}}(3/\varepsilon) \right\rceil \qquad \text{and} \qquad \vartheta = \left\lfloor \log_{\frac{1}{p}}(2) \right\rfloor . \tag{11}$$

By restricting the hypothesis class, on an instance of dimension n, evolution can now take place even when p belongs to the high density region, contrasting Sect. 5.1. Also, no adaptation is needed for any feasible (p, ε) pair.

5.3   Determining $\mu = \min\left\{2p^{q+\vartheta}, \min_{\neq 0}\{\mathcal{A}(u)\}\right\}$

For a specific p, we need to identify the minimum q_m such that $2p^{q_m+\vartheta} < A$. Then, for $\varepsilon < 3p^{q_m-1}$ swaps are more expensive. Thus, q_m satisfies[2]

$$q_m > \log_{\frac{1}{p}}(2p^{\vartheta}/A) = \zeta . \tag{12}$$

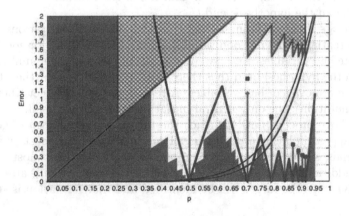

Fig. 3. $\mathcal{H} = \mathcal{C}$ and $n = 8$. Along the line $3p$, $q = 1$ for the lowest possible error at every p. Similarly, the curves $3p^{n-1}$ and $3p^n$ are also drawn. In the top part of the plot, the triangle and the region with the jigsaw frontier that are shaded indicate (p, ε) pairs where evolution needs to adapt a large input ε to a suitable smaller constant ε'; see Sect. 5.1. The shaded region in the lower part of the plot, as well as the individual spikes for the members of $\mathcal{F}$, indicate (p, ε) pairs where swaps determine μ in (8). (When $p < \frac{1}{2}$, the critical p's are obtained by solving numerically $2p^\zeta + 2p - 1 = 0$ for $\zeta \in \mathbb{N}^*$. For $p \geq \frac{1}{2}$, we use stepsize $\Delta p = 10^{-5}$ and for every such p we compute the turning point ε; see Sect. 5.3.) Finally, the smooth boundary that is discussed in Sect. 5.3 is also shown; the separation point for $\mathcal{U}_n$ is $\left(\frac{1}{2}, 3\right)$ and it is not drawn.

---

[2] As p ranges in $(0, 1)$, a natural question in (12) is whether $\zeta \in \mathbb{Z}$; then $q_m = \zeta + 1$, otherwise $q_m = \lceil \zeta \rceil < \zeta + 1$. Equivalently, does $2p^{\zeta+\vartheta} - A = 0$ hold for $\zeta \in \mathbb{Z}$? By Table 1 and the definition of ϑ, for $\mathcal{U}_n$, $\zeta = 1$. Hence, in $\mathcal{U}_n$, when $\frac{3}{2} \leq \varepsilon < 3$ then the two quantities for μ in (8) have the same value *for a range of* ε *values*. Regardless if there are additional integer solutions, q_m can be computed efficiently.

A Smooth Frontier for $\mu = \min\{2p^{q+\vartheta}, \min_{\neq 0}\{\mathcal{A}(u)\}\}$. As q involves rounding, $2p^{q+\vartheta} = 2p^{\lceil \log_{1/p}(3/\varepsilon)\rceil + \lfloor \log_{1/p}(2)\rfloor} \geq 2p^{1+\log_{1/p}(3/\varepsilon)+\log_{1/p}(2)} = \frac{p\varepsilon}{3}$. Thus, by overestimating the required accuracy for swaps, determining μ can be reduced to the simpler $\frac{p\varepsilon}{3} < A \Leftrightarrow \varepsilon < \frac{3A}{p}$. In other words, μ could also be defined as $\mu = \min\{\frac{p\varepsilon}{3}, \min_{\neq 0}\{\mathcal{A}(u)\}\}$ in (8) and in line 6 of Algorithm 1.

Figure 3 presents all the above relationships between p and ε.

6   Convergence

6.1   Short Initial Hypothesis and Short Target

If $U < \frac{1}{2}$, Fig. 1(a) applies. Beneficial mutations can only *add* or *swap* variables. Swaps or additions of good variables increase m. Thus after at most $|c|$ such mutations and at most $q - |h_0|$ additions of bad variables, $U \geq \frac{1}{2}$.

If $U = \frac{1}{2}$, Fig. 1(b) applies. $U = \frac{1}{2} \Rightarrow p \in \mathcal{F}$ for some k. Further, $U = \frac{1}{2} \Rightarrow p^u = \frac{1}{2} \Rightarrow u = \log_{\frac{1}{p}}(2) = k$. Also, $(k = u) \wedge (u \leq |c|) \wedge (|c| \leq q) \Rightarrow k \in \{1, \ldots, q\}$.[3] In one step, the first beneficial swap or addition of good variable brings one more good variable in the hypothesis and $U > \frac{1}{2}$.

When $U > \frac{1}{2}$, corresponding to Fig. 1(c), then beneficial mutations are those that add potentially missing good variables, swap bad variables for good ones, or finally delete bad variables. Each swap or addition increases the number of good variables in the hypothesis and thus there can be $|c|$ of those. Further, there can be at most q removals of bad variables. After we get to the target, there are no beneficial mutations; the only neutral mutation is the target itself.

In the above process, until we reach the target, the number m of good variables that appear in h is non-decreasing. Thus there can be at most $|c|$ additions of good variables and swaps combined. Further, there can be at most $q + q = 2q$ beneficial additions or deletions of bad variables. Hence, overall, after at most $|c| + 2q \leq 3q$ steps the target will be identified and that formation is stable.[4]

---

[3]   A clarification comment is in order here. When $\mathcal{H} = \mathcal{C}$, by Table 2, $q \geq \lceil \log_{\frac{1}{p}}(2)\rceil \geq \log_{\frac{1}{p}}(2) = k$ *always*, and thus, on an instance of dimension n, as p increases in $\mathcal{F}$ for successive values of k, then q increases at least that fast.

The above is not necessarily true when $\mathcal{H} = \mathcal{C}_{\leq q}$. By (11), when $p \in \mathcal{F}$ with $k \geq 3$ (i.e. $p \in \mathcal{F}$ and $p \geq 2^{-1/3}$), for input ε such that $2 > \varepsilon \geq 3p^{k-1} = \frac{3}{2p}$, then $q = \lceil \log_{\frac{1}{p}}(\frac{3}{\varepsilon})\rceil < \lfloor \log_{\frac{1}{p}}(2)\rfloor = \log_{\frac{1}{p}}(2) = k$. However, these distributions and input errors are irrelevant to our discussion as for $|c| \leq q$, $U = p^u \geq p^q \geq p^{k-1} = \frac{1}{2p} > \frac{1}{2}$.

[4]   Diochnos and Turán in [6] gave a bound of $2q$ for $\mathcal{U}_n$. $\mathcal{U}_n$ is once again special, because $p = \frac{1}{2}$ is the *unique* member of $\mathcal{F}$ where in the shrinking phase (Fig. 1(c)), $U > \frac{1}{2} \Rightarrow U = 1 \Rightarrow u = 0$; that is, one needs to argue only about *specializations* of the target. For $p < \frac{1}{2}$, Fig. 1(b) never applies, Fig. 1(c) is again about specializations of the target, and then we can match their $2q$ bound. However, we use $3q$ throughout for uniformity in the analysis.

6.2   Short Initial Hypothesis and Medium Target

Medium targets make sense when $p \geq \frac{1}{2}$ and only when we perform *approximate* learning. Hence, the input error satisfies $\varepsilon \geq 3p^{n-1}$ *always*. Also, for a medium target, $q < |c| = q+j \leq q+\vartheta$, a hypothesis h is a best q-approximation if $m = q$. Then, $u = j \leq \vartheta \Rightarrow U = p^u = p^j \geq p^\vartheta = \frac{1}{2}$.

Thus, starting with a hypothesis h such that $U < \frac{1}{2}$, we have that $m \leq q-1$. Hence, either $|h| \leq q - 1 \Rightarrow N^+ \neq \emptyset$, or $|h| = q \Rightarrow r \geq 1$. In either case, there is at least one beneficial mutation in the neighborhood. As long as $U < \frac{1}{2}$, there can be at most q beneficial additions of variables and at most q beneficial swaps. Therefore, $2q$ generations are enough to form a hypothesis with $U \geq \frac{1}{2}$.

If $U = \frac{1}{2}$, Fig. 1(b) applies. $U = \frac{1}{2} \Rightarrow p \in \mathcal{F}$ for some k. Further, $U = \frac{1}{2} \Rightarrow p^u = \frac{1}{2} \Rightarrow u = \log_{\frac{1}{p}}(2) = k = \vartheta$. In other words, as k increases, we *drill deeper* and thus $\vartheta = k = u$. We distinguish cases.

- If $m = q$, then a best q-approximation is already formed; by the selection of tolerance this formation is stable. By Corollary 1, $\mathrm{Perf}_{\mathcal{B}_n}(h, c) > 1 - \varepsilon$. This case refers to the *longest medium* target; that is, $|c| = q + \vartheta$. For all other medium targets, $m = q$ implies $u < \vartheta$ and thus, $U = p^u > p^\vartheta = \frac{1}{2}$.
  - If $\mathcal{H} = \mathcal{C}$, as medium targets make sense only for approximate learning, by Sect. 5.1, $p < 2^{-\frac{1}{n-1}}$. If $\mathcal{H} = \mathcal{C}_{\leq q}$, then $p \in \mathcal{F}$ and $p < 2^{-\frac{1}{n}}$. To see this, note that $q = \lceil \log_{\frac{1}{p}}(\frac{3}{\varepsilon}) \rceil \Rightarrow q \geq 1$ for any $0 < \varepsilon < 2$. Hence, as $|c| = q + \vartheta \leq n$, it follows that $\vartheta \leq n - 1$ and as a consequence $p < 2^{-\frac{1}{n}}$. Note that $p = 2^{-\frac{1}{n-1}}$ can arise[5,6] under $\mathcal{U}_n$, for $\varepsilon \geq \frac{3}{2}$ and $|c| = n = 2$.
- If $m < q$, since $u = \vartheta$, we are dealing with targets such that $|c| \in \{q+1, \ldots, q+\vartheta - 1\}$. Hence, this case can arise when $p \in \mathcal{F}$ for $k \geq 2$. Since $m < q$, either $|h| = m \Rightarrow N^+ \neq \emptyset$, or $m < |h| \leq q \Rightarrow r \geq 1 \Rightarrow N^{+-} \neq \emptyset$. In either case, in one step, evolution will proceed to the case where $U > \frac{1}{2}$.
  - If $\mathcal{H} = \mathcal{C}$, then again by Sect. 5.1, $p < 2^{-\frac{1}{n}}$. If $\mathcal{H} = \mathcal{C}_{\leq q}$, then $p \leq 2^{-\frac{1}{n}}$; not even the full conjunction can achieve $U = \frac{1}{2}$ for $p > 2^{-\frac{1}{n}}$.

If $U > \frac{1}{2}$, Fig. 1(c) applies. Beneficial mutations either increase good variables with additions or swaps, or redundant bad variables are removed. However, there can be at most q removals of bad variables. Further, the set of good variables can be augmented at most q times through beneficial mutations. Thus, a best q-approximation is formed within at most $2q$ generations.

---

[5] This example reveals another aspect of our approach. There are cases where $q+\vartheta \geq n$, even when $\mathcal{H} = \mathcal{C}$. Then, our method is powerful enough to perform exact learning (there are no long targets). However, only an approximation of the target will be returned, satisfying $\mathrm{Perf}_{\mathcal{B}_n}(h, c) > 1 - \varepsilon$. On the other hand, one can improve the definitions of ϑ in Table 2 and in (11) by setting $\vartheta = \min\{n - q, \lfloor \log_{1/p}(2) \rfloor\}$; we did not do so for simplicity in the presentation.

[6] Also, p can be arbitrarily close to 1. For $k \in \mathbb{N}^*$, $p = 2^{-\frac{1}{k}} \Rightarrow \vartheta = k$. Let, $\varepsilon = \frac{3}{4} \Rightarrow q = \lceil \log_{\frac{1}{p}}(4) \rceil = 2k$. Then, for $n \geq 3k$, we look at the conjunction with size $3k$.

As a summary, in the above process m is non-decreasing. Thus, there can be at most q additions of good variables and swaps combined. Further, there can be at most $q + q = 2q$ beneficial additions or deletions of bad variables. Hence, after at most $q + 2q \leq 3q$ generations, a best q-approximation of a medium target will be formed. That formation is stable. By Corollary 1, $\mathrm{Perf}_{\mathcal{B}_n}(\mathrm{h}, \mathrm{c}) > 1 - \varepsilon$.

6.3   Short Initial Hypothesis and Long Target

$\vartheta = \lfloor \log_{\frac{1}{p}}(2) \rfloor \Rightarrow \vartheta > \log_{\frac{1}{p}}(2p)$. For long targets, $u \geq 1 + \vartheta \Rightarrow U = p^u \leq p^{1+\vartheta} < p^{\log_{1/p}(2)} = 1/2$. Thus, we have $U < \frac{1}{2}$, corresponding to Fig. 1(a). Beneficial mutations are additions of variables or swaps. As long as $|\mathrm{h}| < q$, then $N^+ \neq \emptyset$. Hence, after at most $2q$ generations a hypothesis of size q will be formed. By the selection of tolerance, the mutations in N^- are deleterious. Thus, evolution will wander among hypotheses of size precisely q. By Corollary 2, $\mathrm{Perf}_{\mathcal{B}_n}(\mathrm{h}, \mathrm{c}) > 1 - \varepsilon$.

6.4   Medium or Long Initial Hypothesis

With $\widetilde{\mathcal{O}}(\cdot)$ we ignore polylogarithmic terms; however, we do not ignore q, as q is the frontier of our search and the maximum size of the shortest explanation. As long as $|\mathrm{h}| > q$ the neighborhood is $N = N^- \cup \{\mathrm{h}\}$. Tolerance is t_u from (9); every hypothesis in the neighborhood is neutral. Thus, with probability at least $1 - \delta/4$, in $\widetilde{\mathcal{O}}(n)$ generations we arrive at a hypothesis of size q.

7   Sketch of Complexity Analysis for Evolution

Evolution in $\mathcal{C}_{\leq q}$. Evolution lasts for $3q$ generations. $|N| = \mathcal{O}(nq) \Rightarrow cnq^2$ queries are enough, for some $c > 0$. Table 1 computes $A = \min_{\neq 0}\{\mathcal{A}(u)\}$; by (8), $\mu = \min\{2p^{q+\vartheta}, A\}$. By (8), tolerance is $t = t_\ell$. Requiring $\mathcal{O}\left(\frac{1}{t^2} \cdot \ln(n/\delta)\right)$ samples per hypothesis tested, it follows by Hoeffding's bound and a union bound argument that the performance of each hypothesis in this phase is computed within $\epsilon_s = t$ of its exact value with probability at least $1 - \delta/2$.

Theorem 2. *Let $\mathcal{B}_n$ be a binomial distribution with $0 < p < 1$. Starting with a short initial hypothesis and considering hypotheses in $\mathcal{C}_{\leq q}$, the swapping algorithm, using total sample size $\widetilde{\mathcal{O}}\left(nq^2/t_\ell^2\right)$, in at most $3q$ generations, will evolve a hypothesis h such that $\mathrm{Perf}_{\mathcal{B}_n}(h, c) > 1 - \varepsilon$, with probability at least $1 - \delta/2$.*

Evolution in $\mathcal{C}_{>q}$. With a Chernoff bound argument, selecting from the neutral set when all hypotheses are present there, for $\widetilde{\mathcal{O}}(n)$ generations, then with probability at least $1 - \delta/4$, evolution will form a hypothesis of size q.

On the other hand, by (9), tolerance $t = t_u$. Requiring $\mathcal{O}\left(\frac{1}{t^2}\ln(n/\delta)\right)$ samples per hypothesis tested, with a combination of the Hoeffding bound and a union bound argument, the performance of each hypothesis is computed within $\epsilon_s = t/4$ of its exact value, with probability at least $1 - \delta/4$.

Theorem 3. *Let $\mathcal{B}_n$ be a binomial distribution with $0 < p < 1$. Starting with a long initial hypothesis and considering hypotheses in $\mathcal{C}$, the swapping algorithm, using total sample size $\tilde{O}\left(nq^2/t_\ell^2 + n^2/t_u^2\right)$, in $\tilde{O}(n)$ generations, will evolve a hypothesis h such that $Perf_{\mathcal{B}_n}(h, c) > 1 - \varepsilon$, with probability at least $1 - \delta$.*

8 Further Remarks

With a drilling technique, we examined a local search algorithm for the evolution of monotone conjunctions under binomial distributions. We identified differences between $\mathcal{H} = \mathcal{C}$ and $\mathcal{H} = \mathcal{C}_{\leq q}$ that had to do with the sample size as well as with the overall design and adaptation of the method. Also, on an instance of dimension n, using $\mathcal{H} = \mathcal{C}_{\leq q}$, we are able to cover a wider spectrum of distributions.

Our analysis assumed rational p but can be extended to any real value. We outline the extension; details will be given in the full version. For example, let $p > \frac{1}{2}$ and $p \in [l, r] \subset (2^{-\frac{1}{k}}, 2^{-\frac{1}{k+1}})$, for rational l and r. Setting $q = \lceil \log_{\frac{1}{r}}(3/\varepsilon) \rceil$ and $\theta = \lfloor \log_{\frac{1}{r}}(2) \rfloor = k$, Corollaries 1 and 2 hold for the p of the distribution. Considering $\mathcal{H} = \mathcal{C}_{\leq q}$, we need a lower bound for t_ℓ that works for all $p \in [l, r]$. Notice that $A = \min_{\neq 0}\{\mathcal{A}(u)\} = \min\left\{\left|1 - 2l^k\right|, \left|1 - 2l^{k+1}\right|, \left|1 - 2r^k\right|, \left|1 - 2r^{k+1}\right|\right\}$. Hence, μ in (8) is the minimum between A and $2p^{q+\vartheta}$. Note also that $l = r^b$ for $b \leq 1 + \frac{1}{k}$. Then, for any $p \in [l, r]$, $p^q \geq l^q = r^{bq} \geq \left(r\frac{\varepsilon}{3}\right)^b \geq \left(\frac{\varepsilon}{6}\right)^2$ and $p^\vartheta \geq l^\vartheta = r^{b\vartheta} \geq 2^{-b} \geq \frac{1}{4}$. Thus, $\mu = \min\{\frac{\varepsilon^2}{72}, A\}$. Further, $t_\ell \geq l^{q-1}\mu(1-r) \geq \frac{\varepsilon^2}{36l}\mu(1-r)$. Similar arguments can be made if $p \in \mathcal{F}$ or if $p < \frac{1}{2}$ thus treating uniformly all real values of p in an appropriate interval with rational endpoints.

Concluding, Valiant's model for evolution poses interesting questions even for concept classes that have been studied extensively in learning theory. Perhaps the most distinctive difference between evolvability on one hand and traditional optimization and EAs on the other hand, is that the evolutionary mechanism has access to fitness comparison oracles that have bounded and unbounded precision respectively. Such a distinction on comparison oracles can have independent interest, as for example in [1]. In the case of evolvability, having bounded precision on the comparisons is an artifact of sampling. By trying to understand evolution using local search when the fitness values are corrupted by noise, we have additional results that we will explore in subsequent papers. Finally, studying the method in different computation models or by restricting the parameters on real algebraic numbers might be a problem of independent interest.

Acknowledgement. The author would like to thank György Turán for fruitful discussions on earlier versions of the paper. The author would also like to thank Yanjun Qi and Elias Tsigaridas for some additional interesting discussions.

References

1. Ajtai, M., Feldman, V., Hassidim, A., Nelson, J.: Sorting and selection with imprecise comparisons. ACM Trans. Algorithms **12**(2), 19 (2016)
2. Angelino, E., Kanade, V.: Attribute-efficient evolvability of linear functions. In: ITCS, pp. 287–300 (2014)
3. Angluin, D., Laird, P.D.: Learning from noisy examples. Mach. Learn. **2**(4), 343–370 (1987)
4. Aslam, J.A., Decatur, S.E.: Specification and simulation of statistical query algorithms for efficiency and noise tolerance. J. Comput. Syst. Sci. **56**(2), 191–208 (1998)
5. Bshouty, N.H., Feldman, V.: On using extended statistical queries to avoid membership queries. J. Mach. Learn. Res. **2**, 359–395 (2002)
6. Diochnos, D.I., Turán, G.: On evolvability: the swapping algorithm, product distributions, and covariance. In: SAGA, pp. 74–88 (2009)
7. Droste, S., Jansen, T., Wegener, I.: On the analysis of the (1+1) evolutionary algorithm. Theor. Comput. Sci. **276**(1–2), 51–81 (2002)
8. Feldman, V.: Evolvability from learning algorithms. In: STOC, pp. 619–628 (2008)
9. Feldman, V.: Robustness of Evolvability. In: COLT, pp. 277–292 (2009)
10. Feldman, V.: Distribution-independent evolvability of linear threshold functions. In: COLT, pp. 253–272 (2011)
11. Feldman, V.: A complete characterization of statistical query learning with applications to evolvability. J. Comput. Syst. Sci. **78**(5), 1444–1459 (2012)
12. Goldman, S.A., Sloan, R.H.: Can PAC learning algorithms tolerate random attribute noise? Algorithmica **14**(1), 70–84 (1995)
13. Kanade, V.: Evolution with recombination. In: FOCS, pp. 837–846 (2011)
14. Kanade, V., Valiant, L.G., Vaughan, J.W.: Evolution with drifting targets. In: COLT, pp. 155–167 (2010)
15. Kearns, M.J.: Efficient noise-tolerant learning from statistical queries. In: STOC, pp. 392–401 (1993)
16. Michael, L.: Evolvability via the Fourier transform. Theor. Comput. Sci. **462**, 88–98 (2012)
17. Szörényi, B.: Characterizing statistical query learning: simplified notions and proofs. In: ALT, pp. 186–200 (2009)
18. Valiant, L.G.: A theory of the learnable. Commun. ACM **27**(11), 1134–1142 (1984)
19. Valiant, L.G.: Evolvability. In: Kučera, L., Kučera, A. (eds.) MFCS 2007. LNCS, vol. 4708, pp. 22–43. Springer, Heidelberg (2007). doi:10.1007/978-3-540-74456-6_5
20. Valiant, P.: Distribution free evolvability of polynomial functions over all convex loss functions. In: ITCS, pp. 142–148 (2012)
21. Valiant, P.: Evolvability of real functions. ACM Trans. Comput. Theor. **6**(3), 12:1–12:19 (2014)
22. Wegener, I.: Theoretical aspects of evolutionary algorithms. In: Loeckx, J. (ed.) ICALP 1974. LNCS, vol. 14, pp. 64–78. Springer, Heidelberg (2001). doi:10.1007/3-540-48224-5_6

Exact and Interactive Learning, Complexity of Teaching Models

Exact Learning of Juntas from Membership Queries

Nader H. Bshouty and Areej Costa[✉]

Department of Computer Science, Technion, 3200003 Haifa, Israel
{bshouty,areej.costa}@cs.technion.ac.il

Abstract. In this paper we study adaptive and non-adaptive exact learning of Juntas from membership queries. We use new techniques to find new bounds, narrow some of the gaps between the lower bounds and upper bounds and find new deterministic and randomized algorithms with small query and time complexities.

Some of the bounds are tight in the sense that finding better ones either gives a breakthrough result in some long-standing combinatorial open problem or needs a new technique that is beyond the existing ones.

1 Introduction

Learning from membership queries [1], has flourished due to its many applications in group testing, blood testing, chemical leak testing, chemical reactions, electric shorting detection, codes, multi-access channel communications, molecular biology, VLSI testing and AIDS screening. See many other applications in [6–8,12,13,15]. Many of the new applications raised new models and new problems and in many of those applications the function being learned can be an arbitrary function that depends on few variables. We call this class of functions d-Junta, where d is the number of relevant variables in the function. In some of the applications non-adaptive algorithms are most desirable, where in others adaptive algorithms with limited number of rounds are also useful. Algorithms with high number of rounds can also be useful given that the number of queries they ask is low. In all of the applications, one searches for an algorithm that runs in polynomial time and asks as few queries as possible. In some applications asking queries is very expensive, and therefore, even improving the query complexity by a small non-constant factor is interesting.

In this paper we study adaptive and non-adaptive exact learning of Juntas from membership queries. This problem was studied in [8–11]. In this paper, we find new bounds, tighten some of the gaps between some lower bounds and upper bounds and find new algorithms with better query and time complexities both for the deterministic and the randomized case.

Since learning one term of size d (which is a function in d-Junta) requires at least 2^d queries and asking one query requires time $O(n)$, we cannot expect to learn the class d-Junta in time less than $\Omega(2^d + n)$. We say that the class d-Junta is polynomial time learnable if there is an algorithm that learns d-Junta

© Springer International Publishing Switzerland 2016
R. Ortner et al. (Eds.): ALT 2016, LNAI 9925, pp. 115–129, 2016.
DOI: 10.1007/978-3-319-46379-7_8

in time $poly(2^d, n)$. In this paper we also consider algorithms that run in time $n^{O(d)}$, which is polynomial time for constant d, and algorithms that run in time $poly(d^d, n)$, which is polynomial time for $d = O(\log n / \log \log n)$.

1.1   Results for Non-adaptive Learning

A set $S \subseteq \{0, 1\}^n$ is called an (n, d)-universal set if for every $1 \le i_1 < i_2 < \cdots < i_d \le n$ and $\sigma \in \{0, 1\}^d$ there is an $s \in S$ such that $s_{i_j} = \sigma_j$ for all $j = 1, \ldots, d$. Damaschke, [9], shows that any set of assignments that non-adaptively learns d-Junta must be an (n, d)-universal set. This, along with the lower bound in [14, 21], gives result (1) in Table 1 for the deterministic case. It is not clear that this lower bound is also true for the randomized case. We use the mini-max technique, [20], to show that randomization cannot help reducing the lower bound. See (2) in the table.

Table 1. Results for non-adaptive algorithms

	Lower bound	Upper bound	Time $n^{O(d)}$	Time $poly(d^d, n)$	Time $poly(2^d, n)$
Deterministic					
Previous	$^{(1)}2^d \log n$	$^{(3)}d2^d \log n$ $+d^2 2^d$	$^{(6)}d^2 2^d \log n$	—	$^{(11)}2^{d+O(\log^2 d)} \log^2 n$
Ours	—	$^{(4)}d2^d \log n$	$^{(7)}d2^d \log n$	$^{(9)}d^3 2^d \log n$	$^{(12)}2^{d+O(\log^2 d)} \log n$
Randomized					
Previous	—	—	—	—	—
Ours	$^{(2)}2^d \log n$	$^{(5)}d2^d \log n$	$^{(8)}d2^d \log n$	$^{(10)}d^3 2^d \log n$	$^{(13)}d^2 2^d (\log n + \log(1/\delta))$

Damaschke introduces a graph-theoretic characterization of non-adaptive learning families, called d-wise bipartite connected families [9]. He shows that for an (n, d)-universal set of assignments S, it can non-adaptively learn d-Junta if and only if S is a d-wise bipartite connected family. He then shows, with a non-constructive probabilistic method, that there exists such a family of size $O(d2^d \log n + d^2 2^d)$. See (3) in the table. Then, he shows that a d-wise bipartite connected family of size $O(d^2 2^d \log n)$ can be constructed in $n^{O(d)}$ time [9]. This, along with his deterministic learning algorithm presented in [9] gives result (6) in the table. We further investigate the d-wise bipartite connected families and show that there exists one of size $O(d2^d \log n)$. We then use the technique in [16] to construct one of such size in time $n^{O(d)}$. This gives results (4), (5), (7) and (8), where the results for the randomized algorithms follow from the corresponding deterministic algorithms. We then use the reduction of Abasi et. al in [2] to give a non-adaptive algorithm that runs in $poly(d^d, n)$ time and asks $O(d^3 2^d \log n)$ queries. This is result (9) in the table. Then, result (10) for the randomized case follows.

We also introduce a new simple non-adaptive learning algorithm and apply the same reduction to this algorithm to get result (12) in the table. Result (11) follows from a polynomial time learning algorithm for d-Junta given by Damaschke in [10], which asks $2^{d+O(\log^2 d)} \log^2 n$ queries. Finally, we give a new Monte Carlo randomized polynomial time algorithm that learns d-Junta with $O((d2^d \log n + d^2 2^d) \log(1/\delta))$ queries. Then we present a new reduction for randomized algorithms and apply it to this algorithm to get result (13) in the table.

The significant improvements over the previous results (see the table) are results (2), (4), (9), (12) and (13).

We note here that any improvement in the upper bound would be a breakthrough result since it would give a new result for the size of an (n, d)-universal set, a longstanding open problem. We also believe that to improve the lower bounds one needs a new technique that is beyond the existing ones.

1.2 Results for Adaptive Learning

We summarize here only our new main results for adaptive learning. A more comprehensive presentation of previous and new results, comparison between them (including a summarizing table) and proofs will be introduced in the full version of this paper. The full paper will also include some improvements that were proposed by one of the anonymous reviewers to some of the algorithms.

For deterministic learning we give a two-round algorithm that runs in time $poly(d^d, n)$ and asks $O(d^3 2^d \log n)$ queries. For the randomized case, we present a lower bound of $\Omega(d \log n + 2^d)$ queries. We also give three new Monte Carlo algorithms. The three of them run in $poly(2^d, n, \log 1/\delta)$ time. The first one uses $O(d \log n + d2^d \log d/\delta)$ queries and runs in $O(d \log n)$ rounds. The second one uses $O(d^2 2^d (\log n + \log 1/\delta))$ queries and runs in d rounds. The third algorithms uses $O(d \log n + d^3 2^d \log 1/\delta)$ queries and runs in two rounds. For $poly(d^d, n, \log 1/\delta)$ time randomized algorithms, we refine a previous result of Damaschke and give a two-round algorithm with a query complexity of $O(d \log n + (\log d)d2^d \log 1/\delta)$.

2 Definitions and Preliminary Results

In this section we give some definitions and preliminary results that will be used throughout the paper.

Let n be an integer. We denote $[n] = \{1, \ldots, n\}$. Consider the set of *assignments*, also called *membership queries* or *queries*, $\{0, 1\}^n$. A function f is said to be a *boolean function* on n variables $x = (x_1, \ldots, x_n)$ if $f : \{0, 1\}^n \to \{0, 1\}$. For an assignment $a \in \{0, 1\}^n$, $i \in [n]$ and $\xi \in \{0, 1\}$ we denote by $a|_{x_i \leftarrow \xi}$ the assignment $(a_1, \ldots, a_{i-1}, \xi, a_{i+1}, \ldots, a_n)$. We say that the variable x_i is *relevant* in f if there is an assignment $a = (a_1, \ldots, a_n)$ such that $f(a|_{x_i \leftarrow 0}) \neq f(a|_{x_i \leftarrow 1})$. We say that the variable x_i is *irrelevant* in f if it is not relevant in f.

Given a boolean function f on n variables $x = \{x_1, \ldots, x_n\}$ and an assignment $a \in \{0, 1\}^n$ on the variables, we say that x_i, for $1 \leq i \leq n$, is *sensitive in* f *w.r.t.* a if $f(a|_{x_i \leftarrow 0}) \neq f(a|_{x_i \leftarrow 1})$.

For a boolean function on n variables f, a variable x_i and a value $\xi \in \{0,1\}$ we denote by $f|_{x_i \leftarrow \xi}$ the boolean function on n variables $g(x_1, \ldots, x_n) = f(x_1, \ldots, x_{i-1}, \xi, x_{i+1}, \ldots, x_n)$. In addition, for a boolean variable $x \in \{0,1\}$ we denote $x^1 := x$ and $x^0 := \bar{x}$, where $\bar{x}$ is the negation of x.

For a class of boolean functions C we say that a set of assignment $A \subseteq \{0,1\}^n$ is an *equivalent set for* C if for every two distinct functions $f, g \in C$ there is an $a \in A$ such that $f(a) \neq g(a)$. Obviously, an equivalent set A for C can be used to non-adaptively learn C. Just ask all the the queries in A and find the function $f \in C$ that is consistent with all the answers. By the definition of equivalent set this function is unique.

For integers n and $d \leq n$ we define the set d-Junta as the set of all functions $f : \{0,1\}^n \to \{0,1\}$ with at most d relevant variables.

A set $S \subseteq \{0,1\}^n$ is called a *d-wise independent set* if for every $1 \leq i_1 < i_2 < \cdots < i_d \leq n$ and every $\xi_1, \ldots, \xi_d \in \{0,1\}$, for a random uniform $x = (x_1, \ldots, x_n) \in S$ we have $\Pr[x_{i_1} = \xi_1, \ldots, x_{i_d} = \xi_d] = 1/2^d$. Alon et al. show:

Lemma 1. [3] *There is a construction of a d-wise independent set of size $O((2n)^{\lfloor d/2 \rfloor})$.*

2.1 Universal Sets and d-Wise Bipartite Connected Families

A set $S \subseteq \{0,1\}^n$ is called an *(n,d)-universal set* if for every $1 \leq i_1 < i_2 < \cdots < i_d \leq n$ and $\sigma \in \{0,1\}^d$ there is an $s \in S$ such that $s_{i_j} = \sigma_j$ for all $j = 1, \ldots, d$. Denote by $U(n,d)$ the minimum size of an (n,d)-universal set. It is known, [14,21], that

$$U(n,d) = \Omega(2^d \log n), \quad U(n,d) = O(d2^d \log n). \tag{1}$$

The following result is a folklore result. We give the proof for completeness

Lemma 2. *Let be $m = O(2^d(d \log n + \log(1/\delta)))$. Let be $S = \{s^{(1)}, \ldots, s^{(m)}\}$ where each $s^{(i)}$ is uniformly independently chosen from $\{0,1\}^n$. With probability at least $1 - \delta$ the set S is an (n,d)-universal set.*

Proof. By the union bound and since $s^{(i)}$ are chosen uniformly independently,

$$\Pr[S \text{ is not } (n,d)\text{-US}] = \Pr[(\exists(i_1, \ldots, i_d))(\exists a \in \{0,1\}^d)(\forall j)\ s_{i_1}^{(j)}, \ldots, s_{i_d}^{(j)} \neq a]$$

$$\leq \binom{n}{d} 2^d \left(1 - \frac{1}{2^d}\right)^m \leq \delta.$$

$\square$

This also implies the upper bound of $O(d2^d \log n)$ in (1). Different deterministic constructions for (n,d)-universal sets are useful especially for adaptive learning. The best known constructions are stated in the following:

Lemma 3. *[16,17] There is a deterministic construction for an (n,d)-universal set of size s that runs in time T where*

1. $T = poly(2^d, n)$ and $s = 2^{d+O(\log^2 d)} \log n$.
2. $T = poly(d^d, n)$ and $s = O(d^3 2^d \log n)$. In particular, $T = poly(n)$ for $d = O(\log n / \log \log n)$.
3. $T = n^{O(d)}$ and $s = O(d 2^d \log n)$. In particular, $T = poly(n)$ for $d = O(1)$.

Let $A \subseteq \{0,1\}^n$ be a set of assignments. For non-negative integers d_1, d_2 such that $d = d_1 + d_2$ and for $i = (i_1, \ldots, i_{d_2}), j = (j_1, \ldots, j_{d_1}), k = (k_1, \ldots, k_{d_2})$, with entries that are distinct elements in $[n]$ and $z \in \{0,1\}^{d_1}$, we define a bipartite graph $B := B(i,j,k,z,A)$ as follows. The set of vertices of the left side of the graph is $V_L(B) := \{0,1\}^{d_2} \times \{L\}$ and the set of vertices of the right side of the graph is $V_R(B) := \{0,1\}^{d_2} \times \{R\}$. For $a', a'' \in \{0,1\}^{d_2}$, $\{(a', L), (a'', R)\}$ is an edge in B if and only if there is an assignment $a \in A$ such that $(a_{i_1}, \ldots, a_{i_{d_2}}) = a'$, $(a_{k_1}, \ldots, a_{k_{d_2}}) = a''$ and $(a_{j_1}, \ldots, a_{j_{d_1}}) = z$.

We say that A is a d-wise bipartite connected family if for all d_1, d_2 such that $d = d_1 + d_2$, and for all $i = (i_1, \ldots, i_{d_2}), j = (j_1, \ldots, j_{d_1}), k = (k_1, \ldots, k_{d_2})$ and $z \in \{0,1\}^{d_1}$ as described above, the graph $B(i,j,k,z,A)$ is connected. That is, there is a path from any vertex in the graph to any other vertex. Obviously, any d-wise bipartite connected family is an (n,d)-universal set. Just take $d_1 = 0$ and $d_2 = d$.

In [9], Damaschke proves:

Lemma 4. [9] A is an equivalent set for d-Junta if and only if A is a d-wise bipartite connected family.

3 Deterministic Non-adaptive Algorithms

In this section we study deterministic non-adaptive learning of d-Junta.

3.1 Lower and Upper Bound

For completeness sake, we give a sketch of the proof of the lower bound of $\Omega(2^d \log n)$ on the number of queries in any deterministic adaptive algorithm. This implies the same lower bound for any deterministic non-adaptive algorithm.

The idea of the lower bound is very simple. If the set of asked assignments A is not an (n,d)-universal set and the adversary answers 0 for all the membership queries in A, the learner can't learn the function. This is because, if A is not an (n,d)-universal set, then there are $1 \leq i_1 < i_2 < \cdots < i_d \leq n$ and $\xi_1, \ldots, \xi_d$ such that no assignment $a \in A$ satisfies $(a_{i_1}, \ldots, a_{i_d}) = (\xi_1, \ldots, \xi_d)$. Then, the learner can't distinguish between the zero function and the term $x_{i_1}^{\xi_1} \cdots x_{i_d}^{\xi_d}$. This is because $x_{i_1}^{\xi_1} \cdots x_{i_d}^{\xi_d}$ is also zero on all the assignments of A. Therefore, the learner must ask at least $U(n,d)$ queries which is $\Omega(2^d \log n)$ by (1).

As for the upper bound, Damaschke shows in [9] that a d-wise bipartite connected family is an equivalent set for d-Junta and therefore this family is enough for non-adaptive learning. He then shows that there exists a d-wise bipartite connected family of size $O(d 2^d \log n + d^2 2^d)$. In this section we construct such one of size $O(d 2^d \log n)$. In particular, we have:

Theorem 1. *There is a deterministic non-adaptive algorithm that learns* d-*Junta with* $O(d2^d \log n)$ *queries.*

Proof. We give an algorithmic construction of a d-wise bipartite connected family of size $O(d2^d \log n)$. To prove the result we start with a definition.

For every d_1 and d_2 where $d = d_1 + d_2$ and every $i = (i_1, \ldots, i_{d_2})$, $j = (j_1, \ldots, j_{d_1})$, $k = (k_1, \ldots, k_{d_2})$, with entries that are distinct elements in $[n]$, every $z \in \{0,1\}^{d_1}$ and every set of assignments $A \subseteq \{0,1\}^n$, we define the function $X_{i,j,k,z}$ at A as follows. $X_{i,j,k,z}(A) = t - 1$ where t is the number of connected components in $B(i,j,k,z,A)$. Obviously, if $X_{i,j,k,z}(A) = 0$ then $B(i,j,k,z,A)$ is connected. Consider the function $X(A) = \sum_{i,j,k,z} X_{i,j,k,z}(A)$. The sum here is over all possible d_1, d_2, i, j, k and z as described above. Notice that if $X(A) = 0$ then A is a d-wise bipartite connected family.

We construct A iteratively. At the beginning $A = \varnothing$ and each $B(i,j,k,z,A)$ has 2^{d_2+1} connected components. Therefore, we first have

$$X(\varnothing) = \sum_{d_1+d_2=d} \binom{n}{d_2\ d_2\ d_1\ n-d_1-2d_2} 2^{d_1}(2^{d_2+1}-1) \le (2n)^{2d}.$$

We show that for every $A \subset \{0,1\}^n$ there is an assignment $a \in \{0,1\}^n$ such that

$$X(A \cup \{a\}) \le X(A)\left(1 - 1/2^{d+1}\right). \tag{2}$$

This implies that there is a set A' of size $t = 2d2^{d+1}\ln(2n)$ such that

$$X(A') \le X(\varnothing)\left(1 - 1/2^{d+1}\right)^t \le (2n)^{2d}\left(1 - 1/2^{d+1}\right)^t < (2n)^{2d}e^{-2d\ln(2n)} = 1.$$

Since $X(A')$ is an integer number we get $X(A') = 0$, which implies that A' is a d-wise bipartite connected family of size $t = O(d2^d \log n)$.

We now prove (2). Consider some i, j, k, z, A. Suppose that the number of connected components in $B := B(i,j,k,z,A)$ is t and therefore $X_{i,j,k,z}(A) = t - 1$. Let $C_1, C_2, \ldots, C_t$ be the connected components of B and let s_i and r_i be the number of vertices of the component C_i in $V_L(B)$ and $V_R(B)$ respectively. Consider a random uniform assignment $a \in \{0,1\}^n$. If $(a_{j_1}, \ldots, a_{j_{d_1}}) = z$ then $B(i,j,k,z,A \cup \{a\})$ is the bipartite graph $B(i,j,k,z,A)$ with an addition of a uniform random edge. Therefore the probability that after adding a to A the number of connected components in B reduces by 1 is equal to the probability that $(a_{j_1}, \ldots, a_{j_{d_1}}) = z$ and a uniform random edge in B connects two distinct connected components. This probability is equal to

$$\frac{1}{2^{d_1}}\frac{\sum_{i\neq j} s_i r_j}{2^{2d_2}} = \frac{\left(\sum_i s_i\right)\left(\sum_j r_j\right) - \sum_i s_i r_i}{2^{d_1}2^{2d_2}} = \frac{1}{2^{d_1}} - \frac{\sum_i s_i r_i}{2^{d_1}2^{2d_2}} \tag{3}$$

$$\ge \frac{1}{2^{d_1}} - \frac{2^{2d_2} - (t-1)2^{d_2-1}}{2^{d_1}2^{2d_2}} = \frac{t-1}{2^{d+1}}. \tag{4}$$

Equality (3) is true because $\sum_i s_i = \sum_i r_i = 2^{d_2}$. The inequality (4) is proved later. Therefore $\mathbf{E}_a[X_{i,j,k,z}(A \cup \{a\})] \le (t-1)/2^{d+1} = X_{i,j,k,z}(A)\left(1 - 1/2^{d+1}\right)$.

Since the expectation of a sum is the sum of the expectations, we have $\mathbf{E}_a[X(A \cup \{a\})] \leq X(A)\left(1 - 1/2^{d+1}\right)$. Therefore, for every set A there exists an a such that $X(A \cup \{a\}) \leq X(A)\left(1 - 1/2^{d+1}\right)$.

It remains to prove (4). That is,

$$\max_{r_i, s_i} \sum_{i=1}^{t} s_i r_i \leq 2^{2d_2} - (t-1)2^{d_2-1}. \tag{5}$$

First notice that since the graph B is a bipartite graph we have that $s_i r_i = 0$ if and only if either $s_i = 0$ and $r_i = 1$ or $s_i = 1$ and $r_i = 0$. We first claim that the maximum value in (5) occurs when $r_i s_i = 0$ for all i except for one. If, on the contrary, the maximum value occurs where $s_{i_1} r_{i_1} \neq 0$ and $s_{i_2} r_{i_2} \neq 0$ for some $i_1 \neq i_2$ then by replacing s_{i_1}, r_{i_1} by $0, 1$ and s_{i_2}, r_{i_2} by $s_{i_1} + s_{i_2}, r_{i_1} + r_{i_2} - 1$ we get a larger value and therefore we get a contradiction. Therefore we may assume w.l.o.g that $s_i r_i = 0$ for all $i = 1, \ldots, t-1$, $r_t = 2^{d_2} - t_1$ and $s_t = 2^{d_2} - t_2$ for some $t_1 + t_2 = t - 1$. Then, since $t \leq 2^{d_2+1}$,

$$\max_{r_i, s_i} \sum_{i=1}^{t} s_i r_i = \max_{t_1+t_2=t-1} (2^{d_2} - t_1)(2^{d_2} - t_2) = 2^{2d_2} - (t-1)2^{d_2} + \max_{t_1+t_2=t-1} t_1 t_2$$

$$\leq 2^{2d_2} - (t-1)2^{d_2} + (t-1)^2/4 \leq 2^{2d_2} - (t-1)2^{d_2-1}.$$

$\square$

3.2 Polynomial Time Algorithms

In this section we give three polynomial time algorithms for non-adaptive learning of d-Junta. The first algorithm asks $O(d2^d \log n)$ queries and runs in time $n^{O(d)}$ which is polynomial for constant d. This improves the query complexity $O(d^2 2^d \log n)$ of Damaschke in [9]. The second algorithm asks $O(d^3 2^d \log n)$ queries and runs in time $poly(d^d, n)$ which is polynomial for $d = O(\log n / \log \log n)$. The third algorithm asks $2^{d+O(\log^2 d)} \log n$ queries and runs in polynomial time. This improves the query complexity $2^{d+O(\log^2 d)} \log^2 n$ of Damaschke in [9].

We now present the first algorithm. In the next lemma we show how to construct a d-wise bipartite connected family of size $O(d2^d \log n)$ in time $n^{O(d)}$. We construct a d-wise bipartite connected family A and non-adaptively ask all the queries in A. Then, for every d variables $x_{i_1}, \ldots, x_{i_d}$ we construct a set $M_{i_1, \ldots, i_d} = \{(a_{i_1}, \ldots, a_{i_d}, f(a)) \mid a \in A\}$. We now look for a function $g \in d$-Junta that is consistent with all the answers of the queries in A. If $M_{i_1, \ldots, i_d}$ contains two elements $(a', 0)$ and $(a', 1)$ for some $a' \in \{0, 1\}^d$ then no consistent function exists on those variables. Otherwise, there is a consistent function g and since A is an equivalent set, g is unique and is the target function. After finding the target function, we can then find the relevant variables in g from its truth table if needed.

This algorithm runs in time $n^{O(d)}$. We now show that the construction time of the d-wise bipartite connected family A is $n^{O(d)}$.

In the previous subsection we showed in Theorem 1 an algorithmic construction of a d-wise bipartite connected family of size $O(d2^d \log n)$. We now show that this construction can be performed in $n^{O(d)}$ time.

Lemma 5. *There is an algorithm that runs in time $n^{O(d)}$ and constructs a d-wise bipartite connected family of size $O(d2^d \log n)$.*

Proof. Let $X(A)$ and $X_{i,j,k,z}(A)$ be as in the proof of Theorem 1. $X_{i,j,k,z}(A)$ depends only on $2d_2 + d_1$ entries of the assignments of A. Therefore, the proof of Theorem 1 remains true if the new assignment a is chosen from a $(2d_2 + d_1)$-wise independent set S. By Lemma 1, such set exists and is of size $n^{O(2d_2 + d_1)} = n^{O(d)}$.

Since the number of iterations and the number of random variables $X_{i,j,k,z}(A)$ in the algorithm is at most $(2n)^{2d}$ and each $X_{i,j,k,z}(A)$ can be computed in time $poly(2^{d_2}, |A|) = poly(2^d \log n)$, the result follows. □

Remark. We note here that instead of using a $(2d_2 + d_1)$-wise independent set, one can use a $(1/2^{O(d)})$-almost $(2d_2 + d_1)$-wise independent set of size $poly(2^d, \log n)$, [5]. This is a set of assignments $S \subseteq \{0,1\}^n$ such that for every $1 \le i_1 < i_2 < \cdots < i_d \le n$ and every $B \subseteq \{0,1\}^n$, for a random uniform $x = (x_1, \ldots, x_n) \in S$ we have $|B|/2^d - 1/2^{O(d)} \le \Pr[(x_{i_1}, \ldots, x_{i_d}) \in B] \le |B|/2^d + 1/2^{O(d)}$. The algorithm still needs time $n^{O(d)}$ for computing $X(A)$.

Another approach is the conditional probability method [20]. It follows from the proof of Theorem 1 that for $t = d2^{d+2} \ln(2n)$ i.i.d. random uniform assignments $A = \{a^{(1)}, \ldots, a^{(t)}\}$ we have $\mathbf{E}[X(A)] < 1$. We now construct the bits of $a^{(1)}, \ldots, a^{(t)}$ one at a time while maintaining the property $\mathbf{E}[X(A)|$ *already fixed bits*$] < 1$. At the end all the bits are fixed, say $A = A_0$, and then $X(A_0) = \mathbf{E}[X(A)|A_0] < 1$ which implies (you can see why in the proof of Theorem 1) that A_0 is a d-wise bipartite connected family of size $t = d2^{d+2} \ln(2n)$. In this approach also the algorithm still needs time $n^{O(d)}$ for computing $X(A)$.

Lemma 5 implies:

Theorem 2. *There is a deterministic non-adaptive algorithm that learns d-Junta in time $n^{O(d)}$ with $O(d2^d \log n)$ queries.*

For introducing the second algorithm, we start with presenting a result from [2]. A class of boolean functions C is said to be *closed under variable projection* if for every $f \in C$ and every function $\phi : [n] \rightarrow [n]$ we have $f(x_{\phi(1)}, \ldots, x_{\phi(n)}) \in C$. Obviously, d-Junta is closed under variable projection.

For the second algorithm we apply the reduction described in [2] to the above algorithm. We first give the reduction:

Lemma 6. [2] *Let C be a class of boolean functions that is closed under variable projection. If C is non-adaptively learnable in time $T(n)$ with $Q(n)$ membership queries, then C is non-adaptively learnable in time $O(qd^2 n \log n + d^2 \log n(T(q)$ $+Q(q)n)/\log(q/(d+1)^2))$ with $O\left(d^2 Q(q) \log n / \log(q/(d+1)^2)\right)$ membership queries, where d is an upper bound on the number of relevant variables in $f \in C$ and q is any integer such that $q \ge 2(d+1)^2$.*

We now prove:

Theorem 3. *There is a deterministic non-adaptive algorithm that learns d-Junta in time $poly(d^d, n)$ with $O(d^3 2^d \log n)$ queries.*

Proof. We apply the reduction in Lemma 6 on the result from Theorem 2 with $q = 2d(d+1)^2$, $Q(n) = O(d 2^d \log n)$ and $T(n) = n^{O(d)}$. The time complexity is

$$O(2d(d+1)^2 d^2 n \log n + (d^2 \log n / \log d) \cdot ((2d(d+1)^2)^{O(d)} + d 2^d \log (2d(d+1)^2)n))$$

which is $poly(d^d, n)$, and the query complexity is

$$O(d^2 d 2^d \log (2d(d+1)^2) \log n / \log d) = O(d^3 2^d \log n). \qquad \square$$

We now present the third algorithm. We first give a simple algorithm that runs in polynomial time and uses $2^{d+O(\log^2 d)} n \log n$ queries, and then we use the reduction in Lemma 6.

The algorithm first constructs an (n, d)-universal set U. Then, the algorithm replaces each assignment $a \in U$ by a block of $n + 1$ assignments in the following way: it keeps the original assignment. In addition, for each index $1 \leq i \leq n$, it adds the assignment $a|_{a_i \leftarrow \bar{a}_i}$. Denote the new set of assignments by U'. After asking U', we can find the set of relevant variables by comparing the value of the target function f on the first assignment in each block with the value of f on each one of the other assignments in the block. Since U is a universal set, we can now find the function.

Now, we use a polynomial time construction of an (n, d)-universal set of size $2^{d+O(\log^2 d)} \log n$, as described in Lemma 3, and apply the reduction in Lemma 6 to this algorithm for $q = 2(d+1)^2$. We get a new non-adaptive algorithm that runs in polynomial time and asks $2^{d+O(\log^2 d)} \log n$ queries.

We now give a formal proof:

Theorem 4. *There is a deterministic non-adaptive algorithm that learns d-Junta in polynomial time with $2^{d+O(\log^2 d)} \log n$ queries.*

Proof. Consider the above algorithm. Let f be the target function and let $x_{i_1}, \ldots, x_{i_{d'}}, d' \leq d$, be the relevant variables. Then, we can write $f = g(x_{i_1}, \ldots, x_{i_{d'}})$. For any $1 \leq j \leq d'$, since x_{i_j} is a relevant variable, there is an assignment $b \in \{0,1\}^{d'}$ such that $g(b) \neq g(b|_{b_{i_j} \leftarrow \bar{b}_{i_j}})$. Since U is an (n, d)-universal set, there is an assignment $a \in U$ such that $(a_{i_1}, \ldots, a_{i_{d'}}) = (b_{i_1}, \ldots, b_{i_{d'}})$. Therefore $f(a|_{a_{i_j} \leftarrow \bar{a}_{i_j}}) = g(b|_{b_{i_j} \leftarrow \bar{b}_{i_j}}) \neq g(b) = f(a)$. This shows that the above algorithm can discover all the relevant variables after asking the set of assignments U'.

Since U is an (n, d)-universal set, by looking at the entries of the relevant variables in U we can find all the possible assignments for $x_{i_1}, \ldots, x_{i_{d'}}$. Therefore, g, and consequently f, can be uniquely determined.

This algorithm asks $2^{d+O(\log^2 d)} n \log n$ queries and runs in polynomial time. Finally, we use the reduction in Lemma 3 with $q = 2(d+1)^2$, $T(n) = poly(2^d, n)$ and $Q(n) = 2^{d+O(\log^2 d)} n \log n$, and this completes the proof. $\qquad \square$

4   Randomized Non-adaptive Algorithms

In this section we study randomized non-adaptive learning of d-Junta.

4.1   Lower Bound

The lower bound for deterministic algorithms does not imply the same lower bound for randomized algorithms. To prove a lower bound for randomized non-adaptive algorithms we use the minimax technique. We prove:

Theorem 5. *Let $d < n/2$. Any Monte Carlo randomized non-adaptive algorithm for learning d-Junta must ask at least $\Omega(2^d \log n)$ membership queries.*

Proof. Let $\mathcal{A}(s, Q_f)$ be a Monte Carlo randomized non-adaptive algorithm that learns d-Junta with success probability at least $1/2$, where $s \in \{0,1\}^*$ is the random seed and Q_f is the membership oracle to the target f. Since $\mathcal{A}(s, Q_f)$ is Monte Carlo it stops after some time T and therefore we may assume that $s \in \{0,1\}^T$. Consider the random variable $X(s, f) \in \{0,1\}$ that is equal to 1 if and only if the algorithm $\mathcal{A}(s, Q_f)$ returns f. Then, $\mathbf{E}_s[X(s,f)] \geq 1/2$ for all f.

Consider the set of functions $F = \{f_{\xi,i} := x_1^{\xi_1} \cdots x_{d-1}^{\xi_{d-1}} x_i \mid \xi_1, \ldots, \xi_{d-1} \in \{0,1\}, n \geq i \geq d\}$ and the uniform distribution U_F over F. Then

$$\max_s \mathbf{E}_{U_F}[X(s,f)] \geq \mathbf{E}_s\left[\mathbf{E}_{U_F}[X(s,f)]\right] = \mathbf{E}_{U_F}\left[\mathbf{E}_s[X(s,f)]\right] \geq \frac{1}{2}.$$

Consider any seed s' that maximizes $\mathbf{E}_{U_F}[X(s,f)]$. Then $\mathbf{E}_{U_F}[X(s',f)] \geq 1/2$. Let $A_{s'} = \{a^{(1)}, \ldots, a^{(m_{s'})}\}$ be the queries asked by the algorithm when it uses the seed s'. Note that since the algorithm is non-adaptive, $m_{s'}$ is independent of f. Since the query complexity of a Monte Carlo algorithm is the worst case complexity over all s and f, we have that $m_{s'}$ is a lower bound for the query complexity. So it is enough to show that $m_{s'} = \Omega(2^d \log n)$.

Define for every vector $\xi = (\xi_1, \ldots, \xi_{d-1}) \in \{0,1\}^{d-1}$ the subset $A_{s'}(\xi) = \{a \in A_{s'} \mid (a_1, \ldots, a_{d-1}) = \xi\} \subseteq A_{s'}$. Notice that $\{A_{s'}(\xi)\}_\xi$ are disjoint sets. Suppose that at least $3/4$ fraction of ξ satisfy $|A_{s'}(\xi)| \leq \log(n - d + 1) - 3$. Then, for a random uniform $f_{\xi',i'} \in F$ (and therefore, random uniform ξ'), with probability at least $3/4$ we have $|A_{s'}(\xi')| \leq \log(n - d + 1) - 3$. For any other assignment $a \in A_{s'} \backslash A_{s'}(\xi')$ we have $f_{\xi',i'}(a) = 0$ so no information about i' can be obtained from these assignments. If $|A_{s'}(\xi')| \leq \log(n - d + 1) - 3$ then there are at most $(n-d+1)/8$ distinct values that any $f \in \{f_{\xi',j}\}_j$ can take on $A_{s'}(\xi')$ and therefore the probability to find i' is at most $1/4$. Therefore, if at least $3/4$ fraction of ξ satisfy $|A_{s'}(\xi)| \leq \log(n-d+1)-3$, then the probability of success, $\mathbf{E}_{U_F}[X(s',f)]$, is at most $1 - (3/4)^2 = 7/16 < 1/2$. This gives a contradiction. Therefore for at least $1/4$ fraction of ξ we have $|A_{s'}(\xi)| > \log(n-d+1)-3$. Then $m_{s'} = |A_{s'}| = \sum_{\xi \in \{0,1\}^{d-1}} |A_{s'}(\xi)| \geq (2^{d-1}/4)(\log(n-d+1)-3) = \Omega(2^d \log n)$. $\square$

4.2   Upper Bound and Polynomial Time Algorithms

If randomization is allowed, for some cases the performance of our deterministic non-adaptive algorithms is satisfying when comparing with algorithms that take advantage of the randomization. This applies for $n^{O(d)}$ time algorithms, where the algorithm from Lemma 5 gives good results. This algorithm provides also the upper bound of $O(d2^d \log n)$ queries for randomized non-adaptive learning. In addition, for $poly(d^d, n)$ time algorithms, we can apply the algorithm from Theorem 3 that asks $O(d^3 2^d \log n)$ queries.

But, this is not the case for $poly(2^d, n)$ time algorithms. In this case we can improve over the deterministic result for certain values of the failure probability δ. We next present a new Monte Carlo non-adaptive algorithm that runs in $poly(2^d, \log n, \log(1/\delta))$ time.

We first prove:

Lemma 7. *Let be $1 \le i_1 \le i_2 \le \cdots \le i_d \le n$ and $B \subseteq \{0,1\}^d$. If we randomly uniformly choose $2^d(\ln|B| + \ln(1/\delta))$ assignments $A \subseteq \{0,1\}^n$, then with probability at least $1 - \delta$ we have: for every $b \in B$ there is an $a \in A$ such that $(a_{i_1}, \ldots, a_{i_d}) = b$.*

Proof. The probability of failure is at most $|B|(1 - 2^{-d})^{|A|} \le |B|e^{-|A|/2^d} = \delta$. $\square$

We say that the variable x_i *is sensitive in f with respect to* an assignment a if $f(a|_{x_i \leftarrow 0}) \ne f(a|_{x_i \leftarrow 1})$. Obviously, if x_i is relevant in f then there is an assignment a where x_i is sensitive in f with respect to a. If x_i is irrelevant then x_i is not sensitive in f with respect to any assignment.

We now prove:

Lemma 8. *Let f be a d-Junta function. Let a be an assignment that x_j is sensitive in f with respect to, and let x_i be an irrelevant variable in f. Let b be a random assignment where each entry $b_\ell \in \{0,1\}$ is independently chosen to be 1 with probability $1/(3d)$. Then $\Pr_b[f(a+b) = f(a) \text{ and } b_i = 1] \ge 0.2/d$, and $\Pr_b[f(a+b) = f(a) \text{ and } b_j = 1] \le 0.1/d$.*

Proof. If $b_\ell = 0$ for all the relevant variables x_ℓ of f then $f(a+b) = f(a)$. Therefore $\Pr_b[f(a+b) = f(a) \text{ and } b_i = 1]$ is greater or equal to the probability that $b_\ell = 0$ for all the relevant variables x_ℓ of f and $b_i = 1$. This probability is at least $(1 - 1/3d)^d /3d \ge 0.2/d$.

If x_j is sensitive in f with respect to a then the probability that $f(a+b) = f(a)$ when $b_j = 1$ is less or equal than the probability that $b_\ell = 1$ for some other relevant variable x_ℓ of f. Therefore $\Pr_b[f(a+b) = f(a) \text{ and } b_j = 1] = \Pr[b_j = 1]\Pr[f(a+b) = f(a)|b_j = 1] \le 1/3d \cdot \left(1 - (1 - 1/3d)^{d-1}\right) \le 0.1/d$. $\square$

From Chernoff bound it follows that:

Lemma 9. *Let f be a d-Junta function. Let a be any assignment. There is an algorithm that asks $O(d(\log n + \log 1/\delta))$ membership queries and with probability at least $1 - \delta$ finds all the sensitive variables in f with respect to a (and maybe other relevant variables of f as well).*

Now we give the algorithm. We first choose $t = O(2^d(\log d + \log(1/\delta)))$ random uniform assignments A. To find the relevant variables we need for each one an assignment that the variable is sensitive in f with respect to it. Therefore, we need at most d such assignments. By Lemma 7, with probability at least $1 - (\delta/3)$, for every relevant variable in f there is an assignment a in A such that this variable is sensitive in f with respect to it. Now, by Lemma 9, for each $a \in A$ there is an algorithm that asks $O(d(\log n + \log(t/\delta)))$ membership queries and with probability at least $1 - (\delta/(3t))$ finds all the variables that are sensitive to it. Therefore, there is an algorithm that asks $O(dt(\log n + \log(t/\delta)))$ membership queries and with probability at least $1 - (\delta/3)$ finds every variable that is sensitive in f with respect to some assignment in A. This gives all the relevant variables of f. Now again by Lemma 7, with another $O(2^d(d + \log(1/\delta)))$ assignments we can find, with probability at least $1 - (\delta/3)$, the value of the function in all the possible assignments for the relevant variables. At the end, we can output the set of relevant variables and a representation of the target function as a truth table.

This algorithm runs in time $poly(2^d, n, \log(1/\delta))$ and asks

$$O(2^d(\log d + \log(1/\delta)) \cdot d(\log n + \log(1/\delta) + d) + 2^d(d + \log(1/\delta)))$$

$$= O(d2^d(\log d + \log(1/\delta)) \log n + d2^d(d + \log(1/\delta))(\log d + \log(1/\delta))).$$

membership queries. For $\delta = 1/d$ we have the complexity $O((\log d)d2^d(\log n + d))$. We can repeat this algorithm $O(\log(1/\delta)/\log d)$ times (non-adaptively) to get a success probability of at least $1 - \delta$ and a query complexity of $O((d2^d \log n + d^2 2^d) \log(1/\delta))$.

Notice that if $\delta = 1/poly(n)$ then the query complexity becomes quadratic in $\log n$. In the next section we solve this problem by giving a reduction that changes the query complexity to $O(d^2 2^d(\log n + \log(1/\delta)))$.

4.3 A Reduction for Randomized Non-adaptive Algorithms

In this section we give a reduction for randomized non-adaptive algorithms. Using this reduction we prove:

Theorem 6. *There is a Monte Carlo randomized non-adaptive algorithm for learning d-Junta in $poly(2^d, n, \log(1/\delta))$ time that asks $O(d^2 2^d(\log n + \log(1/\delta)))$ membership queries.*

This result improves over the result from the previous subsection for values of δ small enough.

We start with some definitions. Let C be a class of functions and let H be a class of representations of functions. We say that C is a *subclass* of H if for every function f in C there is at least one representation h in H. We say that C is *non-adaptively learnable from H* if there is a non-adaptive learning algorithm that for every function $f \in C$ outputs a function $h \in H$ that is equivalent to the target function f. We say that C is *closed under distinct variable projection* if for any function $f \in C$ and any permutation $\phi : [n] \to [n]$ we have $f(x_{\phi(1)}, \ldots, x_{\phi(n)}) \in C$. We now prove (Fig. 1):

Algorithm reduction

$\mathcal{A}(n,d,\delta)$ is a non-adaptive learning algorithm for C from H.
    $\mathcal{A}$ also outputs the set of relevant variables of its input.
    $\mathcal{A}$ runs in time $T(d,n,\delta)$.
1) $P \leftarrow$ Choose $O(\log(n/\delta))$ random hash functions $h : [n] \rightarrow [q]$.
2) For all $h \in P$ in parallel
    Run $\mathcal{A}(q,d,1/8)$ to learn $f_h := f(x_{h(1)}, \ldots, x_{h(n)})$
        and find its relevant variables.
    Stop after $T(d,q,1/8)$ time.
    For processes that do not stop output the function 0,
        and an empty set of relevant variables.
    Let $f'_h \in H$ be the output of $\mathcal{A}$ on f_h.
    Let V_h be the set of relevant variables that $\mathcal{A}$ outputs on f_h.
3) $W_h = \{x_i \mid x_{h(i)} \in V_h\}$.
    $W \leftarrow$ Variables appearing in more than $1/2$ of the $\{W_h\}_{h \in P}$.
4) $T \leftarrow$ All $h \in H$ such that for each $v_i \in V_h$ there is
        exactly one variable $x_j \in W$ for which $h(j) = i$.
5) For each $h \in T$: $g_h \leftarrow$ Replace each relevant variable v_i in f'_h
        by $x_j \in W$ where $h(j) = i$.
6) Output $\mathrm{Popular}(\{g_h\}_{h \in T})$.

Fig. 1. Algorithm reduction

Theorem 7. *Let C be a class of boolean functions that is closed under distinct variable projection. Let d be an upper bound on the number of relevant variables in any $f \in C$. Let H be a class of representations of boolean functions. Let h_1, h_2 be functions in H with at most d relevant known variables each. Suppose there is a deterministic algorithm $\mathcal{B}(d,n)$ that for such input h_1 and h_2 decides whether the two functions are equivalent in time $E(d,n)$.*

Let $\mathcal{A}(d,n,\delta)$ be a Monte Carlo non-adaptive algorithm that with failure probability at most δ, learns C from H and finds the set of relevant variables of the input in time $T(d,n,\delta)$ with $Q(d,n,\delta)$ membership queries. Then, C is Monte Carlo non-adaptively learnable from H in time $O((T(d,q,1/8)n + E(d,q)(\log n + \log(1/\delta)))(\log n + \log(1/\delta)))$ with $O(Q(d,q,1/8)(\log n + \log(1/\delta)))$ membership queries, where δ is the failure probability and q is any integer such that $q \geq 8d^2$.

Proof. Let $\mathcal{A}(d,n,\delta)$ and $\mathcal{B}(d,n)$ be as stated in the theorem above. Let $f \in C$ be the target function and suppose the relevant variables in f are $x_{i_1}, \ldots, x_{i_{d'}}$ where $d' \leq d$. Let be $I = \{i_1, \ldots, i_{d'}\}$.

We first choose $O(\log(n/\delta))$ random hash functions $h : [n] \rightarrow [q]$ and put them in P. For each hash function $h \in P$ we define the following events. The event A_h is true if $h(I) := \{h(i_1), \ldots, h(i_{d'})\}$ are not distinct. The event $B_{h,j}$, $j \in [n] \backslash I$, is true if $h(j) \in h(I)$. For any $h \in P$, the probability that A_h is true is at most $\sum_{i=1}^{d-1} i/q = d(d-1)/(2q) \leq 1/16$. For any $h \in P$ and $j \in [n] \backslash I$, the probability that $B_{h,j}$ is true is at most $d/q \leq 1/8$.

By Chernoff bound, with failure probability at most $\delta/(3n)$, we have that at least $7/8$ of $\{A_h\}_{h \in P}$ are false. Therefore, with failure probability at most $\delta/(3n)$, at least $7/8$ of $\{f_h := f(x_{h(1)}, \ldots, x_{h(n)})\}_{h \in P}$ are still in C. This is true because C is closed under distinct variable projection.

Let V_h be the set of relevant variables that $\mathcal{A}$ outputs when running with the input f_h. Since hashing can not raise the number of relevant variables in the function, we can now for each $h \in P$ run in parallel $\mathcal{A}(d, q, 1/8)$ to learn f_h and find V_h. Let $S \subseteq P$ denote the set of $h \in P$ where the corresponding processes finish after $T(d, q, 1/8)$ time. For each $h \in S$, denote the output of $\mathcal{A}$ on f_h by f'_h. With failure probability at most $\delta/(3n)$, it holds that $|S| \geq 7/8|P|$. For any other $h \in P$, we stop its process and set $f'_h = 0$, $V_h = \varnothing$. Applying Chernoff bound on $\{f'_h\}_{h \in S}$ yields that for at least $6/7$ of them $\mathcal{A}$ succeeds, with failure probability at most $\delta/(3n)$. Therefore, with failure probability at most $2\delta/(3n) \leq 2\delta/3$ we have that for at least $7/8 \times 6/7 = 3/4$ of $h \in P$ it holds that $h(I)$ are distinct, $f'_h = f_h$ and the set of relevant variables V_h is correct.

For each $h \in P$ define the set $W_h = \{x_i \mid x_{h(i)} \in V_h\}$ and let W be the set of all the variables appearing in more than $1/2$ of the $\{W_h\}_{h \in P}$. We now find the probability that a relevant variable of f is in W_h and compare it with the probability that an irrelevant variable of f is in W_h. The probability that $x_{i_1} \notin W_h$ is at most the probability that A_h is true or that $\mathcal{A}$ fails. This probability is at most $1/16 + 1/8 = 3/16$. Therefore the probability that a relevant variable is in W_h is at least $13/16$. The probability that an irrelevant variable x_j, $j \notin I$, is in W_h is at most the probability that A_h or $B_{h,j}$ is true or that $\mathcal{A}$ fails. This is bounded by $1/16 + 1/8 + 1/8 = 5/16$.

Therefore, by Chernoff bound, when running the algorithm for $O(\log(n/\delta))$ random hash functions, W contains all the relevant variables of the target function f and only them, with probability at least $1 - \delta/3$.

Let T be the set of all $h \in H$ such that for each $v_i \in V_h$ there is exactly one variable $x_j \in W$ for which $h(j) = i$. For each $h \in T$, let be $V_h = \{v_{i_1}, \ldots, v_{i_{l_h}}\}$, where $l_h \leq d'$, and $f'_h := f'_h(v_{i_1}, \ldots, v_{i_{l_h}})$. Define $g_h := f'_h(x_{j_1}, \ldots, x_{j_{l_h}})$ where $h(j_k) = i_k$. Now, with failure probability at most $2\delta/3 + \delta/3 = \delta$, it holds that at least $3/4$ of $\{g_h\}_{h \in T}$ are identical to the target function f. Therefore, at the end we use $\mathcal{B}(d, q)$ to find Popular($\{g_h\}_{h \in T}$) and output it.   □

Let C be d-Junta and let H be a class of representations of boolean functions as truth tables. Let $\mathcal{A}$ be the algorithm described in the previous subsection. This algorithm learns d-Junta from H and outputs the set of relevant variables. Let $\mathcal{B}$ be the simple algorithm that given two functions in H with at most d known relevant variables, decides if they are identical by comparing the values of the two functions on the $O(2^d)$ relevant entries in the tables. This algorithm runs in $poly(2^d, n)$ time. We can now apply the reduction and get the result from Theorem 6.

Acknowledgments. We would like to thank the anonymous reviewers for their insightful comments. Improvements to some algorithms have been suggested and they will be included in the full version of this paper.

References

1. Angluin, D.: Queries and concept learning. Mach. Learn. **2**(4), 319–342 (1987)
2. Abasi, H., Bshouty, N.H., Mazzawi, H.: Non-adaptive learning a hidden hipergraph. In: ALT 2015, pp. 89–101 (2015)
3. Alon, N., Babai, L., Itai, A.: A fast and simple randomized parallel algorithm for the maximal independent set problem. J. Algorithms **7**(4), 567–583 (1986)
4. Blum, A.: Learning a function of r relevant variables. In: COLT 2003, pp. 731–733 (2003)
5. Bshouty, N.H.: Derandomizing Chernoff bound with union bound with an application to k-wise independent sets. ECCC. TR16-083
6. Ciccalese, F.: Group testing. In: Ciccalese, F. (ed.) Fault-Tolerant Search Algorithms. Monographs in Theoretical Computer Science. An EATCS Series 2013, pp. 139–173. Springer, Heidelberg (2013)
7. Biglieri, E., Gyorfi, L.: Multiple Access Channels: Theory and Practice. IOS Press, Amsterdam (2007)
8. De Bonis, A., Gasieniec, L., Vaccaro, U.: Optimal two-stage algorithms for group testing problems. SIAM J. Comput. **34**(5), 1253–1270 (2005)
9. Damaschke, P.: Adaptive versus nonadaptive attribute-efficient learning. Mach. Learn. **41**(2), 197–215 (2000)
10. Damaschke, P.: Computational aspects of parallel attribute-efficient learning. In: Richter, M.M., Smith, C.H., Wiehagen, R., Zeugmann, T. (eds.) ALT 1998. LNCS, vol. 1501, pp. 103–111. Springer, Heidelberg (1998)
11. Damaschke, P.: On parallel attribute-efficient learning. J. Comput. Syst. Sci. **67**(1), 46–62 (2003)
12. Du, D., Hwang, F.K.: Combinatorial Group Testing and Its Applications. World Scientific Pub Co Inc., Hong Kong (2000)
13. Du, D., Hwang, F.K.: Pooling Design and Nonadaptive Group Testing: Important-Tools for DNA Sequencing. World Scientific Pub Co Inc., Hong Kong (2006)
14. Kleitman, D.J., Spencer, J.: Families of k-independent sets. Discret. Math. **6**(3), 255–262 (1972)
15. Ngo, H.Q., Du, D.-Z.: A survey on combinatorial group testing algorithms with applications to DNA library screening. DIMACS Series in Discrete Mathematics and Theoretical Computer Science (2000)
16. Naor, M., Schulman, L.J., Srinivasan, A.: Splitters and near-optimal derandomization. In: FOCS 1995, pp. 182–191 (1995)
17. Bshouty, N.H.: Linear time constructions of some d-Restriction problems. In: Paschos, V.T., Widmayer, P. (eds.) CIAC 2015. LNCS, vol. 9079, pp. 74–88. Springer, Heidelberg (2015)
18. Bshouty, N.H., Hellerstein, L.: Attribute-efficient learning in query and mistake-bound models. In: COLT 1996, pp. 235–243 (1996)
19. Mossel, E., O'Donnell, R., Servedio, R.A.: Learning juntas. In: STOC 2003, pp. 206–212 (2003)
20. Motwani, R., Raghavan, P.: Randomized Algorithms. Cambridge University Press, Cambridge (1995)
21. Seroussi, G., Bshouty, N.H.: Vector sets for exhaustive testing of logic circuits. IEEE Trans. Inf. Theor. **34**(3), 513–522 (1988)
22. Abasi, H., Bshouty, N.H., Mazzawi, H.: On exact learning monotone DNF from membership queries. In: Auer, P., Clark, A., Zeugmann, T., Zilles, S. (eds.) ALT 2014. LNCS, vol. 8776, pp. 111–124. Springer, Heidelberg (2014)

Submodular Learning and Covering with Response-Dependent Costs

Sivan Sabato$^{(\boxtimes)}$

Ben-Gurion University of the Negev, Beersheba, Israel
sabatos@cs.bgu.ac.il

Abstract. We consider interactive learning and covering problems, in a setting where actions may incur different costs, depending on the response to the action. We propose a natural greedy algorithm for response-dependent costs. We bound the approximation factor of this greedy algorithm in active learning settings as well as in the general setting. We show that a different property of the cost function controls the approximation factor in each of these scenarios. We further show that in both settings, the approximation factor of this greedy algorithm is near-optimal among all greedy algorithms. Experiments demonstrate the advantages of the proposed algorithm in the response-dependent cost setting.

Keywords: Interactive learning · Submodular functions · Outcome costs

1 Introduction

We consider interactive learning and covering problems, a term introduced in [7]. In these problems, there is an algorithm that interactively selects actions and receives a response for each action. Its goal is to achieve an objective, whose value depends on the actions it selected, their responses, and the state of the world. The state of the world, which is unknown to the algorithm, also determines the response to each action. The algorithm incurs a cost for every action it performs. The goal is to have the total cost incurred by the algorithm as low as possible.

Many real-world problems can be formulated as interactive learning and covering problems. These include pool-based active learning problems [2,12], maximizing the influence of marketing in a social network [7], interactive sensor placement [4] and document summarization [11] with interactive user feedback. Interactive learning and covering problems cannot be solved efficiently in general [14,19]. Nevertheless, many such problems can be solved near-optimally by efficient algorithms, when the functions that map the sets of actions to the total reward are *submodular*. It has been shown in several settings, that a simple greedy algorithm pays a near-optimal cost when the objective function is submodular (e.g., [1,4,7]). Many problems naturally lend themselves to a submodular formulation. These include covering objectives, objectives promoting diversity [13] and active learning [2,4,6,7].

© Springer International Publishing Switzerland 2016
R. Ortner et al. (Eds.): ALT 2016, LNAI 9925, pp. 130–144, 2016.
DOI: 10.1007/978-3-319-46379-7_9

Interactive learning and covering problems have so far been studied mainly under the assumption that the cost of the action is known to the algorithm before the action is taken. In this work we study the setting in which the costs of actions depend on the outcome of the action, which is only revealed by the observed response. This is the case in many real-world scenarios. For instance, consider an active learning problem, where the goal is to learn a classifier that predicts which patients should be administered a specific drug. Each action in the process of learning involves administering the drug to a patient and observing the effect. In this case, the cost (poorer patient health) is higher if the patient suffers adverse effects. Similarly, when marketing in a social network, an action involves sending an ad to a user. If the user does not like the ad, this incurs a higher cost (user dissatisfaction) than if they like the ad.

We study the achievable approximation guarantees in the setting of response-dependence costs, and characterize the dependence of this approximation factor on the properties of the cost function. We propose a natural generalization of the greedy algorithm of [7] to the response-dependent setting, and provide two approximation guarantees. The first guarantee holds whenever the algorithm's objective describes an active learning problem. We term such objectives *learning objectives*. The second guarantee holds for general objectives, under a mild condition. In each case, the approximation guarantees depend on a property of the cost function, and we show that this dependence is necessary for any greedy algorithm. Thus, this fully characterizes the relationship between the cost function and the approximation guarantee achievable by a greedy algorithm. We further report experiments that demonstrate the achieved cost improvement.

Response-dependent costs has been previously studied in specific cases of active learning, assuming there are only two possible labels [15–18]. In [8] this setting is also mentioned in the context of active learning. Our work is more general: First, it addresses general objective functions and not only specific active learning settings. Our results indicate that the active learning setting and the general setting are inherently different. Second, it is not limited to settings with two possible responses. As we show below, previous guarantees for two responses do not generalize to tight guarantees for cases with more than two responses. We thus develop new proof techniques that allow deriving these tighter bounds.

2 Definitions and Preliminaries

For an integer n, denote $[n] := \{1, \ldots, n\}$. A set function $f : 2^{\mathcal{Z}} \to \mathbb{R}$ is *monotone* (non-decreasing) if $\forall A \subseteq B \subseteq \mathcal{Z}$, $f(A) \leq f(B)$. Let $\mathcal{Z}$ be a domain, and let $f : 2^{\mathcal{Z}} \to \mathbb{R}_+$ be a set function. Define, for any $z \in \mathcal{Z}, A \subseteq \mathcal{Z}$, $\delta_f(z \mid A) := f(A \cup \{z\}) - f(A)$. f is *submodular* if $\forall z \in \mathcal{Z}, A \subseteq B \subseteq \mathcal{Z}$, $\delta_f(z \mid A) \geq \delta_f(z \mid B)$.

Assume a finite domain of actions $\mathcal{X}$ and a finite domain of responses $\mathcal{Y}$. For simplicity of presentation, we assume that there is a one-to-one mapping between world states and mappings from actions to their responses. Thus the states of the world are represented by the class of possible mappings $\mathcal{H} \subseteq \mathcal{Y}^{\mathcal{X}}$. Let $h^* \in \mathcal{H}$ be the true, unknown, mapping from actions to responses. Let $S \subseteq \mathcal{X} \times \mathcal{Y}$ be a set of action-response pairs.

We consider algorithms that iteratively select a action $x \in \mathcal{X}$ and get the response $h^*(x)$, where $h^* \in \mathcal{H}$ is the true state of the world, which is unknown to the algorithm. For an algorithm $\mathcal{A}$, let $S^h[\mathcal{A}]$ be the set of pairs collected by $\mathcal{A}$ until termination if $h^* = h$. Let $S_t^h[\mathcal{A}]$ be the set of pairs collected by $\mathcal{A}$ in the first t iterations if $h^* = h$. In each iteration, $\mathcal{A}$ decides on the next action to select based on responses to previous actions, or it decides to terminate. $\mathcal{A}(S) \in \mathcal{X} \cup \{\bot\}$ denotes the action that $\mathcal{A}$ selects after observing the set of pairs S, where $\mathcal{A}(S) = \bot$ if $\mathcal{A}$ terminates after observing S.

Each time the algorithm selects an action and receives a response, it incurs a cost, captured by a cost function $\mathsf{cost} : \mathcal{X} \times \mathcal{Y} \to \mathbb{R}_+$. If $x \in \mathcal{X}$ is selected and the response $y \in \mathcal{Y}$ is received, the algorithm pays $\mathsf{cost}(x, y)$. Denote $\mathsf{cost}(S) = \sum_{(x,y)\in S} \mathsf{cost}(x,y)$. The total cost of a run of the algorithm when the state of the world is h^*, is thus $\mathsf{cost}(S^{h^*}[\mathcal{A}])$. For a given $\mathcal{H}$, define the *worst-case cost* of $\mathcal{A}$ by $\mathsf{cost}(\mathcal{A}) := \max_{h \in \mathcal{H}} \mathsf{cost}(S^h[\mathcal{A}])$. Let $Q > 0$ be a threshold, and let $f : 2^{\mathcal{X} \times \mathcal{Y}} \to \mathbb{R}_+$ be a monotone non-decreasing submodular objective function. We assume that the goal of the interactive algorithm is to collect pairs S such that $f(S) \geq Q$, while minimizing $\mathsf{cost}(\mathcal{A})$.

Guillory and Bilmes [7] consider a setting in which instead of a single global f, there is a set of monotone non-decreasing objective functions $\mathcal{F}_{\mathcal{H}} = \{f_h : 2^{\mathcal{X} \times \mathcal{Y}} \to \mathbb{R}_+ \mid h \in \mathcal{H}\}$, and the value $f_h(S)$, for $S \subseteq \mathcal{X} \times \mathcal{Y}$, represents the reward obtained by the algorithm if $h^* = h$. They show that obtaining $f_{h^*}(S) \geq Q$ is equivalent to obtaining $\bar{F}(S) \geq Q$, where $\bar{F} : 2^{\mathcal{X} \times \mathcal{Y}} \to \mathbb{R}_+$ is defined by

$$\bar{F}(S) := \frac{1}{|\mathcal{H}|}\left(Q|\mathcal{H} \setminus \mathrm{VS}(S)| + \sum_{h \in \mathrm{VS}(S)} \min(Q, f_h(S))\right). \tag{1}$$

Here $\mathrm{VS}(S)$ is the *version space* induced by S on $\mathcal{H}$, defined by $\mathrm{VS}_{\mathcal{H}}(S) = \{h \in \mathcal{H} \mid \forall(x,y) \in S, y = h(x)\}$. It is shown in [7] that if all the functions in $\mathcal{F}_{\mathcal{H}}$ are monotone and submodular then so is $\bar{F}$. Thus our setting of a single objective function can be applied to the setting of [7] as well.

Let $\alpha \geq 1$. An interactive algorithm $\mathcal{A}$ is an α-*approximate greedy algorithm* for *utility function* $u : \mathcal{X} \times 2^{\mathcal{X} \times \mathcal{Y}} \to \mathbb{R}_+$, if the following holds: For all $S \subseteq \mathcal{X} \times \mathcal{Y}$, if $f(S) \geq Q$ then $\mathcal{A}(S) = \bot$, and otherwise, $\mathcal{A}(S) \in \mathcal{X}$ and $u(\mathcal{A}(S), S) \geq \frac{1}{\alpha}\max_{x \in \mathcal{X}} u(x, S)$. As shown below, consistently with previous works, (e.g. [4]), competitive guarantees are better for α-approximate-greedy algorithms with $\alpha = 1$ or α close to 1. However, due to efficiency of computation or other practical considerations, it is not always feasible to implement a 1-greedy algorithm. Thus, for full generality, we analyze also α-greedy algorithms for $\alpha > 1$. Let $\mathrm{OPT} := \min_{\mathcal{A}} \mathsf{cost}(\mathcal{A})$, where the minimum is taken over all interactive $\mathcal{A}$ that obtain $f(S) \geq Q$ at termination, for all for all possible $h^* \in \mathcal{H}$. If no such $\mathcal{A}$ exist, define $\mathrm{OPT} = \infty$.

In [7] it is assumed that costs are not response-dependent, thus $\mathsf{cost}(x, y) \equiv \mathsf{cost}(x)$, and a greedy algorithm is proposed, based on the following utility function:

$$u(x, S) := \frac{\delta_{\bar{F}}((x, h(x)) \mid S)}{\mathsf{cost}(x)}. \tag{2}$$

It is shown that for integral functions, this algorithm obtains an integer Q with a worst-case cost of at most $GCC(\ln(Q|\mathcal{H}|) + 1) \cdot OPT$, where GCC is a lower bound on OPT. In [4], a different greedy algorithm and analysis guarantees a worst-case cost of $\alpha(\ln(Q) + 1) \cdot OPT$ for adaptive submodular objectives and α-approximate greedy algorithms. It is well known that the factor of $\ln(Q)$ cannot be substantially improved by an efficient algorithm, even for non-interactive problems [3,19].

The results of [7] can be trivially generalized to the response-dependent cost setting using the *cost ratio* of the problem:

$$r_{cost} := \max_{x \in \mathcal{X}} \frac{\max_{y \in \mathcal{Y}} cost(x, y)}{\min_{y \in \mathcal{Y}} cost(x, y)}.$$

Consider a generalized version of u:

$$u(x, S) := \frac{\delta_{\bar{F}}((x, h(x)) \mid S)}{\min_{y \in \mathcal{Y}} cost(x, y)}. \tag{3}$$

Setting $\overline{cost}(x) := \min_{y \in \mathcal{Y}} cost(x, y)$, we have $cost \leq r_{cost} \cdot \overline{cost}$. Using this fact, it is easy to derive an approximation guarantee of $r_{cost} \cdot OPT(\ln(Q|\mathcal{H}|) + 1)$, for a greedy algorithm which uses the utility function in Eq. (3) with a response-dependent cost, or $r_{cost} \cdot \alpha(\ln(Q) + 1)OPT$ when applied to the setting of [4]. However, in this work we show that this trivial derivation is loose, since our new approximation bounds can be finite even if r_{cost} is infinite.

3   A Greedy Algorithm for Response-Dependent Costs

We provide approximation guarantee for two types of objective functions. The first type captures active learning settings, while the second type is more general. Our results show that objective functions for active learning have better approximation guarantees than general objective functions. For both types of objective functions, we analyze a greedy algorithm that selects an element maximizing (or approximately maximizing) the following utility function:

$$u^f(x, S) := \min_{h \in VS(S)} \frac{\delta_{\min(f,Q)}((x, h(x)) \mid S)}{cost(x, h(x))}.$$

Note that $u^{\bar{F}}$ is equal to the function u defined in Eq. (3). We employ the following standard assumption in our results (see e.g. [5]):

Assumption 1. *Let $f : 2^{\mathcal{X} \times \mathcal{Y}} \to \mathbb{R}_+$, $Q > 0$, $\eta > 0$. Assume that f is submodular and monotone, $f(\emptyset) = 0$, and that for any $S \subseteq \mathcal{X} \times \mathcal{Y}$, if $f(S) \geq Q - \eta$ then $f(S) \geq Q$.*

In Sect. 3.1 we show an approximation guarantee for objectives meant for active learning, which we term *learning objectives*. In Sect. 3.2 we consider general monotone submodular objective functions. Our guarantees hold for objective functions f that satisfy the following property, which we term *consistency-aware*. This property requires that the function gives at least Q to any set of action-response pairs that are inconsistent with $\mathcal{H}$.

Definition 1. *A function* $f : 2^{\mathcal{X} \times \mathcal{Y}} \to \mathbb{R}_+$ *is* consistency-aware *for threshold* $Q > 0$ *if for all* $S \subseteq \mathcal{X} \times \mathcal{Y}$ *such that* $\mathrm{VS}_{\mathcal{H}}(S) = \emptyset$, $f(S) \geq Q$.

Note that the definition is concerned with the value of f only on inconsistent sets S, which the algorithm never encounters. Therefore, it suffices that there exist an extension of f to these sets that is consistent with all the other requirements from f. The function $\bar{F}$ defined in Eq. (1) is consistency-aware. In addition, a similar construction to $\bar{F}$ with non-uniform weights for mappings is also consistency-aware. Such a construction is sometimes more efficient to compute than the uniform-weight construction. For instance, as shown in [6], non-uniform weights allow a more efficient computation when the mappings represent linear classifiers with a margin. In general, any objective f can be made consistency aware using a simple transformation such as $\bar{F}$. Thus our results are relevant to a diverse class of problems.

3.1   Guarantees for Learning Objectives

Active learning is an important special case of interactive learning. In active learning, the only goal is to discover information on the identity of h^*. We term functions that represent such a goal *learning objectives*.

Definition 2. *A function* $f : 2^{\mathcal{X} \times \mathcal{Y}} \to \mathbb{R}_+$ *is a* learning objective *for* $\mathcal{H}$ *if* $f(S) = g(\mathrm{VS}_{\mathcal{H}}(S))$ *where* g *is a monotone non-increasing function*.

It is easy to see that all learning objectives $S \mapsto f(S)$ are monotone non-decreasing in S. In many useful cases, they are also submodular. In noise-free active learning, where the objective is to exactly identify the correct mapping h^*, one can use the learning objective $f(S) := 1 - |\mathrm{VS}(S)|/|\mathcal{H}|$, with $Q = 1 - 1/|\mathcal{H}|$. This is the *version space reduction* objective function [4,7]. In [5] noise-aware active learning and its generalization to the problem of Equivalence Class Determination is considered. In this generalization, there is some partition of $\mathcal{H}$, and the goal is to identify the class to which h^* belongs. The objective function proposed by [5] measures the weight of pairs in $\mathrm{VS}(S)$ in which each mapping belongs to a different class. This function is also a learning objective. In [1] the *total generalized version space reduction* function is proposed. This function is also a learning objective. More generally, consider a set of structures $\mathcal{G} \subseteq 2^{\mathcal{H}}$, where the goal is to disqualify these structures from the version space, by proving that at least one of the mappings in this structure cannot be the true h^*. In this case one can define the submodular learning objective $f(S) := w(\mathcal{G}) - w(\mathcal{G} \cap 2^{\mathrm{VS}(S)})$, where w is a modular weight function on $\mathcal{G}$, and $Q = w(\mathcal{G})$. For instance, if $\mathcal{G}$ is the set of pairs from different equivalence classes in $\mathcal{H}$, this is the Equivalence Class Determination objective. If $\mathcal{G}$ is a set of triplets from different equivalence classes, this encodes an objective of reducing the uncertainty on the identity of h^* to at most two equivalence classes.

We show that for learning objectives, the approximation factor for a greedy algorithm that uses u^f depends on a new property of the cost function, which we term the *second-smallest cost ratio*, denoted by $r_{cost}^{[2]}$. For $x \in \mathcal{X}$, let $\phi(x)$ be the second-smallest value in the multiset $\{\text{cost}(x,y) \mid y \in \mathcal{Y}\}$. Define

$$r_{cost}^{[2]} := \max_{x \in \mathcal{X}, y \in \mathcal{Y}} \frac{\text{cost}(x,y)}{\phi(x)}.$$

Theorem 1. *Let* $f : 2^{\mathcal{X} \times \mathcal{Y}} \to \mathbb{R}_+, Q > 0, \eta > 0$ *such that Assumption 1 holds. Let $\mathcal{A}$ be an α-approximate greedy algorithm for the utility function u^f. If f is a learning objective, then* $\text{cost}(\mathcal{A}) \leq r_{cost}^{[2]} \cdot \alpha(\ln(Q/\eta) + 1)\text{OPT}.$

The ratio between the trivial bound that depends on the cost ratio r_{cost}, mentioned in Sect. 2, and this new bound, is $r_{cost}/r_{cost}^{[2]}$, which is unbounded in the general case: for instance, if each action has one response which costs 1 and the other responses cost $M \gg 1$, then $r_{cost} = M$ but $r_{cost}^{[2]} = 1$. Whenever $|\mathcal{Y}| = 2$, $r_{cost}^{[2]} = 1$. Thus, the approximation factor of the greedy algorithm for any binary active learning problem is independent of the cost function. This coincides with the results of [17,18] for active learning with binary labels. If $|\mathcal{Y}| > 2$, then the bound is smallest when $r_{cost}^{[2]} = 1$, which would be the case if for each action there is one preferred response which has a low cost, while all other responses have the same high cost. For instance, consider a marketing application, in which the action is to recommend a product to a user, and the response is either buying the product (a preferred response), or not buying it, in which case additional feedback could be provided from the user, but the cost (user dissatisfaction) remains the same regardlesss of that feedback.

To prove Theorem 1, we use the following property of learning objectives: For such objectives, there exists an optimal algorithm (that is, one that obtains OPT) that only selects actions for which at least two responses are possible given the action-response pairs observed so far. Formally, we define *bifurcating* algorithms. Denote the set of possible responses for x given the history S by $\mathcal{Y}_{\mathcal{H}}(x, S) := \{h(x) \mid h \in \text{VS}_{\mathcal{H}}(S)\}$. We omit the subscript $\mathcal{H}$ when clear from context.

Definition 3. *An interactive algorithm $\mathcal{A}$ is* bifurcating *for $\mathcal{H}$ if for all t and $h \in \mathcal{H}$, $|\mathcal{Y}_{\mathcal{H}}(\mathcal{A}(S_t^h[\mathcal{A}]), S_t^h[\mathcal{A}])| \geq 2$.*

Lemma 1. *For any learning objective f for $\mathcal{H}$ with an optimal algorithm, then there exists a bifurcating optimal algorithm for $f, \mathcal{H}$.*

Proof. Let $\mathcal{A}$ be an optimal algorithm for f. Suppose there exists some t, h such that $\mathcal{Y}(x_0, S_{t-1}^h[\mathcal{A}]) = \{y_0\}$ for some $y_0 \in \mathcal{Y}$, where $x_0 := \mathcal{A}(S_{t-1}^h[\mathcal{A}])$. Let $\mathcal{A}'$ be an algorithm that selects the same actions as $\mathcal{A}$, except that it skips the action x_0 it if has collected the pairs $S_{t-1}^h[\mathcal{A}]$. That is, $\mathcal{A}'(S) = \mathcal{A}(S)$ for $S \not\supseteq S_{t-1}^h[\mathcal{A}]$, and $\mathcal{A}'(S) = \mathcal{A}(S \cup \{(x_0, y_0)\})$ for $S \supseteq S_{t-1}^h$. Since $\text{VS}(S) = \text{VS}(S \cup \{(x_0, y_0)\})$, and $\mathcal{A}$ is a learning objective, $\mathcal{A}'$ obtains Q as well, at the same cost of $\mathcal{A}$ or less. By repeating this process a finite number of steps, we can obtain an optimal algorithm for $\mathcal{H}$ which is bifurcating. $\square$

The following lemma is the crucial step in proving Theorem 1, and will also be used in the proof for the more general case below. The lemma applies to general consistency-aware functions. It can be used for learning objectives, because all learning objectives with a finite OPT are consistency-aware: Suppose that f is a learning objective, and let $S \subseteq \mathcal{X} \times \mathcal{Y}$ such that $\mathrm{VS}_{\mathcal{H}}(S) = \emptyset$. For any $h \in \mathcal{H}$, denote $S_*^h := \{(x, h(x)) \mid x \in \mathcal{X}\}$. We have $\mathrm{VS}(S_*^h) \supseteq \mathrm{VS}(S)$, therefore, since f is a learning objective, $f(S) \geq f(S_*^h)$. Since OPT is finite, $f(S_*^h) \geq Q$. Therefore $f(S) \geq Q$. Thus f is consistency-aware.

Lemma 2. *Let f, Q, η which satisfy Assumption 1 such that f is consistency-aware. Let $\mathcal{A}$ be an interactive algorithm that obtains $f(S) \geq Q$ at termination. Let $\gamma = r_{\mathsf{cost}}^{[2]}$ if $\mathcal{A}$ is bifurcating, and let $\gamma = r_{\mathsf{cost}}$ otherwise. then*

$$\exists x \in \mathcal{X} \ s.t. \ u^f(x, \emptyset) \geq \frac{Q}{\gamma \cdot \mathsf{cost}(\mathcal{A})}.$$

Proof. Denote for brevity $\delta \equiv \delta_{\min(f,Q)}$. Define $\bar{\mathcal{H}} := \mathcal{Y}^{\mathcal{X}}$. Consider an algorithm $\bar{\mathcal{A}}$ such that for any S that is consistent with some $h \in \mathcal{H}$ (that is $\mathrm{VS}_{\mathcal{H}}(S) \neq \emptyset$), $\bar{\mathcal{A}}(S) = \mathcal{A}(S)$, and $\bar{\mathcal{A}}(S) = \perp$ otherwise. Since f is consistency-aware, we have $f(S^h[\bar{\mathcal{A}}]) \geq Q$ for all $h \in \bar{\mathcal{H}}$.

Consider a run of $\bar{\mathcal{A}}$, and denote the pair in iteration t of this run by (x_t, y_t). Denote $S_t = \{(x_i, y_i) \mid i \leq t\}$. Choose the run such that in each iteration t, the response y_t is in $\mathrm{argmin}_{y \in \mathcal{Y}} \delta(x_t, y \mid S_{t-1})/\mathsf{cost}(x_t, y)$. Let T be the length of the run until termination. Denote $\psi := \max_{h \in \bar{\mathcal{H}}} \mathsf{cost}(S^h[\bar{\mathcal{A}}])$, the worst-case cost of $\bar{\mathcal{A}}$ over $\bar{\mathcal{H}}$. We have

$$Q/\psi \leq f(S_T)/\mathsf{cost}(S_T) = \frac{\sum_{t \in [T]} (f(S_t) - f(S_{t-1}))}{\sum_{t \in [T]} \mathsf{cost}(x_t, y_t)}$$

$$= \frac{\sum_{t \in [T]} \delta((x_t, y_t) \mid S_{t-1})}{\sum_{t \in [T]} \mathsf{cost}(x_t, y_t)} \leq \max_{t \in [T]} \left(\delta((x_t, y_t) \mid S_{t-1})/\mathsf{cost}(x_t, y_t) \right),$$

where we used $f(\emptyset) = 0$ in the second line. Thus there exists some $t \in [T]$ such that $Q/\psi \leq \delta((x_t, y_t) \mid S_{t-1})/\mathsf{cost}(x_t, y_t)$. Therefore

$$u(x_t, \emptyset) = \min_{y \in \mathcal{Y}} \delta((x_t, y) \mid \emptyset)/\mathsf{cost}(x_t, y) \geq \min_{y \in \mathcal{Y}} \delta((x_t, y) \mid S_{t-1})/\mathsf{cost}(x_t, y)$$

$$= \delta((x_t, y_t) \mid S_{t-1})/\mathsf{cost}(x_t, y_t) \geq Q/\psi. \tag{4}$$

The second line follows from the submodularity of f. The third line follows from the definition of y_t. To prove the claim, we have left to show that $\psi \leq r^{[2]} \cdot \mathsf{cost}(\mathcal{A})$. Consider again a run of $\bar{\mathcal{A}}$. If all observed pairs are consistent with some $h \in \mathcal{H}$, $\bar{\mathcal{A}}$ and $\mathcal{A}$ behave the same. Hence $\mathsf{cost}(S^h[\bar{\mathcal{A}}]) = \mathsf{cost}(S^h[\mathcal{A}])$. Now, consider $h \in \bar{\mathcal{H}} \setminus \mathcal{H}$. By the definition of $\bar{\mathcal{A}}$, $S^h[\bar{\mathcal{A}}]$ is a prefix of $S^h[\mathcal{A}]$. Let $T = |S^h[\bar{\mathcal{A}}]|$ be the number of iterations until $\bar{\mathcal{A}}$ terminates. Then $S_{T-1}^h[\bar{\mathcal{A}}]$ is consistent with some $h' \in \mathcal{H}$.

Let x_T be the action that $\mathcal{A}$ and $\bar{\mathcal{A}}$ select at iteration T, and let $h' \in \mathcal{H}$ which is consistent with $S_{T-1}^h[\bar{\mathcal{A}}]$, and incurs the maximal possible cost in iteration T. Formally, h' satisfies $h'(x_T) \in \text{argmax}_{y \in \mathcal{Y}_\mathcal{H}(x_T, S^h_{T-1}[\mathcal{A}])} \text{cost}(x_T, y)$. Now, compare the run of $\bar{\mathcal{A}}$ on h to the run of $\mathcal{A}$ on h'. In the first $T-1$ iterations, the algorithms observe the same pairs. In iteration T, they both select x_T. $\bar{\mathcal{A}}$ observes $h(x_T)$, while $\mathcal{A}$ observes $h'(x_T)$. $\bar{\mathcal{A}}$ terminates after iteration T. Hence $\text{cost}(S^h[\bar{\mathcal{A}}]) = \text{cost}(S^h_{T-1}[\mathcal{A}]) + \text{cost}(x_T, h(x_T)) = \text{cost}(S^h_T[\mathcal{A}]) - \text{cost}(x_T, h'(x_T)) + \text{cost}(x_T, h(x_T))$. Consider two cases: (a) $\mathcal{A}$ is not bifurcating. Then $\gamma = r$, and so $\text{cost}(x_T, h(x_T)) \le \gamma \text{cost}(x_T, h'(x_T))$. (b) $\mathcal{A}$ is bifurcating. Then there are at least two possible responses in $\mathcal{Y}_\mathcal{H}(x_T, S^h_{T-1}[\mathcal{A}])$. Therefore $\text{cost}(x_T, h'(x_T)) \ge \phi(x_T)$. By the definition of $r^{[2]}_{\text{cost}}$, $\text{cost}(x_T, h(x_T)) \le r^{[2]}_{\text{cost}} \cdot \phi(x_T)$. Therefore $\text{cost}(x_T, h(x_T)) \le r_{\text{cost}} \text{cost}(x_T, h'(x_T))) = \gamma \text{cost}(x_T, h'(x_T)))$.

In both cases, $\text{cost}(x_T, h(x_T)) - \text{cost}(x_T, h'(x_T)) \le (\gamma - 1)\text{cost}(x_T, h'(x_T))$. Therefore $\text{cost}(S^h[\bar{\mathcal{A}}]) \le \text{cost}(S^h_T[\mathcal{A}]) + (\gamma - 1)\text{cost}(x_T, h'(x_T)) \le \gamma \text{cost}(S^h_T[\mathcal{A}])$, where the last inequality follows since $\text{cost}(S^h_T[\mathcal{A}]) \le \text{cost}(S^{h'}_T[\mathcal{A}])$. Thus for all $h \in \bar{\mathcal{H}}$, $\text{cost}(S^h[\bar{\mathcal{A}}]) \le \gamma \cdot \text{cost}(\mathcal{A})$, hence $\psi \le \gamma \cdot \text{cost}(\mathcal{A})$. Combining this with Eq. (4), the proof is concluded.   □

In the proof of Theorem 1 we further use the following lemmas, which can be proved using standard techniques (see e.g. [4,7]). The proofs are omitted due to lack of space.

Lemma 3. *Let $\beta, \alpha \ge 1$. Let f, Q, η such that Assumption 1 holds. If for all $S \subseteq \mathcal{X} \times \mathcal{Y}$, $\max_{x \in \mathcal{X}} u^f(x, S) \ge \frac{Q - f(S)}{\beta \text{OPT}}$, then for any α-approximate greedy algorithm with u^f, $\text{cost}(\mathcal{A}) \le \alpha\beta(\log(Q/\eta) + 1)\text{OPT}$.*

Lemma 4. *Let f, Q, η such that Assumption 1 holds and f is consistency-aware. Let $S \subseteq \mathcal{X} \times \mathcal{Y}$. Define $f' : 2^{\mathcal{X} \times \mathcal{Y}} \to \mathbb{R}_+$ by $f'(T) := f(T \cup S) - f(S)$. Let $Q' = Q - f(S)$. Then*

1. *f' is submodular, monotone and consistency-aware, with $f'(\emptyset) = 0$.*
2. *Let $\mathcal{A}$ be an interactive algorithm for f', Q'. Let $\beta \ge 1$. If $\max_{x \in \mathcal{X}} u^{f'}(x, \emptyset) \ge \frac{Q'}{\beta \text{OPT}'}$, where OPT' is the optimal cost for f', Q', then for any $S \subseteq \mathcal{X} \times \mathcal{Y}$, $\max_{x \in \mathcal{X}} u^f(x, S) \ge \frac{Q - f(S)}{\beta \text{OPT}}$.*

Proof (of Theorem 1). Let f', Q', OPT' as in Lemma 4, and let $\mathcal{A}^*$ be an optimal algorithm for f', Q'. Let $\mathcal{A}^*$ be an optimal algorithm for f', Q'. Since f is a learning objective, then so is f', and by Lemma 1 we can choose $\mathcal{A}^*$ to be bifurcating. Combining this with the first part of Lemma 4, the conditions of Lemma 2 hold. Therefore $u^{f'}(x, \emptyset) \ge Q'/\text{cost}(\mathcal{A}^*) \ge Q'/(r^{[2]}_{\text{cost}} \cdot \text{OPT}')$. By the second part of Lemma 4, $\forall S \subseteq \mathcal{X} \times \mathcal{Y}, u^f(x, S) \ge \frac{Q - f(S)}{r^{[2]}_{\text{cost}} \cdot \text{OPT}}$. Therefore, by Lemma 3, $\text{cost}(\mathcal{A}) \le \alpha(\log(Q/\eta) + 1) \cdot r^{[2]}_{\text{cost}} \cdot \text{OPT}$.   □

Next, we show that a linear dependence of the approximation guarantee on $r^{[2]}_{\text{cost}}$ is necessary for any greedy algorithm. To show the lower bound, we must exclude greedy algorithms that choose the utility function according to the set of available actions $\mathcal{X}$. Formally, define *local* greedy algorithms as follows. Assume there is a super-domain of all possible actions $\bar{\mathcal{X}}$, and consider an algorithm which receives as input a subset $\mathcal{X} \subseteq \bar{\mathcal{X}}$ of available actions. We say that such an algorithm is *local* greedy if it greedily selects the next action out of $\mathcal{X}$ using a fixed utility function $u : \bar{\mathcal{X}} \times 2^{\bar{\mathcal{X}} \times \mathcal{Y}} \to \mathbb{R}_+$, which does not depend on $\mathcal{X}$. The following lower bound shows that there exists a learning objective such that the approximation guarantee of any local greedy algorithm grows with $r_{\text{cost}}^{[2]}$ or is trivially bad.

Theorem 2. *Let f be the version-space reduction objective function with the corresponding $Q = 1 - 1/|\mathcal{H}|$ and $\eta = 1/|\mathcal{H}|$. For any value of OPT, $r_{\text{cost}}^{[2]} > 1$, and integer size of Q/η, there exist $\bar{\mathcal{X}}, \mathcal{H}$, and cost such that $\text{cost}(x,y)$ depends only on y, and such that for any local greedy algorithm $\mathcal{A}$, there exists an input domain $\mathcal{X} \subseteq \bar{\mathcal{X}}$ such that, for η as in Theorem 1,*

$$\text{cost}(\mathcal{A}) \geq \min\left(\frac{r_{\text{cost}}^{[2]}}{\log_2(Q/\eta)}, \frac{Q/\eta}{\log_2(Q/\eta)}\right) \cdot \text{OPT}.$$

Here $\text{cost}(\mathcal{A})$ and OPT refer to the costs for the domain $\mathcal{X}$.

Proof. Define $\mathcal{Y} = \{1,2,3\}$. Let $\bar{\mathcal{X}} = \{a_i \mid i \in [k]\} \cup \{b_j^t \mid j \in [k], t \in [[\lceil \log_2(k-2) \rceil]]\}$. Let $\text{cost}(x,y) = c_y$ for all $x \in \mathcal{X}$, where $c_3 \geq c_2 > c_1 = 0$. Set c_2, c_3 such that $r_{\text{cost}}^{[2]} = c_3/c_2$. Let $\mathcal{H} = \{h_i \mid i \in [k]\}$, where h_i is defined as follows: for a_j,

$$h_i(a_j) := \begin{cases} 1 & i = j \\ 2 & i \neq j. \end{cases}$$

For b_j^t and $i \neq j$, let $l_{i,j}$ be the location of i in $(1, \ldots, j-1, j+1, \ldots, k)$, where the locations range from 0 to $k-2$. Denote by $l_{i,j}^t$ the t'th most significant bit in the binary expansion of $l_{i,j}$ to $\lceil \log_2(k-2) \rceil$ bits. Define

$$h_i(b_j^t) := \begin{cases} 1 & i \neq j \wedge l_{i,j}^t = 0 \\ 2 & i \neq j \wedge l_{i,j}^t = 1 \\ 3 & i = j \end{cases}$$

Fix an index $n \in [k]$. Let $\mathcal{X}_n = \{a_i \mid i \in [k]\} \cup \{b_n^t \mid t \in [[\lceil \log_2(k-2) \rceil]]\}$. We now show an interactive algorithm for $\mathcal{X}_n$ and bound its worst-case cost. On the first iteration, the algorithm selects action a_n. If the result is 1, then $\text{VS}(S) = \{h_n\}$, hence $f(S) \geq Q$. In this case the cost is $c_1 = 0$. Otherwise, the algorithm selects all actions in $\{b_n^t \mid t \in [[\lceil \log_2(k-2) \rceil]]\}$. The responses reveal the binary expansion of $l_{j,n}$, thus limiting the version space to the a single h_i, hence $f(S) \geq Q$. In this case the total cost is at most $c_2 \lceil \log_2(k-2) \rceil$.

Now, consider a local greedy algorithm. Let $\sigma : [k] \to [k]$ be a permutation that represents the order in which $a_1, \ldots, a_k$ would be selected by the utility function if only a_i were available, and their response was always 2. Formally, $\sigma(i) = \text{argmax}_{i \in [k]} u(a_{\sigma(i)}, \{(a_{\sigma(i')}, 2) \mid i' \in [i-1]\})$.[1]

---

[1] We may assume without loss of generality that $u(x, S) = 0$ whenever $(x, y) \in S$.

Suppose the input to the algorithm is $\mathcal{X}_{\sigma(k)}$. Denote $S_i = \{(a_{\sigma(i')}, 2) \mid i' \in [i-1]\}$, and suppose $h^* = h_{\sigma(k)}$. First, assume that $\max_t u(b^t_{\sigma(k)}, S_{k-1}) < u(a_{\sigma(k)}, S_{k-1})$. Then all of $a_{\sigma(1)}, \ldots, a_{\sigma(k-1)}$ are selected before any of $b^t_{\sigma(k)}$, and the version space is reduced to a singleton only after these $k-1$ actions. Therefore the cost of the run is at least $c_2(k-1)$. Second, assume that this assumption does not hold. Then there exists an integer i' such that $\max_t u(b^t_{\sigma(k)}, S_{i'-1}) > u(a_{\sigma(i)}, S_{i'-1})$. Let i' be the smallest such integer. Then, the algorithm receives 2 on each of the actions $a_{\sigma(1)}, \ldots, a_{\sigma(i'-1)}$, and its next action is $b^t_{\sigma(k)}$ for some t. Hence the cost of the run is at least c_3.

To summarize, the worst-case cost of every local greedy algorithm is at least $\min\{c_3, c_2(k-1)\}$ for at least one of the inputs $\mathcal{X}_n$. The worst-case cost of the optimal algorithm for each $\mathcal{X}_n$ is at most $c_2\lceil \log_2(k-2) \rceil$. The statement of the theorem follows. $\qquad\square$

3.2   Guarantees for General Objectives

In the previous section we showed that for learning objectives, the achievable approximation guarantee for greedy algorithms is characterized by $r^{[2]}_{\text{cost}}$. We now turn to general consistency-aware objective functions. We show that the factor of approximation for this class depends on a different property of the cost function, which is lower bounded by $r^{[2]}_{\text{cost}}$. Define $\text{cost}_{\max} := \max_{(x,y)\in\mathcal{X}\times\mathcal{Y}} \text{cost}(x,y)$, and let

$$\phi_{\min} := \min_{x\in\mathcal{X}} \phi(x), \qquad gr^{[2]}_{\text{cost}} := \frac{\text{cost}_{\max}}{\phi_{\min}}.$$

We term the ratio $gr^{[2]}_{\text{cost}}$ the *Global second smallest cost ratio*. As we show below, the approximation factor is best when $gr^{[2]}_{\text{cost}}$ is equal to 1. This is the case if there is at most one preferred response for every action, and in addition, all the non-preferred responses for all actions have the same cost.

Theorem 3. *Let $f : 2^{\mathcal{X}\times\mathcal{Y}} \to \mathbb{R}_+, Q > 0, \eta > 0$ such that Assumption 1 holds and f is consistency-aware. Let $\mathcal{A}$ be an α-approximate greedy algorithm for the utility function u^f. Then $\text{cost}(\mathcal{A}) \leq 2\min(gr^{[2]}_{\text{cost}}, r_{\text{cost}}) \cdot \alpha \cdot (\log(Q/\eta)+1) \cdot \text{OPT}$.*

Like Theorem 1 for learning objectives, this result for general objectives is a significant improvement over the trivial bound, mentioned in Sect. 2, which depends on the cost ratio, since the ratio $gr^{[2]}_{\text{cost}}/r_{\text{cost}}$ can be unbounded. For instance, consider a case where each action has one response with a cost of 1 and all other responses have a cost of $M \gg 1$. Then $r_{\text{cost}} = M$ but $gr^{[2]}_{\text{cost}} = 1$.

The proof of Theorem 3 hinges on two main observations: First, any interactive algorithm may be "reordered" without increasing its cost, so that all actions with only one possible response (given the history so far) are last. Second, there are two distinct cases for the optimal algorithm: In one case, for all $h \in \mathcal{H}$, the optimal algorithm obtains a value of at least $Q/2$ before performing actions with a single possible response. In the other case, there exists at least one mapping h for which actions with a single possible response obtain at least $Q/2$ of the value. We start with the following lemma, which handles the case where OPT $< \phi_{\min}$.

Lemma 5. *Let* $f : 2^{\mathcal{X} \times \mathcal{Y}} \to \mathbb{R}_+$, $Q > 0$. *Suppose that* f *is submodular, and* $f(\emptyset) = \emptyset$. *If* $\mathrm{OPT} < \phi_{\min}$, *then* $\max_{x \in \mathcal{X}} u^f(x, \emptyset) \geq Q/OPT$.

Proof. For every action $x \in \mathcal{X}$ there is at most a single y with $\mathsf{cost}(x, y) < \phi_{\min}$. Denote this response by $y(x)$. Let $\mathcal{A}$ be an optimal algorithm for f, Q. For any value of $h^* \in \mathcal{H}$, $\mathcal{A}$ only receives responses with costs less than $\phi_{\min}$. Therefore for any x that $\mathcal{A}$ selects, it receives the response $y(x)$, regardless of the identity of h^*. In other words, for all $h \in \mathcal{H}$, in every iteration t, $\mathcal{A}$ selects an action x such that $\mathcal{Y}(x, S_{t-1}^h[\mathcal{A}]) = \{y(x)\}$. It follows that for all t, $S_t^h[\mathcal{A}]$ is the same for all $h \in \mathcal{H}$. Therefore, there is a fixed set of actions that $\mathcal{A}$ selects during its run, regardless of h^*. Let $\mathcal{X}' \subseteq \mathcal{X}$ be that set. Then for all $h \in \mathcal{H}, x \in X'$, $h(x) = y(x)$. For a set $A \subseteq \mathcal{X}$, denote $A^{[y(x)]} = \{(x, y(x)) \mid x \in A\}$. We have $f(\mathcal{X}'^{[y(x)]}) \geq Q$ and $\mathsf{cost}(\mathcal{X}'^{[y(x)]}) = \mathrm{OPT}$. By the submodulrity of f, and since $f(\emptyset) = 0$, we have $Q/\mathrm{OPT} \leq f(\mathcal{X}'^{[y(x)]})/\mathrm{OPT} \leq \sum_{x \in \mathcal{X}'} f((x, y(x)))/\sum_{x \in \mathcal{X}'} \mathsf{cost}(x, y(x))$. Therefore there exists some $x \in \mathcal{X}'$ with $f((x, y(x)))/\mathsf{cost}(x, y(x)) \geq Q/\mathrm{OPT}$. Moreover, for this x we have $\mathcal{Y}(x, \emptyset) = \{y(x)\}$. Therefore $u^f(x, \emptyset) = f((x, y(x)))/\mathsf{cost}(x, y(x)) \geq Q/\mathrm{OPT}$. $\square$

We now turn to the main lemma, to address the two cases described above.

Lemma 6. *Let* $f : 2^{\mathcal{X} \times \mathcal{Y}} \to \mathbb{R}_+$, $Q > 0$. *Suppose that* f *is submodular, and* $f(\emptyset) = \emptyset$. *Assume that* f *is consistency-aware. There exists* $x \in \mathcal{X}$ *such that* $u^f(x, \emptyset) \geq \dfrac{Q}{2 \min(gr_{cost}^{[2]}, r_{cost})\mathrm{OPT}}$.

Proof. If $\mathrm{OPT} < \phi_{\min}$, the statement holds by Lemma 5. Suppose that $\mathrm{OPT} \geq \phi_{\min}$. Let $\mathcal{A}^*$ be optimal algorithm for f, Q. We may assume without loss of generality, that for any $h^* \in \mathcal{H}$, if $\mathcal{A}^*$ selects an action that has only one possible response (given the current version space) at some iteration t, then all actions selected after iteration t also have only one possible response. This does not lose generality: let t be the first iteration such that the action at iteration t has one possible response, and the action at iteration $t + 1$ has two possible responses. Consider an algorithm which behaves the same as $\mathcal{A}^*$, except that at iteration t it selects the second action, and at iteration $t + 1$ it selects the first action (regardless of the response to the first action). This algorithm has the same cost as $\mathcal{A}^*$.

For $h \in \mathcal{H}$, define $\mathrm{val}(h) := f(S_{t_h}^h[\mathcal{A}^*])$, where t_h is the last iteration in which an action with more than one possible response (given the current version space) is selected, if $h^* = h$. Consider two cases: (a) $\min_{h \in \mathcal{H}} \mathrm{val}(h) \geq Q/2$ and (b) $\exists h \in \mathcal{H}, \mathrm{val}(h) < Q/2$. In case (a), there is a bifurcating algorithm that obtains $f(S) \geq Q/2$ at cost at most OPT: This is the algorithm that selects the same actions as $\mathcal{A}^*$, but terminates before selecting the first action that has a single response given the current version space. We also have $r_{cost}^{[2]} \leq \min(gr_{cost}^{[2]}, r_{cost})$. By Lemma 2, there exists some $x \in \mathcal{X}$ such that $u^f(x, \emptyset) \geq \dfrac{Q}{2 \min(gr_{cost}^{[2]}, r_{cost})\mathrm{OPT}}$.

In case (b), let $h \in \mathcal{H}$ such that $\mathrm{val}(h) < Q/2$. Denote $S_t := S_t^h[\mathcal{A}^*]$. Let $(x_t, h(x_t))$ be the action and the response received in iteration t if $h^* = h$. Then $f(S_{t_h}) < Q/2$. Let $S' = \{(x_t, h(x_t)) \mid t > t_h\}$. Then $f(S_{t_h} \cup S') \geq Q$.

Since $f(\emptyset) = 0$ and f is submodular, $f(S') = f(S') - f(\emptyset) \geq f(S_{t_h} \cup S') - f(S_{t_h}) \geq Q - \mathrm{val}(h) \geq Q/2$. In addition, $f(S') \leq \sum_{t > t_h} f(\{(x_t, h(x_t))\})$. Hence $\frac{Q}{2\mathrm{OPT}} \leq \frac{f(S')}{\mathrm{OPT}} \leq \frac{\sum_{t > t_h} f(\{(x_t, h(x_t))\})}{\sum_{t > t_h} \mathrm{cost}(x_t, y_t)}$. Therefore there is some t' such that $\frac{f(\{(x_{t'}, h(x_{t'}))\})}{\mathrm{cost}(x_{t'}, h(x_{t'}))} \geq \frac{Q}{2\mathrm{OPT}}$. Therefore,

$$u^f(x_{t'}, \emptyset) = \min_{y \in \mathcal{Y}(x_{t'}, \emptyset)} \min\{f(\{(x_{t'}, y)\}), Q\}/\mathrm{cost}(x_{t'}, y)$$

$$\geq \min\{Q/\mathrm{cost}_{\max}, \min_{y \in \mathcal{Y}} f(\{(x_{t'}, y)\})/\mathrm{cost}(x_{t'}, y)\}$$

$$\geq \min\{Q/\mathrm{cost}_{\max}, \frac{Q}{2\mathrm{OPT}}, \min_{y \in \mathcal{Y} \setminus \{h(x_{t'})\}} f(\{(x_{t'}, y)\})/\mathrm{cost}(x_{t'}, y)\}.$$

Now, $\mathrm{cost}_{\max} \leq gr_{\mathrm{cost}}^{[2]} \cdot \phi_{\min} \leq gr_{\mathrm{cost}}^{[2]} \cdot \mathrm{OPT}$, from our assumption that $\mathrm{OPT} \geq \phi_{\min}$, and $\mathrm{cost}_{\max} \leq r_{\mathrm{cost}}\mathrm{cost}(x_{t'}, h(x_{t'})) \leq r_{\mathrm{cost}} \cdot \mathrm{OPT}$. Therefore

$$u^f(x_{t'}, \emptyset) \geq \min\left\{ \frac{Q}{2\min(gr_{\mathrm{cost}}^{[2]}, r_{\mathrm{cost}})\mathrm{OPT}}, \min_{y \in \mathcal{Y} \setminus \{h(x_{t'})\}} \frac{f(\{(x_{t'}, y)\})}{\mathrm{cost}(x_{t'}, y)} \right\}.$$

We have left to show a lower bound on $\min_{y \in \mathcal{Y} \setminus \{h(x_{t'})\}} \frac{f(\{(x_{t'}, y)\})}{\mathrm{cost}(x_{t'}, y)}$. By the choice of t', $x_{t'}$ has only one possible response given the current version space, that is $|\mathcal{Y}(x_{t'}, S_{t'-1})| = 1$. Since the same holds for all $t > t_h$, we have $\mathrm{VS}(S_{t'-1}) = \mathrm{VS}(S_{t_h})$, hence also $\mathcal{Y}(x_{t'}, S_{t_h}) = \{h(x_{t'})\}$. It follows that for $y \in \mathcal{Y} \setminus \{h(x_{t'})\}$, the set $S_{t_h} \cup \{(x_{t'}, y)\}$ is not consistent with any $h \in \mathcal{H}$. Since f is consistency-aware, it follows that $f(S_{t_h} \cup \{(x_{t'}, y)\}) \geq Q$. Therefore $f(\{(x_{t'}, y)\}) = f(\{(x_{t'}, y)\}) - f(\emptyset) \geq f(S_{t_h} \cup \{(x_{t'}, y)\}) - f(S_{t_h}) \geq Q - \mathrm{val}(h) \geq Q/2$. Hence $\frac{f(\{(x_{t'}, y)\})}{\mathrm{cost}(x_{t'}, y)} \geq \frac{Q}{2c_{\max}} \geq \frac{Q}{2\min(gr_{\mathrm{cost}}^{[2]}, r_{\mathrm{cost}})\mathrm{OPT}}$. It follows that $u^f(x_t, \emptyset) \geq \frac{Q}{2c_{\max}} \geq \frac{Q}{2\min(gr_{\mathrm{cost}}^{[2]}, r_{\mathrm{cost}})\mathrm{OPT}}$ also in case (b). $\square$

Using the lemmas above, the proof of Theorem 3 is straight forward.

Proof (of Theorem 3). Let f', Q', OPT' as in Lemma 4, and let $\mathcal{A}^*$ be an optimal algorithm for f', Q'. Let $\mathcal{A}^*$ be an optimal algorithm for f', Q'. From the first part of Lemma 4, the conditions of Lemma 6 hold for f', Q'. Therefore $u^{f'}(x, \emptyset) \geq \frac{Q'}{2\min(gr_{\mathrm{cost}}^{[2]}, r_{\mathrm{cost}})\mathrm{OPT}'}$. By the second part of Lemma 4, $\forall S \subseteq \mathcal{X} \times \mathcal{Y}$, $u^f(x, S) \geq \frac{Q - f(S)}{2\min(gr_{\mathrm{cost}}^{[2]}, r_{\mathrm{cost}})\mathrm{OPT}}$. Therefore, by Lemma 3, $\mathrm{cost}(\mathcal{A}) \leq 2\alpha \min(gr_{\mathrm{cost}}^{[2]}, r_{\mathrm{cost}})(\log(Q/\eta) + 1) \cdot \mathrm{OPT}$. $\square$

The guarantee of Theorem 3 for general objectives is weaker than the guarantee for learning objectives given in Theorem 1: The ratio $\min(gr_{\mathrm{cost}}^{[2]}, r_{\mathrm{cost}})/r_{\mathrm{cost}}^{[2]}$ is always at least 1, and can be unbounded. For instance, if there are two actions that have two responses each, and all action-response pairs cost 1, except for one action-response pair which costs $M \gg 1$, then $r_{\mathrm{cost}} = 1$ but $gr_{\mathrm{cost}}^{[2]} = M$. Nonetheless, the following theorem shows that for general functions, a dependence on $\min(gr_{\mathrm{cost}}^{[2]}, r_{\mathrm{cost}})$ is essential in any greedy algorithm.

Theorem 4. *For any values of* $gr^{[2]}_{\text{cost}}, r_{\text{cost}} > 0$, *there exist* $\bar{\mathcal{X}}, \mathcal{Y}, \mathcal{H}, \text{cost}$ *with* $|\mathcal{Y}| = 2$ *and* $r^{[2]}_{\text{cost}} = 1$, *and a submodular monotone f which is consistency-aware, with $Q/\eta = 1$, such that for any local greedy algorithm $\mathcal{A}$, there exists an input domain $\mathcal{X} \subseteq \bar{\mathcal{X}}$ such that* $\text{cost}(\mathcal{A}) \geq \frac{1}{2}\min(gr^{[2]}_{\text{cost}}, r_{\text{cost}}) \cdot \text{OPT}$, *where* $\text{cost}(\mathcal{A})$ *and* OPT *refer to the costs of an algorithm running on the domain $\mathcal{X}$.*

Proof. Define $\mathcal{Y} := \{0,1\}$. Let $g, r > 0$ be the desired values for $gr^{[2]}_{\text{cost}}, r_{\text{cost}}$. Let $c_1 > 0$, $c_2 := c_1 \min(g, r)$. If $g < r$, define $c_3 := c_1/r$, $c_4 := c_1$. Otherwise, set $c_4 := c_3 := c_2/g$. Define $k := \lceil c_2/c_1 \rceil + 1$. Let $\bar{\mathcal{X}} = \{a_i \mid i \in [k]\} \cup \{b_i \mid i \in [k]\} \cup \{c\}$. Let $\bar{\mathcal{H}} = \{h_i \mid i \in [k]\}$ where h_i is defined as follows: $\forall i, j \in [k], h_i(a_j) = h_i(b_j)$, and equal to 1 if and only if $i = j$, and zero otherwise, and $\forall i \in [k], h_i(c) = i \bmod 2$. Let the cost function be as follows, where $c_2 \geq c_1 > 0$, and $c_3, c_4 > 0$: $\text{cost}(a_i, y) = c_1$, $\text{cost}(b_i, y) = c_{y+1}$, and $\text{cost}(c, y) = c_{y+3}$. Then $gr^{[2]}_{\text{cost}} = g$, $r_{\text{cost}} = r$ as desired.

Define f such that $\forall S \subseteq \mathcal{X} \times \mathcal{Y}$, $f(S) = Q$ if there exists in S at least one of $(a_i, 1)$ for some $i \in [k]$ or (b_i, y) for some $i \in [k], y \in \mathcal{Y}$. Otherwise, $f(S) = 0$. Note that (f, Q) is consistency-aware. Fix an index $n \in [k]$. Let $\mathcal{X}_n = \{a_i \mid i \in [k]\} \cup \{b_n\}$. We have $\text{OPT} = 2c_1$: An interactive algorithm can first select a_n, and then, only if the response is $y = 0$, select b_n. Now, consider a local greedy algorithm with a utility function u. Let $\sigma : [k] \to [k]$ be a permutation that represents the order in which $a_1, \ldots, a_k$ would be selected by the utility function if only a_i were considered, and their response was always $y = 0$. Formally, $\sigma(i) = \text{argmax}_{i \in [k]} u(a_{\sigma(i)}, \{(a_{\sigma(i')}, 0) \mid i' \in [i-1]\})$ (See footnote 1).

Now, suppose the input to the algorithm is $\mathcal{X}_{\sigma(k)}$. Denote $S_i = \{(a_{\sigma(i')}, 0) \mid i' \in [i-1]\}$, and suppose that there exists an integer i' such that $u(b_{\sigma(k)}, S_{i'-1}) > u(a_{\sigma(i)}, S_{i'-1})$, and let i' be the smallest such integer. Then, if the algorithm receives 0 on each of the actions $a_{\sigma(1)}, \ldots, a_{\sigma(i'-1)}$, its next action will be $b_{\sigma(k)}$. In this case, if $h^* = h_{\sigma(k)}$, then $b_{\sigma(k)}$ is queried before $a_{\sigma(k)}$ is queried and the response $y = 1$ is received. Thus the algorithm pays at least c_2 in the worst-case. On the other hand, if such an integer i' does not exist, then if $h^* = h_{\sigma(k)}$, the algorithm selects actions $a_{\sigma(1)}, \ldots, a_{\sigma(k-1)}$ before terminating. In this case the algorithm receives $k - 1$ responses 0, thus its cost is at least $c_1(k - 1)$. To summarize, every local greedy algorithm pays at least $\min\{c_2, c_1(k - 1)\}$ for at least one of the inputs $\mathcal{X}_n$, while $\text{OPT} = 2c_1$. By the definition of k, $\min\{c_2, c_1(k - 1)\} \geq c_2$. Hence the cost of the local greedy algorithm is at least $\frac{c_2}{2c_1}\text{OPT}$. $\square$

To summarize, for both learning objectives and general objectives, we have shown that the factors $r^{[2]}_{\text{cost}}$ and $gr^{[2]}_{\text{cost}}$, respectively, characterize the approximation factors obtainable by a greedy algorithm.

4   Experiments

We performed experiments to compare the worst-case costs of a greedy algorithm that uses the proposed u^f, to a greedy algorithm that ignores

Table 1. Results of experiments. Left: Facebook dataset, right: GR-QC dataset

Test parameters			Results: $cost(\mathcal{A})$			Test parameters			Results: $cost(\mathcal{A})$		
Test	$r_{cost}^{[2]}$	$gr_{cost}^{[2]}$	u^f	u_2^f	u_3^f	Test	$r_{cost}^{[2]}$	$gr_{cost}^{[2]}$	u^f	u_2^f	u_3^f
f=edge users, 3 communities	5	5	52	255	157	f=edge users, 3 communities	5	5	51	181	123
	100	100	148	5100	2722		100	100	147	3503	1833
	1	100	49	52	2821		1	100	51	53	2526
f=edge users, 10 communities	5	5	231	256	242	f=edge users, 10 communities	5	5	246	260	245
	100	100	4601	5101	4802		100	100	4901	5200	4900
	1	100	50	52	2915		1	100	49	52	3217
f=v. reduction, 3 communities	5	5	13	20	15	f=v. reduction, 3 communities	5	5	10	20	15
	100	100	203	400	300		100	100	106	400	300
	1	100	3	4	201		1	100	3	400	300
f=v. reduction, 10 communities	5	5	8	20	15	f=v. reduction, 10 communities	5	5	15	16	15
	100	100	105	400	300		100	100	300	301	300
	1	100	101	103	201		1	100	3	201	300

response-dependent costs, and uses instead variant of u^f, notated u_2^f, that assumes that responses for the same action have the same cost, which was set to be the maximal response cost for this action. We also compared to u_3^f, a utility function which gives the same approximation guarantees as given in Theorem 3 for u^f. Formally, $u_2^f(x, S) := \min_{h \in \mathrm{VS}(S)} \frac{\delta_{\min(f,Q)}((x,h(x))|S)}{\max_{y \in \mathcal{Y}} cost(x,y)}$ and $u_3^f(x, S) := \min_{h \in \mathrm{VS}(S)} \frac{\delta_{\min(f,Q)}((x,h(x))|S)}{\min\{cost(x,h(x)),\phi_{\min}\}}$. We tested these algorithms on a social network marketing objective, where users in a social network are partitioned into communities. Actions are users, and a response identifies the community the user belongs to. We tested two objective functions: "edge users" counts how many of the actions are users who have friends not from their community, assuming that these users can be valuable promoters across communities. The target value Q was set to 50. The second objective function was the version-reduction objective function, and the goal was to identify the true partition into communities out of the set of possible partitions, which was generated by considering several possible sets of "center users", which were selected randomly. We compared the worst-case costs of the algorithms under several configurations of number of communities and the values of $r_{cost}^{[2]}$, $gr_{cost}^{[2]}$. The cost ratio r_{cost} was infinity in all experiments, obtained by always setting a single response to have a cost of zero for each action. Social network graphs were taken from a friend graph from Facebook[2] [10], and a collaboration graph from Arxiv GR-QC community[3] [9]. The results are reported in Table 1. We had $|\mathcal{H}| = 100$ for all tests with 3 communities, and $|\mathcal{H}| = 500$ for all tests with 10 communities. The results show an overall preference to the proposed u^f.

Acknowledgements. This work was supported in part by the Israel Science Foundation (grant No. 555/15).

---

[2] http://snap.stanford.edu/data/egonets-Facebook.html.
[3] http://snap.stanford.edu/data/ca-GrQc.html.

References

1. Cuong, N., Lee, W., Ye, N.: Near-optimal adaptive pool-based active learning with general loss. In: 30th Conference on Uncertainty in Artificial Intelligence (2014)
2. Dasgupta, S.: Analysis of a greedy active learning strategy. In: NIPS 17, pp. 337–344 (2004)
3. Feige, U.: A threshold of ln(n) for approximating set cover. J. ACM (JACM) **45**(4), 634–652 (1998)
4. Golovin, D., Krause, A.: Adaptive submodularity: theory and applications in active learning and stochastic optimization. J. Artif. Intell. Res. **42**, 427–486 (2011)
5. Golovin, D., Krause, A., Ray, D.: Near-optimal bayesian active learning with noisy observations. In: NIPS, pp. 766–774 (2010)
6. Gonen, A., Sabato, S., Shalev-Shwartz, S.: Efficient active learning of halfspaces: an aggressive approach. J. Mach. Learn. Res. **14**, 2487–2519 (2013)
7. Guillory, A., Bilmes, J.A.: Interactive submodular set cover. In: Proceedings of the 27th International Conference on Machine Learning (ICML), pp. 415–422 (2010)
8. Kapoor, A., Horvitz, E., Basu, S.: Selective supervision: guiding supervised learning with decision-theoretic active learning. In: Proceedings of IJCAI (2007)
9. Leskovec, J., Kleinberg, J., Faloutsos, C.: Graph evolution: densification and shrinking diameters. ACM Trans. Knowl. Disc. Data (TKDD) **1**(1), 2 (2007)
10. Leskovec, J., Mcauley, J.J.: Learning to discover social circles in ego networks. In: Advances in Neural Information Processing Systems, pp. 539–547 (2012)
11. Lin, H., Bilmes, J.: A class of submodular functions for document summarization. In: Proceedings of the 49th Annual Meeting of the Association for Computational Linguistics: Human Language Technologies, HLT 2011, vol. 1, pp. 510–520 (2011)
12. McCallum, A.K., Nigam, K.: Employing em and pool-based active learning for text classification. In: ICML (1998)
13. Mirzasoleiman, B., Karbasi, A., Sarkar, R., Krause, A.: Distributed submodular maximization: identifying representative elements in massive data. In: NIPS, pp. 2049–2057 (2013)
14. Nemhauser, G.L., Wolsey, L.A., Fisher, M.L.: An analysis of approximations for maximizing submodular set functions. Math. Program. **14**(1), 265–294 (1978)
15. Sabato, S., Sarwate, A.D., Srebro, N.: Auditing: active learning with outcome-dependent query costs. In: Advances in Neural Information Processing Systems 26 (NIPS) (2013)
16. Sabato, S., Sarwate, A.D., Srebro, N.: Auditing: active learning with outcome-dependent query costs. arXiv preprint arXiv:1306.2347v4 (2015)
17. Saettler, A., Laber, E., Cicalese, F.: Trading off worst and expected cost in decision tree problems and a value dependent model. arXiv preprint arXiv:1406.3655 (2014)
18. Saettler, A., Laber, E., Cicalese, F.: Approximating decision trees with value dependent testing costs. Inf. Process. Lett. **115**(68), 594–599 (2015)
19. Wolsey, L.: An analysis of the greedy algorithm for the submodular set covering problem. Combinatorica **2**(4), 385–393 (1982)

Classifying the Arithmetical Complexity
of Teaching Models

Achilles A. Beros[1], Ziyuan Gao[2]([✉]), and Sandra Zilles[2]

[1] Department of Mathematics, University of Hawaii, Honolulu, USA
beros@math.hawaii.edu
[2] Department of Computer Science, University of Regina,
Regina S4S 0A2, SK, Canada
{gao257,zilles}@cs.uregina.ca

Abstract. This paper classifies the complexity of various teaching models by their position in the arithmetical hierarchy. In particular, we determine the arithmetical complexity of the index sets of the following classes: (1) the class of uniformly r.e. families with finite teaching dimension, and (2) the class of uniformly r.e. families with finite positive recursive teaching dimension witnessed by a uniformly r.e. teaching sequence. We also derive the arithmetical complexity of several other decision problems in teaching, such as the problem of deciding, given an effective coding $\{\mathcal{L}_0, \mathcal{L}_1, \mathcal{L}_2, \ldots\}$ of all uniformly r.e. families, any e such that $\mathcal{L}_e = \{L_0^e, L_1^e, \ldots, \}$, any i and d, whether or not the teaching dimension of L_i^e with respect to $\mathcal{L}_e$ is upper bounded by d.

1 Introduction

A fundamental problem in computational learning theory is that of characterising identifiable classes in a given learning model. Consider, for example, Gold's [5] model of learning from positive data, in which a learner is fed piecewise with all the positive examples of an unknown target language – often coded as a set of natural numbers – in an arbitrary order; as the learner processes the data, it outputs a sequence of hypotheses that must converge syntactically to a correct conjecture. Of particular interest to inductive inference theorists is the learnability of classes of recursively enumerable (r.e.) languages. Angluin [1] demonstrated that a uniformly recursive family is learnable in Gold's model if and only if it satisfies a certain "tell-tale" condition. As a consequence, the family of nonerasing pattern languages over any fixed alphabet and the family of regular expressions over $\{0, 1\}$ that contain no operators other than concatenation and Kleene plus are both learnable in the limit. On the other hand, even a relatively simple class such as one consisting of an infinite set L and all the finite subsets of L cannot be learnt in the limit [5, Theorem I.8]. Analogous characterisations of learnability have since been discovered for uniformly r.e. families as well as for other learning models such as behaviourally correct learning [2,8].

Intuitively, the structural properties of learnable families seem to be related to the "descriptive complexity" of such families. By fixing a system of describing families of sets, one may wish to compare the relative descriptive complexities of

© Springer International Publishing Switzerland 2016
R. Ortner et al. (Eds.): ALT 2016, LNAI 9925, pp. 145–160, 2016.
DOI: 10.1007/978-3-319-46379-7_10

families identifiable under different criteria. One idea, suggested by computability theory and the fact that many learnability criteria may be expressed as first-order formulae, is to analyse the quantifier complexity of the formula defining the class of learnable families. In other words, one may measure the descriptive complexity of identifiable classes that are first-order definable by determining the position of their corresponding index sets in the *arithmetical hierarchy*. This approach to measuring the complexity of learnable classes was taken by Brandt [4], Klette [9], and later Beros [3]. More specifically, Brandt [4, Corollary 1] showed that every identifiable class of partial-recursive functions is contained in another identifiable class with an index set that is in $\Sigma_3 \cap \Pi_3$, while Beros [3] established the arithmetical complexity of index sets of uniformly r.e. families learnable under different criteria.

The purpose of the present work is to determine the arithmetical complexity of various decision problems in *algorithmic teaching*. Teaching may be viewed as a natural counterpart of learning, where the goal is to find a sample efficient learning *and* teaching protocol that guarantees learning success. Thus, in contrast to a learning scenario where the learner has to guess a target concept based on labelled examples from a truthful but arbitrary (possibly even adversarial) source, the learner in a cooperative teaching-learning model is presented with a sample of labelled examples carefully chosen by the teacher, and it decodes the sample according to some pre-agreed protocol. We say that a family is "teachable" in a model if and only if the associated teaching parameter – such as the *teaching dimension* [6], the *extended teaching dimension* [7] or the *recursive teaching dimension* [14] – of the family is finite. Due to the ubiquity of number-able families of r.e. sets in theoretical computer science and the naturalness of such families, our work will focus on the class of uniformly r.e. families. Our main results classify the arithmetical complexity of index sets of uniformly r.e. families that are teachable under the teaching dimension model and a few variants of the recursive teaching dimension model.

From the viewpoint of computability theory, our work provides a host of natural examples of complete sets, thus supporting Rogers' view that many "arithmetical sets with intuitively simple definitions ... have proved to be Σ_n^0-complete or Π_n^0-complete (for some n)" [12, p. 330]. From the viewpoint of computational learning theory, our results shed light on the recursion-theoretic structural properties of the classes of uniformly r.e. families that are teachable in some well-studied models.

2   Preliminaries

The notation and terminology from computability theory adopted in this paper follow in general the book of Rogers [12].

$\forall^\infty x$ denotes "for almost every x", $\exists^\infty x$ denotes "for infinitely many x" and $\exists! x$ denotes "for exactly one x". $\mathbb{N}$ denotes the set of natural numbers, $\{0, 1, 2, \ldots\}$, and $\mathcal{R}$ $(= 2^{\mathbb{N}})$ denotes the power set of $\mathbb{N}$. For any function f, $\mathrm{ran}(f)$ denotes the range of f. For any set A, $\mathbb{1}_A$ will denote the characteristic function of A, that is, $\mathbb{1}_A(x) = 1$ if $x \in A$, and $\mathbb{1}_A(x) = 0$ if $x \notin A$. For any sets A and B, $A \times B = \{\langle a, b \rangle : a \in A \wedge b \in B\}$ and $A \oplus B$, the *join of A and B*, is the set $\{2x : x \in A\} \cup \{2y + 1 : y \in B\}$. Analogously, for any class $\{A_i : i \in \mathbb{N}\}$ of sets, $\bigoplus_{i \in \mathbb{N}} A_i$ is the infinite join of the A_i's. For any set A, A^* denotes the set of all finite sequences of elements of A. Given a sequence of families, $\{\mathcal{F}_i\}_{i \in \mathbb{N}}$, we define $\bigsqcup_{i \in \mathbb{N}} \mathcal{F}_i = \bigcup_{i \in \mathbb{N}} \{F \oplus \{i\} : F \in \mathcal{F}_i\}$. If there are only finitely many families, $\mathcal{F}_0, \ldots, \mathcal{F}_n$, we denote this by $\mathcal{F}_0 \sqcup \ldots \sqcup \mathcal{F}_n$. We call this the *disjoint union of $\mathcal{F}_1, \ldots, \mathcal{F}_n$*.

For any $\sigma, \tau \in \{0, 1\}^*$, $\sigma \preceq \tau$ if and only if σ is a prefix of τ, $\sigma \prec \tau$ if and only if σ is a proper prefix of τ, and $\sigma(n)$ denotes the element in the nth position of σ, starting from $n = 0$.

Let $W_0, W_1, W_2, \ldots$ be an acceptable numbering of all r.e. sets, and let $D_0, D_1, D_2, \ldots$ be a canonical numbering of all finite sets such that $D_0 = \emptyset$ and for any pairwise distinct numbers $x_1, \ldots, x_n$, $D_{2^{x_1} + \ldots + 2^{x_n}} = \{x_1, \ldots, x_n\}$. For all e and j, define $L_j^e = \{x : \langle j, x \rangle \in W_e\}$ and $\mathcal{L}_e = \{L_j^e : j \in \mathbb{N}\}$. L_j^e is the jth column of W_e. $L_{j,s}^e$ denotes the sth approximation of L_j^e which, without loss of generality, we assume is a subset of $\{0, \ldots, s\}$. Note that $\{\mathcal{L}_e : e \in \mathbb{N}\}$ is the class of all uniformly r.e. (u.r.e.) families, each of which is encoded as an r.e. set. Let COINF denote the index set of the class of all coinfinite r.e. sets, and let COF denote the index set of the class of all cofinite r.e. sets. Let INF denote the index set of the class of all infinite r.e. sets and FIN denote the index set of the class of all finite sets.

Definition 1. [12] A set $A \subseteq \mathbb{N}$ is in $\Sigma_0 (= \Pi_0 = \Delta_0)$ iff A is recursive. A is in Σ_n^0 iff there is a recursive relation R such that

$$x \in B \leftrightarrow (\exists y_1)(\forall y_2) \ldots (Q_n y_n) R(x, y_1, y_2, \ldots, y_n) \tag{1}$$

where $Q_n = \forall$ if n is even and $Q_n = \exists$ if n is odd. A set $A \subseteq \mathbb{N}$ is in Π_n^0 iff its complement $\overline{A}$ is in Σ_n^0. $(\exists y_1)(\forall y_2) \ldots (Q_n y_n)$ is known as a Σ_n^0 prefix; $(\forall y_1)(\exists y_2) \ldots (Q_n, y_n)$, where $Q_n = \exists$ if n is even and $Q_n = \forall$ if n is odd, is known as a Π_n^0 prefix. The formula on the right-hand side of (1) is called a Σ_n^0 *formula* and its negation is called a Π_n^0 *formula*. A set A is in Δ_n^0 iff A is in Σ_n^0 and A is in Π_n^0. Sets in Σ_n^0 (Π_n^0, Δ_n^0) are known as Σ_n^0 *sets* (Π_n^0 *sets*, Δ_n^0 *sets*). For any $n \geq 1$, a set A is Σ_n^0-*hard* (Π_n^0-*hard*) iff every Σ_n^0 (Π_n^0) set B is many-one reducible to it, that is, there exists a recursive function f such that $x \in B \leftrightarrow f(x) \in A$. A is Σ_n^0-*complete* (Π_n^0-*complete*) iff A is definable with a Σ_n^0 (Π_n^0) formula and A is Σ_n^0-hard (Π_n^0-hard).

The following proposition collects several useful equivalent forms of Σ_n^0 or Π_n^0 formulas (for any n).

Proposition 2

(I) *For every Σ_{n+1}^0 set A, there is a Π_n^0 predicate P such that for all x,*

$$x \in A \leftrightarrow (\forall^\infty a) P(a, x) \leftrightarrow (\exists a) P(a, x).$$

(II) *For every Σ_{n+1}^0 set B, there is a Π_n^0 predicate Q such that for all x,*

$$x \in B \leftrightarrow (\exists! a) Q(a, x) \leftrightarrow (\exists a) Q(a, x).$$

3   Teaching

Goldman and Kearns [6] introduced a variant of the on-line learning model in which a helpful teacher selects the instances presented to the learner. They considered a combinatorial measure of complexity called the *teaching dimension*, which is the mininum number of labelled examples required for any consistent learner to uniquely identify any target concept from the class.

Let $\mathcal{L}$ be a family of subsets of $\mathbb{N}$. Let $L \in \mathcal{L}$ and T be a subset of $\mathbb{N} \times \{+, -\}$. Furthermore, let $T^+ = \{n : (n, +) \in T\}$, $T^- = \{n : (n, -) \in T\}$ and $X(T) = T^+ \cup T^-$. A subset L of $\mathbb{N}$ is said to be *consistent* with T iff $T^+ \subseteq L$ and $T^- \cap L = \emptyset$. T is a *teaching set* for L with respect to $\mathcal{L}$ iff T is consistent with L and for all $L' \in \mathcal{L} \setminus \{L\}$, T is not consistent with L'. If T is a teaching set for L with respect to $\mathcal{L}$, then $X(T)$ is known as a *distinguishing set* for L with respect to $\mathcal{L}$. Every element of $\mathbb{N} \times \{+, -\}$ is known as a *labelled example*.

Definition 3. [6,13] Let $\mathcal{L}$ be any family of subsets of $\mathbb{N}$. Let $L \in \mathcal{L}$ be given. The size of a smallest teaching set for L with respect to $\mathcal{L}$ is called the *teaching dimension of L with respect to $\mathcal{L}$*, denoted by $\mathrm{TD}(L, \mathcal{L})$. The *teaching dimension of $\mathcal{L}$* is defined as $\sup\{\mathrm{TD}(L, \mathcal{L}) : L \in \mathcal{L}\}$ and is denoted by $\mathrm{TD}(\mathcal{L})$. If there is a teaching set for L with respect to $\mathcal{L}$ that consists of only positive examples, then the *positive teaching dimension of L with respect to $\mathcal{L}$* is defined to be the smallest possible size of such a set, and is denoted by $\mathrm{TD}^+(L, \mathcal{L})$. If there is no teaching set for L w.r.t. $\mathcal{L}$ that consists of only positive examples, then $\mathrm{TD}^+(L, \mathcal{L})$ is defined to be ∞. A teaching set for L with respect to $\mathcal{L}$ that consists of only positive examples is known as a *positive teaching set* for L with respect to $\mathcal{L}$. The *positive teaching dimension of $\mathcal{L}$* is defined as $\sup\{\mathrm{TD}^+(L, \mathcal{L}) : L \in \mathcal{L}\}$.

Another complexity parameter recently studied in computational learning theory is the recursive teaching dimension. It refers to the maximum size of teaching sets in a series of nested subfamilies of the family.

Definition 4. (Based on [10,14]). Let $\mathcal{L}$ be any family of subsets of $\mathbb{N}$. A *teaching sequence for $\mathcal{L}$* is any sequence $\mathrm{TS} = ((\mathcal{F}_0, d_0), (\mathcal{F}_1, d_1), \ldots)$ where (i) the families $\mathcal{F}_i$ form a partition of $\mathcal{L}$ with each $\mathcal{F}_i$ nonempty, and (ii) $d_i = \sup\{\mathrm{TD}(L, \mathcal{L} \setminus \bigcup_{0 \leq j < i} \mathcal{F}_j) : L \in \mathcal{F}_i\}$ for all i. $\sup\{d_i : i \in \mathbb{N}\}$ is called the *order of TS*, and is denoted by $\mathrm{ord}(\mathrm{TS})$. The *recursive teaching dimension of $\mathcal{L}$* is defined as $\inf\{\mathrm{ord}(\mathrm{TS}) : \mathrm{TS}$ is a teaching sequence for $\mathcal{L}\}$ and is denoted by $\mathrm{RTD}(\mathcal{L})$.

We shall also briefly consider the *extended teaching dimension* (XTD) of a class. This parameter may be viewed as a generalisation of the teaching dimension; it expresses the complexity of unique specification with respect to a concept class $\mathcal{C}$ for *every* concept (not just members of $\mathcal{C}$) over a given instance space X. As Hegedüs [7] showed, the extended teaching dimension of a concept class $\mathcal{C}$ is closely related to the query complexity of learning $\mathcal{C}$.

Definition 5. [7] Let $\mathcal{L}$ be a family of subsets of $\mathbb{N}$, and let L be a subset of $\mathbb{N}$. A set $S \subseteq \mathbb{N}$ is a *specifying set for L with respect to $\mathcal{L}$* iff there is at most one concept L' in $\mathcal{L}$ such that $L \cap S = L' \cap S$. Define the *extended teaching dimension* (XTD) of $\mathcal{L}$ as $\inf\{d :$ for every set $L \subseteq \mathbb{N}$ there exists an at most d-element specifying set with respect to $\mathcal{L}\}$.

A set $S \subseteq L$ is a *positive specifying set for L with respect to $\mathcal{L}$* iff there is at most one concept in $\mathcal{L}$ that contains S. Define the *positive extended teaching dimension* (XTD$^+$) of $\mathcal{L}$ as $\inf\{d :$ for every set $\emptyset \neq L \subseteq \mathbb{N}$ there exists an at most d-element positive specifying set with respect to $\mathcal{L}\}$. If there is a nonempty set L that does not have a positive specifying set w.r.t $\mathcal{L}$, define XTD$^+(\mathcal{L}) = \infty$.

The next series of definitions will introduce various subsets of $\mathbb{N}$, each of which is a set of codes for u.r.e. families that satisfy some notion of teachability.

(I) $I_{TD}^{\forall} = \{e : (\forall L \in \mathcal{L}_e)[\mathrm{TD}(L, \mathcal{L}_e) < \infty]\}$.
(II) $I_{TD} = \{e : \mathrm{TD}(\mathcal{L}_e) < \infty\}$.
(III) $I_{TD}^{\forall\infty} = \{e : (\forall^{\infty} L \in \mathcal{L}_e)[\mathrm{TD}(L, \mathcal{L}_e) < \infty]\}$.
(IV) $I_{TD^+}^{\forall} = \{e : (\forall L \in \mathcal{L}_j)[\mathrm{TD}^+(L, \mathcal{L}_e) < \infty]\}$.
(V) $I_{TD^+} = \{e : \mathrm{TD}^+(\mathcal{L}_e) < \infty\}$.
(VI) $I_{TD^+}^{\forall\infty} = \{e : (\forall^{\infty} L \in \mathcal{L}_e)[\mathrm{TD}^+(L, \mathcal{L}_e) < \infty]\}$.
(VII) $I_{XTD} = \{e : \mathrm{XTD}(\mathcal{L}_e) < \infty\}$.
(VIII) $I_{XTD^+} = \{e : \mathrm{XTD}^+(\mathcal{L}_e) < \infty\}$.

Owing to space constraints, many proofs will be omitted or sketched.

4   Teaching Dimension

In this section we study the arithmetical complexity of the class of u.r.e. families with finite teaching dimension; several related decision problems will also be considered.

Before proceeding with the main theorems on the arithmetical complexity of the teaching dimension model and its variants, a series of preparatory results will be presented. Theorem 7 addresses the question: how hard (arithmetically) is it to determine whether or not, given $e \in \mathbb{N}$ and a finite set D, D can distinguish an r.e. set $L_j^e \in \mathcal{L}_e$ from the other members of $\mathcal{L}_e$?

Definition 6. DS $:= \{\langle e, x, u \rangle : (\forall y)[L_x^e \neq L_y^e \rightarrow L_x^e \cap D_u \neq L_y^e \cap D_u]\}$.[1]

Theorem 7. *DS is Π_2^0-complete.*

Proof. By the definition of DS, $\langle e, x, u \rangle \in$ DS $\leftrightarrow (\forall y)(\forall t)(\exists v > t)[[L_{x,v}^e \cap D_u \neq L_{y,v}^e \cap D_u] \vee (\forall p)(\forall a)(\exists b > a)[\mathbb{1}_{L_{x,b}^e}(p) = \mathbb{1}_{L_{y,b}^e}(p)]]$. Thus DS has a Π_2^0 description. Now, since INF is Π_2^0-complete [12], it suffices to show that INF is many-one reducible to DS. Let g be a one-one recursive function with

$$W_{g(e,i)} = \begin{cases} \mathbb{N} & \text{if } i = 0 \vee (\exists j > i)[j \in W_e]; \\ \{0, 1, \dots, i\} & \text{otherwise.} \end{cases}$$

---

[1] DS stands for "distinguishing set."

Let f be a recursive function such that $\mathcal{L}_{f(e)} = \{W_{g(e,i)} : i \in \mathbb{N}\}$ and $L_0^{f(e)} = W_{g(e,0)}$. Recall that $D_0 = \emptyset$. It is readily verified that $e \in \text{INF} \leftrightarrow \langle f(e), 0, 0 \rangle \in \text{DS}$. ∎

The expectation that the arithmetical complexity of determining if a finite D is a *smallest* possible distinguishing set for some W_x belonging to $\mathcal{L}_e$ is at most one level above that of DS is confirmed by Theorem 9.

Definition 8. $\text{MDS} := \{\langle e, x, u \rangle \in \text{DS} : (\forall u')[|D_{u'}| < |D_u| \to \langle e, x, u' \rangle \notin \text{DS}]\}$.[2]

Theorem 9. *MDS is Π_3^0-complete.*

Proof. (Sketch.) By the definition of MDS,

$$\langle e, x, u \rangle \in \text{MDS} \leftrightarrow \langle e, x, u \rangle \in \text{DS} \wedge (\forall u')[(|D_{u'}| \geq |D_u|) \vee \langle e, x, u' \rangle \notin \text{DS}].$$

By Theorem 7, DS has a Π_2^0 description and $\overline{\text{DS}}$ has a Σ_2^0 description. Thus MDS has a Π_3^0 description. We omit the proof that MDS is Π_3^0-hard. ∎

Another problem of interest is the complexity of determining whether or not the teaching dimension of some W_x w.r.t. a class $\mathcal{L}_e$ is upper-bounded by a given number d. For $d = 0$, this problem is just as hard as DS (see Proposition 11); for $d > 0$, however, the complexity of the problem is exactly one level above that of DS (see Theorem 12). We omit the proofs.

Definition 10. $\text{TDDP} := \{\langle e, x, d \rangle : d \geq 1 \wedge (\exists u)[|D_u| \leq d \wedge \langle e, x, u \rangle \in \text{DS}]\}$.[3]

Proposition 11. $\{\langle e, x \rangle : \mathcal{L}_e = \{L_x^e\}\}$ *is Π_2^0-complete.*

Theorem 12. *TDDP is Σ_3^0-complete.*

Our first main result states that the class of all u.r.e. families $\mathcal{L}$ such that any finite subclass $\mathcal{L}' \subseteq \mathcal{L}$ has finite teaching dimension with respect to $\mathcal{L}$ is Π_4-complete.

Theorem 13. $I_{TD}^{\forall}$ *is Π_4^0-complete.*

Proof. First, note the following equivalent conditions:

$$e \in I_{TD}^{\forall} \leftrightarrow (\forall i)[\text{TD}(L_i^e, \mathcal{L}_e) < \infty)] \leftrightarrow (\forall i)(\exists u)[\langle e, i, u \rangle \in \text{DS}].$$

By Theorem 7, $\langle e, i, u \rangle \in \text{DS}$ may be expressed as a Π_2^0 predicate, so $I_{TD}^{\forall}$ is Π_4^0.

Now consider any Π_4^0 unary predicate $P(e)$; $P(e)$ is of the form $(\forall x)[Q(e,x)]$, where Q is a Σ_3^0 predicate. Since COF is Σ_3^0-complete [12], there is a recursive function $g(e,x)$ such that $P(e) \leftrightarrow (\forall x)[Q(e,x)] \leftrightarrow (\forall x)[g(e,x) \in \text{COF}]$ holds. For each triple $\langle e, x, i \rangle$, define

$$H_{\langle e,x,i \rangle} = \begin{cases} \{\langle e, x \rangle\} \oplus (W_{g(e,x)} \cup \{i\}) & \text{if } i > 0; \\ \{\langle e, x \rangle\} \oplus W_{g(e,x)} & \text{if } i = 0 \end{cases}$$

Let h be a recursive function such that for all e, $\mathcal{L}_{h(e)} = \{H_{\langle e,x,i \rangle} : x, i \in \mathbb{N}\}$.

---

[2] MDS stands for "minimal distinguishing set."
[3] TDDP stands for "Teaching dimension decision problem."

Case (i): $P(e)$ holds. Then for all x, $W_{g(e,x)}$ is cofinite. Thus for all x and each $i > 0$ such that $i \notin W_{g(e,x)}$, $H_{\langle e,x,i \rangle}$ has the teaching set $\{(2\langle e,x \rangle, +), (2i + 1, +)\}$ with respect to $\mathcal{L}_{h(e)}$. Furthermore, for all x and each i such that either $i \neq 0 \wedge i \in W_{g(e,x)}$ or $i = 0$, $H_{\langle e,x,i \rangle}$ has the teaching set $\{(2\langle e,x \rangle, +)\} \cup \{(2j + 1, -) : j \notin W_{g(e,x)} \wedge j > 0\}$ with respect to $\mathcal{L}_{h(e)}$. Therefore $\text{TD}(H_{\langle e,x,i \rangle}, \mathcal{L}_{h(e)}) < \infty$ for every pair $\langle x, i \rangle$, so that $h(e) \in I_{TD}^\vee$.

Case (ii): $\neg P(e)$ holds. Then $W_{g(e,x)}$ is coinfinite for some x. Fix such an x. Then $\mathcal{L}_{h(e)}$ contains $\mathcal{L}' = \{H_{\langle e,x,i \rangle} : i \in \mathbb{N}\}$. Furthermore, for each positive $i \notin W_{g(e,x)}$, since $\{\langle e, x \rangle\} \oplus (W_{g(e,x)} \cup \{i\}) \in \mathcal{L}'$, any teaching set for $H_{\langle e,x,0 \rangle}$ w.r.t. $\mathcal{L}_{h(e)}$ must contain $(2i + 1, -)$. Hence $\text{TD}(H_{\langle e,x,0 \rangle}, \mathcal{L}_{h(e)}) = \infty$, so that $h(e) \notin I_{TD}^\vee$.

Thus $I_{TD}^\vee$ is Π_4^0-complete. ∎

Extending $I_{TD}^\vee$ to include u.r.e. families $\mathcal{L}$ for which there is a *cofinite* subclass $\mathcal{L}' \subseteq \mathcal{L}$ belonging to $I_{TD}^\vee$ increases the arithmetical complexity of $I_{TD}^\vee$ to Σ_5^0.

Theorem 14. $I_{TD}^{\vee\infty}$ *is* Σ_5^0-complete.

Proof. (Sketch.) For any e, s, let g be a recursive function such that $\mathcal{L}_{g(e,s)} = \{L_i^e : i > s\}$. Note that the expression for $I_{TD}^{\vee\infty}$ can be re-written as $e \in I_{TD}^{\vee\infty} \leftrightarrow (\exists t)(\forall s > t)[g(e,s) \in I_{TD}^\vee]$. Since $\mathcal{L}_e \setminus \mathcal{L}_{g(e,t)}$ is finite, it follows that $e \in I_{TD}^{\vee\infty}$. By Theorem 13, the predicate $g(e,s) \in I_{TD}^\vee$ has a Π_4^0 description, so that $I_{TD}^{\vee\infty}$ is definable with a Σ_5^0 predicate.

Now let P be any Σ_5^0 predicate. By Proposition 2 and the Σ_3^0-completeness of COF [12], there is a recursive function h such that $P(e) \leftrightarrow (\forall^\infty a)(\forall b)[h(e,a,b) \in \text{COF}] \leftrightarrow (\exists a)(\forall b)[h(e,a,b) \in \text{COF}]$ and $(\exists b)[h(e,a,b) \in \text{COINF}] \leftrightarrow (\exists! b)[h(e,a,b) \in \text{COINF}]$. Now let g be a one-one recursive function such that

$$W_{g(e,a,b,i)} = \begin{cases} \{\langle e,a,b \rangle\} \oplus (W_{h(e,a,b)} \cup \{i\}) & \text{if } i > 0; \\ \{\langle e,a,b \rangle\} \oplus W_{h(e,a,b)} & \text{if } i = 0. \end{cases}$$

Let f be a one-one recursive function such that $\mathcal{L}_{f(e)} = \{W_{g(e,a,b,i)} : a, b, i \in \mathbb{N}\}$. Note that for all $a, b, i \in \mathbb{N}$, $\text{TD}(W_{g(e,a,b,i)}, \mathcal{L}_{f(e)}) < \infty \leftrightarrow h(e,a,b) \in \text{COF}$. One can show as in the proof of Theorem 13 that $\text{TD}(L, \mathcal{L}_{f(e)}) < \infty$ holds for almost all $L \in \mathcal{L}_{f(e)}$ iff $P(e)$ is true. ∎

The next theorem shows that the index set of the class consisting of all u.r.e. families with finite teaching dimension is Σ_5^0-complete.

Theorem 15. I_{TD} *is* Σ_5^0-complete.

Proof. (Sketch.) From $\text{TD}(\mathcal{L}_e) < \infty \leftrightarrow (\exists a)(\forall b)[\langle e, b, a \rangle \in \text{TDDP}]$ and the fact that TDDP is Σ_3^0-complete by Theorem 12 we have that I_{TD} is Σ_5^0.

To prove that I_{TD} is Σ_5^0-hard, consider any Σ_5^0 predicate $P(e)$. There is a binary recursive function g such that $P(e) \rightarrow (\exists a)(\forall b)[g(e,a,b) \in \text{COF}]$ and

$\neg P(e) \rightarrow (\forall a)(\forall^\infty b)[g(e, a, b) \in \text{COINF}]$. Now fix $e, b \in \mathbb{N}$. For each a, let $\{H_{\langle a,0\rangle}, H_{\langle a,1\rangle}, H_{\langle a,2\rangle}, \ldots\}$ be a numbering of the union of two u.r.e. families $\{C_{\langle a,0\rangle}, C_{\langle a,1\rangle}, C_{\langle a,2\rangle}, \ldots\}$ and $\{L_{\langle a,0\rangle}, L_{\langle a,1\rangle}, L_{\langle a,2\rangle}, \ldots\}$, which are defined as follows. (For simplicity, the notation for dependence on e and b is dropped.)

1. $\{C_{\langle a,0\rangle}, C_{\langle a,1\rangle}, C_{\langle a,2\rangle}, \ldots\}$ is a numbering of $\{X \subseteq \mathbb{N} : |\mathbb{N} \setminus X| = a\}$.
2. Let $E_{\langle a,0\rangle}, E_{\langle a,1\rangle}, E_{\langle a,2\rangle}, \ldots$ be a one-one enumeration of $\{X \subseteq \mathbb{N} : |\mathbb{N} \setminus X| < \infty \wedge |\mathbb{N} \setminus X| \neq a\}$. Let f be a recursive function such that for all $n, s \in \mathbb{N}$, $f(n, s)$ is the $(n + 1)$st element of $\mathbb{N} \setminus W_{g(e,a,b),s}$. For all $n, s \in \mathbb{N}$, define

$$L_{\langle a, \langle n,s\rangle\rangle} = \begin{cases} E_{\langle a,n\rangle} & \text{if } (\forall t \geq s)[f(n, s) = f(n, t)]; \\ \mathbb{N} & \text{if } (\exists t > s)[f(n, s) \neq f(n, t)]. \end{cases}$$

Note that $\{H_{\langle a,0\rangle}, H_{\langle a,1\rangle}, H_{\langle a,2\rangle}, \ldots\} \neq \emptyset$. Now construct a u.r.e. family $\{G_{\langle e,b,0\rangle}, G_{\langle e,b,1\rangle}, \ldots\}$ with the following properties:

(I) For all s, $G_{\langle e,b,s\rangle}$ is of the form $\{b\} \oplus \{s\} \oplus \bigoplus_{j \in \mathbb{N}} H_{\langle j,i_j\rangle}$.

(II) For every nonempty finite set $\{H_{\langle c_0,d_0\rangle}, \ldots, H_{\langle c_k,d_k\rangle}\}$ with $c_0 < \ldots < c_k$, there is at least one t for which $G_{\langle e,b,t\rangle} = \{b\} \oplus \{t\} \oplus \bigoplus_{i \in \mathbb{N}} A_i$, where $A_{c_i} = H_{\langle c_i,d_i\rangle}$ for all $i \in \{0, \ldots, k\}$ and $A_i \in \{H_{\langle i,j\rangle} : j \in \mathbb{N}\}$ for all $i \notin \{c_0, \ldots, c_k\}$.

(III) For every t such that $G_{\langle e,b,t\rangle} = \{b\} \oplus \{t\} \oplus \bigoplus_{j \in \mathbb{N}} H_{\langle j,i_j\rangle}$, there are infinitely many $t' \neq t$ such that $G_{\langle e,b,t'\rangle} = \{b\} \oplus \{t'\} \oplus \bigoplus_{j \in \mathbb{N}} H_{\langle j,i_j\rangle}$.

The family $\{G_{\langle e,i,j\rangle} : i, j \in \mathbb{N}\}$ may be thought of as an infinite matrix M in which each row represents a set parametrised by $g(e, a, b)$ for a fixed b and a ranging over $\mathbb{N}$. Furthermore, if there exists an a such that $W_{g(e,a,b)}$ is cofinite for all b, then the ath column of M contains all cofinite sets with complement of size a plus a finite number of other cofinite sets; if no such a exists then every column of M contains all cofinite sets. Let h be a recursive function with $\mathcal{L}_{h(e)} = \{\{b\} \oplus \bigoplus_{i \in \mathbb{N}} \emptyset : b \in \mathbb{N}\} \cup \{G_{\langle e,b,s\rangle} : b, s \in \mathbb{N}\}$. Showing that $P(e)$ iff $h(e) \in I_{TD}$ proves I_{TD} to be Σ_5^0-complete. ∎

To conclude our discussion on the general teaching dimension, we demonstrate that the criterion for a u.r.e. family to have finite extended teaching dimension is so stringent that only finite families have a finite XTD.

Theorem 16. I_{XTD} is Σ_3-complete.

Proof. We show $\text{XTD}(\mathcal{L}_e) < \infty \leftrightarrow |\mathcal{L}_e| < \infty$. First, suppose $\mathcal{L}_e = \{L_1, \ldots, L_k\}$. As $\mathcal{L}_e$ is finite, $\text{TD}(L_i, \mathcal{L}_e) \leq k - 1$ for all $i \in \{1, \ldots, k\}$. Consider any $L \notin \mathcal{L}_e$. For each $i \in \{1, \ldots, k\}$, fix y_i with $\mathbb{1}_L(y_i) \neq \mathbb{1}_{L_i}(y_i)$. Then $\{(y_i, +) : 1 \leq i \leq k \wedge y_i \in L\} \cup \{(y_j, -) : 1 \leq j \leq k \wedge y_j \notin L\}$ is a specifying set for L with respect to $\mathcal{L}_e$ of size k.

Second, suppose $|\mathcal{L}_e| = \infty$. Let $T = \{\sigma \in \{0,1\}^* : (\exists^\infty L \in \mathcal{L}_e)(\forall x < |\sigma|)[\sigma(x) = \mathbb{1}_L(x)]\}$. Note that $|\mathcal{L}_e| = \infty$ implies T is an infinite binary tree.

Thus by König's lemma, T has at least one infinite branch, say B. Then for all $n \in \mathbb{N}$, there exist infinitely many $L \in \mathcal{L}_e$ such that $\mathbb{1}_B(x) = \mathbb{1}_L(x)$ for all $x < n$. Therefore B has no finite specifying set with respect to $\mathcal{L}_e$ and so $\mathrm{XTD}(\mathcal{L}_e) = \infty$. Consequently, $\mathrm{XTD}(\mathcal{L}_e) < \infty \leftrightarrow |\mathcal{L}_e| < \infty \leftrightarrow (\exists a)(\forall b)(\exists c \le a)(\forall x)(\forall s)(\exists t > s)[\mathbb{1}_{L_{b,t}^e}(x) = \mathbb{1}_{L_{c,t}^e}(x)]$; as any two quantifiers, at least one of which is bounded, may be permuted, it follows that the last expression is equivalent to a Σ_3^0 formula. To show that I_{XTD} is Σ_3^0-hard, consider any Σ_3^0 predicate P, and let g be a recursive function such that

$$P(e) \leftrightarrow g(e) \in \mathrm{COF}.$$

Let f be a recursive function with $\mathcal{L}_{f(e)}$ equal to $\{W_{g(e)} \cup D : \mathbb{N} \supseteq |D| < \infty\}$, the class of all sets consisting of the union of $W_{g(e)}$ and a finite set of natural numbers. Then

$$P(e) \leftrightarrow g(e) \in \mathrm{COF} \leftrightarrow |\mathcal{L}_{f(e)}| < \infty \leftrightarrow \mathrm{XTD}(\mathcal{L}_{f(e)}) < \infty,$$

and so I_{XTD} is indeed Σ_3^0-complete. ■

5 Positive Teaching Dimension

We now consider the analogues of the results in the preceding section for the positive teaching dimension model. In studying the process of child language acquisition, Pinker [11, p. 226] points to evidence in prior research that children are seldom "corrected when they speak ungrammatically", and "when they are corrected they take little notice". It seems likely, therefore, that children learn languages mainly from positive examples. Thus, as a model for child language acquisition, the positive teaching dimension model is probably closer to reality than the general teaching dimension model in which negative examples are allowed. The next two results are the analogues of Theorems 13 and 15 for the positive teaching dimension model. It is noteworthy that $I_{TD}^\vee$ and $I_{TD+}^\vee$ have equal arithmetical complexity; that is to say, restricting the teaching sets of each $L \in \mathcal{L}_e$ with $e \in I_{TD}^\vee$ to positive teaching sets has no overall effect on the arithmetical complexity of $I_{TD}^\vee$.

Theorem 17. $I_{TD+}^\vee$ is Π_4^0-complete.

Proof. (Sketch.) Observe that

$$(\forall L \in \mathcal{L}_e)[\mathrm{TD}^+(L, \mathcal{L}_e) < \infty] \leftrightarrow (\forall i)(\exists u)[(\exists s)[D_u \subseteq L_{i,s}^e]$$

$$\wedge (\forall j)[(\forall x)(\forall s)(\exists t > s)[\mathbb{1}_{L_{j,t}^e}(x) = \mathbb{1}_{L_{i,t}^e}(x)] \vee \forall s'[D_u \not\subseteq L_{j,s'}^e]]],$$

Since the right-hand side simplifies to a Π_4^0 predicate, we know that $I_{TD+}^\vee$ is Π_4^0.

For the proof that $I_{TD+}^\vee$ is Π_4^0-hard, take any Π_4^0 predicate P, and let g be a recursive function such that $P(e) \leftrightarrow (\forall a)[g(e, a) \in \mathrm{COF}]$. Define a u.r.e. family

$\mathcal{L} = \{L_i\}_{i \in \mathbb{N}}$ as follows. (For notational simplicity, the notation for dependence on e is dropped.) For all $a, i \in \mathbb{N}$,

$$L_{\langle a,0 \rangle} = \{a\} \oplus W_{g(e,a)},$$

$$L_{\langle a,i+1 \rangle} = \begin{cases} \{a\} \oplus W_{g(e,a)} & \text{if } i \in W_{g(e,a)}; \\ \{a\} \oplus (\{i\} \cup \{x : x < i \wedge x \in W_{g(e,a)}\}) & \text{if } i \notin W_{g(e,a)}. \end{cases}$$

Let f be a recursive function for which $\mathcal{L}_{f(e)} = \mathcal{L}$. One can show that $\mathrm{TD}^+(L, \mathcal{L}_{f(e)}) < \infty$ holds for all $L \in \mathcal{L}_{f(e)}$ iff $P(e)$ is true. ∎

Theorem 18. I_{TD^+} *is* Σ_5^0-*complete.*

Proof. (Sketch.) For any e, one has $\mathrm{TD}^+(\mathcal{L}_e) < \infty \leftrightarrow (\exists a)(\forall b)[\mathrm{TD}^+(L_b^e, \mathcal{L}_e) < a]$ and

$$\mathrm{TD}^+(L_b^e, \mathcal{L}_e) < a \leftrightarrow (\exists u)(\forall c)[(\exists s')[D_u \subseteq L_{b,s'}^e] \wedge |D_u| < a$$

$$\wedge [(\forall x)(\forall s)(\exists t > s)[\mathbb{1}_{L_b^e}(x) = \mathbb{1}_{L_c^e}(x)] \vee (\forall t')[D_u \not\subseteq L_{c,t'}^e]]].$$

Simplifying the last equivalence yields a Σ_5^0-predicate for $\mathrm{TD}^+(\mathcal{L}_e) < \infty$.

The proof that I_{TD^+} is Σ_5^0-hard is similar to the earlier proof that I_{TD} is Σ_5^0-hard (but requires some additional ideas). Given any Σ_5^0 formula P, let R be a recursive predicate such that $P(e) \rightarrow (\exists a)(\forall b)(\forall^\infty c)(\forall d)(\exists l)[R(e, a, b, c, d, l)]$ and $\neg P(e) \rightarrow (\forall a)(\forall^\infty b)(\forall c)(\exists d)(\forall l)[\neg R(e, a, b, c, d, l)]$. Now fix any $e, b \in \mathbb{N}$. Let $(\langle a_0^0, b_0^0, c_0^0 \rangle, \ldots, \langle a_{k_0}^0, b_{k_0}^0, c_{k_0}^0 \rangle), (\langle a_0^1, b_0^1, c_0^1 \rangle, \ldots, \langle a_{k_1}^1, b_{k_1}^1, c_{k_1}^1 \rangle), \ldots$ be a one-one enumeration of all non-empty finite sequences of triples such that $a_0^j < \ldots < a_{k_j}^j$ for all $j \in \mathbb{N}$. Define the set (dropping the notation for dependence on e)

$$S_b := \{i : (\exists j \in \{0, \ldots, k_i\})[|D_{b_j^i}| \neq a_j^i + 1 \wedge (\exists l)[R(e, a_j^i, b, b_j^i, c_j^i, l)]]\},$$

using our fixed numbering $D_0, D_1, D_2, \ldots$ of all finite sets. For each b (with e fixed), construct a u.r.e. family $\{G_{\langle b, -1 \rangle}\} \cup \bigcup_{s \in \mathbb{N}} \{G_{\langle b, s \rangle}\}$ as follows.

$$G_{\langle b, s \rangle} = \{b\} \oplus S_b \oplus \bigoplus_{i \in \mathbb{N}} \mathbb{N} \quad \text{if } s = -1 \text{ or } s \in S_b.$$

If $s \notin S_b$, set $G_{\langle b, s \rangle} = \{b\} \oplus (S_b \cup \{s\}) \oplus \bigoplus_{i \in \mathbb{N}} H_i$, where

$$H_i = \begin{cases} \emptyset & \text{if } i \notin \{a_0^s, \ldots, a_{k_s}^s\}; \\ D_{b_i^s} & \text{if } i \in \{a_0^s, \ldots, a_{k_s}^s\}. \end{cases}$$

Let f be a recursive function such that $\mathcal{L}_{f(e)} = \bigcup_{b \in \mathbb{N}}(\{G_{\langle b, -1 \rangle}\} \cup \bigcup_{s \in \mathbb{N}} \{G_{\langle b, s \rangle}\})$ (note again that the notation for dependence on e in the definition of $G_{\langle b, s \rangle}$ has been dropped). We omit the proof that $P(e)$ holds iff $\mathrm{TD}^+(\mathcal{L}_{f(e)}) < \infty$. ∎

Theorem 19. $I_{TD^+}^{\forall^\infty}$ *is* Σ_5^0-*complete.*

Proof. (Sketch.) The condition

$$(\forall^\infty i)[\text{TD}^+(L_i^e, \mathcal{L}_e) < \infty] \leftrightarrow (\exists i)(\forall j \geq i)(\exists a)[\text{TD}^+(L_j^e, \mathcal{L}_e) < a)],$$

together with the fact that $\text{TD}^+(L_j^e, \mathcal{L}_e) < a$ is a Σ_3^0 predicate (as shown in the proof of Theorem 18), shows that $I_{TD^+}^{\forall^\infty}$ is Σ_5^0. The proof that $I_{TD^+}^{\forall^\infty}$ is Σ_5^0-hard is very similar to that of Theorem 14. ∎

Finally, consider the positive extended teaching dimension. Like the u.r.e. families with finite extended teaching dimension, those with finite *positive* extended teaching dimension have a particularly simple structure.

Theorem 20. I_{XTD^+} *is* Π_2^0-*complete.*

Proof. (Sketch.) One may verify directly that

$$\text{XTD}^+(\mathcal{L}_e) < \infty \leftrightarrow (\forall i, j)[L_i^e = L_j^e \vee L_i^e \cap L_j^e = \emptyset].$$

Note that $(\forall i, j)[L_i^e = L_j^e \vee L_i^e \cap L_j^e = \emptyset]$ iff

$$(\forall i, j)[(\forall x, s)(\exists t > s)[\mathbb{1}_{L_{i,t}^e}(x) = \mathbb{1}_{L_{j,t}^e}(x)] \vee (\forall s')[L_{i,s'}^e \cap L_{j,s'}^e = \emptyset]],$$

and that the latter expression reduces to a Π_2^0 predicate. Hence I_{XTD^+} is Π_2^0.

To establish that I_{XTD^+} is Π_2^0-hard, take any Π_2^0 predicate P, and let g be a recursive function such that $P(e) \leftrightarrow g(e) \in \text{INF}$. Let f be a recursive function such that $\mathcal{L}_{f(e)} = \{\mathbb{N}, G\}$, where

$$G = \begin{cases} \mathbb{N} & \text{if } |W_{g(e)}| = \infty; \\ \{0\} \cup \{x : x < m\} & \text{if } m \text{ is the least number} \\ & \text{such that } (\forall s \geq m)[W_{g(e),s} = W_{g(e),s+1}]. \end{cases}$$

Then $P(e) \leftrightarrow g(e) \in \text{INF} \leftrightarrow \text{XTD}^+(\mathcal{L}_{f(e)}) < \infty$. ∎

6   Recursive Teaching Dimension

Although the classical teaching dimension model is quite succinct and intuitive, it is rather restrictive. For example, let $\mathcal{L}$ be the concept class consisting of the empty set $L_0 = \emptyset$ and all singleton sets $L_i = \{i\}$ for positive $i \in \mathbb{N}$. Then $\text{TD}(L_i, \mathcal{L}) = 1$ for all $i \in \mathbb{N} \setminus \{0\}$ but $\text{TD}(L_0, \mathcal{L}) = \infty$. Thus $\text{TD}(\mathcal{L}) = \infty$ even though the class $\mathcal{L}$ is relatively simple. One deficiency of the teaching dimension model is that it fails to exploit any property of the learner other than the learner being *consistent*. The *recursive teaching model* [10,14], on the other hand, uses inherent structural properties of concept classes to define a teaching-learning protocol in which the learner is not just consistent, but also "cooperates" with the teacher by learning from a sequence of samples that is defined relative to the given concept class. In this section, we shall consider the arithmetical complexity of the index set of the class of all u.r.e. families that are teachable in some variants of the recursive teaching model. The complexities of interesting problems relating to the original recursive teaching model remain open.

Definition 21. A *positive teaching plan for* $\mathcal{L}$ is is any sequence TP $=$ $((L_0, d_0), (L_1, d_1), \ldots)$ where (i) the families $\{L_i\}$ form a partition of $\mathcal{L}$, and (ii) $d_i = \text{TD}^+(L_i, \mathcal{L} \setminus \bigcup_{0 \le j < i}\{L_j\})$ for all i. $\text{RTD}_1^+(\mathcal{L})$ is defined to be inf $\{\text{ord(TS)} : \text{TS is a positive teaching plan for } \mathcal{L}\}$. Since this paper only considers positive teaching plans, positive teaching plans will simply be called "teaching plans". Note that a positive teaching plan for $\mathcal{L}$ is essentially a teaching sequence for $\mathcal{L}$ that employs only positive examples and partitions $\mathcal{L}$ into singletons.

We begin with a lemma; the proof is omitted.

Lemma 22. *Let* $\{\mathcal{F}_i\}_{i \in \mathbb{N}}$ *be any sequence of families. If* $\sup\{RTD_1^+(\mathcal{F}_i) : i \in \mathbb{N}\} < \infty$, *then* $\sup\{RTD_1^+(\mathcal{F}_i) : i \in \mathbb{N}\} \le RTD_1^+(\bigsqcup_{i \in \mathbb{N}} \mathcal{F}_i) \le \sup\{RTD_1^+(\mathcal{F}_i) : i \in \mathbb{N}\} + 1$; *otherwise,* $RTD_1^+(\bigsqcup_{i \in \mathbb{N}} \mathcal{F}_i) = \infty$.

Definition 23. We denote by R_1^+ the set of codes for u.r.e. families, $\mathcal{L}$, such that $\text{RTD}_1^+(\mathcal{L})$ is finite.

Theorem 24. R_1^+ *is* Σ_4^0-*complete.*

Proof. To show that R_1^+ is Σ_4^0, fix e, a code for a u.r.e. family, $\mathcal{F}$. Given $n \in \mathbb{N}$, whether or not $\text{RTD}_1^+(\mathcal{F}) \le n$ can be decided by executing the following algorithm. Find the $F \in \mathcal{F}$ of least index such that $\text{TD}^+(F, \mathcal{F} \setminus \{F\}) \le n$ and call this set F_0. Having defined $F_0, \ldots, F_i$, let F_{i+1} be the set of least index in $\mathcal{F} \setminus \{F_0, \ldots, F_i\}$ such that $\text{TD}^+(F_{i+1}, \mathcal{F} \setminus \{F_0, \ldots, F_{i+1}\}) \le n$. If there is a teaching plan for $\mathcal{F}$ with order at most n, then the above algorithm will produce such a teaching plan because $\text{RTD}_1^+(\mathcal{F} \setminus \{F_0, \ldots, F_i\}) \le n$ for all $i \in \mathbb{N}$. Conversely, if there is no such teaching plan, then clearly the algorithm must initiate a non-terminating search at some stage.

Observe that the statement $\text{TD}^+(F_{i+1}, \mathcal{F} \setminus \{F_0, \ldots, F_{i+1}\}) \le n$ is Σ_2^0, therefore $\text{RTD}_1^+(\mathcal{F}) \le n$ is Π_3^0. Finally, $\text{RTD}_1^+(\mathcal{F}) < \infty$ is equivalent to $(\exists n)$ $\left(\text{RTD}_1^+(\mathcal{F}) \le n\right)$; hence, it is Σ_4^0. It remains to show that R_1^+ is Σ_4^0-hard.

Every Σ_4^0 predicate is of the form $(\forall^\infty n)\left(g(e, n) \in \text{COINF}\right)$, where g is a computable function. Fix such a predicate, P, and computable function g.

For $k \in \mathbb{N}$, let $f_k : \mathbb{N} \to \mathbb{N}$ be a uniformly computable sequence of functions such that (1) $D_{f_k(n)} \subsetneq [0, k]$, (2) $(\forall S \subsetneq [0, k])(\exists n)\left(D_{f_k(n)} = S\right)$ and (3) $(\forall n \in \text{ran}(f_k))\left(|f_k^{-1}(n)| = \infty\right)$. Define $\mathcal{L}_k = \{[0, k] \oplus \emptyset\} \cup \{D_{f_k(n)} \oplus \{n\} : n \in \mathbb{N}\}$. We will denote the members of $\mathcal{L}_k$ by L_i^k, where $L_0^k = [0, k] \oplus \emptyset$ and $L_{n+1}^k = D_{f_k(n)} \oplus \{n\}$. Observe that $\text{RTD}_1^+(\mathcal{L}_k) = k + 1$. However, for any finite $\mathcal{L}' \subset \mathcal{L}_k$, $\text{RTD}_1^+(\mathcal{L}') = 1$.

Let m be a computable function such that $m(a, n, s)$ is the number of $t < s$ such that the nth element of the complement of $W_{a,t}$ differs from the nth element of the complement of $W_{a,t+1}$. Define $\mathcal{G}_{a,i} = \bigsqcup_{n \in \mathbb{N}} \{L_j^i : (\exists s)(j < m(a, n, s))\}$ and let $\mathcal{F}_e = \bigsqcup_{i \in \mathbb{N}} \mathcal{G}_{g(e,i),i}$. By the construction, there is a computable function, h, such that $W_{h(e)} = \mathcal{F}_e$. We omit the proof that $\text{RTD}_1^+(\mathcal{F}_e) < \infty$ iff $P(e)$. ∎

Definition 25. Given $\sigma \in \mathbb{N}^*$ and $S = \{s_0, \ldots, s_n\} \subset \mathbb{N}$ with $s_0 < \cdots < s_n < |\sigma|$, define $\sigma[S] \in \mathbb{N}^*$ by $\sigma[S](i) = \sigma(s_i)$ for $i \le n$.

We now consider a "semi-effective" version of the recursive teaching model in which the teacher presents only positive teaching sets to the learner.

Definition 26. Let $\mathcal{L}$ be any family of subsets of $\mathbb{N}$. A *positive teaching sequence for* $\mathcal{L}$ is any sequence TS $= ((\mathcal{F}_0, d_0), (\mathcal{F}_1, d_1), \dots)$ such that (i) the families $\mathcal{F}_i$ form a partition of $\mathcal{L}$ with each $\mathcal{F}_i$ nonempty, and (ii) for all i and all $L \in \mathcal{F}_i$, there is a subset $S_L \subseteq L$ with $|S_L| = d_i < \infty$ such that for all $L' \in \bigcup_{j>i} \mathcal{F}_j$, it holds that $S_L \subseteq L' \to L = L'$. $\sup\{d_i : i \in \mathbb{N}\}$ is called the *order of* TS, and is denoted by ord(TS). The *positive recursive teaching dimension of* $\mathcal{L}$ is defined as $\inf\{\text{ord(TS)} : \text{TS is a positive teaching sequence for } \mathcal{L}\}$ and is denoted by $\text{RTD}^+(\mathcal{L})$.

We denote by R_{ure}^+ the set of codes for u.r.e. families, $\mathcal{L}$, such that $\text{RTD}^+(\mathcal{L})$ is finite and witnessed by a u.r.e. teaching sequence. In this section, a "teaching sequence" will always mean a u.r.e. teaching sequence.

Our last major result is that R_{ure}^+ is Σ_5^0-complete, which we establish in the following three theorems.

Theorem 27. R_{ure}^+ *is* Σ_5^0-*hard.*

Proof. Fix a Σ_5^0-predicate P. As in the proof of Theorem 4.4 from [3], let g be computable such that $P(e) \to (\exists x)\Big((\forall x' > x)(\forall y)(g(e, x', y) \in \text{COF}) \wedge (\forall x' \leq x)(\exists^{\leq 1} y)(g(e, x', y) \in \text{COINF})\Big)$ and $\neg P(e) \to (\forall x)(\exists! y)\Big(g(e, x, y) \in \text{COINF}\Big)$.

As in the proof of Theorem 24, let $\{f_k\}_{k \in \mathbb{N}}$ be a uniformly computable sequence of functions such that for all $k, n \in \mathbb{N}$, (1) $D_{f_k(n)} \subsetneq [0, k]$, (2) $(\forall S \subsetneq [0, k])(\exists m)(D_{f_k(m)} = S)$ and (3) $(\forall m \in \text{ran}(f_k))(|f_k^{-1}(m)| = \infty)$.

Fix $a, n \in \mathbb{N}$ and $\sigma \in \mathbb{N}^*$. Define $H_a(x, \sigma) = \{\langle \sigma, x \rangle\}$ if $x \in W_a$ and $H_a(x, \sigma) = \emptyset$ otherwise. Using this notation, we define the set $A^n = \bigoplus_{j \in \mathbb{N}}([0, n] \oplus \emptyset)$ and the sets $A_{i,\sigma}^{a,n} = \Big(D_{f_n(\sigma(0))} \oplus H_a(i, \sigma)\Big) \oplus \Big(\bigoplus_{j<i}([0, n] \oplus \emptyset)\Big) \oplus \Big(\bigoplus_{1 \leq j < |\sigma|}(D_{f_n(\sigma(j))} \oplus \{\langle \sigma[\{0\} \cup [j, |\sigma|)] \rangle\})\Big)$. Using the above sets, we define the following families: $\mathcal{G}_i^{a,n} = \{A_{i,\sigma}^{a,n} : \sigma \in \mathbb{N}^* \wedge |\sigma| \geq 2\}$; $\mathcal{G}^{a,n} = \{A^n\} \cup \bigcup_{i \in \mathbb{N}} \mathcal{G}_i^{a,n}$.

Suppose that $a \in \text{COF}$ and let $x_0, \dots x_k$ be an increasing enumeration of $\overline{W}_a$. The following is a teaching sequence for $\mathcal{G}^{a,n}$: $\Big((\bigcup_{i \in W_a} \mathcal{G}_i^{a,n}, 1), (\mathcal{G}_{x_0}^{a,n}, 1), \dots, (\mathcal{G}_{x_k}^{a,n}, 1), (\{A^{a,n}\}, 1)\Big)$. Thus, $\text{RTD}^+(\mathcal{G}^{a,n}) = 1$. Now suppose that $a \in \text{COINF}$ and let $x_0, x_1, \dots$ be an increasing enumeration of $\overline{W}_a$. Suppose TS $= ((\mathcal{L}_0, d_0), (\mathcal{L}_1, d_1), \dots)$ is a teaching sequence for $\mathcal{G}^{a,n}$ and ord(TS) $\leq n$. Consider an arbitrary $A_{x_i,\sigma}^{a,n} \in \mathcal{G}_{x_i}^{a,n}$ for $i \geq 1$. $A_{x_i,\sigma}^{a,n} \notin \mathcal{L}_0$, because $n + 1$ points are needed to distinguish $A_{x_i,\sigma}^{a,n}$ from every member of $\mathcal{G}_{x_0}^{a,n}$. Since $\mathcal{G}_{x_0}^{a,n} \cap \mathcal{L}_0 = \emptyset$, we know that $A^n \notin \mathcal{L}_0$. Now suppose that $\mathcal{G}_{x_k}^{a,n} \subseteq \bigcup_{i \geq k} \mathcal{L}_i$ and $A^n \notin \bigcup_{i < k} \mathcal{L}_i$, then $\mathcal{L}_k$ cannot contain any member of $\mathcal{G}_{x_i}^{a,n}$ for $i \geq k + 1$ because $n + 1$ points are needed to distinguish the members of $\mathcal{G}_{x_k}^{a,n}$ from the members of $\mathcal{G}_{x_i}^{a,n}$. As before, this also implies $A^n \notin \mathcal{L}_{k+1}$. By induction, we conclude that $A^n \notin \mathcal{L}_i$ for any $i \in \mathbb{N}$. This is a contradiction, so $\text{RTD}^+(\mathcal{G}^{a,n}) \geq n + 1$. Since

$$\text{TS} = \Big((\{A^n\}, n+1), (\mathcal{G}_0^{a,n}, 1), (\mathcal{G}_1^{a,n}, 1), \dots \Big)$$

is a teaching sequence for $\mathcal{G}^{a,n}$ and $\text{ord}(\text{TS}) = n + 1$, we conclude that $\text{RTD}^+(\mathcal{G}^{a,n}) = n+1$. Finally, define

$$\mathcal{F}_{e,x} = \bigsqcup_{y \in \mathbb{N}} \mathcal{G}^{g(e,x,y),x} \text{ and } \mathcal{F}_e = \bigsqcup_{x \in \mathbb{N}} \mathcal{F}_{e,x}.$$

We wish to prove that $\text{RTD}^+(\mathcal{F}_e) < \infty$ if and only if $P(e)$. First, suppose $P(e)$. For all but finitely many x, $g(e,x,y) \in \text{COF}$ for all y. This means that $\text{RTD}^+(\mathcal{F}_{e,x}) = 1$ for all but finitely many x. For each x for which $\text{RTD}^+(\mathcal{F}_{e,x}) \neq 1$ the dimension is still finite, hence, there is a uniform bound, n, on the recursive teaching dimension of all the $\mathcal{F}_{e,x}$. We conclude that $\text{RTD}^+(\mathcal{F}_e) < \infty$.

On the other hand, if $\neg P(e)$, then for every x there is exactly one y such that $g(e,x,y) \in \text{COINF}$. Hence, $\mathcal{F}_e$ is the disjoint union of families whose RTD is unbounded. We have thus reduced an arbitrary Σ_5^0-predicate to R_{ure}^+.   ∎

Theorem 28. R_{ure}^+ *is* Σ_5^0.

Proof. Let $\{S_n\}_{n \in \mathbb{N}}$ enumerate the u.r.e. teaching sequences, with $S_n = ((\mathcal{L}_0^n, d_0^n), (\mathcal{L}_1^n, d_1^n), \dots)$. Consider a u.r.e. family, $\mathcal{F} = \{F_0, F_1, \dots\}$ coded by e.

$$e \in R_{ure}^+ \leftrightarrow (\exists a, n)\Big(\text{ord}(\text{TS}_a) \leq n \wedge \text{TS}_a \text{ is a teaching sequence for } \mathcal{F} \Big)$$

$$\text{ord}(\text{TS}_a) \leq n \leftrightarrow (\forall i)\Big(d_i \leq n \Big)$$

To say that S_a is a teaching sequence for $\mathcal{F}$ is equivalent to (1) $\{\mathcal{L}_0, \mathcal{L}_1, \dots\}$ is a partition of $\mathcal{F}$ and (2) $(\forall i \in \mathbb{N}, L \in \mathcal{L}_i)(\text{TD}^+(L, \bigcup_{j \geq i} \mathcal{L}_j) \leq d_i)$.

Since the statement $\text{TD}^+(L, \bigcup_{j \geq i} \mathcal{L}_j) \leq d_i$ is Σ_3^0, we know that the statement $(\forall i \in \mathbb{N}, L \in \mathcal{L}_i)(\text{TD}^+(L, \bigcup_{j \geq i} \mathcal{L}_j) \leq d_i)$ is Π_3^0. The statement that $\{\mathcal{L}_0, \mathcal{L}_1, \dots\}$ is a partition of $\mathcal{F}$ is equivalent to

$$(\forall i \in \mathbb{N}, F \in \mathcal{L}_i)(\exists F' \in \mathcal{F})\Big(F = F' \Big) \wedge (\forall F \in \mathcal{F})(\exists i \in \mathbb{N}, F' \in \mathcal{L}_i)\Big(F = F' \Big)$$

$$\wedge (\forall i, j \in \mathbb{N}, F \in \mathcal{L}_i, F' \in \mathcal{L}_j)\Big(i \neq j \to F \neq F' \Big),$$

which is Π_4^0. Thus, $e \in R_{ure}^+$ is Σ_5^0.   ∎

Theorem 29. R_{ure}^+ *is* Σ_5^0-*complete*.

7   Conclusion

This paper studied the arithmetical complexity of index sets of classes of u.r.e. families that are teachable under various teaching models. Our main results are summarised in Table 1. While u.r.e. families constitute a very special case of families of sets, many of our results may be extended to the class of families of

Table 1. Summary of main results on u.r.e. families. The notation $\mathrm{TD}^{(+)}$ indicates that the result holds for both TD and TD^+.

Index set	Arithmetical complexity
$\{e : (\forall L \in \mathcal{L}_e)[\mathrm{TD}^{(+)}(L, \mathcal{L}_e) < \infty]\}$	Π_4^0-complete (Theorems 13 and 17)
$\{e : \mathrm{TD}^{(+)}(\mathcal{L}_e) < \infty\}$	Σ_5^0-complete (Theorems 15 and 18)
$\{e : \mathrm{XTD}(\mathcal{L}_e) < \infty\}$; $\{e : \mathrm{XTD}^+(\mathcal{L}_e) < \infty\}$	Σ_3^0-complete (Theorem 16); Π_2^0-complete (Theorem 20)
$\{e : \mathrm{RTD}_1^+(\mathcal{L}_e) < \infty\}$	Σ_4^0-complete (Theorem 24)
$\{e : \mathrm{RTD}_{ure}^+(\mathcal{L}_e) < \infty\}$	Σ_5^0-complete (Theorem 29)

countably many sets; more precisely, if we define $C_j^X = \{x : \langle j, x \rangle \in X\}$ and $\mathcal{C}_X = \{C_j^X : j \in \mathbb{N}\}$ for any $X \subseteq \mathbb{N}$, it is not difficult to apply our results to determine the position of $\{X : I(\mathcal{C}_X) < \infty\}$ for different teaching parameters I in the *hierarchy of sets of sets* [12, Sect. 15.1]. We also determined first-order formulas with the least possible quantifier complexity defining some fundamental decision problems in algorithmic teaching. Our work may be extended in several directions. For example, it might be interesting to investigate the arithmetical complexity of index sets of classes of general – even non-u.r.e. – families that are teachable under the teaching models considered in the present paper. In particular, it may be asked whether the arithmetical complexity of the class of teachable families with one-one numberings is less than that of the class of teachable families that do not have one-one numberings.

References

1. Angluin, D.: Inductive inference of formal languages from positive data. Inf. Control **45**(2), 117–135 (1980)
2. Baliga, G., Case, J., Jain, S.: The synthesis of language learners. Inf. Comput. **152**, 16–43 (1999)
3. Beros, A.: Learning theory in the arithmetic hierarchy. J. Symbolic Logic **79**(3), 908–927 (2014)
4. Brandt, U.: The position of index sets of identifiable sets in the arithmetical hierarchy. Inf. Control **68**, 185–195 (1986)
5. Gold, E.M.: Language identification in the limit. Inf. Control **10**, 447–474 (1967)
6. Goldman, S.A., Kearns, M.J.: On the complexity of teaching. J. Comput. Syst. Sci. **50**(1), 20–31 (1995)
7. Hegedüs, T.: Generalised teaching dimensions and the query complexity of learning. In: Proceedings of the COLT, pp. 108–117 (1995)
8. de Jongh, D., Kanazawa, M.: Angluin's theorem for indexed families of r.e. sets and applications. In: Proceedings of the COLT, pp. 193–204 (1996)
9. Klette, R.: Indexmengen und Erkennung Rekursiver Funktionen. Z. Math. Logik Grundlag. Math. **22**, 231–238 (1976)
10. Mazadi, Z., Gao, Z., Zilles, S.: Distinguishing pattern languages with membership examples. In: Dediu, A.-H., Janoušek, J., Martín-Vide, C., Truthe, B. (eds.) LATA 2016. LNCS, vol. 9618, pp. 528–540. Springer, Heidelberg (2014). doi:10.1007/978-3-319-04921-2_43

11. Pinker, S.: Formal models of language learning. Cognition **7**, 217–283 (1979)
12. Rogers Jr., H.: Theory of Recursive Functions and Effective Computability. MIT Press, Cambridge (1987)
13. Shinohara, A., Miyano, S.: Teachability in computational learning. New Gener. Comput. **8**(4), 337–347 (1991)
14. Zilles, S., Lange, S., Holte, R., Zinkevich, M.: Models of cooperative teaching and learning. J. Mach. Learn. Res. **12**, 349–384 (2011)

Inductive Inference

Learning Finite Variants of Single Languages from Informant

Klaus Ambos-Spies[✉]

Institut für Informatik, Heidelberg University, 69120 Heidelberg, Germany
ambos@math.uni-heidelberg.de

Abstract. We show that the family $S_L^+ = \{L \cup \{x\} : x \in \omega\} \cup \{L\}$ consisting of the languages obtained from a given language (i.e., computably enumerable set) L by adding at most one additional element can be explanatorily learned from informant (i.e., is InfEx-learnable) if and only if L is autoreducible. Similarly, the subfamily $\hat{S}_L^+ = \{L \cup \{x\} : x \notin L\}$ of S_L^+ consisting of the languages obtained from L by adding exactly one additional element can be learned from informant without mind changes (i.e., is InfFin-learnable) if and only if L is autoreducible.

1 Introduction

In the setting of Gold's [8] model of learning languages (i.e., computably enumerable sets) in the limit, numerous natural learnability criteria have been studied. Such a criterion is determined by the mode of representation of the languages to be identified and by the mode of convergence of the hypotheses produced by the learner. The most common way for representing a language L is either by *text (Txt)* (i.e., by a – not necessarily effective – list of the elements of L; see [8]) or by *informant (Inf)* (i.e., by the characteristic sequence $L(0)L(1)L(2)\ldots$ of L; see [7]), where the former reveals only positive information on L to the learner whereas the latter provides negative information too. The most common convergence criterion is *explanatory (Ex)* learning [8] where the hypotheses produced by the learner eventually converge to a fixed computably enumerable index of L (w.r.t. some acceptable numbering of the family of c.e. sets). More restrictive is the setting of *finite (Fin)* learning [3] where, once a hypothesis is produced, the learner is not allowed to change its mind anymore; more relaxed is the setting of *behaviourally correct (BC)* learning [2,5] where eventually all hypotheses have to be indices of the language to be identified but not necessarily the same index. For more background on the theory of inductive inference and for precise definitions of the learnability criteria see e.g. the monograph [10] by Jain et al. or Chap. 12 of Stephan's lecture notes [14].

Though numerous interesting learnability criteria are studied in the literature (see [10] and [14]), one may argue that the criteria TxtFin, TxtEx, TxtBC, InfFin, InfEx, and InfBC obtained along these lines are the most fundamental learnability criteria in the theory of inductive inference. It is well known that these criteria are mutually different (and that – besides the trivial implications among these concepts – the only nontrivial implication is InfFin $\subseteq$ TxtEx).

© Springer International Publishing Switzerland 2016
R. Ortner et al. (Eds.): ALT 2016, LNAI 9925, pp. 163–173, 2016.
DOI: 10.1007/978-3-319-46379-7_11

The goal of this note is to give a uniform separation for five of these six classes. For this sake we consider, for a given language L, the class

$$S_L^+ = \{L \cup \{x\} : x \in \omega\} \cup \{L\} \tag{1}$$

consisting of the languages obtained by adding at most one element to L (note that, for nonempty L, the second part in the definition may be omitted since $L \in \{L \cup \{x\} : x \in \omega\}$); and, for any learning criterion given above, we specify the languages L for which S_L^+ is identifiable under this criterion.

For the criteria TxtFin, TxtEx, TxtBC, InfFin and InfBC, these characterizations are straightforward and well known. Still, for the sake of completeness, we present these results in Sect. 2 below. The characterization of the languages L for which the class S_L^+ can be InfEx-identified, however, is less straightforward. It turns out that, for some noncomputable languages L – for instance for the halting problem K – the class S_L^+ can be InfEx-identified, but there are also languages L for which this is not the case. A complete characterization of the languages for which S_L^+ can be InfEx-identified is provided by Trahtenbrot's notion of autoreducibility [15] which was intensively studied in other areas of computability theory but has not been previously linked to the theory of inductive inference. This characterization, which is the main result of this note, is presented in Sect. 3.

In Sect. 4 we look at learnability of the variant

$$\hat{S}_L^+ = \{L \cup \{x\} : x \notin L\}$$

of S_L^+ consisting of the languages obtained by adding *exactly* one element to L. Since $\hat{S}_L^+$ is a subclass of S_L^+, namely $\hat{S}_L^+ = S_L^+ \setminus \{L\}$, the positive learnability results for the classes S_L^+ carry over to the classes $\hat{S}_L^+$. In fact, since all languages in $\hat{S}_L^+$ are incomparable, we get stronger results here. In particular, for autoreducible L, the class $\hat{S}_L^+$ can be InfFin-learned (not just InfEx-learned as in case of S_L^+).

Finally, in Sect. 5 we shortly summarize our results, and we give some straightforward extensions as well as some directions for further research.

Our notation is standard. For unexplained notation and for the basic definitions on inductive inference we refer the reader to the monograph [10] by Jain et al. A general reference for basic facts and notation from computability theory is Soare [13].

2 Learnability of the Classes S_L^+: Basic results

For the learning criteria TxtFin, TxtEx, TxtBC, InfFin and InfBC, the characterization of the languages L for which the classes S_L^+ (defined in (1) above) are identifiable is straightforward and, probably, well known. (For instance, for the case of learning from text, see Example 12.6 in [14] where it is shown that, for noncomputable L, $S_L^+ \in$ TxtBC $\setminus$ TxtEx.)

Theorem 1 (Folklore)

(i) The class S_L^+ is TxtFin-*learnable if and only if* $L = \omega$.

(ii) The class S_L^+ is TxtEx-*learnable if and only if* L *is computable*.

(iii) The class S_L^+ is TxtBC-*learnable (hence* InfBC-*learnable) for all c.e. sets* L.

(iv) The class S_L^+ is InfFin-*learnable if and only if* L *is cofinite*.

Proof. In the following let L be a fixed language, let $W_{e_0} = L$, and let f be a computable function such that $W_{f(x)} = L \cup \{x\}$.

(i). First assume that $L = \omega$. Then $S_L^+ = \{L\}$. So, since any singleton class of c.e. sets is TxtFin-learnable, S_L^+ is TxtFin-learnable.

For the (contraposition of the) converse implication, assume that $L \neq \omega$, say $x_0 \notin L$. Then, for $L_0 = L$ and $L_1 = L \cup \{x_0\}$, $L_0, L_1 \in S_L^+$ and $L_0 \subset L_1$. But, obviously, for any languages L_0 and L_1 such that $L_0 \subset L_1$ and for any learner M which TxtFin-learns L_0, M does not TxtFin-learn L_1 since any finite initial segment of a text for L_0 can be extended to a text for L_1. So, for $L \neq \omega$, $S_L^+ \notin$ TxtFin.

(ii). First assume that L is computable. Then M TxtEx-learns S_L^+ where M is defined by

$$M(\sigma) = e_0$$

if $content(\sigma) \subseteq L$ and

$$M(\sigma) = f(\mu\, x\, (x \in content(\sigma) \setminus L))$$

otherwise.

For the converse implication assume that M TxtEx-learns S_L^+. By Blum and Blum [4] and by $L \in S_L^+$, fix a locking sequence σ for M on L (i.e., a sequence σ such that $content(\sigma) \subseteq L$, $W_{M(\sigma)} = L$, and, for any τ with $content(\tau) \subseteq L$, $M(\sigma\tau) = M(\sigma)$). Then, for any computable text T for L and for any $x \in \omega$,

$$x \notin L \;\Leftrightarrow\; \exists\, n\, (M(\sigma\, x\, T(0) \ldots T(n)) \neq M(\sigma)).$$

(Note that the implication "$\Rightarrow$" follows from the fact that M TxtEx learns S_L^+ since, for $x \notin L$, $\sigma x T$ is a text for the language $L \cup \{x\}$, while the (contraposition of the) implication "$\Leftarrow$" holds by the fact that σ is a locking sequence.) So $\overline{L}$ is c.e., hence L is computable.

(iii). Let $\{L_s\}_{s \geq 0}$ be a computable enumeration of L. The learner M defined by

$$M(\sigma) = \begin{cases} e_0 & \text{if } content(\sigma) \subseteq L_{|\sigma|} \\ f(\mu\, x\, (x \in content(\sigma) \setminus L_{|\sigma|})) & \text{otherwise} \end{cases}$$

TxtBC-learns S_L^+.

(iv). First assume that L is cofinite. If $L = \omega$ then S_L^+ is TxtFin-learnable (by (i)) hence InfFin-learnable. So w.l.o.g. we may assume that $\overline{L} = \{x_0, \ldots, x_n\}$ where $n \geq 0$ and $x_0 < x_1 < \cdots < x_n$. Then M InfFin-learns S_L^+ where M is defined by

$$M(i(0)\dots i(m)) = \begin{cases} ? & \text{if } m < x_n \\ e_0 & \text{if } m \geq x_n \text{ and } i(x_j) = 0 \text{ for all } j \leq n \\ f(x_j) & \text{if } m \geq x_n \text{ and } j \leq n \text{ is minimal} \\ & \text{such that } i(x_j) = 1. \end{cases}$$

For the converse implication, assume that S_L^+ is InfFin-learnable. Fix M such that M InfFin-learns S_L^+. Then there are a unique number n_0 and a unique index e of L such that $M(L(0)\dots L(n_0)) = e$ and $M(L \restriction n) = ?$ for $n \leq n_0$. It follows that $\overline{L} \subseteq \{0,\dots n_0\}$. (Namely, otherwise, fix $n_1 \notin L$ such that $n_0 < n_1$. Then M does not InfFin-learn $L \cup \{n_1\}$.)

This completes the proof of Theorem 1.

3   InfEx-Learnable Classes S_L^+

The languages L for which the classes S_L^+ can be explanatorily learned from informant are just the autoreducible languages. Here a language L is autoreducible if there is an effective procedure for deciding the membership problem for any number x in L provided that the information about membership in L for all other numbers is given.

Definition 1 (Trahtenbrot [15]). *A set $A \subseteq \omega$ is autoreducible if there is a Turing functional Ψ such that, for all $x \in \omega$, $A(x) = \Psi^{A \setminus \{x\}}(x)$.*

Theorem 2. *Let L be a language. The following are equivalent.*

(i) The class S_L^+ is InfEx-learnable.
(ii) L is autoreducible.

Proof. $(i) \Rightarrow (ii)$. Assume that $S_L^+ \in$ InfEx, and fix an InfEx-learner M of S_L^+. By $L \in S_L^+$, M InfEx-learns L whence there are a number n_0 and an index e_0 of L such that

$$\forall\, n \geq n_0 \; (M(L(0)\dots L(n)) = e_0).$$

Since M InfEx-learns S_L^+, it follows that, for all $n \geq n_0$,

$$L(n+1) = 0 \;\Leftrightarrow\; \exists\, n' > n+1 \; \Big(M(L(0)\dots L(n)1L(n+2)\dots L(n')) \neq e_0 \Big) \quad (2)$$

holds. Using a computable enumeration $\{L_s\}_{s \geq 0}$ of L this allows to compute $L(n+1)$ from $L \setminus \{n+1\}$ for $n \geq n_0$ as follows. Find $s > n+1$ minimal such that $n+1 \in L_s$ or $M(L(0)\dots L(n)1L(n+2)\dots L(s)) \neq e_0$ holds. By (2) such an s must exist, and $n+1 \in L$ iff $n+1 \in L_s$. So L is autoreducible.

$(ii) \Rightarrow (i)$. Assume that L is autoreducible, let Ψ be a Turing functional such that $L(x) = \Psi^{L \setminus \{x\}}(x)$ for all $x \geq 0$, and let $\{L_s\}_{s \geq 0}$ and $\{\Psi_s\}_{s \geq 0}$ be computable enumerations of L and Ψ, respectively. The definition of a learner M which InfEx-identifies S_L^+ is based on the following observation.

Claim 1. Let $\hat{L} \in S_L^+$ and let x be any number. The following are equivalent.

(a) $\hat{L} = L \cup \{x\}$ and $x \notin L$.
(b) There are numbers y and z such that $x < y < z$ and the following hold.
  (α) $\hat{L}(x) = 1$.
  (β) $L_z(x) = 0$.
  (γ) $\Psi_z^{L_z \setminus \{x\}}(x) = 0$ and the use of this computation (i.e., the greatest oracle query in this computation) is less than y.
  (δ) $\forall x' < y \ (x' \neq x \Rightarrow \hat{L}(x') = L_z(x'))$.

Proof of Claim 1. (a) $\Rightarrow$ (b). Assume that $\hat{L} = L \cup \{x\}$ and $L(x) = 0$. Then (α) is immediate and (β) and (δ) hold for all sufficiently large numbers z. Moreover, by choice of Ψ, $0 = L(x) = \Psi^{L \setminus \{x\}}(x)$. It follows that, for the least $y > x$ which strictly bounds the oracle queries in the computation of $\Psi^{L \setminus \{x\}}(x)$, ($\gamma$) holds for all sufficiently large z. (Namely, it suffices to choose z such that $\Psi^{L \setminus \{x\}}(x)$ converges in $\leq z$ steps and $L_z \upharpoonright y = L \upharpoonright y$.) So there is a number $z > y$ such that (β), (γ) and (δ) hold.

(b) $\Rightarrow$ (a). Fix numbers y and z such that $x < y < z$ and (α)–(δ) hold. Note that, by $\hat{L} \in S_L^+$, L is a subset of $\hat{L}$. Hence, for any numbers x' and z, $L_z(x') \leq L(x') \leq \hat{L}(x')$. So, by ($\delta$), $L_z(x') = L(x')$ for all $x' < y$ with $x' \neq x$. It follows, by (γ) and by the Use Principle, that $\Psi^{L \setminus \{x\}}(x) = 0$. So, by choice of Ψ, $L(x) = 0$. Finally, by $\hat{L} \in S_L^+$ and (α), this implies that $\hat{L} = L \cup \{x\}$.

This completes the proof of Claim 1.

Now, given a c.e. index e_0 of L and a computable function f such that

$$W_{f(x)} = L \cup \{x\}$$

for all $x \geq 0$, a learner M which InfEx-learns S_L^+ works as follows.

Given a finite binary sequence $i(0) \ldots i(n)$, for the value of $M(i(0) \ldots i(n))$ distinguish the following two cases. If there are numbers x, y and z such that $x < y < z \leq n$ and such that conditions (α)–(δ) in Claim 1 hold if we replace $\hat{L}(0) \ldots \hat{L}(n)$ by $i(0) \ldots i(n)$ there, then, for the least such x, let

$$M(i(0) \ldots i(n)) = f(x).$$

Otherwise, let
$$M(i(0) \ldots i(n)) = e_0.$$

For the correctness of M, fix $\hat{L} \in S_L^+$. It suffices to show that the values $M(\hat{L}(0) \ldots \hat{L}(n))$ converge to an index of $\hat{L}$ (for $n \to \infty$). Distinguish the following two cases. First assume that $\hat{L} = L$. Then, by Claim 1, there are no numbers x, y and z such that $x < y < z$ and (α)–(δ) hold. So $M(\hat{L}(0) \ldots \hat{L}(n)) = e_0$ for all $n \geq 0$, where by choice of e_0 and by assumption, $W_{e_0} = L = \hat{L}$. Second, assume that $\hat{L} = L \cup \{x\}$ where $x \notin L$. Then, again by Claim 1, x is unique such that there are numbers y and z satisfying $x < y < z$ and the conditions (α)–(δ). So, for the least n_0 such that there are such numbers y and z with $z \leq n_0$,

$$M(\hat{L}(0) \ldots \hat{L}(n)) = e_0$$

for $n < n_0$ and

$$M(\hat{L}(0) \dots \hat{L}(n)) = f(x)$$

for $n \geq n_0$ and $\hat{L} = W_{f(x)}$.

For later use note that the above shows that, for $\hat{L} = L \cup \{x\} \in S_L^+ \setminus \{L\}$, M does not only converge to an index $f(x)$ of $\hat{L}$ but it changes its mind exactly once: namely it switches from the hypothesis L (expressed by the fixed index e_0 of L) to the hypothesis $\hat{L} = L \cup \{x\}$ (expressed by the fixed index $f(x)$) at stage n_0.

This completes the proof of Theorem 2.

Note that any computable set is autoreducible. Moreover, for any noncomputable language (i.e., noncomputable c.e. set) L,

$$L \oplus L = \{2x : x \in L\} \cup \{2x + 1 : x \in L\}$$

is a noncomputable language and autoreducible. So there are noncomputable autoreducible languages (in fact, any c.e. many-one degree – hence any c.e. Turing degree – contains a c.e. autoreducible set). Further obvious examples of autoreducible sets are cylinders. Since any many-one complete set is a cylinder, it follows that the halting set K is autoreducible. By Theorem 2, these observations imply

$$\{L : L \text{ is computable}\} \subset \{L : S_L^+ \in \text{InfEx}\} \tag{3}$$

(where $\subset$ denotes a strict inclusion) and $S_K^+ \in \text{InfEx}$.

On the other hand, Ladner [11] has shown that there are languages L which are not autoreducible whence, by Theorem 2,

$$\{L : S_L^+ \in \text{InfEx}\} \subset \{L : L \text{ is c.e.}\}, \tag{4}$$

By combining (3) and (4) we obtain

$$\{L : L \text{ is computable}\} \subset \{L : S_L^+ \in \text{InfEx}\} \subset \{L : L \text{ is c.e.}\}.$$

Together with Theorem 1 this implies

Corollary 1. $\{L : S_L^+ \in \text{TxtEx}\} \subset \{L : S_L^+ \in \text{InfEx}\} \subset \{L : S_L^+ \in \text{TxtBC}\}$.

Non-autoreducible sets and their (Turing) degrees have been extensively studied. Ladner [11] has shown that a language L is autoreducible if and only if L is *mitotic* (where a language A, i.e., a c.e. set A, is mitotic if A is the disjoint union of c.e. sets A_0 and A_1 which are Turing equivalent to A); and in [12] Ladner has shown that there is a noncomputable c.e. Turing degree $\mathbf{a}$ such that all c.e. sets in $\mathbf{a}$ are autoreducible. On the other hand, Ingrassia [9] has shown that the non-autoreducible c.e. degrees, i.e., the c.e. degrees containing non-autoreducible languages, are dense in the c.e. degrees. For more information on the distribution of the non-autoreducible c.e. degrees among all c.e. degrees see Downey and Slaman [6].

4   Learnability of the Classes $\hat{S}_L^+$

We now consider the subclass $\hat{S}_L^+ = \{L \cup \{x\} : x \notin L\}$ of S_L^+ consisting of the languages obtained by adding (exactly) one element to L, i.e., $\hat{S}_L^+ = S_L^+ \setminus \{L\}$, and we characterize the languages L for which these classes can be identified under the criteria considered here. In the following cases these characterizations are straightforward.

Theorem 3 (Folklore)

(i) The class $\hat{S}_L^+$ is TxtFin-*learnable if and only if* L *is computable.*

(ii) For any language L, the class $\hat{S}_L^+$ is TxtEx-*learnable (hence* InfEx-, TxtBC-*and* InfBC-*learnable).*

Proof. Let L be a fixed language, let $W_{e_0} = L$, and let f be a computable function such that $W_{f(x)} = L \cup \{x\}$.

(i). First assume that L is computable. We have to show that there is a learner M which TxtFin-learns $\hat{S}_L^+$. For $L = \omega$ this is obvious (since $\hat{S}_L^+$ is empty). So w.l.o.g. we may assume that $L \neq \omega$. Then M TxtFin-learns $\hat{S}_L^+$ where

$$M(\sigma) = \begin{cases} ? & \text{if } content(\sigma) \subseteq L \\ f(\mu x(x \in content(\sigma) \setminus L)) & \text{otherwise.} \end{cases}$$

For the converse direction, assume that $\hat{S}_L^+$ is TxtFin-learnable. We have to show that L is computable. If L is cofinite then this is trivially true. So w.l.o.g. we may assume that L is coinfinite. Fix a learner M which TxtFin-learns $\hat{S}_L^+$, and let T be a computable text for L. Then, for any $x \geq 0$,

$$x \notin L \iff \exists n \, (M(x(T \upharpoonright n)) \downarrow). \tag{5}$$

Namely, for a proof of the direction $\Rightarrow$, assume that $x \notin L$. Then xT is a text of the language $L \cup \{x\} \in \hat{S}_L^+$ whence M outputs a fixed index of $L \cup \{x\}$ on all sufficiently long initial segments of xT. For a proof of the direction $\Leftarrow$, for a contradiction assume that there are numbers n and e such that $M(x(T \upharpoonright n)) = e$ and $x \in L$. By coinfinity of L choose $x' \notin L$ such that $L \cup \{x'\} \neq W_e$. Then $x(T \upharpoonright n)x'T$ is a text for $L \cup \{x'\}$ and M does not TxtFin-learn $L \cup \{x'\}$ from this text. Since $L \cup \{x'\}$ is in the class $\hat{S}_L^+$, this contradicts the choice of M.

Now, by (5), the complement $\overline{L}$ of L is c.e., hence L is computable.

(ii). For $L = \omega$, $\hat{S}_L^+$ is empty hence TxtEx-learnable. So w.l.o.g. we may assume that $L \neq \omega$. Then, given a computable enumeration $\{L_s\}_{s \geq 0}$ of L, a learner M which TxtEx-learns $\hat{S}_L^+$ is as follows. If there is a number x such that $x \in content(\sigma) \setminus L_{|\sigma|}$ then let $M(\sigma) = f(x)$ for the least such x. Otherwise let $M(\sigma) = ?$.

This completes the proof of Theorem 3.

Theorem 3 leaves open the case of InfFin-learnability which we settle here.

Theorem 4. *Let L be a language. The following are equivalent.*

(i) The class $\hat{S}_L^+$ is InfFin-learnable.
(ii) L is autoreducible.

Proof. The proof is a variant of the proof of Theorem 2. So we refer to some of the relevant facts established there.

$(i) \Rightarrow (ii)$. Assume that $\hat{S}_L^+ \in$ InfFin. Since, for computable L, the claim is trivial, w.l.o.g. we may assume that L is noncomputable (hence coinfinite). Let M be an InfFin-learner of $\hat{S}_L^+$. It suffices to show that

$$L(n) = 0 \Leftrightarrow \exists n' > n \left(M(L(0) \ldots L(n-1)1L(n+1) \ldots L(n')) \neq ? \right) \quad (6)$$

holds. Then, as in the proof of the corresponding direction of Theorem 2, we may argue that L is autoreducible.

Now, the implication $\Rightarrow$ in (6) is immediate since, for $n \notin L$,

$$L(0) \ldots L(n-1)1L(n+1)L(n+2)L(n+3) \ldots$$

is the characteristic sequence of the language $L \cup \{n\} \in \hat{S}_L^+$. Since M InfFin-learns $\hat{S}_L^+$ – hence $L \cup \{n\}$ – it follows that $M(L(0) \ldots L(n-1)1L(n+1) \ldots L(n'))$ is an index of $L \cup \{n\}$ for all sufficiently large n'.

For a proof of the implication $\Leftarrow$, for a contradiction assume that

$$M(L(0) \ldots L(n-1)1L(n+1) \ldots L(n')) \neq ?$$

for some $n' > n$ and that $n \in L$. Fix e such that

$$M(L(0) \ldots L(n-1)1L(n+1) \ldots L(n')) = e,$$

and, by coinfinity of L, fix $x > n'$ such that $x \notin L$ and $L \cup \{x\} \neq W_e$. Then $L(0) \ldots L(n-1)1L(n+1) \ldots L(n')$ is an initial segment of the characteristic sequence of $L \cup \{x\}$, and $L \cup \{x\} \in \hat{S}_L^+$. So M does not InfFin-learn $L \cup \{x\}$ hence does not InfFin-learn $\hat{S}_L^+$ contrary to assumption.

$(ii) \Rightarrow (i)$. Assume that L is autoreducible, and let M be the learner defined in the second part of the proof of Theorem 2, which InfEx-learns S_L^+. As observed there, on a language $\hat{L} = L \cup \{x\} \neq L$ in S_L^+, M makes just one mind change switching from the hypothesis L (expressed by index e_0) to the hypothesis $\hat{L}$ (expressed by index $f(x)$). So M can be converted into a learner which learns $\hat{S}_L^+ = S_L^+ \setminus \{L\}$ from informant without mind changes by replacing the clause $M(i(0) \ldots i(n)) = e_0$ in the definition of M by $M(i(0) \ldots i(n)) = ?$.

This completes the proof of Theorem 4.

5 Summary

The results of this note can be summarized as follows where L is assumed to be a language (i.e., a c.e. set), S_L^+ is the class of languages obtained by adding at most one element to L, and $\hat{S}_L^+$ is the class of languages obtained by adding (exactly) one element to L.

Criterion	S_L^+ is learnable if and only if	$\hat{S}_L^+$ is learnable if and only if
TxtFin	$L = \omega$	L is computable
TxtEx	L is computable	L is any language
TxtBC	L is any language	L is any language
InfFin	L is cofinite	L is autoreducible
InfEx	L is autoreducible	L is any language
InfBC	L is any language	L is any language

(7)

Since the class of autoreducible languages splits the class of the noncomputable languages, the results in the first column show that the major learning criteria TxtFin, InfFin, TxtEx, InfEx and TxtBC (or InfBC in place of TxtBC) can be pairwise distinguished by considering the languages L for which the classes

$$S_L^+ = \{L \cup \{x\} : x \geq 0\} \cup \{L\} = \{\hat{L} : L \subseteq \hat{L} \ \& \ |\hat{L} \setminus L| \leq 1\}$$

can be learned under the respective criterion.

In fact, for the incomparable learnability criteria TxtBC and InfEx, the classes S_L^+ provide a witness for the non-inclusion TxtBC $\not\subseteq$ InfEx. The separation of the criteria TxtBC and InfBC is not witnessed by learnability of the classes S_L^+, however, whence, in particular, the non-inclusion InfEx $\not\subseteq$ TxtBC cannot be shown this way.

We obtain witnesses for these separations, however, by considering the dual classes

$$S_L^- = \{L \setminus \{x\} : x \geq 0\} \cup \{L\} = \{\hat{L} : \hat{L} \subseteq L \ \& \ |L \setminus \hat{L}| \leq 1\}$$

consisting of the languages $\hat{L}$ obtained from a given language L by omitting at most one element. Namely, for infinite L, $S_L^- \notin$ TxtBC. (This easily follows from the fact that, for any learner M which TxtBC-learns L, there is a TxtBC-locking sequence for M on L, i.e., a finite sequence σ such that $content(\sigma) \subseteq L$, $W_{M(\sigma)} = L$, and, for any finite sequence τ with $content(\tau) \subseteq L$, $W_{M(\sigma\tau)} = L$; see Exercise 6–9 in [10] for the existence of such sequences.) On the other hand, for any language L, $S_L^- \in$ InfEx. (Namely, given a computable enumeration $\{L_n\}_{n \geq 0}$ of L, a learner M which uses the following strategy InfEx-learns S_L^-: given a finite initial segment $\hat{L} \upharpoonright n$ of the characteristic sequence of a language $\hat{L}$, distinguish the following two cases. If there is a number $x < n$ such that $x \in L_n$ and $(\hat{L} \upharpoonright n)(x) = 0$ then, for the least such x, M guesses that $\hat{L} = L \setminus \{x\}$. Otherwise, M guesses that $\hat{L} = L$.) So, for any infinite language L, $S_L^- \in$ InfEx $\setminus$ TxtBC thereby giving the desired separation.

More systematically, we obtain the following classification of the learnability of classes S_L^- and the corresponding classes

$$\hat{S}_L^- = \{L \setminus \{x\} : x \in L\} = \{\hat{L} : \hat{L} \subset L \ \& \ |L \setminus \hat{L}| = 1\}$$

consisting of the languages obtained by omitting (exactly) one element of a given language L.

Criterion	S_L^- is learnable if and only if	$\hat{S}_L^-$ is learnable if and only if
TxtFin	$L = \emptyset$	L is finite
TxtEx	L is finite	L is any language
TxtBC	L is finite	L is any language
InfFin	L is finite	L is any language
InfEx	L is any language	L is any language
InfBC	L is any language	L is any language

(8)

We omit the proofs of the claims in (8) since they are standard.

There are obvious ways to extend our results. First, one may consider learnability of the above classes under other learning criteria. Second, one may consider other classes of finite variants of a single language L. Here we have considered only the cases where (at most) one element is added or (at most) one element is deleted. For instance one may consider the case of finite extensions of the given language L, i.e., the class

$$\tilde{S}_L^+ = \{L \cup F : F \text{ is finite}\}.$$

For these classes, one of the referees pointed out the following interesting result on InfEx-learnability: the class $\tilde{S}_L^+$ can be InfEx-learned if and only if the language L is "upward autoreducible" in the sense that there is a Turing functional Ψ such that

$$L(x) = \Psi^{L \setminus \{0, \ldots, x\}}(x)$$

for all numbers x. (Note that any upward autoreducible set is autoreducible but the converse is not true in general. For instance, for a non-autoreducible set A, the selfjoin $A \oplus A$ of A is autoreducible but not upward autoreducible.)

Acknowledgements. We would like to thank Wolfgang Merkle for some very helpful discussions and one of the anonymous referees for his very useful comments and suggestions.

References

1. Angluin, D.: Inductive inference of formal languages from positive data. Inf. Control **45**(2), 117–135 (1980)
2. Barzdiņš, Ja.M.: Two theorems on the identification in the limit of functions. (Russian) Latviisk. Gos. Univ. Učen. Zap. 210 Teorija Algoritmov i Programm No. 1, pp. 82–88 (1974)
3. Barzdiņš, J., Freivalds, R.: On the prediction of general recursive functions. Sov. Math. Doklady **13**, 1224–1228 (1972)
4. Blum, L., Blum, M.: Toward a mathematical theory of inductive inference. Inf. Control **28**, 125–155 (1975)
5. Case, J., Lynes, C.: Machine inductive inference and language identification. Automata, Languages and Programming. LNCS, vol. 140, pp. 107–115. Springer, Berlin (1982)
6. Downey, R.G., Slaman, T.A.: Completely mitotic R.E. degrees. Ann. Pure Appl. Logic **41**(2), 119–152 (1989)
7. Freivalds, R.V., Wiehagen, R.: Inductive inference with additional information. Elektron. Informationsverarb. Kybernet. **15**(4), 179–185 (1979)
8. Gold, E.M.: Language identification in the limit. Inf. Control **10**, 447–474 (1967)
9. Ingrassia, M.A.: P-genericity for recursively enumerable sets. Thesis (Ph.D.), University of Illinois at Urbana-Champaign, ProQuest LLC, Ann Arbor, MI, 162 p. (1981)
10. Jain, S., Osherson, D., Royer, J.S., Sharma, A.: Systems That Learn, An Introduction to Learning Theory, 2nd edn. The MIT Press, Cambridge (1999)
11. Ladner, R.E.: Mitotic recursively enumerable sets. J. Symb. Logic **38**, 199–211 (1973)
12. Ladner, R.E.: A completely mitotic nonrecursive R.E. degree. Trans. Am. Math. Soc. **184**, 479–507 (1973, 1974)
13. Soare, R.I.: Recursively Enumerable Sets and Degrees. A Study of Computable Functions and Computably Generated Sdets. Perspectives in Mathematical Logic. Springer, Berlin (1987). xviii+437
14. Stephan, F.: Recursion Theory. National University of Singapore, Lecture Notes (2012). http://www.comp.nus.edu.sg/~fstephan/recursiontheory-pstopdf.pdf
15. Trahtenbrot, B.A.: Autoreducibility. (Russian). Dokl. Akad. Nauk SSSR **192**, 1224–1227 (1970)

Intrinsic Complexity of Partial Learning

Sanjay Jain[1](✉) and Efim Kinber[2]

[1] School of Computing, National University of Singapore,
Singapore 117417, Singapore
sanjay@comp.nus.edu.sg
[2] Department of Computer Science, Sacred Heart University,
Fairfield, CT 06432-1000, USA
kinbere@sacredheart.edu

Abstract. A partial learner in the limit [16], given a representation of
the target language (a text), outputs a sequence of conjectures, where
one correct conjecture appears infinitely many times and other conjectures each appear a finite number of times. Following [5,14], we define
intrinsic complexity of partial learning, based on reducibilities between
learning problems. Although the whole class of recursively enumerable
languages is partially learnable (see [16]) and, thus, belongs to the complete learnability degree, we discovered a rich structure of incomplete
degrees, reflecting different types of learning strategies (based, to some
extent, on topological structures of the target language classes). We also
exhibit examples of complete classes that illuminate the character of the
strategies for partial learning of the hardest classes.

1 Introduction

In his seminal paper [8], E.M. Gold introduced the framework for algorithmic learning of languages in the limit from their representations (texts), which
became the standard for exploring learnability of languages in the limit (see,
for example [16]). In this model (we will refer to it as **TxtEx**), a learner outputs an infinite sequence of conjectures stabilizing on a correct grammar for
the target language. However, Gold himself was the first one to notice that the
TxtEx model has a strong limitation: whereas the class of all finite languages
is easily learnable within this framework, no class $\mathcal{L}$ containing just one infinite
language and all its finite subsets is **TxtEx**-learnable. In particular, the class of
all regular languages cannot be learnt in the limit just from positive data. To
capture the extent to which the aforementioned class $\mathcal{L}$ would still be learnable
from positive data, Osherson, Stob and Weinstein [16] introduced the concept
of *partial learning* in the limit: a learner outputs an infinite sequence of conjectures, where one correct grammar of the target language occurs infinitely many
times, whereas all other conjectures occur at most a finite number of times. The

---

S. Jain—Supported in part by NUS grant numbers C252-000-087-001 and R146-000-
181-112.

E. Kinber—Supported by URCG grant from Sacred Heart University.

R. Ortner et al. (Eds.): ALT 2016, LNAI 9925, pp. 174–188, 2016.
DOI: 10.1007/978-3-319-46379-7_12

aforementioned class $\mathcal{L}$ containing an infinite recursive language L and all its finite subsets is easily learnable in this model by a simple strategy that, every time when a new datum appears on the input, conjectures a grammar for L, and conjectures some standard code for the input seen so far, otherwise. Yet, as it was noted in [16], partial learning, without any other constraints, is very powerful: the whole class of all recursively enumerable languages turns out to be partially learnable — albeit by a much more complex strategy than the one trivially learning the aforementioned class $\mathcal{L}$.

Partial learning, under various natural constraints, has attracted a lot of attention recently (see, for example, [6,7,9,10,15]). Though partial learning can be done for the whole class of recursively enumerable sets, partial learning with constraints gives interesting results. Although partial learning does not seem to be as natural as Gold's classical model of inductive inference, one can hope that partial learning strategies for important classes of languages (like the class of regular languages) not learnable within Gold's framework can shed new light on the general problem of learnability of such classes (and, perhaps, their important subclasses) from positive data and, possibly, additional information. For example, if a relatively simple partial learning strategy for the class of regular languages from positive data is found, one can try to look at what kind of reasonable additional information could be sufficient for converting such partial strategy to a more realistic learning strategy for this class (perhaps, it could be different from Angluin's classical strategy for learning regular languages from membership queries and counterexamples to conjectures [2]). We hope that our paper can be a start for this line of research.

One of the potential issues is to understand exactly how partial learning happens and what is involved in it, as, at no particular instant, one can say what is the current "planned" hypothesis of the learner. To understand more about partial learning, we consider reductions between different learning problems (classes of languages). Reductions gave an interesting structure in explanatory learning (see [4,5,12–14]), and we hope to be able to understand much more about partial learning using reductions between different classes, which would, in some sense, highlight the ease/difficulty of partial learning of various subsets of the full class of all recursively enumerable languages.

Thus, our main goal in the current research is to find natural, yet non-**TxtEx**-learnable, classes (in particular, *indexed* classes [1], with decidable membership problem) and — whenever it would be possible — corresponding natural partial learning strategies that would be simpler than that for the class of all recursively enumerable languages. The concept of *reducibility* between partial learnability problems that we introduce in this paper is based on similar models defined first for learning in the limit of classes of recursive functions in [5] and then, for **TxtEx**-learnability in [14] (see also [4] for a related but different concept of complexity of learning).

A partial learnability problem (a class of languages $\mathcal{L}$) is reducible to another partial learning problem (a class of languages $\mathcal{L}'$) if there exist two computable operators, Θ and Ψ, such that (a) Θ translates every text for a language in $\mathcal{L}$ to a text for a language in $\mathcal{L}'$ and (b) Ψ translates every sequence of conjectures where a grammar for a language $L' \in \mathcal{L}'$ occurs infinitely many times and all other conjectures occur at most a finite number of times (we will call such sequences of conjectures *admissible*) back to an admissible sequence of conjectures for the language $L \in \mathcal{L}$ such that some text for L is translated by Θ to a text for L'. We make a distinction between *strong* reducibility, where Θ translates every text for the same language in $\mathcal{L}$ to a text for the same language in $\mathcal{L}'$ and *weak* reducibility, where Θ may translate different texts of the same language $L \in \mathcal{L}$ to texts of different languages in $\mathcal{L}'$. Based on this concept of reducibility, one can naturally define degrees of learnability and the complete degree (which contains the class of all recursively enumerable languages).

Firstly, we found two relatively simple and transparent classes that are complete for weak and, respectively, strong reducibilities — these classes illuminate the character of the partial learning strategies for the hardest problems. We also show (Theorem 13) that the class of all recursive languages is not strongly complete. In particular, it means that all indexed classes, including the class of all regular languages, are not strongly complete.

A major accomplishment of our research is the discovery of a rich structure of incomplete classes under the degree of the class of all regular languages — based on a number of classes representing certain natural partial learning strategies. In particular, we define the class **iCOINIT**, which contains an infinite chain $L_1, L_2, \ldots$ of infinite recursive subsets of a recursive infinite language, and all their finite subsets, where, for every i, $L_{i+1} \subset L_i$. The natural strategy to learn this class, when choosing an infinite language as its conjecture, immediately finds out an upper bound on the possible number of infinite languages that may be conjectured in the future. We also define the counterpart of **iCOINIT**, the class **iINIT**, which also contains an enumerable, but indefinitely growing chain of infinite recursive languages and all their finite subsets. The natural learning strategy for **iINIT**, when choosing an infinite language as its conjecture, also faces a bound on the number of infinite languages that can be conjectured in the future, but unlike the case of **iCOINIT**, this bound is not known to the learner. We show that **iCOINIT** is weakly reducible to **iINIT** (Theorem 15), yet it is not strongly reducible to **iINIT** (Theorem 16); also, **iINIT** is not even weakly reducible to **iCOINIT** (Theorem 17).

We also introduce the class **iRINIT**, which contains an infinitely growing chain of recursive languages and all their finite subsets, yet, unlike the case of **iINIT**, the enumeration of members of the chain is based on the set of all rational numbers between 0 and 1. In particular, for any two infinite languages $L, L' \in$ **iRINIT**, $L \subset L'$, there is another language in **iRINIT** between L and L'. We show that **iRINIT** is not weakly complete (Corollary 19), yet all variants and generalizations formed using **iINIT** and **iCOINIT** (as defined in this paper) are strongly reducible to **iRINIT** (Theorem 26), and **iRINIT** is strongly reducible to none of them (Theorem 26). On the other hand, **iRINIT** itself turns out to be weakly reducible to **iINIT** (Theorem 22). **iRINIT** is also strictly under the degree of all regular languages (Theorems 18 and 24).

We also define a variant $\mathbf{iCOINIT}_k$ of $\mathbf{iCOINIT}$, which, in addition to every infinite language L in the chain, contains also all languages extending L by at most k additional elements. A natural strategy learning an infinite target language $L \in \mathbf{iCOINIT}_k$, when a new datum appears on the input, first conjectures an infinite language $M \in \mathbf{iCOINIT}$, and when up to k new elements $x \notin M$ appear on the input, conjectures appropriate finite variants of M, before moving to the next M' in the chain when the number of new data not in M exceeds k. Similarly the variant $\mathbf{iINIT}_k$ is defined for $\mathbf{iINIT}$. Interestingly, though, all classes $\mathbf{iCOINIT}_k, k \geq 1$ turn out to be strongly reducible to $\mathbf{iCOINIT}_1$ and, respectively, all classes $\mathbf{iINIT}_k, k \geq 1$ are strongly reducible to $\mathbf{iINIT}_1$. Yet, surprisingly, $\mathbf{iINIT}_1$ is not strongly reducible to $\mathbf{iINIT}$ and $\mathbf{iCOINIT}_1$ is not even weakly reducible to $\mathbf{iCOINIT}$ (see Theorem 25). All these classes, though, are weakly reducible to $\mathbf{iINIT}$ (as $\mathbf{iRINIT}$ is weakly reducible to $\mathbf{iINIT}$, see above).

Lastly, based on similar multidimensional classes of languages defined in [13], one can define classes of "multidimensional" languages, where partial learning of one "dimension" aids in learning next "dimension". For example, one can, using cylindrification, define the class $(\mathbf{iINIT}, \mathbf{iCOINIT})$, where the conjecture that is output infinitely many times for the first "dimension" can be used to partially learn the second "dimension". We have extended this idea to any arbitrary sequence Q of $\mathbf{iINIT}$ and $\mathbf{iCOINIT}$ and have shown that if a sequence Q is a proper subsequence of Q', then the class corresponding to Q is strongly reducible to the one corresponding to Q', but not vice versa. Due to space constraints, results on multi-dimensional languages are not described in this paper but will be given in the full paper.

Our result on the incompleteness of any indexed class suggests that there may exist natural, relatively simple, strategies that can partially learn an indexed class. This can shed a new light on the potential of learnability of many important classes of languages from positive data.

2 Preliminaries

Any unexplained recursion theoretic notation is from [17]. N denotes the set of natural numbers $\{0, 1, 2, \ldots\}$. A language is any subset of N. We let $\emptyset, \subseteq, \subset, \supseteq, \supset$ denote empty set, subset, proper subset, superset and proper superset respectively. $A \Delta B$ denotes the symmetric difference of sets A and B, that is $A \Delta B = (A - B) \cup (B - A)$. $\overline{L} = N - L$ denotes the complement of L. We let $\mathrm{card}(S)$ denote the cardinality of a set S. For $S \subseteq N$, let $\max(S), \min(S)$ respectively denote maximum and minimum of a set S, where $\min(\emptyset) = \infty$ and $\max(\emptyset) = 0$. We sometimes use sets of rational numbers. In this case, we use $\max(S)$ to denote the least upper bound of the rational numbers in the set S.

A finite set $S \subseteq N$ can be coded as $\mathrm{code}(S) = \sum_{x \in S} 2^x$. D_i denotes the finite set A with $\mathrm{code}(A) = i$.

φ denotes a fixed standard acceptable numbering [17]. φ_i denotes the i-th program in the acceptable numbering φ. Let $W_i = \mathrm{domain}(\varphi_i)$. Thus, W_i is the language/set enumerated by the i-th grammar in the acceptable programming system $W_0, W_1, \ldots$. Let Φ be a Blum complexity measure [3] for the φ programming system. Let

$$\varphi_{i,s}(x) = \begin{cases} \varphi_i(x), & \text{if } x < s \text{ and } \Phi_i(x) < s; \\ \uparrow, & \text{otherwise} \end{cases}$$

Let $W_{i,s} = \text{domain}(\varphi_{i,s})$. Intuitively, $W_{i,s+1} - W_{i,s}$ can be thought of as the elements enumerated by W_i in $(s+1)$-th step. For purposes of this paper, one can assume without loss of generality that $W_{i,s+1} - W_{i,s}$ contains at most one element.

$\mathcal{R}$ denotes the class of all recursive languages. $\mathcal{E}$ denotes the class of all recursively enumerable sets.

An indexed family is a family $(L_i)_{i \in N}$ of languages such that there exists a recursive function f uniformly deciding the membership question for L_i, that is, for all i, x, $f(i, x) = 1$ iff $x \in L_i$.

Let $\langle \cdot, \cdot \rangle$ denote a fixed recursive bijection from $N \times N$ to N. $\langle \cdot, \cdot \rangle$ can be extended to pairing of n-ary tuples by taking $\langle x_1, x_2, \ldots, x_n \rangle = \langle x_1, \langle x_2, \ldots, x_n \rangle \rangle$. For notation convenience we let $\langle x \rangle = x$. Let $\pi_i^n(\langle x_1, x_2, \ldots, x_n \rangle) = x_i$, where we drop the superscript in case $n = 2$.

Let $pad(\cdot, \cdot)$ be a 1–1 recursive function, increasing in both its arguments, such that for all i and j, $\varphi_{pad(i,j)} = \varphi_i$. Note that there exists such a padding function pad (see [17]).

$RAT_{0,1}$ denotes the set of rational numbers between 0 and 1 (both inclusive). Let ntor be a recursive bijection from N to $RAT_{0,1}$. Let rton be the inverse of ntor. Left r.e. real means a real number which is approximable from below using rational numbers enumerated by a recursive procedure. That is, a real number r is called a left r.e. real iff there exists a recursive function f mapping N to the set of rational numbers such that: for all i, $f(i) \leq f(i+1)$, and $\lim_{i \in N} f(i) = r$.

We now give some concepts from language learning theory. Let $\#$ be a special pause symbol. A *finite sequence* σ is a mapping from an initial segment of N to $(N \cup \{\#\})$. Let Λ denote the empty sequence. SEQ denotes the set of all finite sequences. A *text* is a mapping from N to $(N \cup \{\#\})$. Let $|\sigma|$ denote the length of sequence σ. Let $T[n]$ denote the initial segment of length n of the text T. For $n \leq |\sigma|$, $\sigma[n]$ denotes the initial segment of length n of the sequence σ. The concatenation of sequences σ and τ is denoted by $\sigma \diamond \tau$. The content of T, denoted content(T), is the set of the numbers in the range of T, that is, $\{T(n) : n \in N\} - \{\#\}$. Let content($\sigma$) be defined similarly. We say that T is a text for a language L iff content(T) = L.

A *language learning machine* is a partial computable function which maps SEQ to N. We let **M**, with or without decorations, range over learning machines.

Definition 1. [16]

(a) **M Part**-learns L iff for all texts T for L,

(i) for all n, $\mathbf{M}$ is defined on $T[n]$,

(ii) there exists a unique p such that $p = \mathbf{M}(T[n])$ for infinitely many n, and

(iii) for p as in (ii) above, $W_p = L$.

(b) $\mathbf{M}$ **Part**-learns a class $\mathcal{L}$ iff it **Part**-learns each $L \in \mathcal{L}$.

(c) $\mathbf{Part} = \{\mathcal{L} : (\exists \mathbf{M})[\mathbf{M}\ \mathbf{Part}\text{-learns}\ \mathcal{L}\}$.

It can be shown that $\mathcal{E}$ is **Part**-learnable [16]. If $\mathbf{M}$ **Part**-learns a class $\mathcal{L}$ then we say that $\mathbf{M}$ witnesses **Part**-learnability of $\mathcal{L}$. If an infinite sequence $p_0 p_1 \ldots$ satisfies the following two requirements:

(i) there exists a unique p such that $p = p_n$ for infinitely many n, and

(ii) for p as in (i) above, $W_p = L$,

then, we say that the sequence $p_0 p_1 \ldots$ witnesses **Part**-learnability of L.

An *enumeration operator* (or just operator) Θ is an algorithm mapping from SEQ to SEQ such that for all $\sigma, \tau \in SEQ$, if $\sigma \subseteq \tau$, then $\Theta(\sigma) \subseteq \Theta(\tau)$. We let $\Theta(T) = \bigcup_{n \in N} \Theta(T[n])$.

We further assume that $\lim_{n \to \infty} |\Theta(T[n])| = \infty$, that is texts are mapped to texts by the operator Θ. Note that any operator Θ can be modified to satisfy the above property without violating the content of its output on infinite texts.

We will also use Θ as an operator on languages (rather than individual texts representing them, as above). Note that, in general, for different texts T, T' of a language L, Θ may produce texts $\Theta(T)$ and $\Theta(T')$ of different languages. Thus, we define $\Theta(L)$ as a collection of languages: $\Theta(L) = \{\text{content}(\Theta(T)) : T \text{ is a text for } L\}$, and, accordingly, the image $\Theta(\mathcal{L}) = \bigcup_{L \in \mathcal{L}} \Theta(L)$. In the special case (important for our strong reductions, defined below), when $\Theta(L)$ is a singleton $\{L'\}$, we abuse notation and say simply $\Theta(L) = L'$. (Note that if $\Theta(L) = \{L'\}$, then $L' = \bigcup_{\sigma : \text{content}(\sigma) \subseteq L} \text{content}(\Theta(\sigma))$.)

We let Θ and Ψ range over operators, where for ease of notation, we assume that for Ψ the input and output sequences contain only elements of N (and thus do not contain $\#$). We view Ψ as mapping sequences of grammars to sequences of grammars. Again, as in the definition of operator Θ, we assume that Ψ maps infinite sequences to infinite sequences. This can be easily done without changing the set of grammars which appear infinitely often in the sequence.

The following two definitions are based on the corresponding reductions for explanatory function learning [5] and explanatory language learning [14]. In these definitions, we view operators Θ as mapping texts to texts, as well as mapping languages to collections of languages (as discussed above).

Definition 2. We say that $\mathcal{L} \leq_{\mathbf{Part}}^{Weak} \mathcal{L}'$ iff there exist operators Θ and Ψ such that

(a) for all $L \in \mathcal{L}$, $\Theta(L) \subseteq \mathcal{L}'$.

(b) for all $L \in \mathcal{L}$, for all $L' \in \Theta(L)$, if $p_0 p_1 \ldots$ is a sequence witnessing **Part**-learnability of L', then $\Psi(p_0 p_1 \ldots)$ witnesses **Part**-learnability of L.

Intuitively, Θ reduces a text for a language $L \in \mathcal{L}$ to a text for a language $L' \in \mathcal{L}'$. Ψ then converts sequences witnessing **Part**-learnability of L' to sequences witnessing **Part**-learnability of L.

However, as we noted above, different texts for L may be mapped by Θ to texts for different languages in $\mathcal{L}'$. If we require that the mapping should be to the texts of the same language, then we get strong reduction.

Definition 3. We say that $\mathcal{L} \leq_{\mathbf{Part}}^{Strong} \mathcal{L}'$, iff there exist operators Θ and Ψ such that

(a) Θ and Ψ witness that $\mathcal{L} \leq_{\mathbf{Part}}^{Weak} \mathcal{L}'$ and
(b) for all $L \in \mathcal{L}$, $\mathrm{card}(\Theta(L)) = 1$.

For ease of notation, when considering strong reductions, as discussed above, we consider Θ as directly mapping languages to languages, rather than considering it as a mapping from languages to a set containing just one language.

We say that $\mathcal{L} <_{\mathbf{Part}}^{Weak} \mathcal{L}'$ if $\mathcal{L} \leq_{\mathbf{Part}}^{Weak} \mathcal{L}'$ but $\mathcal{L}' \not\leq_{\mathbf{Part}}^{Weak} \mathcal{L}$. Similarly, $\mathcal{L} \equiv_{\mathbf{Part}}^{Weak}$ if $\mathcal{L} \leq_{\mathbf{Part}}^{Weak} \mathcal{L}'$ and $\mathcal{L}' \leq_{\mathbf{Part}}^{Weak} \mathcal{L}$.

Similarly, we can define $\mathcal{L} <_{\mathbf{Part}}^{Strong} \mathcal{L}'$ and $\mathcal{L} \equiv_{\mathbf{Part}}^{Strong} \mathcal{L}'$.

Definition 4. We say that $\mathcal{L}$ is $\leq_{\mathbf{Part}}^{Weak}$-complete if

(a) $\mathcal{L} \in \mathbf{Part}$ and
(b) For all $\mathcal{L}' \in \mathbf{Part}$, $\mathcal{L}' \leq_{\mathbf{Part}}^{Weak} \mathcal{L}$.

$\leq_{\mathbf{Part}}^{Strong}$-completeness can be defined similarly.

We now define some languages and classes which are often used in the paper. We used the names **iINIT, iCOINIT, iRINIT** for the classes defined below as the infinite languages in these classes are obtained by cylindrification of the languages in $INIT$, $COINIT$ and $RINIT$ used in the literature (the class $INIT$ contains languages $\{1, 2, \ldots, i\}$ and $COINIT$ contains languages $\{i, i + 1, i + 2, \ldots\}$; $RINIT$ is similar to $INIT$ and contains, for each $r \in R_{0,1}$, the language having (the representatives of) rational numbers below r). Additionally the classes **iINIT, iCOINIT, iRINIT** contain all the finite languages. For $i \in N, r \in RAT_{0,1}$,

(a) $INIT_i = \{\langle x, y \rangle : x, y \in N \text{ and } x \leq i\}$,
(b) $COINIT_i = \{\langle x, y \rangle : x, y \in N \text{ and } x \geq i\}$,
(c) $RINIT_r = \{\langle x, y \rangle : x, y \in N \text{ and } \mathrm{ntor}(x) \leq r\}$,
(d) $INIT_{i,s} = D_s \cup INIT_i$,
(e) $COINIT_{i,s} = D_s \cup COINIT_i$,
(f) $\mathcal{FIN} = \{L : L \text{ is finite}\}$,
(g) **iINIT** $= \{INIT_i : i \in N\} \cup \mathcal{FIN}$,
(h) **iCOINIT** $= \{COINIT_i : i \in N\} \cup \mathcal{FIN}$,
(i) **iRINIT** $= \{RINIT_r : r \in RAT_{0,1}\} \cup \mathcal{FIN}$,
(j) **iINIT$_k$** $=$ **iINIT** $\cup \{INIT_{i,s} : \mathrm{card}(D_s) \leq k, i \in N\}$,
(k) **iCOINIT$_k$** $=$ **iCOINIT** $\cup \{COINIT_{i,s} : \mathrm{card}(D_s) \leq k, i \in N\}$.

A natural partial learning strategy for the languages in **iINIT** is as follows: when a new datum $\langle j, x \rangle$, where j is larger than all m for all pairs $\langle m, y \rangle$ seen so far, appears on the input, the learner, for the first time, outputs the conjecture $INIT_j$. Now, as long as no new (not previously seen) datum appears on the input, the learner conjectures the finite set representing the input seen so far; if a new datum (not previously seen) from $INIT_j$ appears on the input, the learner repeats the conjecture $INIT_j$. This continues as long as no datum outside $INIT_j$ is seen. Clearly, the correct language $INIT_j$ or some finite input set is the only one that will be conjectured infinite number of times. A similar strategy works for **iRINIT**.

For **iCOINIT** a similar strategy chooses a new infinite conjecture $COINIT_j$ when a new pair $\langle j, x \rangle$, where j is smaller than all m for all pairs $\langle m, y \rangle$ seen so far, appears on the input. Otherwise, the strategy is identical to the one for **iINIT**.

For **iINIT$_k$** the above **iINIT**-learning strategy can be adjusted as follows: the learner keeps track of the smallest j such that the set $E = \{\langle x, y \rangle : x > j$ and $\langle x, y \rangle$ is seen in the input so far$\}$ has at most k elements. Then, the strategy for learning **iINIT$_k$** is similar to that of **iINIT**, except that whenever the strategy for **iINIT** outputs an infinite conjecture, strategy for **iINIT$_k$** outputs an infinite conjecture for $INIT_{j,s}$, where $D_s = E$, where j, E are as described above. Similar modification to the strategy for **iCOINIT** works for **iCOINIT$_k$**.

3 Basic Properties of Reductions

In this section, we establish a number of technical facts used in many proofs of our results.

Lemma 5. *Suppose Θ witnesses part (a) of Definition 2 for $\mathcal{L} \leq_{\text{Part}}^{\text{weak}} \mathcal{L}'$. Suppose F_1, F_2 are computable functions such that, for any $L \in \mathcal{L}$ and $L' \in \Theta(L)$, the following three properties hold:*

(i) if L' is finite, then $F_1(L')$ is a grammar for L.
(ii) if L' is infinite and p is a grammar for L', then $\lim_{t \to \infty} F_2(p, t)$ exists and is a grammar for L.
(iii) if L' is infinite, then for any sequence of finite sets $S_1, S_2, \ldots$ such that $S_0 \subset S_1 \subset S_2 \ldots$ and $\bigcup_{i \in N} S_i = L'$, for all t, for all but finitely many t', $F_1(S_t) \neq F_1(S_{t'})$.

*Then, there exists a Ψ such that, for all $L \in \mathcal{L}$, for any sequence of grammars $p_0 p_1 \ldots$ witnessing **Part**-learnability of $L' \in \Theta(L)$, $\Psi(p_0 p_1 \ldots) = q_0 q_1 \ldots$ witnesses **Part**-learnability of L.*

The above lemma is useful in simplifying the construction of Ψ in many of the proofs: we can just give the relevant F_1 and F_2.

Proposition 6. *There exists an operator Ψ such that for any sequence $q_0 q_1 \ldots$, $\Psi(q_0 q_1 \ldots) = q_0' q_1', \ldots$ such that:*

(a) at most one grammar appears in $q'_0 q'_1 \ldots$ infinitely often,

(b) if q is the least grammar which appears infinitely often in $q_0 q_1 \ldots$, then q' appears infinitely often in $q'_0 q'_1 \ldots$, where $W_{q'} = W_q$.

Proof. Let $q'_i = pad(q_i, j)$, where $j = \mathrm{card}(\{i' : i' \leq i \text{ and } q_{i'} < q_i\})$. It is easy to verify that the above sequence satisfies the requirements of the proposition. ∎

The above proposition is useful to simplify some of the constructions for Ψ in our proofs.

Proposition 7. *Suppose $\mathcal{L} \leq^{weak}_{\mathbf{Part}} \mathcal{L}'$ as witnessed by Θ and Ψ. Then, for all distinct $L, L' \in \mathcal{L}$, $\Theta(L) \cap \Theta(L') = \emptyset$.*

Proposition 8. *For any operator Θ, if $L \subseteq L'$, $\Theta(L) = \{X\}$ and $\Theta(L') = \{X'\}$, then $X \subseteq X'$.*

Proposition 9. *Suppose L is infinite and $\mathcal{L}$ contains L and all finite subsets of L. Suppose further that $\mathcal{L} \leq^{weak}_{\mathbf{Part}} \mathcal{L}'$ as witnessed by Θ and Ψ. Then, for all finite sets S such that $S \subseteq L'$ for some $L' \in \Theta(L)$, there exists an infinite superset of S in $\Theta(L)$ (in particular, $\Theta(L)$ contains an infinite language).*

4 Complete Classes and the Class $\mathcal{R}$

As $\mathcal{E} \in \mathbf{Part}$, we trivially have that $\mathcal{E}$ is $\leq^{Strong}_{\mathbf{Part}}$-complete. The following results give some simple classes which are complete.

Let $\mathbf{iCOINIT}_* = \{COINIT_i \cup A : i \in N \text{ and } A \text{ is finite}\}$. We first show that every text for every recursively enumerable language can be appropriately "encoded" as a text for some language in $\mathbf{iCOINIT}_*$ — thus showing that $\mathbf{iCOINIT}_*$ is weakly complete.

Theorem 10. $\mathbf{iCOINIT}_*$ *is $\leq^{weak}_{\mathbf{Part}}$-complete.*

Proof. To show that $\mathcal{E} \leq^{weak}_{\mathbf{Part}} \mathbf{iCOINIT}_*$, define Θ and Ψ as follows.

Suppose T is a given text. Let $C_{p,T} = \max(\{t : W_{p,t} \subseteq \mathrm{content}(T) \text{ and } \mathrm{content}(T[t]) \subseteq W_p\})$. Note that $C_{p,T}$ can be approximated from below (that is, there exists a recursive function f such that $f(p, T[n]) \leq f(p, T[n+1]) \leq C_{p,T}$ and $\lim_{n \to \infty} f(p, T[n]) = C_{p,T}$.

Define Θ as follows. $\Theta(T) = T'$ such that $\mathrm{content}(T') = \{\langle q, x \rangle : (\exists q' \leq q)[x \leq C_{q',T}]\}$.

Now, suppose p is the least grammar for W_p, T is a text for W_p, and $T' = \Theta(T)$. Then, it is easy to verify that $\mathrm{content}(T') = COINIT_{p,s}$, for some s, as $C_{p,T} = \infty$, but $C_{p',T} < \infty$ for $p' < p$.

We define Ψ as follows.

$\Psi'(p_0 p_1 \ldots) = q_0 q_1 \ldots$, where

$q_i = pad(j, p_i)$, where, for some x, $\langle j, x \rangle$ is the last new element enumerated in $W_{p_i, k}$ and k is the number of times p_i appears in $p_0 p_1 \ldots p_i$.

Claim 11. *Suppose $L \in \mathcal{E}$, and $L' \in \Theta(L)$.*

*If $p_0 p_1 \ldots$ witnesses **Part**-learnability of L', then $q_0 q_1 \ldots$ is a sequence satisfying:*

there exists a minimal q such that q appears infinitely often in $q_0 q_1 \ldots$ and this $q = pad(j, p)$, for minimal grammar j for L and some p.

To see the claim, suppose $L' \in \Theta(L)$ and $p_0, p_1 \ldots$ is a sequence witnessing **Part**-learnability of L' and $\Psi'(p_0 p_1 \ldots) = q_0 q_1 \ldots$. Suppose p is the only grammar which appears infinitely often in $p_0 p_1 \ldots$. Then only q_i with $p_i = p$ can possibly appear infinitely often in the sequence $q_0 q_1 \ldots$ as q_i used p_i in its padding. Furthermore, $pad(j, p)$, with $j \geq \min(\{e : W_e = L\})$ appear infinitely often in the sequence $q_0 q_1 \ldots$. The claim follows.

Now the theorem follows using Proposition 6. ∎

We now consider a $\leq_{\mathbf{Part}}^{Strong}$-complete class.

Let $V(L) = \frac{1}{4} + \sum_{x \in L} 4^{-x-1}$.

Intuitively, V maps languages to real numbers where the mapping is monotonic in L. Furthermore, if $L \neq L'$ and $\min(L \Delta L') \in L$, then $V(L) > V(L')$.

The reason for choosing the additive part "$\frac{1}{4}$" is just to make sure that $V(L)$ is non-zero.

Let $L_{r_0, r_1, \ldots, r_k} = \{\langle i, x \rangle : i < k \text{ and } ntor(x) < r_i \text{ or } i \geq k \text{ and } ntor(x) < r_k\}$.

Let $\mathcal{STRCOMP} = \{L_{r_0, r_1, \ldots, r_k} : k \in N, r_0 \leq r_1 \leq \ldots \leq r_k, \text{ and } r_0, r_1, \ldots, r_k \text{ are left r.e. reals}\}$.

$\mathcal{STRCOMP}$ denotes "strong complete class". The languages in $\mathcal{STRCOMP}$ can be thought of as follows: in the i-th cylinder we keep rational numbers $< r_i$. The r_i's are monotonically non-decreasing left r.e. real numbers and the sequence $r_0, r_1, \ldots$ converges (that is, for some k, for all $i \geq k$, $r_i = r_k$). We suggest the reader to contrast this class with the previously defined class **iRINIT** (which, in the sequel, will be shown to be incomplete).

Theorem 12. $\mathcal{STRCOMP}$ *is* $\leq_{\mathbf{Part}}^{Strong}$-*complete.*

Proof. For any index i and any language L, let

$X_{i,L} = \{\frac{x}{y} : x, y \in N, y \neq 0, \frac{x}{y} < V(W_i \cap L) \text{ and } x, y \leq \min(\{t : W_{i,t} - L \neq \emptyset\})\}$.

Intuitively, $sup(X_{i,L})$ gives a value to how much W_i and L are similar to each other:

(P1) If $W_i = L$, then $sup(X_{i,L})$ is $V(L)$ (as $\min(\{t : W_{i,t} - L \neq \emptyset\})$ is infinite).

(P2) If $W_i \neq L$, then $sup(X_{i,L}) < V(L)$. To see this note that if $L \nsubseteq W_i$, then clearly $V(W_i \cap L) < V(L)$ and thus $sup(X_{i,L}) \leq V(W_i \cap L) < V(L)$. On the other hand, if $W_i \nsubseteq L$, then $sup(X_{i,L}) < V(W_i \cap L) \leq V(L)$, as $X_{i,L}$ is a finite set and the supremum of a finite set of rational numbers $< r$ is $< r$ for any positive real number r.

Note that $X_{i,L}$ depends only on i and L and not on the particular presentation of L. This allows us to construct Θ (and corresponding Ψ) which give a strong reduction from $\mathcal{E}$ to $\mathcal{STRCOMP}$.

$\Theta(L) = \bigcup_{i \in N}[\{\langle i, y \rangle : (\exists i' \leq i)(\exists s \in X_{i',L})[\text{ntor}(y) \leq s]\}]$.

Intuitively, Θ just collects all the members of $X_{i,L}$ in the i-th cylinder and then does an upward closure.

Note that $\Theta(L)$ can be enumerated from a text T for L. Moreover, $\Theta(L)$ depends only on L and not on the particular presentation T, and thus the reduction is a strong reduction.

We say that m improves at time step j in the enumeration of W_p, if, for some x, the following two conditions are satisfied:

(i) $\langle m, x \rangle$ gets enumerated at time step j in W_p.
(ii) Suppose $j' < j$ is the largest earlier time step when m improved, if any, in the enumeration of W_p (take $j' = 0$, if m did not improve earlier). Then, for any y such that $\langle m', y \rangle \in W_{p,j'}$, $\text{ntor}(x) > \text{ntor}(y)$.

Note that for any grammar p for $L' = L_{r_0, r_1, \ldots, r_k}$, with $r_0 \leq r_1 \leq r_2 \ldots \leq r_{k-1} < r_k$, k improves at infinitely many steps j in the enumeration of W_p, but $0, 1, \ldots, k-1$ improve only for finitely many steps j in the enumeration of W_p.

We define an operator Ψ' as follows. This can then be converted to the desired Ψ using Proposition 6. $\Psi'(p_0 p_1 \ldots) = q_0 q_1 \ldots$, where q_i is defined below. Note that if $L \in \mathcal{E}$, $\Theta(L) = L'$ and $p_0, p_1, \ldots$ witnessed **Part**-learning of L', then we want $\Psi(p_0 p_1 \ldots)$ to witness **Part**-learning of L.

Without loss of generality assume that, for any k, W_k enumerates at most one element at any step. For any fixed i, suppose p_i appears j times in $p_0 p_1 \ldots p_i$, and m_i improves at step j in the enumeration of W_{p_i} (if no such m_i exists, then take m_i to be $i + 1$). Then, let $q_i = pad(m_i, p_i)$.

Now suppose $L \in \mathcal{E}$ and $L' = \Theta(L)$ and i is the minimal grammar for L. Then, by the properties (P1) and (P2), for all $j < i$, $sup(X_{i,L}) > sup(X_{j,L})$, and for all $j \geq i$, $sup(X_{i,L}) \geq sup(X_{j,L})$. Thus, L' is of the form $L_{r_0, r_1, \ldots, r_i}$, where $r_0 \leq r_1 \leq r_{i-1} < r_i$.

Now, suppose p appears infinitely often in $p_0 p_1 \ldots$, which witnesses **Part**-learnability of L'. Then for $\Psi'(p_0 p_1 \ldots) = q_0 q_1 \ldots$ only q_j with $p_j = p$ could possibly appear infinitely often in the sequence $q_0 q_1 \ldots$ as q_j used p_j in its padding. Furthermore, i is the minimal number which improves at infinitely many steps in the enumeration of W_p. It follows that $pad(i, p)$ is the minimal element which appears infinitely often in the sequence $q_0 q_1 \ldots$.

Now Ψ' can be converted to the required Ψ using Proposition 6. The theorem follows.   ∎

Our next result states that the class $\mathcal{R}$ of all recursive languages is not strongly complete. In particular, this means that all indexed classes of languages, including the class of all regular languages, are incomplete. This opens a possibility of creating partial learning strategies for these classes that would be simpler than the general strategy for partial learning of all recursively enumerable languages.

Theorem 13. $\mathcal{R}$ *is not* $\leq_{\mathbf{Part}}^{Strong}$*-complete.*

Proof. Suppose Θ (and Ψ) witness that $\mathcal{E} \leq_{\mathbf{Part}}^{Strong} \mathcal{R}$. Let K denote the halting set $\{i : i \in N$ and $\varphi_i(i) \downarrow\}$. Now, by Propositions 7 and 8, for all x, $\Theta(K) \subset \Theta(K \cup \{x\})$. Let $S = \Theta(K)$. Note that, by assumption, S is recursive. Now, $x \in K$ iff $\Theta(\{x\} \cup K) \subseteq S$. As S is recursive, this would imply that $\overline{K}$ is recursively enumerable, which is in contradiction to a known fact that K is not recursively enumerable [17]. ∎

5   Relationship Between iINIT, iCOINIT and iRINIT Classes

In this section, we explore the relationships between the classes **iINIT**, **iCOINIT** (and some of their variants), and **iRINIT**. We also establish that their degrees are strictly under the degree of all regular languages.

Proposition 14. *Fix* $n \in N$. *Suppose* $\mathcal{L} \subseteq \mathcal{E}$ *contains only* n *infinite languages. Then,*

(a) $\mathcal{L} \leq_{\mathbf{Part}}^{Strong}$ **iINIT**.
(b) $\mathcal{L} \leq_{\mathbf{Part}}^{Strong}$ **iCOINIT**.

First, we explore the relationship between **iINIT** and **iCOINIT**. We begin with establishing that **iCOINIT** is reducible to **iINIT**, but only weakly. Perhaps, this fact and the fact that **iINIT** is not reducible to **iCOINIT** (see below) are not surprising, as the chain of infinite languages in **iINIT** is growing indefinitely, whereas every growing chain of infinite languages in **iCOINIT** is finite.

Theorem 15. **iCOINIT** $\leq_{\mathbf{Part}}^{weak}$ **iINIT**.

Theorem 16. **iCOINIT** $\nleq_{\mathbf{Part}}^{Strong}$ **iINIT**.

On the other hand, **iINIT** is not reducible to **iCOINIT** even weakly.

Theorem 17. **iINIT** $\nleq_{\mathbf{Part}}^{weak}$ **iCOINIT**.

The class **iRINIT** is similar to **iINIT** in that it features an infinitely growing chain of infinite languages. However, unlike **iINIT**, between any two infinite languages in **iRINIT**, there is always another language. We show that the $\leq_{\mathbf{Part}}^{Strong}$-degree of **iRINIT** is strictly above the degrees of the classes **iINIT** and **iCOINIT**. However, first we show that **iRINIT** is not even weakly complete. Let $\mathcal{REG}$ denote the class of all regular sets [11] (we assume some standard recursive bijection between strings and N, so that regular sets can be considered as subsets of natural numbers). Topologically, $\mathcal{REG}$ is much more complex than containing just one growing chain of infinite languages **iRINIT** (plus finite sets), and this translates into greater complexity of partial learning of $\mathcal{REG}$, as the following theorem indicates.

Theorem 18. $\mathcal{REG} \not\leq_{\mathbf{Part}}^{weak}$ **iRINIT**.

Proof. Suppose by way of contradiction that Θ and Ψ witness that $\mathcal{REG} \leq_{\mathbf{Part}}^{\mathrm{Weak}}$ **iRINIT**.

Inductively define σ_i as follows.

$\sigma_0 = \Lambda$.

σ_{i+1} is an extension of $\sigma_i \diamond i$ such that $\max(\{\mathrm{ntor}(x) : \langle x, y \rangle \in \Theta(\sigma_{i+1})\}) > \max(\{\mathrm{ntor}(x) : \langle x, y \rangle \in \Theta(\sigma_i)\})$.

If all σ_{i+1} get defined, then for $T = \bigcup_{i \in N} \sigma_i$, T is a text for N (which is regular), but $sup(\{\mathrm{ntor}(x) : \langle x, y \rangle \in \mathrm{content}(\Theta(T))\})$ does not belong to $\{\mathrm{ntor}(x) : \langle x, y \rangle \in \mathrm{content}(\Theta(T))\}$ (as if it belonged, then it would belong to $\{\mathrm{ntor}(x) : \langle x, y \rangle \in \mathrm{content}(\Theta(\sigma_j))\}$, for some j, and that would violate the definitions of σ_i's).

If some σ_{i+1} does not get defined, then let $r = \max(\{\mathrm{ntor}(x) : \langle x, y \rangle \in \mathrm{content}(\Theta(\sigma_i))\})$. Now, for all infinite regular languages L containing $\mathrm{content}(\sigma_i)$, $RINIT_r \in \Theta(L)$ (as $\Theta(L)$ contains an infinite language containing $\langle \mathrm{rton}(r), y \rangle$, for some y, and σ_{i+1} did not get defined). A contradiction to Proposition 7, as there are infinitely many (in particular at least two) infinite regular languages which contain $\mathrm{content}(\sigma_i)$. ∎

Corollary 19. **iRINIT** *is not* $\leq_{\mathbf{Part}}^{weak}$*-complete.*

Our next result shows that both **iINIT** and **iCOINIT** are strongly reducible to **iRINIT**. Let $0 < r_0, r_1, \ldots$ in $RAT_{0,1}$ be a strictly increasing sequence of rational numbers. $\{INIT_i : i \in N\}$ can be naturally embedded into **iRINIT**, by mapping $INIT_i$ to $RINIT_{r_i}$. Note that, by our convention on coding of finite sets, $D_s \subset D_{s'}$ implies $s < s'$. For any s, let $k_s = \max(\{x : \langle x, y \rangle \in D_s\})$. Now, mapping finite sets D_s to $RINIT_{r_{k_s} + (r_{k_s+1} - r_{k_s}) * r_s}$ ensures that **iINIT** $\leq_{\mathbf{Part}}^{Strong}$ **iRINIT**. A similar method works to show that **iCOINIT** $\leq_{\mathbf{Part}}^{Strong}$ **iRINIT**.

Theorem 20.

(a) **iINIT** $\leq_{\mathbf{Part}}^{Strong}$ **iRINIT**.
(b) **iCOINIT** $\leq_{\mathbf{Part}}^{Strong}$ **iRINIT**.

Next we show that **iRINIT** is neither strongly reducible to **iINIT**, nor even weakly reducible to **iCOINIT**. Yet, it is weakly reducible to **iINIT**. The latter fact is quite interesting: every text for a language in **iRINIT** can be encoded as a text for a language in **iINIT**, yet the corresponding languages in **iINIT** for such texts may be different for different texts of the same language in **iRINIT**.

Theorem 21. **iRINIT** $\not\leq_{\mathbf{Part}}^{Strong}$ **iINIT**.

Proof. Suppose by way of contradiction otherwise, as witnessed by Θ and Ψ. Note that, by Proposition 9, $\Theta(RINIT_{0.2})$ cannot be a finite set. Suppose $\Theta(RINIT_{0.2}) = INIT_k$.

Then for two different values of $r < 0.2$, $\Theta(RINIT_r) = INIT_i$, for same i, as for all $r < 0.2$, $\Theta(RINIT_r) \subseteq INIT_k$. A contradiction. ∎

Theorem 22. $\textbf{iRINIT} \leq_{\textbf{Part}}^{weak} \textbf{iINIT}$.

Proof. Let $r_S = \max(\{\mathrm{ntor}(x) : \langle x, y \rangle \in S\})$. Define $m_{T[n]}$ as follows.

(i) $m_\Lambda = \langle rton(0), 0 \rangle$.
(ii) $m_{T[n+1]} = m_{T[n]}$, if $r_{\mathrm{content}(T[n+1])} = r_{\mathrm{content}(T[n])}$; otherwise $m_{T[n+1]} = \langle rton(r_{\mathrm{content}(T[n+1])}), m_{T[n]} + 1 \rangle$.

Now, let $\Theta(T[n]) = \{\langle x, y \rangle : x \leq m_{T[n]}, y \leq \mathrm{code}(\mathrm{content}(T[n]))\}$.
It is easy to verify that,

(1) For a finite set L, $\Theta(L) \subseteq \{\{\langle x, y \rangle : x \leq i, y \leq \mathrm{code}(L)\} : i \in N\}$.
(2) $\Theta(RINIT_r) \subseteq \{INIT_{\langle \mathrm{rton}(r), w \rangle} : w \in N\}$.

We can define the operator Ψ for the reduction using Lemma 5, where F_1 and F_2 are defined as follows.

$F_1(S) = $ canonical grammar for D_w, where $w = \max(\{y : \langle 0, y \rangle \in S\})$.
$F_2(p, t) = $ canonical grammar for $RINIT_{\mathrm{ntor}(j)}$, where, for some w, $\langle j, w \rangle = \max(\{x : \langle x, y \rangle \in W_{p,t}\})$.

It is now easy to verify using Lemma 5 that Θ and Ψ (as given by Lemma 5) witness that $\textbf{iRINIT} \leq_{\textbf{Part}}^{Weak} \textbf{iINIT}$. ∎

Theorem 23. $\textbf{iRINIT} \not\leq_{\textbf{Part}}^{weak} \textbf{iCOINIT}$.

The next result shows that $\textbf{iRINIT}$ is strongly reducible to $\mathcal{REG}$, the class of all regular languages. As we noted above, $\mathcal{REG}$ is not reducible to $\textbf{iRINIT}$ (even weakly), thus, the degree of $\textbf{iRINIT}$ is strictly below the degree of $\mathcal{REG}$.

Theorem 24. $\textbf{iRINIT} \leq_{\textbf{Part}}^{Strong} \mathcal{REG}$.

Now we turn our attention to the classes $\textbf{iINIT}_k$ and $\textbf{iCOINIT}_k$. The infinite languages in these classes do not form simple strict chains, as, for every infinite language L in the chain, both classes contain its variants having up to k extra elements. Interestingly, though, it turns out that, whereas adding one such extra element to infinite languages in the chain makes the partial learning problem harder, the difficulty of the partial learning problem does not increase when more elements are added.

Theorem 25. *For all $k > 0$,*

(a) $\textbf{iINIT}_k \leq_{\textbf{Part}}^{Strong} \textbf{iINIT}_1$.
(b) $\textbf{iCOINIT}_k \leq_{\textbf{Part}}^{Strong} \textbf{iCOINIT}_1$.
(c) $\textbf{iINIT}_1 \not\leq_{\textbf{Part}}^{Strong} \textbf{iINIT}$.
(d) $\textbf{iCOINIT}_1 \not\leq_{\textbf{Part}}^{weak} \textbf{iCOINIT}$.
(e) $\textbf{iINIT}_1 \leq_{\textbf{Part}}^{weak} \textbf{iINIT}$.

Theorem 26. *For all $k > 0$,*

(a) $\textbf{iINIT}_k \leq_{\textbf{Part}}^{Strong} \textbf{iRINIT}$.
(b) $\textbf{iCOINIT}_k \leq_{\textbf{Part}}^{Strong} \textbf{iRINIT}$.
(c) $\textbf{iRINIT} \not\leq_{\textbf{Part}}^{Strong} \textbf{iINIT}_k$.
(d) $\textbf{iRINIT} \not\leq_{\textbf{Part}}^{weak} \textbf{iCOINIT}_k$.

Acknowledgements. We thank Frank Stephan and the referees for several helpful comments, which improved the presentation of the paper.

References

1. Angluin, D.: Inductive inference of formal languages from positive data. Inf. Control **45**, 117–135 (1980)
2. Angluin, D.: Learning regular sets from queries and counter-examples. Inf. Comput. **75**, 87–106 (1987)
3. Blum, M.: A machine-independent theory of the complexity of recursive functions. J. ACM **14**(2), 322–336 (1967)
4. Case, J., Kötzing, T.: Computability-theoretic learning complexity. Phil. Trans. R. Soc. London **370**, 3570–3596 (2011)
5. Freivalds, R., Kinber, E., Smith, C.: On the intrinsic complexity of learning. Inf. Comput. **123**(1), 64–71 (1995)
6. Gao, Z., Jain, S., Stephan, F.: On conservative learning of recursively enumerable languages. In: Bonizzoni, P., Brattka, V., Löwe, B. (eds.) CiE 2013. LNCS, vol. 7921, pp. 181–190. Springer, Heidelberg (2013)
7. Gao, Z., Jain, S., Stephan, F., Zilles, S.: A survey on recent results on partial learning. In: Proceedings of the Thirteenth Asian Logic Conference, pp. 68–92. World Scientific (2015)
8. Gold, E.M.: Language identification in the limit. Inf. Control **10**(5), 447–474 (1967)
9. Gao, Z., Stephan, F.: Confident and consistent partial learning of recursive functions. In: Bshouty, N.H., Stoltz, G., Vayatis, N., Zeugmann, T. (eds.) ALT 2012. LNCS, vol. 7568, pp. 51–65. Springer, Heidelberg (2012)
10. Gao, Z., Stephan, F., Zilles, S.: Combining models of approximation with partial learning. In: Chaudhuri, K., Gentile, C., Zilles, S. (eds.) ALT 2015. LNSC (LNAI), vol. 9355, pp. 56–70. Springer, Heidelberg (2015)
11. Hopcroft, J., Ullman, J.: Introduction to Automata Theory, Languages, and Computation. Addison-Wesley, Boston (1979)
12. Jain, S., Kinber, E., Papazian, C., Smith, C., Wiehagen, R.: On the intrinsic complexity of learning recursive functions. Inf. Comput. **184**(1), 45–70 (2003)
13. Jain, S., Kinber, E., Wiehagen, R.: Language learning from texts: degrees of intrinsic complexity and their characterizations. J. Comput. Syst. Sci. **63**, 305–354 (2001)
14. Jain, S., Sharma, A.: The intrinsic complexity of language identification. J. Comput. Syst. Sci. **52**, 393–402 (1996)
15. Jain, S., Stephan, F.: Consistent partial learning. In: Proceedings of the Twenty Second Annual Conference on Learning Theory (2009)
16. Osherson, D., Stob, M., Weinstein, S.: Systems that Learn: An Introduction to Learning Theory for Cognitive and Computer Scientists. MIT Press, Cambridge (1986)
17. Rogers, H.: Theory of Recursive Functions and Effective Computability. McGraw-Hill, New York (1967). Reprinted by MIT Press in 1987

Learning Pattern Languages over Groups

Rupert Hölzl[1], Sanjay Jain[2]([⊠]), and Frank Stephan[2,3]

[1] Institute 1, Faculty of Computer Science, Universität der Bundeswehr München,
Werner-Heisenberg-Weg 39, 85577 Neubiberg, Germany
r@hoelzl.fr
[2] School of Computing, National University of Singapore,
Singapore 117417, Republic of Singapore
{sanjay,fstephan}@comp.nus.edu.sg
[3] Department of Mathematics, National University of Singapore,
Singapore 119076, Republic of Singapore

Abstract. This article studies the learnability of classes of pattern languages over automatic groups. It is shown that the class of bounded unions of pattern languages over finitely generated Abelian automatic groups is explanatorily learnable. For patterns in which variables occur at most n times, it is shown that the classes of languages generated by such patterns as well as their bounded unions are, for finitely generated automatic groups, explanatorily learnable by an automatic learner. In contrast, automatic learners cannot learn the unions of up to two arbitrary pattern languages over the integers. Furthermore, there is an algorithm which, given an automaton describing a group G, generates a learning algorithm M_G such that either M_G explanatorily learns all pattern languages over G or there is no learner for this set of languages at all, not even a non-recursive one. For some automatic groups, non-learnability results of natural classes of pattern languages are provided.

1 Introduction

Gold [10] introduced inductive inference, a model for learning classes of languages $\mathcal{L}$ (a language is a subset of Σ^* for some alphabet Σ) from positive data; this model was studied extensively in the subsequent years [1,2,8,9,24]. Inductive inference can be described as follows: The learner reads, one by one as input, elements of a language L from a class $\mathcal{L}$ of languages; these elements are provided in an arbitrary order and with arbitrarily many repetitions and pauses; such a presentation of data is called a *text* for the language. While reading the data items from the text, the learner conjectures a sequence of hypotheses (grammars), one hypothesis at a time; subsequent data may lead to revision of earlier hypotheses. The learner is considered to have learnt the target language L if the sequence

S. Jain is supported in part by NUS grants R146-000-181-112, R252-000-534-112 and C252-000-087-001.

F. Stephan is supported in part by NUS grants R146-000-181-112 and R252-000-534-112.

© Springer International Publishing Switzerland 2016
R. Ortner et al. (Eds.): ALT 2016, LNAI 9925, pp. 189–203, 2016.
DOI: 10.1007/978-3-319-46379-7_13

of hypotheses converges syntactically to a grammar for L. The learner is said to learn the class $\mathcal{L}$ of languages if it learns each language in $\mathcal{L}$. The above model of learning is often referred to as explanatory learning (**Ex**-learning) or learning in the limit; see Sect. 2 for the formal details.

Angluin [2] introduced one important concept studied in learning theory, namely the concept of (non-erasing) pattern languages, generated by patterns in which variables can be replaced by non-empty strings. Shinohara [29] generalised it to the concept of *erasing pattern languages* in which the variables are allowed to be substituted by empty strings. Suppose Σ is an alphabet set (usually finite) and X is an infinite set of variables. A pattern is a string over $\Sigma \cup X$. A substitution is a mapping from X to Σ^*. Using different substitutions for variables, different strings can be generated from a pattern. The language generated by a pattern is the set of strings that can be obtained from the pattern using some substitution. Angluin showed that when variables can be substituted by non-empty strings then the class of pattern languages is **Ex**-learnable; Lange and Wiehagen [18] provided a polynomial-time learner for non-erasing pattern languages. On the other hand, Reidenbach [27] showed that if arbitrary strings (including empty strings) are allowed for substitutions, then the class of pattern languages is not **Ex**-learnable if the alphabet size is 2, 3 or 4. Shinohara and Arimura [30] consider learning of unbounded unions of pattern languages from positive data.

This paper explores learnability of pattern languages over groups; pattern languages and verbal languages (that are languages generated by patterns without constants) have been used in group theory extensively, for example in the work showing the decidability of the theory of the free group [14,15]. Miasnikov and Romankov [20] studied when verbal languages are regular. In the following, consider a group $(G, \circ)$. The elements of G are then used as constants in patterns and the substitutions map variables to group elements; for example, if $L(xax^{-1}yyab) = \{x \circ a \circ x^{-1} \circ y \circ y \circ ab : x, y \in G\}$ then letting $x = b$ and $y = a^{-2}$ produces $bab^{-1}a^{-3}b \in L(xax^{-1}yyab)$. That is, concatenation in the pattern is replaced by the group operation $\circ$, producing a subset of G. Note that variables may be replaced by G's identity element.

This paper considers when pattern languages over groups and their bounded unions are **Ex**-learnable, where the focus is on automatic groups. Informally, an automatic group or more generally an automatic structure can be defined as follows (see Sect. 2 for formal details). Consider a structure $(A, R_1, R_2, \ldots)$, where $A \subseteq \Sigma^*$ and $R_1, R_2, \ldots$ are relations over Σ^* (an n-ary function can be considered as a relation over $n+1$ arguments — n inputs and one output). The structure is said to be automatic if the set A is regular and each of the relations is regular, where multiple arguments are given to the automata in parallel with shorter inputs being padded by some special symbol to make the lengths of all inputs the same. An automatic group is an automatic structure (A, R), where A is a representation of G (that is, there exists a one-one and onto mapping rep from G to A) and $R = \{(rep(x), rep(y), rep(z)) : x, y, z \in G \text{ and } x \circ y = z\}$. Automatic groups in this paper follow the original approach by Hogdson [11,12] and later by Khoussainov and Nerode [17] and Blumensath and Grädel [5,6]; they have also been studied by Nies, Oliver, Thomas and Tsankov [21–23,31].

Some automatic groups allow to represent the class of all pattern languages over the group or some natural subclass of it as an automatic family, $(L_e)_{e \in E}$, which is given by an automatic relation $\{(e, x) : e \in E \wedge x \in L_e\}$ for some regular index set E. Automatic families allow to implement learners which are themselves automatic; such learners satisfy some additional complexity bound and results in a restriction though many complexity bounds in learning theory are not restrictive [7,13,26]. The use of automatic structures and families has the further advantage that the first-order theory of these structures is decidable and that first-order definable functions and relations are automatic [11,17]; see the surveys of Khoussainov and Minnes [16] and Rubin [28] for more information. Theorem 6 below strengthens Angluin's characterisation [1] result on the learnability of indexed families of languages by showing that for the class of pattern languages over an automatic group to be learnable, it is already sufficient that they satisfy Angluin's tell-tale condition non-effectively (see Sect. 3 for definition of the tell-tale condition). It follows from Angluin's work that this non-effective version of the tell-tale condition is necessary. Note that for general indexed families, this non-effective version of tell-tale condition is not sufficient and gives rise only to a behaviourally correct learner [3].

Section 4 explores the learnability of the class of pattern languages when the number of occurrences of variables in the pattern is bounded by some constant. Let $\mathbf{Pat}_n(G)$ denote the class of pattern languages over group G where the number of occurrences of the variables in the pattern is bounded by n. Then, Theorem 7 shows that $\mathbf{Pat}_1(G)$ is $\mathbf{Ex}$-learnable for all automatic groups G, though $\mathbf{Pat}_2(G)$ is not $\mathbf{Ex}$-learnable for some automatic group G. This group G has infinitely many generators. Theorem 10 shows that $\mathbf{Pat}_n(G)$ is $\mathbf{Ex}$-learnable for all finitely generated automatic groups G (in fact, for any fixed m, even the class of unions of up to m such pattern languages is $\mathbf{Ex}$-learnable).

Sections 5 and 6 consider learnability of the class of all pattern languages. Theorem 16 shows that for some automatic group G generated by two elements, $\mathbf{Pat}(G)$, the class of all pattern languages over G, is not $\mathbf{Ex}$-learnable. On the other hand, Theorem 20 shows that for finitely generated Abelian groups G, $\mathbf{Pat}(G)$ as well as the class of unions of up to m pattern languages over G is $\mathbf{Ex}$-learnable, for any fixed m. Theorem 14 shows that for the class of pattern languages over finitely generated Abelian groups the learners can even be made automatic using a suitable representation of the group and hypothesis space (see Sect. 2 for the definition of automatic learners), though this hypothesis space cannot in general be an automatic family. However, the class of unions of up to two pattern languages over the integers with group operation $+$ is not automatically $\mathbf{Ex}$-learnable (see Theorem 18).

2   Preliminaries

The symbol $\mathbb{N}$ denotes the set of natural numbers $\{0, 1, 2, \ldots\}$ and the symbol $\mathbb{Z}$ denotes the set of integers $\{\ldots, -2, -1, 0, 1, 2, \ldots\}$. For any alphabet set Σ, Σ^* denotes a set of strings over Σ. For the purpose of the study of algebraic groups, let Σ^{-1} denote the set of inverses of the elements of Σ. The length of a string w is denoted by $|w|$ and $w(i)$ denotes the $(i+1)$-th character of the string, that is $w = w(0)w(1)w(2)\ldots w(|w|-1)$, where each $w(i)$ is in Σ (or $\Sigma \cup \Sigma^{-1}$ depending on the context).

The convolution of two strings u and v is defined as follows: Let $m = \max(\{|u|,|v|\})$ and let $\Diamond \notin \Sigma$ be a special symbol used for padding words. If $i < |u|$ then let $u'(i) = u(i)$ else let $u'(i) = \Diamond$; similarly, if $i < |v|$ then let $v'(i) = v(i)$ else let $v'(i) = \Diamond$. Now, $conv(u,v) = w$ is the string of length m such that, for $i < m$, $w(i) = (u'(i), v'(i))$. The convolution over n-tuples of strings is defined similarly. The convolution is useful when considering relations with two or more inputs like the graph of a function.

A function f is *automatic* if $\{conv(x, f(x)): x \in \mathrm{dom}(f)\}$ is regular. An n-ary relation R is *automatic* if $\{conv(x_1, x_2, \ldots, x_n): (x_1, x_2, \ldots, x_n) \in R\}$ is regular. A structure $(A, R_1, R_2, \ldots, f_1, f_2, \ldots)$ is said to be automatic if A is regular and, for all i, f_i is an automatic function from A^{k_i} to A for some k_i and R_i is an automatic relation over A^{h_i} for some h_i. A class $\mathcal{L}$ of languages over alphabet Σ is said to be an *automatic family* if there exists a regular index set I and there exist languages L_α, $\alpha \in I$, such that $\mathcal{L} = \{L_\alpha: \alpha \in I\}$ and the set $\{conv(\alpha, x): \alpha \in I, x \in \Sigma^*, x \in L_\alpha\}$ is regular. For $x, y \in \Sigma^*$ for a finite alphabet Σ, let $x <_{ll} y$ iff $|x| < |y|$ or $|x| = |y|$ and x is lexicographically before y, where some fixed ordering among symbols in Σ is assumed. Let $\leq_{ll}$, $>_{ll}$ and $\geq_{ll}$ be defined analogously. The relation $<_{ll}$ is called the length-lexicographical order on Σ. The following fact is useful for showing that various relations or functions are automatic.

Fact 1 (Blumensath and Grädel [6], Hodgson [11,12], Khoussainov and Nerode [17]). *Any relation or function that is first-order definable from existing automatic relations and functions is automatic.*

A group is a set of elements G along with an operation $\circ$ such that the following conditions hold:

Closure: for all $a, b \in G$, $a \circ b \in G$;
Associativity: for all $a, b, c \in G$, $(a \circ b) \circ c = a \circ (b \circ c)$;
Identity: there is an $\varepsilon \in G$ such that for all $a \in G$, $a \circ \varepsilon = \varepsilon \circ a = a$;
Inverse: for all $a \in G$ exists an $a^{-1} \in G$ such that $a \circ a^{-1} = a^{-1} \circ a = \varepsilon$.

Often when referring to the group G, the group operation $\circ$ is implicit. A group $(G, \circ)$ is said to be *Abelian* if for all $a, b \in G$, $a \circ b = b \circ a$. When considering groups, a string over G is identified with the element of G obtained by replacing concatenation with $\circ$: thus for $a, b, c \in G$, $ab^{-1}c$ represents the group element $a \circ b^{-1} \circ c$.

Σ is a set of generators for a group $(G, \circ)$ if all elements of G can be written as a finite string over elements of Σ and their inverses, where the concatenation operation is replaced by $\circ$. Note that, in general, the set of generators may be finite or infinite. An Abelian group $(G, \circ)$ is said to be a *free Abelian group* generated by a finite set $\{a_1, a_2, \ldots, a_n\}$ of generators iff

$$G = \{a_1^{m_1} \circ a_2^{m_2} \circ \ldots \circ a_n^{m_n} : m_1, m_2, \ldots, m_n \in \mathbb{Z}\}$$

and for each group element the choice of $m_1, m_2, \ldots, m_n$ is unique.

Usually a group $(G, \circ)$ is represented using a set of representatives over a finite alphabet Σ via a one-one function rep from G to Σ^*. Then, $rep(\alpha)$ is said to be the representative of $\alpha \in G$. For $L \subseteq G$, $rep(L) = \{rep(a): a \in L\}$. Often a group element is identified with its representative, and thus $L \subseteq G$ is also identified with $rep(L)$.

A group $(G, \circ)$ is said to be automatic if there exists a one-one function rep from G to Σ^*, where $rep(\alpha)$ is the representative of $\alpha \in G$, such that the following conditions hold:

- $A = \{rep(\alpha): \alpha \in G\}$ is a regular subset of Σ^*;
- The function $f(rep(\alpha), rep(\beta)) = rep(\alpha \circ \beta)$ is automatic.

In this case $(A, \circ)$ is called an automatic presentation of the group $(G, \circ)$; in the following, for the ease of notation, the groups are identified with their automatic presentation. Note that the second clause above implies that the function mapping $\alpha \mapsto \alpha^{-1}$, computing inverses, is also automatic, as it can be defined using a first order formula over automatic functions. Without loss of generality it is assumed that $\varepsilon \in \Sigma^*$ is the representative of $\varepsilon \in G$. An example of an automatic group is $(\mathbb{Z}, +)$, where $+$ denotes addition. The representation used for this automatic group is the reverse binary representation of numbers where the leftmost bit is the least significant bit (the sign of the number can be represented using a special symbol). In the above group, the order given by $<$ is also automatic and so the entire automatic structure $(\mathbb{Z}, +, <)$ is often used.

Angluin [2] introduced the concept of pattern languages to the subject of learning theory; the corresponding notions for pattern languages over a group $(G, \circ)$ are defined as follows. A *pattern* π is a string over $G \cup \{x_1, x_2, \ldots\} \cup \{x_1^{-1}, x_2^{-1}, \ldots\}$, where $X = \{x_1, x_2, \ldots\}$ is a set of variables. Sometimes the symbols x, y, z are also used for variables. The elements of G appearing in a pattern are called constants. A substitution is a mapping from X to G. Note that the substitution of variables by ε is allowed. Let $sub(\pi)$ denote the string formed from π by using the substitution sub, that is, by replacing every variable x by $sub(x)$ and x^{-1} by $(sub(x))^{-1}$ in the pattern π. The language generated by π, denoted $L(\pi)$, is the set $L(\pi) = \{sub(\pi): sub \text{ is a substitution}\}$. Two patterns π_1 and π_2 are said to be *equivalent* (with respect to the group G) iff $L(\pi_1) = L(\pi_2)$. A *pattern language* (over a group G) is a language generated by some pattern π. In case the pattern π does not contain any constants, then the language $L(\pi)$ generated by π is called a *verbal* language. Let $\mathbf{Pat}(G)$ denote the class of all pattern languages over the group $(G, \circ)$ and $\mathbf{Pat}^m(G)$ denote the class of all unions of up to m pattern languages over the group G. Let $\mathbf{Pat}_n(G)$ denote the class of pattern languages generated by patterns containing up to n occurrences of variables or inverted variables and correspondingly $\mathbf{Pat}_n^m(G)$ denote the class of all unions of up to m pattern languages generated by patterns containing up to n occurrences of variables or inverted variables.

Proposition 2. *The classes* $\mathbf{Pat}_n(G)$ *and* $\mathbf{Pat}_n^m(G)$ *are automatic families for all automatic groups* $(G, \circ)$ *and* $m, n \in \mathbb{N}$.

Gold [10] introduced the model of learning in the limit which is described below. Fix a group $(G, \circ)$. A *text* T is a mapping from $\mathbb{N}$ to $rep(G) \cup \{\#\}$, where as mentioned earlier elements of G are identified with $rep(G)$, and thus a text can be viewed as a sequence of elements of $G \cup \{\#\}$. A finite sequence is an initial segment of a text. Let $|\sigma|$ denote the length of sequence σ. The content of a text T, denoted content(T), is the set of elements of G in the range of T, that is, content(T)$= \{T(i) : T(i) \neq \#\}$. Similarly, the content of a finite sequence σ is $\{\sigma(i) : i < |\sigma| \text{ and } \sigma(i) \neq \#\}$ and denoted by content(σ). Intuitively, $\#$'s denote pauses in the presentation of data. T is a *text for* $L \subseteq G$ iff content(T)$= L$. Let $T[n]$ denote the initial segment of T of length n.

Intuitively, a *learner* reads from the input, one element at a time, some text T for a target language L. Based on this new element, the learner updates its memory and conjecture. The learner has some initial memory and conjecture before it has received any data. Note that a text denotes only positive data being presented to the learner; the learner is never given information about what is not in the target language L. The learner uses some hypothesis space $\mathcal{H} = \{H_\alpha : \alpha \in J\}$ for its conjectures. It is assumed for this paper that the hypothesis space is uniformly recursive, that is, $\{(\alpha, x) : \alpha \in J, x \in H_\alpha\}$ is recursive. More formally, a learner is defined as follows. Parts (d) to (f) of the definition give a basic learning criterion called explanatory learning.

Definition 3 (Based on Gold [10]). Fix a group $(G, \circ)$. Suppose I and J are some index sets (regular sets of strings over some finite alphabet).

Suppose $\mathcal{L} = \{L_\alpha : \alpha \in I\}$ is a class to be learnt and $\mathcal{H} = \{H_\beta : \beta \in J\}$ is a hypothesis space, with $L_\alpha, H_\beta \subseteq G$ for all $\alpha \in I$ and $\beta \in J$.

Suppose Δ is a finite alphabet used for storing memory by the learner and ? is a special symbol not in $J \cup \Delta^*$ used for the null hypothesis as well as for null memory.

(a) A *learner* is a recursive mapping from the set $(\Delta^* \cup \{?\}) \times (G \cup \{\#\})$ to the set $(\Delta^* \cup \{?\}) \times (J \cup \{?\})$. It has initial memory $mem_0 \in \Delta^* \cup \{?\}$ and initial conjecture $hyp_0 \in J \cup \{?\}$.

(b) Suppose a learner $\mathbf{M}$ with initial memory mem_0 and initial hypothesis hyp_0 is given. Suppose a text T for a language $L \subseteq G$ is given.
  – Let $mem_0^T = mem_0$ and $hyp_0^T = hyp_0$.
  – Let $(mem_{n+1}^T, hyp_{n+1}^T) = \mathbf{M}(mem_n^T, T(n))$.
  Intuitively, mem_{n+1}^T and hyp_{n+1}^T denote the memory and conjecture of the learner $\mathbf{M}$ after receiving input $T[n+1]$.

(c) M converges on T to a hypothesis hyp iff, for all but finitely many n, $hyp_n^T = hyp$.

(d) $\mathbf{M}$ is said to **Ex**-learn a language L with respect to the hypothesis space $\mathcal{H}$ iff for all texts T for L, $\mathbf{M}$ converges on T to a hypothesis hyp such that $H_{hyp} = L$. Note that the memory of the learner need not converge.

(e) **M** is said to **Ex**-learn a class $\mathcal{L}$ of languages with respect to the hypothesis space $\mathcal{H}$ iff **M Ex**-learns each $L \in \mathcal{L}$ with respect to the hypothesis space $\mathcal{H}$. In this case **M** is said to be an **Ex**-learner for $\mathcal{L}$.

(f) $\mathcal{L}$ is said to be **Ex**-learnable iff there exists a recursive learner **M** and a hypothesis space $\mathcal{H}$ such that **M Ex**-learns $\mathcal{L}$ with respect to the hypothesis space $\mathcal{H}$.

The notation **Ex** in the above definition stands for "explanatory learning" and was first introduced by Gold [10]. Often, reference to the hypothesis space $\mathcal{H}$ is dropped and is implicit. Furthermore, in some cases when learning automatic families, some automatic family is used as the hypothesis space.

A learner **M** *makes a mind change* [8,9] at $T[n]$, if $? \neq hyp_n^T \neq hyp_{n+1}^T$ as defined in the above definition. A learner makes at most m mind changes on a text T iff $|\{n : ? \neq hyp_n^T \neq hyp_{n+1}^T\}| \leq m$.

A learner is said to be *automatic* if its updating function is automatic, that is, if the relation $\{conv(mem, x, mem', hyp) : \mathbf{M}(mem, x) = (mem', hyp)\}$ is automatic.

For ease of presentation of proofs, sometimes it is informally described how the learner updates its memory and hypothesis when a new datum is presented. Furthermore, sometimes a language is directly used as hypothesis rather than an index; the index conjectured is implicit in such a case.

The following lemma by Gold is useful to show some of the results below.

Lemma 4 (Gold [10]). *Suppose $L_0 \subset L_1 \subset \ldots$ and $L = \bigcup_{i \in \mathbb{N}} L_i$. Then the class $\{L\} \cup \{L_0, L_1, \ldots\}$ is not **Ex**-learnable.*

3   A Characterisation

A *tell-tale set* [1] for a language L with respect to a class $\mathcal{L}$, is a finite subset D of L such that, for every $L' \in \mathcal{L}$, $D \subseteq L' \subseteq L$ implies $L' = L$. A class $\mathcal{L}$ satisfies Angluin's tell-tale condition non-effectively iff every language L in $\mathcal{L}$ has a tell-tale set with respect to $\mathcal{L}$. A class $\mathcal{L} = \{L_i : i \in I\}$ satisfies Angluin's tell-tale condition effectively iff for each $i \in I$, a tell-tale set D for L_i with respect to $\mathcal{L}$ can be enumerated effectively in i. Furthermore, $\mathcal{L}$ is called an indexed family iff there exists an indexing $\mathcal{L} = \{L_i : i \in I\}$, where I is a recursive set, such that $\{conv(i, x) : i \in I, x \in L_i\}$ is recursive.

Proposition 5 (Angluin [1]). *An indexed family $\mathcal{L}$ is **Ex**-learnable iff it satisfies Angluin's tell-tale condition effectively.*

Baliga, Case and Jain [3] gave a similar characterisation for behaviourally correct learning (using an acceptable numbering as hypothesis space) for indexed families satisfying Angluin's tell-tale condition non-effectively. Behaviourally correct learning, see [4,8,25], is a weaker learning notion where a successful learner for L on text T for L is required to eventually only output grammars that produce L; the learner is not required to converge to a single such grammar.

Angluin's result is now used to obtain a characterisation for learnability of $\mathbf{Pat}^m(G)$, for any automatic group G.

Theorem 6. *Given an automatic group G, if $\mathbf{Pat}^m(G)$ satisfies Angluin's tell-tale condition non-effectively, then $\mathbf{Pat}^m(G)$ is $\mathbf{Ex}$-learnable. The learner can be found effectively given m and the automaton describing G.*

Thus, $\mathbf{Pat}^m(G)$ is $\mathbf{Ex}$-learnable iff it satisfies Angluin's tell-tale condition non-effectively. The above result also holds when considering $\mathbf{Pat}^m_n(G)$ instead of $\mathbf{Pat}^m(G)$. The above result implies that the $\mathbf{Ex}$-learnability of $\mathbf{Pat}^m(G)$ or $\mathbf{Pat}^m_n(G)$ does not depend on the automatic representation chosen for the group; however, the learnability by an automatic learner might still depend on it.

4   Learning Patterns with up to n Variable Occurrences

In this section it is shown that the class $\mathbf{Pat}^m_n(G)$ is $\mathbf{Ex}$-learnable for all finitely generated automatic groups G. For other automatic groups, while $\mathbf{Pat}_1(G)$ is always $\mathbf{Ex}$-learnable, in general $\mathbf{Pat}_2(G)$ is not.

Theorem 7. *For every automatic group $(G, \circ)$, $\mathbf{Pat}_1(G)$ can be $\mathbf{Ex}$-learnt by an automatic learner which makes at most one mind change. There is, however, an automatic group $(G, \circ)$ such that $\mathbf{Pat}_2(G)$ cannot be $\mathbf{Ex}$-learnt by any learner.*

In the following, it will be shown that for finitely generated automatic groups, for all $n \in \mathbb{N}$, the class $\mathbf{Pat}_n(G)$ has an automatic learner. For this result, it is necessary to recall some facts from group theory:

Oliver and Thomas [23] showed that a finitely generated group is automatic iff it is Abelian by finite, that is, it has an Abelian subgroup of finite index. They furthermore noted [23, Remark 3] that if any group has an Abelian subgroup of finite index, then it also has an Abelian normal subgroup of finite index. Also note that if a finitely generated group has a normal subgroup of finite index, then this normal subgroup is finitely generated.

Thus, given an automatic finitely generated group $(G, \circ)$, it can be assumed without loss of generality that there is a finite subset H of G and generators $b_1, \ldots, b_m$ of a normal Abelian subgroup H' of G, such that every group element of G is of the form $a \circ b_1^{\ell_1} \circ b_2^{\ell_2} \circ \ldots \circ b_m^{\ell_m}$ where $a \in H$ and $\ell_1, \ell_2, \ldots, \ell_m \in \mathbb{Z}$. Furthermore, for each $a \in H$ and generator b_i, there are $j_{a,1}, \ldots, j_{a,m}$ with

$$b_i \circ a = a \circ b_1^{j_{a,1}} \circ b_2^{j_{a,2}} \circ \ldots \circ b_m^{j_{a,m}}.$$

Therefore it can be assumed without loss of generality that the group is represented as a convolution of $a \in H$ and $\ell_1, \ldots, \ell_m \in \mathbb{Z}$ in some automatic presentation of $(\mathbb{Z}, +, <)$. Then this allows to automatically carry out various group operations such as $\circ$, testing membership in H', and finding, for any element c of H' and any fixed group elements $d_1, \ldots, d_h$ that generate H', a tuple $(k_1, \ldots, k_h) \in \mathbb{Z}^h$ such that $c = d_1^{k_1} \circ \cdots \circ d_h^{k_h}$.

For G, H, H' as above, let S be a finite set of generators of H'; note that for each $S' \subseteq S$, the set generated by the elements of S' is a regular subset of H'. Furthermore, let $\mathcal{R}$ be a finite set of regular subsets of G with the property that for every $U \in \mathcal{R}$ and $\beta, \beta' \in H'$, either $U \circ \beta = U \circ \beta'$ or $U \circ \beta \cap U \circ \beta' = \emptyset$. Now the following family is automatic for each constant n:

$$\mathcal{Q}_n(\mathcal{R}, H') = \left\{ (U_1 \circ \beta_1) \cup \ldots \cup (U_h \circ \beta_h): \begin{array}{c} h \leq n \text{ and } U_1, \ldots, U_h \in \mathcal{R} \\ \text{and } \beta_1, \ldots, \beta_h \in H' \end{array} \right\}.$$

Here, an element of $\mathcal{Q}_n(\mathcal{R}, H')$ can be represented as follows. As $\mathcal{R}$ is finite, its elements can be represented using finitely many symbols. Each pair (U, β) can be represented using $conv(u, \beta)$, where u is a single symbol representing the element U of $\mathcal{R}$ and β is a representation of the corresponding group element. Each element of $\mathcal{Q}_n(\mathcal{R}, H')$ can now be represented using a convolution of the representation of up to n pairs (U, β).

Proposition 8. *For every n, the class $\mathcal{Q}_n(\mathcal{R}, H')$ has an automatic **Ex**-learner using $\mathcal{Q}_n(\mathcal{R}, H')$ as the hypothesis space.*

Proof. The learner in its memory maintains a set of candidate conjectures, where each candidate conjecture is of the form

$$(U_1 \circ \beta_1) \cup (U_2 \circ \beta_2) \cup \ldots \cup (U_h \circ \beta_h),$$

with $h \leq n$, each $U_i \in \mathcal{R}$ and each $\beta_i \in H'$. It will be the case that the number of candidates in the set is bounded by some constant c. The set of candidates can be memorised using a convolution of the representation of each of the candidates.

The conjecture of the learner at any stage is the minimal candidate (subset-wise) among all the candidates. In case of several minimal candidates, the candidate with the length-lexicographically smallest representation is used. Note that the above is an automatic operation.

Initially the learner has only one candidate which is the empty union. At any stage, when a new input w is processed, the following is done for each candidate in the current set. If a candidate $(U_1 \circ \beta_1) \cup (U_2 \circ \beta_2) \cup \ldots \cup (U_h \circ \beta_h)$ with $h < n$ does not contain w, then the candidate is replaced by a set of candidates of the form

$$(U_1 \circ \beta_1) \cup (U_2 \circ \beta_2) \cup \ldots \cup (U_h \circ \beta_h) \cup (U_{h+1} \circ \beta_{h+1}),$$

which satisfy that U_{h+1} is an element of $\mathcal{R}$ and that β_{h+1} exists and is the length-lexicographically least element of H' such that $w \in U_{h+1} \circ \beta_{h+1}$. If no such U_{h+1}, β_{h+1} exist, then $(U_1 \circ \beta_1) \cup (U_2 \circ \beta_2) \cup \ldots \cup (U_h \circ \beta_h)$ is simply dropped from the candidate set. Note that all the above operations are automatic. Furthermore, note that for each replaced candidate, at most $|\mathcal{R}|$ new candidates are added in. As none of these unions has more than n terms, the overall number of candidates considered through the runtime of the algorithm is at most $1 + |\mathcal{R}| + \ldots + |\mathcal{R}|^n$; if c is chosen equal to this number then never more than c candidates need to be memorised.

Now, suppose T is a text for $L \in \mathcal{Q}_n(\mathcal{R}, H')$. For all candidates which are not supersets of the language L to be learnt, the learner will eventually see a counter example and remove the candidate from the candidate set. Thus only candidates containing all elements of L will survive in the limit in the candidate set maintained by the learner.

Suppose $L = (U_1 \circ \beta_1) \cup \ldots \cup (U_h \circ \beta_h)$, where $h \leq n$ is minimal. Without loss of generality assume that, for each $i \in \{1, 2, \ldots, h\}$, β_i is the length-lexicographically least element β_i' of H' such that $U_i \circ \beta_i = U_i \circ \beta_i'$. Then eventually, $(U_1 \circ \beta_1) \cup \ldots \cup (U_h \circ \beta_h)$ is added to the candidate set by the learner. This can be shown by induction as follows. When the first datum different from $\#$ is observed by the learner, it adds $(U_{i_1} \circ \beta_{i_1})$ to the candidate set, for some $i_1 \in \{1, 2, \ldots, h\}$. Now, suppose the learner has placed $(U_{i_1} \circ \beta_{i_1}) \cup \ldots \cup (U_{i_k} \circ \beta_{i_k})$ in the candidate set with $k < h$ and $\{i_1, i_2, \ldots, i_k\} \subset \{1, 2, \ldots, h\}$. Then, eventually, it will add $(U_{i_1} \circ \beta_{i_1}) \cup \ldots \cup (U_{i_{k+1}} \circ \beta_{i_{k+1}})$ to the candidate set where $\{i_1, \ldots, i_k\} \subset \{i_1, \ldots, i_k, i_{k+1}\} \subseteq \{1, 2, \ldots, h\}$. This happens as soon as the learner is first presented some $w \in L - [(U_{i_1}, \beta_{i_1}) \cup \ldots \cup (U_{i_k}, \beta_{i_k})]$. Thus, by induction the learner will eventually put $(U_1 \circ \beta_1) \cup \ldots \cup (U_h \circ \beta_h)$, in the candidate set. As $(U_1, \beta_1) \cup \ldots \cup (U_h, \beta_h)$ is never dropped from the candidate set, eventually the learner only outputs the length-lexicographically least correct conjecture among the conjectures in its candidate set. Thus, it **Ex**-learns L. As L was arbitrary in $\mathcal{Q}_n(\mathcal{R}, H')$, it follows that the learner **Ex**-learns $\mathcal{Q}_n(\mathcal{R}, H')$. $\square$

Proposition 9. *Every automatic finitely generated group $(G, \circ)$ has a normal Abelian subgroup H' of finite index such that, for each n, there is a set $\mathcal{R}$ of finitely many regular languages and an n' such that $\mathbf{Pat}_n(G) \subseteq \mathcal{Q}_{n'}(\mathcal{R}, H')$.*

Note that if an automatic learner can learn an automatic family $\mathcal{L}$ using $\mathcal{L}$ as the hypothesis space and $\mathcal{L}' \subseteq \mathcal{L}$ is an automatic subfamily, then there is another automatic learner which learns $\mathcal{L}'$ using $\mathcal{L}'$ as the hypothesis space. This holds as there is an automatic function which translates any index in $\mathcal{L}$ for a language $L \in \mathcal{L}'$ into an index for L in $\mathcal{L}'$ (see [13]); furthermore the domain of this function is regular, as a finite automaton can determine whether an index of an element of $\mathcal{L}$ is for a language in $\mathcal{L}'$. Thus Propositions 8 and 9 give the following theorem.

Theorem 10. *Let $(G, \circ)$ be a finitely generated automatic group. Then, for each n, there is an automatic **Ex**-learner for the class $\mathbf{Pat}_n(G)$. Furthermore, for each $m, n \in \mathbb{N}$, there is an automatic **Ex**-learner for the class $\mathbf{Pat}_n^m(G)$ using $\mathbf{Pat}_n^m(G)$ itself as the hypothesis space.*

5   Automatic Learning of All Patterns

If $(G, \circ)$ is Abelian automatic group, then, in some cases, the class of all pattern languages is learnable by an automatic learner. For this, sometimes a non-automatic family must be considered as the hypothesis space, as the class of pattern languages may not always form an automatic family.

Example 11. *The group $(\mathbb{Z}, +)$ does not have any presentation $(A, +)$ in which the family $\{A_i : i \in I\}$ of $\mathbf{Pat}(A)$ is automatic.*

The following two propositions will be crucial in this and the subsequent section.

Proposition 12. *Suppose L is a pattern language over an Abelian group $(G, \circ)$. Then L is generated by a pattern of the form αx^n for some $\alpha \in G$ and $n \in \mathbb{N}$. Furthermore, α can be chosen as any element of L.*

Proposition 13. *Suppose L is a pattern language over a finitely generated free Abelian group and β_1, β_2 are two distinct elements of L. Then, effectively from β_1 and β_2, a finite set of patterns can be found, one of which generates L.*

In the case that a group is finite, the class of its pattern languages is obviously **Ex**-learnable simply by maintaining a list of all elements observed. So for the following result, only the case where the group is infinite is interesting.

Theorem 14. *Let $(G, \circ)$ be a finitely generated Abelian automatic group. Then the class $\mathbf{Pat}(G)$ of its pattern languages has, in a suitable representation of the group, an automatic **Ex**-learner which uses a suitable hypothesis space.*

In the above theorem, if another hypothesis space or representation of the group G is chosen, then the learner might fail to be automatic.

Remark 15. Note that the learning algorithm of Theorem 14 works for every automatic Abelian group of the form $(\mathbb{Z}, +) \times (A, \bullet)$, independently of what the second part of the direct product is. This gives a more general learning algorithm which covers many automatic groups, but not all of them; for example, the Prüfer group is not of this form. Furthermore, for some Abelian groups like

$$\left(\left\{ \tfrac{m}{n} : n > 0 \text{ and no prime factor of } n \text{ occurs twice} \right\}, + \right),$$

the class of pattern languages is not learnable. However, it is unknown whether this group is automatic; most likely, this group is not automatic.

Theorem 16. *There exists an automatic group G generated by two elements such that $\mathbf{Pat}(G)$ is not **Ex**-learnable.*

Proof. Consider the group with two generators a, b and the following equations for the group operation $\circ$:

- $a \circ b = b^{-1} \circ a$;
- $a \circ a = \varepsilon$.

Thus every group element is either of the form b^i or ab^i for some $i \in \mathbb{Z}$. Furthermore, consider the pattern languages

$$
\begin{aligned}
L(x_1 a x_1^{-1} a) &= \{ b^{2i} : i \in \mathbb{Z} \}; \\
L(b x_1 b^{-1} x_1^{-1}) &= \{ \varepsilon, b^2 \}; \\
L_i = L(b^{-2i} &\circ (b x_1 b^{-1} x_1^{-1}) \circ \ldots \circ (b x_{2i} b^{-1} x_{2i}^{-1})) \\
&= \{ b^{-2i}, b^{-2i+2}, \ldots, b^{-2}, \varepsilon, b^2, \ldots, b^{2i-2}, b^{2i} \}.
\end{aligned}
$$

It is easy to see that $L_1 \subset L_2 \subset L_3 \subset \ldots$ and $\bigcup_{i \in \mathbb{N}} L_{i+1} = L(x_1 a x_1^{-1} a)$. Thus, $\mathbf{Pat}(G)$ is not **Ex**-learnable by Lemma 4. □

The class of verbal languages over the group used in the above theorem is **Ex**-learnable, however for some group G, the class of verbal languages is not **Ex**-learnable.

Proposition 17. *There is a finitely generated automatic group where the class of verbal languages is not* **Ex***-learnable.*

6 Learning Bounded Unions of Patterns

Recall that the Prüfer group is the group of all rationals of the form $m/2^n$ with $0 \le m < 2^n$ and the rule that when $x + y \ge 1$ for group elements, one identifies this number with $x + y - 1$. In the Prüfer group, all patterns are either singletons or the full group; thus the bounded unions of the pattern languages have an automatic learner. In contrast to this, the automatic learnability of unions of two pattern languages fails already for the group of the integers.

Theorem 18. *Consider the group $(\mathbb{Z}, +)$. The class* **Pat**$^2(\mathbb{Z})$ *of unions of up to two pattern languages over the integers does not have an automatic* **Ex***-learner for any automatic representation of the group.*

Though there is no automatic **Ex**-learner for unions of two pattern languages over $\mathbb{Z}$, the next results will show that for finitely generated Abelian groups, there is a recursive (non-automatic) **Ex**-learner for unions of pattern languages.

Theorem 19. *Suppose G is a finitely generated free Abelian group. Then the class* **Pat**$^k(G)$ *is* **Ex***-learnable.*

Proof. Let Σ be the finite set of generators for G. Suppose T is a text for $L \in$ **Pat**$^k(G)$. Let P_G be a set of at most k patterns such that $L = \bigcup_{\pi \in P_G} L(\pi)$. Without loss of generality assume that the number of patterns in P_G generating at least two elements is minimised. The learner **M** keeps the following memory:

(a) the full input sequence $T[n]$ seen so far and
(b) a finite labeled tree; the labels on the nodes of the tree are patterns that generate at least two elements of G, except for the root which has an empty label; the tree is finitely branching and has depth of at most k.

For any leaf z in the tree, the language S_z associated with the leaf is the union of the languages generated by the labels on the nodes in the simple path from the root to z (excluding the root, but including z if it is not the root).

Initially, the tree in the memory of the learner **M** has only one node, the root. On input $T(n)$, the learner updates the tree as described in (A).

(A) For any leaf z in the tree, let $X_z = \text{content}(T[n+1]) - S_z$. If z is at depth $r < k$, where the root is at depth 0, and $\text{card}(X_z) \ge k - r + 1$, then using Proposition 13 find the finite number of possible patterns which contain a pattern equivalent to any pattern that generates at least two elements of X_z. Each of these finitely many patterns is added as a child of z in the tree.

As the tree in the memory of the learner is of bounded depth, has a finite branching degree and only leaf nodes can be expanded, the tree stabilizes as n goes to infinity. The learner's conjecture is computed as follows:

(B) For the tree as above after receiving $T[n+1]$, define the language Y_z for any leaf z at depth r of the tree as follows. Let $X_z = \text{content}(T[n+1]) - S_z$. Let $Y_z = S_z \cup X_z$, if $\text{card}(X_z) \leq k - r$. Otherwise, $Y_z = \Sigma^*$ (in this case, Y_z can be considered spoiled).

(C) Conjecture the minimal Y_z where z is a leaf of the tree (minimal is taken subset wise, as observed for strings in $\Sigma^0 \cup \Sigma^1 \cup \ldots \cup \Sigma^n$). Here a minimal language Y_z is not computable, so the learner just considers each language Y_z on strings (over Σ) of length at most n, and then chooses the minimal based on this. In case of multiple such minimal Y_z, choose the z which is leftmost in the tree. Note that the set of at most k patterns generating the language Y_z can be correspondingly computed.

To see that **M Ex**-learns $\mathbf{Pat}^k(G)$, note that, by induction, for each n, after seeing input $T[n+1]$, for each leaf z in the tree, for r, Y_z, X_z as defined in (B):

(i) $\text{content}(T[n+1]) \subseteq Y_z$;
(ii) if $\text{card}(X_z) > k - r$, then the depth of z is k (as otherwise, children would have been added to the leaf z in (A));
(iii) the pattern labels on the simple path from the root to z generate pairwise distinct languages;
(iv) for some leaf z', the simple path from the root to z' (excluding the root) is labeled only using patterns equivalent to some patterns in P_G.

Thus, for large enough n, for z' as given by item (iv) above, $Y_{z'} \subseteq L \subseteq Y_{z'}$. It follows that, for large enough n, the conjecture of **M** is L.   □

As every finitely generated Abelian group has a normal divisor of the form $(\mathbb{Z}^m, +)$ of finite index, the following theorem can be shown.

Theorem 20. *Consider any finitely generated Abelian group, not necessarily the free Abelian group. Then this group has an automatic presentation $(G, +)$ such that $\mathbf{Pat}^m(G)$ is **Ex**-learnable for all m using some suitable hypothesis space.*

7   Conclusions

In this paper the learnability of pattern languages over groups was studied. It was shown that for every finitely generated automatic group G, the class of pattern languages over G generated by patterns having a bounded number of variable occurrences is **Ex**-learnable. The same holds for bounded unions of such languages. Furthermore, for finitely generated Abelian groups G, the class of all pattern languages over G (and their bounded unions) is **Ex**-learnable. However, for some non-Abelian automatic group G generated by two elements, the class of pattern languages over G is not **Ex**-learnable. Similarly, for some infinitely generated group G, even the class of pattern languages over G generated by patterns having at most two occurrences of variables is not **Ex**-learnable.

Wiehagen [32] called a learner *iterative* if its memory is identical to the most recent hypothesis. Proposition 8, Theorems 10, 14, 19 and 20 can be shown using iterative learners which use class-preserving hypothesis spaces [19]. It is an open question whether for every automatic group it holds that if the class of all pattern languages over this group is **Ex**-learnable then it is also iteratively learnable.

Acknowledgements. The authors would like to thank the referees for detailed comments that helped to improve the presentation of this article.

References

1. Angluin, D.: Inductive inference of formal languages from positive data. Inf. Control **45**, 117–135 (1980)
2. Angluin, D.: Finding patterns common to a set of strings. J. Comput. Syst. Sci. **21**, 46–62 (1980)
3. Baliga, G., Case, J., Jain, S.: The synthesis of language learners. Inf. Comput. **152**, 16–43 (1999)
4. Bārzdiņš, J.: Two theorems on the limiting synthesis of functions. In: Theory of Algorithms and Programs, vol. 1, pp. 82–88. Latvian State University (1974). (in Russian)
5. Blumensath, A.: Automatic structures. Diploma thesis, RWTH Aachen (1999)
6. Blumensath, A., Grädel, E.: Automatic structures. In: Fifteenth Annual IEEE Symposium on Logic in Computer Science, Santa Barbara, LICS 2000, pp. 51–62. IEEE Computer Society Press, Los Alamitos (2000)
7. Case, J., Jain, S., Ong, Y.S., Semukhin, P., Stephan, F.: Automatic learners with feedback queries. J. Comput. Syst. Sci. **80**, 806–820 (2014)
8. Case, J., Lynes, C.: Machine inductive inference and language identification. In: Nielsen, M., Schmidt, E.M. (eds.) ICALP 1982. LNCS, vol. 140, pp. 107–115. Springer, Heidelberg (1982). doi:10.1007/BFb0012761
9. Case, J., Smith, C.: Comparison of identification criteria for machine inductive inference. Theor. Comput. Sci. **25**, 193–220 (1983)
10. Gold, E.M.: Language identification in the limit. Inf. Control **10**, 447–474 (1967)
11. Hodgson, B.R.: Théories décidables par automate fini. Ph.D. thesis, University of Montréal (1976)
12. Hodgson, B.R.: Décidabilité par automate fini. Ann. Sci. Math. Qué. **7**(1), 39–57 (1983)
13. Jain, S., Ong, Y.S., Shi, P., Stephan, F.: On automatic families. In: Proceedings of the Eleventh Asian Logic Conference in Honour of Professor Chong Chi Tat on his Sixtieth Birthday, pp. 94–113. World Scientific (2012)
14. Kharlampovich, O., Myasnikov, A.: Elementary theory of free non-abelian groups. J. Algebra **302**(2), 451–552 (2006)
15. Kharlampovich, O., Myasnikov, A.: Definable subsets in a hyperbolic group. Int. J. Algebra Comput. **23**(1), 91–110 (2013)
16. Khoussainov, B., Minnes, M.: Three lectures on automatic structures. In: Proceedings of Logic Colloquium 2007, Lecture Notes in Logic, vol. 35, pp. 132–176 (2010)

17. Khoussainov, B., Nerode, A.: Automatic presentations of structures. In: Leivant, D. (ed.) LCC 1994. LNCS, vol. 960, pp. 367–392. Springer, Heidelberg (1995). doi:10.1007/3-540-60178-3_93
18. Lange, S., Wiehagen, R.: Polynomial time inference of arbitrary pattern languages. New Gener. Comput. **8**, 361–370 (1991)
19. Lange, S., Zeugmann, T.: Incremental learning from positive data. J. Comput. Syst. Sci. **53**, 88–103 (1996)
20. Myasnikov, A., Romankov, V.: On rationality of verbal subsets in a group. Theory Comput. Syst. **52**(4), 587–598 (2013)
21. Nies, A.: Describing groups. Bull. Symb. Log. **13**, 305–339 (2007)
22. Nies, A., Thomas, R.M.: FA-presentable groups and rings. J. Algebra **320**, 569–585 (2008)
23. Oliver, G.P., Thomas, R.M.: Automatic presentations for finitely generated groups. In: Diekert, V., Durand, B. (eds.) STACS 2005. LNCS, vol. 3404, pp. 693–704. Springer, Heidelberg (2005). doi:10.1007/978-3-540-31856-9_57
24. Osherson, D., Stob, M., Weinstein, S.: Systems that Learn: An Introduction to Learning Theory for Cognitive and Computer Scientists. Bradford - The MIT Press, Cambridge (1986)
25. Osherson, D., Weinstein, S.: Criteria for language learning. Inf. Control **52**, 123–138 (1982)
26. Pitt, L.: Inductive inference, DFAs, and computational complexity. In: Jantke, K.P. (ed.) AII 1989. LNCS, vol. 397, pp. 18–44. Springer, Heidelberg (1989). doi:10.1007/3-540-51734-0_50
27. Reidenbach, D.: A non-learnable class of E-pattern languages. Theor. Comput. Sci. **350**, 91–102 (2006)
28. Rubin, S.: Automata presenting structures: a survey of the finite string case. Bull. Symb. Log. **14**, 169–209 (2008)
29. Shinohara, T.: Polynomial time inference of extended regular pattern languages. In: Goto, E., Furukawa, K., Nakajima, R., Nakata, I., Yonezawa, A. (eds.) RIMS Symposium on Software Science and Engineering 1982. LNCS, vol. 147, pp. 115–127. Springer, Heidelberg (1983). doi:10.1007/3-540-11980-9_19
30. Shinohara, T., Arimura, H.: Inductive inference of unbounded unions of pattern languages from positive data. In: Arikawa, S., Sharma, A.K. (eds.) ALT 1996. LNCS, vol. 1160, pp. 256–271. Springer, Heidelberg (1996). doi:10.1007/3-540-61863-5_51
31. Tsankov, T.: The additive group of the rationals does not have an automatic presentation. J. Symb. Log. **76**(4), 1341–1351 (2011)
32. Wiehagen, R.: Limes-Erkennung rekursiver Funktionen durch spezielle Strategien. J. Inf. Process. Cybern. (EIK) **12**(1–2), 93–99 (1976)

Online Learning

The Maximum Cosine Framework for Deriving Perceptron Based Linear Classifiers

Nader H. Bshouty[1] and Catherine A. Haddad-Zaknoon[2(✉)]

[1] Department of Computer Science, Technion, Haifa, Israel
[2] Department of Computer Science, University of Haifa, Haifa, Israel
catherin.haddad@gmail.com

Abstract. In this work, we introduce a mathematical framework, called the *Maximum Cosine Framework* or *MCF*, for deriving new linear classifiers. The method is based on selecting an appropriate bound on the cosine of the angle between the target function and the algorithm's. To justify its correctness, we use the MCF to show how to regenerate the update rule of Aggressive ROMMA [5]. Moreover, we construct a cosine bound from which we build the *Maximum Cosine Perceptron* algorithm or, for short, the *MCP* algorithm. We prove that the MCP shares the same mistake bound like the Perceptron [6]. In addition, we demonstrate the promising performance of the MCP on a real dataset. Our experiments show that, under the restriction of single pass learning, the MCP algorithm outperforms PA [1] and Aggressive ROMMA.

Keywords: Online learning · Linear classifiers · Perceptron

1 Introduction

Large-scale classification problems are characterized by huge datasets, high dimension and sparse examples. Moreover, the feature space is normally unknown to the learner. Therefore, an efficient learning algorithm should comply with two main requirements: (1) single pass over the examples dataset such that, for each example **x**, the time complexity for processing the example and adapting the algorithm hypothesis is linear in number of the non-zero features in **x**, and (2) space complexity is linear in the number of the relevant features. Consequently, in real world applications, classification via linear classifiers has gained a lot of attention due to their efficiency in time and memory.

The roots of many papers discussing linear classifiers date back to the Perceptron algorithm [6]. The Perceptron algorithm gained its popularity due to its efficiency in time and space as well as its polynomial mistake bound. The perceptron update rule complies naturally with the space and time requirements. Many algorithms introduced later followed the perceptron update paradigm including ALMA [3], NORMA [4] and PA [1].

In this work, we introduce a mathematical framework, called the *Maximum Cosine Framework* or *MCF*, for deriving new algorithms that follow the perceptron update scheme. That is, the algorithm observes examples in a sequence of

© Springer International Publishing Switzerland 2016
R. Ortner et al. (Eds.): ALT 2016, LNAI 9925, pp. 207–222, 2016.
DOI: 10.1007/978-3-319-46379-7_14

rounds. On round i, it constructs its classification hyperplane $\mathbf{w}_i$ incrementally each time the online algorithm makes a prediction mistake or its confidence in the prediction is inadequately low. It updates its classification scheme using an update rule of the form $\mathbf{w}_{i+1} = \mathbf{w}_i + y_i\lambda_i\mathbf{a}_i$, where $\mathbf{a}_i$ is the current observed example and λ_i is a parameter. To calculate the parameter λ_i, we formulate an upper bound on the cosine of the angle between the target hyperplane and the algorithm's one. Then we choose λ_i to be the value that optimizes the cosine bound. We argue that the tighter the cosine bound is, the closer we progress towards the target function. To justify the usefulness of the method, we use the MCF to regenerate the update rule of Aggressive ROMMA. In addition, using the MCF, we build a new linear classifier called the *Maximum Cosine Perceptron* or *MCP* for binary classification. We prove that the MCP shares the same mistake bound like the Perceptron. In addition, we demonstrate the promising performance of the MCP on a real dataset. Our experiments show that, under the restrictions of memory and single pass learning, the MCP algorithm outperforms PA and Aggressive ROMMA.

This paper is organized as follows. In Sect. 2 we bring a formal definition for the classification problem via linear classifiers. Moreover, we define the cosine bound concept and develop some preliminaries and useful lemmas that will be used along the discussion on algorithm construction under the MCF. In Sect. 3, we use the MCF to develop another algorithm called New Aggressive ROMMA (NAROMMA). We prove equivalence of NAROMMA to the well known Aggressive ROMMA in terms of the cosine of the angle between the target hyperplane and the algorithm's hypothesis. Section 4 outlines in details the construction of the local cosine bound from which the MCP algorithm generates its update rule. Furthermore, we discuss the mistake bound of the MCP algorithm and prove formally that it has the same mistake bound like the Perceptron, PA and Aggressive ROMMA. In Sect. 5 we describe the experiments we made to compare the performance of the MCP vs. the well known PA and Aggressive ROMMA.

2   The Maximum Cosine Framework

2.1   Problem Settings

In the binary classification setting, a *linear classifier* is an n-dimensional hyperplane that splits the space into two, where the points on the different sides correspond to the positive and negative labels. The target hyperplane is described by an n-dimensional vector called the *weight* vector and denoted by $\mathbf{w} \in \mathbb{R}^n$. Along the discussion we assume that $\|\mathbf{w}\|_2 = 1$. Our goal is to learn a prediction function, normally denoted by an n-dimensional vector $\mathbf{w}_i \in \mathbb{R}^n$, from a sequence of training examples $\{(\mathbf{a}_1, y_1), \cdots, (\mathbf{a}_T, y_T)\}$ where $\mathbf{a}_i \in \mathbb{R}^n$ and $\|\mathbf{a}_i\|_2 \leq R$ for some $R > 0$. In addition, $y_i \in \{-1, +1\}$ is the class label assigned to $\mathbf{a}_i$. In the *online learning model*, the learning process proceeds in *trials*. On trial i, the learning algorithm observes an example $\mathbf{a}_i$ and *predicts* the classification $\hat{y}_i \in \{-1, +1\}$ such that $\hat{y}_i = sign(\mathbf{w}_i^T \cdot \mathbf{a}_i)$. We say that the algorithm made a *mistake* if $\hat{y}_i \neq y_i$. The magnitude $|\mathbf{w}_i^T \mathbf{a}_i|$ is interpreted as the degree of confidence in the prediction. We refer to the term $y_i(\mathbf{w}_i^T \mathbf{a}_i)$ as the (signed) *margin* attained at round i. Let $\gamma > 0$, and let S be a set of binary labeled examples. We say that $\mathbf{w}$ seperates S with margin γ, if for all $\mathbf{a} \in S$, $|\mathbf{w}^T \mathbf{a}| \geq \gamma$. The margin γ is unknown to the algorithm. An online learning algorithm is *conservative* if it updates its weight vector $\mathbf{w}_i$ only on mistake, and *non-conservative* if it performs its update when the margin did not achieve a predefined threshold.

Let $\mathcal{A}$ be some online algorithm for binary classification that is introduced to some examples set. We say that $\mathcal{A}$ follows the *perceptron algorithm scheme*, if it maintains some hypothesis $h_i = sign(\mathbf{w}_i^T \mathbf{x})$ initialized to some value $\mathbf{w}_1$, normally chosen to be $\mathbf{0}$. On each example $\mathbf{a}_i$, the algorithm decides to update its hypothesis according to some predefined condition using the following update rule $\mathbf{w}_{i+1} = \mathbf{w}_i + \lambda_i(y_i \mathbf{a}_i)$.

From now on, we will use the terms *perceptron-like algorithm* and *perceptron algorithm* alternately to point the fact that the algorithm follows the perceptron algorithm scheme.

Along the discussion in this paper, we restrict our analysis to the case where the algorithm uses 0 *bias* hypothesis. That is, the target hypothesis is of the form $\mathbf{w} + \theta$ where we assume $\theta = 0$. It is well known in the literature that, for the variable bias case, we can get analogues theorems to the ones we prove in this work that are a constant factor worse than the original bounds [2].

2.2   The Cosine Bound

Let θ_i be the angle between the target hypothesis $\mathbf{w}$ and $\mathbf{w}_i$. Recall that $\|\mathbf{w}\|_2 = 1$. Our target is to choose λ_i such that

$$\alpha_{i+1} \triangleq \cos\theta_{i+1} = \frac{\mathbf{w}^T \mathbf{w}_{i+1}}{\|\mathbf{w}_{i+1}\|_2} \tag{1}$$

is maximal. The incentive of this choice is that we want to choose $\mathbf{w}_{i+1}$ to be as close as possible to $\mathbf{w}$. We start by choosing $\mathbf{w}_1 = cy_0\mathbf{a}_0$ for some $c > 0$. Under the separability assumption, this choice will guarantee that $\alpha_1 > 0$. Since the target hypothesis is unknown to the algorithm, it is obviously clear that we cannot find an accurate value for λ_i that maximizes the expression in (1). Instead, we will formulate a lower bound for $\cos\theta_i$ that we will call the *cosine bound* on which optimality can be achieved. Then, we find the value of λ_i that maximizes it. The optimal value of λ_i defines the algorithm's update on each trial. It is needless to say that choosing different cosine bounds will derive different update rules, namely different algorithms. To achieve a better classifier, we aim to maximize the value of α_i at each round i. For that purpose, we will start with some lemmas that will assist us develop cosine bounds from which we can derive new perceptron-like algorithms. For clarity, we bring the proofs of the following three lemmas in Appendix A.

Lemma 1. *Let* $\{(\mathbf{a}_0, y_0), \cdots, (\mathbf{a}_T, y_T)\}$ *be a sequence of examples where* $\mathbf{a}_i \in \mathbb{R}^n$ *and* $y_i \in \{-1, +1\}$ *for all* $0 \leq i \leq T$. *Let* $\gamma > 0$ *and* $\mathbf{w} \in \mathbb{R}^n$ *with* $\|\mathbf{w}\|_2 = 1$, *such that for all $0 \leq i \leq T$, $|\mathbf{w}^T \mathbf{a}_i| \geq \gamma$. Let $\mathbf{w}_1 = c\mathbf{a}_0$ for any constant $c > 0$. Let $\lambda_i > 0$ and $\mathbf{w}_{i+1} = \mathbf{w}_i + \lambda_i(y_i \mathbf{a}_i)$ be the update we use after the ith example. Let*

$$x_i = \frac{\lambda_i}{\|\mathbf{w}_i\|_2} \geq 0. \tag{2}$$

Then,

$$\alpha_{i+1} \geq \frac{\alpha_i + \gamma x_i}{\sqrt{1 + \|\mathbf{a}_i\|_2^2 x_i^2 + 2\frac{y_i(\mathbf{w}_i^T \mathbf{a}_i)}{\|\mathbf{w}_i\|_2} x_i}}. \tag{3}$$

The following two lemmas will assist us in optimizing the value of λ_i.

Lemma 2. *Let*

$$\Phi(x) = \frac{r + p \cdot x}{\sqrt{s + q \cdot x^2}}, \tag{4}$$

where $s, q > 0$. If $r \neq 0$ the optimal (maximal or minimal) value of $\Phi(x)$ is in $x^* = (ps)/(rq)$ and is equal to

$$sign(r)\sqrt{\frac{r^2}{s} + \frac{p^2}{q}}. \tag{5}$$

The point x^* is minimal if $r < 0$ and maximal if $r > 0$. If $r = 0$, the function $\Phi(x)$ is monotone increasing if $p > 0$ and monotone decreasing if $p < 0$.

Lemma 3. *Let*

$$\Phi(x) = \frac{r + p \cdot x}{\sqrt{s + q \cdot x^2 + 2tq \cdot x}} \tag{6}$$

where $s + q \cdot x^2 + 2tq \cdot x > 0$ for all x. Let Φ^* be the maximal value of $\Phi(x)$ over $\mathbb{R}^+$, i.e., $\Phi^* = \max_{x \in \mathbb{R}^+} \Phi(x)$. Then we have the following cases,

1. *If $r - pt = 0$ and $p > 0$ then, $\Phi^* = \Phi(\infty) = p/\sqrt{q}$.*
2. *If $r - pt = 0$ and $p < 0$ then, $\Phi^* = \Phi(0) = r/\sqrt{s}$.*
3. *If $r - pt > 0$ and $ps - rtq \geq 0$, let $x^* = (ps - rtq)/(rq - ptq) \geq 0$. Then, $\Phi^* = \Phi(x^*) = \sqrt{(r - pt)^2/(s - qt^2) + p^2/q}$.*
4. *If $r - pt > 0$ and $ps - tqr < 0$ then, $\Phi^* = \Phi(0) = r/\sqrt{s}$.*
5. *If $r - pt < 0$ and $ps - tqr \geq 0$ then, $\Phi^* = \Phi(\infty) = p/\sqrt{q}$.*
6. *If $r - pt < 0$ and $ps - tqr < 0$ then,*
   $\Phi^* = \max(\Phi(\infty), \Phi(0)) = \max(p/\sqrt{q}, r/\sqrt{s})$.

3   New Aggressive ROMMA

We use the MCF to construct a non-conservative algorithm - the NAROMMA algorithm. We start by formulating a local cosine bound from which we derive the best choice of λ_i at each trial. Along the discussion in this section and Sect. 4, we assume that the prerequisites of Lemma 1 apply, that is, for a sequence of examples $\{(\mathbf{a}_0, y_0), \cdots, (\mathbf{a}_T, y_T)\}$ there is $\gamma > 0$ and a separating hyperplane $\mathbf{w} \in \mathbb{R}^n$ with $\|\mathbf{w}\|_2 = 1$ such that $|\mathbf{w}^T \mathbf{a}_i| > \gamma$ for all i.

3.1   The Local Cosine Bound for the NAROMMA Algorithm

In this algorithm, $\mathbf{w}_1 = y_0 \mathbf{a}_0$ where $\mathbf{a}_0$ is the first example received by the algorithm, and the update follows the perceptron paradigm, that is $\mathbf{w}_{i+1} = \mathbf{w}_i + \lambda_i(y_i \mathbf{a}_i)$. Let

$$\gamma_i = \frac{y_i(\mathbf{w}_i^T \mathbf{a}_i)}{\|\mathbf{w}_i\|_2} \tag{7}$$

which is the projection of $y_i \mathbf{a}_i$ on the direction of $\mathbf{w}_i$. Let

$$\delta_i \triangleq \frac{\cos(\theta_i)}{\gamma} = \frac{\mathbf{w}^T \mathbf{w}_i}{\gamma \|\mathbf{w}_i\|_2} = \frac{\alpha_i}{\gamma}, \tag{8}$$

where α_i is as defined in (1). Then, by Lemma 1 we have,

$$\delta_{i+1} \geq \frac{\delta_i + x_i}{\sqrt{1 + \|\mathbf{a}_i\|_2^2 x_i^2 + 2\gamma_i x_i}} \tag{9}$$

for all $x_i \geq 0$. Since we cannot find an accurate evaluation for δ_i, we will look alternatively for some lower bound ℓ_i for δ_i, which will act as a local cosine bound for the algorithm, such that

$$\delta_i \geq \ell_i \tag{10}$$

for all i. We start by choosing ℓ_1,

$$\delta_1 = \frac{\cos(\theta_1)}{\gamma} = \frac{y_0(\mathbf{w}^T \mathbf{a}_0)}{\gamma \|\mathbf{a}_0\|_2} \geq \frac{1}{\|\mathbf{a}_0\|_2} = \ell_1. \tag{11}$$

Assuming that $\delta_i \geq \ell_i$ for some i, then by (9) we get,

$$\delta_{i+1} \geq \frac{\ell_i + x_i}{\sqrt{1 + \|\mathbf{a}_i\|_2^2 x_i^2 + 2\gamma_i x_i}}. \tag{12}$$

Inequality (12) formulates a local cosine bound for the NAROMMA algorithm. Therefore, since it holds for all $x_i \geq 0$, our next step is to find x_i^* that maximizes the right-hand side of (12) and get an optimal lower bound for δ_i. To achieve this, we start by the following lemma.

Lemma 4. *Let* $\{(\mathbf{a}_0, y_0), \cdots, (\mathbf{a}_T, y_T)\}$ *be a sequence of examples where* $\mathbf{a}_i \in \mathbb{R}^n$ *and* $y_i \in \{-1, +1\}$ *for all* $0 \leq i \leq T$. *Let* $\gamma > 0$ *and* $\mathbf{w} \in \mathbb{R}^n$ *with* $\|\mathbf{w}\|_2 = 1$, *such that for all* $0 \leq i \leq T$, $|\mathbf{w}^T \mathbf{a}_i| \geq \gamma$. *Let,*

$$\ell_{i+1}(x_i, \ell_i) \triangleq \frac{\ell_i + x_i}{\sqrt{1 + \|\mathbf{a}_i\|_2^2 x_i^2 + 2\gamma_i x_i}}, \tag{13}$$

and

$$x_i^*(\ell_i) \triangleq \arg\max_{x_i \geq 0} \frac{\ell_i + x_i}{\sqrt{1 + \|\mathbf{a}_i\|_2^2 x_i^2 + 2\gamma_i x_i}} \tag{14}$$

and finally,

$$\ell_{i+1}^* \triangleq \ell_{i+1}(x_i^*(\ell_i^*), \ell_i^*), \tag{15}$$

where $\ell_1^* = \ell_1 = \frac{1}{\|\mathbf{a}_0\|_2}$. Then for all $i \geq 1$,

$$\delta_i \geq \ell_i^*. \tag{16}$$

Proof. We prove by induction. For $i = 1$ we have, $\delta_1 = \frac{\cos(\theta_1)}{\gamma} = \frac{y_0(\mathbf{w}^T\mathbf{a}_0)}{\gamma\|\mathbf{a}_0\|_2} \geq \frac{1}{\|\mathbf{a}_0\|_2} = \ell_1 = \ell_1^*$. Assume that the lemma holds for $i = j$, i.e. $\delta_j \geq \ell_j^*$, we prove for $i = j + 1$. By (9) and the induction assumption we get for all $x_j \geq 0$, $\delta_{j+1} \geq \frac{\ell_j^* + x_j}{\sqrt{1 + \|\mathbf{a}_j\|_2^2 x_j^2 + 2\gamma_j x_j}}$. Specifically, the above inequality holds for $x_j = x_j^*(\ell_j^*)$. Using (13) and (15) we get, $\delta_{j+1} \geq \frac{\ell_j^* + x_j^*(\ell_j^*)}{\sqrt{1 + \|\mathbf{a}_j\|_2^2 x_j^*(\ell_j^*)^2 + 2\gamma_j x_j^*(\ell_j^*)}} = \ell_{j+1}^*$, which proves our lemma. $\qquad\square$

The above discussion implicitly assumes that (13) and (14) are well defined for all $x_i \geq 0$, and that, given some ℓ_i^*, we can easily determine $x_i^*(\ell_i^*)$ by solving the optimization problem implied by (14). Therefore, for completeness of the discussion, we first need to show that $1 + \|\mathbf{a}_i\|_2^2 x_i^2 + 2\gamma_i x_i > 0$. Second, we need to propose some direct solution from which the optimal x_i^* can be obtained. By calculating the discriminant of $1 + \|\mathbf{a}_i\|_2^2 x_i^2 + 2\gamma_i x_i$, one can conclude that it has a negative value for all $x_i \geq 0$ when $|\mathbf{w}_i^T \mathbf{a}_i|/(\|\mathbf{w}_i\|_2\|\mathbf{a}_i\|_2) < 1$. It is evident that $|\mathbf{w}_i^T \mathbf{a}_i|/\|\mathbf{w}_i\|_2\|\mathbf{a}_i\|_2 = 1$ if and only if $\mathbf{a}_i$ and $\mathbf{w}_i$ are linearly dependent. However, in this case the update will not change the direction of $\mathbf{w}_{i+1}$ regardless of the choice of λ_i and hence such examples will be disregarded by the algorithm. The following lemma provides a direct way to calculate ℓ_i^* and x_i^* for all i.

Lemma 5. *Let $\{(\mathbf{a}_0, y_0), \cdots, (\mathbf{a}_T, y_T)\}$ be a sequence of examples where $\mathbf{a}_i \in \mathbb{R}^n$ and $y_i \in \{-1, +1\}$ for all $0 \leq i \leq T$. Let $\gamma > 0$ and $\mathbf{w} \in \mathbb{R}^n$ with $\|\mathbf{w}\|_2 = 1$, such that for all $0 \leq i \leq T$, $|\mathbf{w}^T\mathbf{a}_i| \geq \gamma$. Let $\Phi(x_i) = \frac{\ell_i^* + x_i}{\sqrt{1 + \|\mathbf{a}_i\|_2^2 x_i^2 + 2\gamma_i x_i}}$, and let $x_i^* = \arg\max_{x_i \geq 0} \Phi(x_i)$. If $\gamma_i > 0$ then,*

$$x_i^* = \begin{cases} \infty & , \ell_i^* \leq \frac{\gamma_i}{\|\mathbf{a}_i\|_2^2} \\ \frac{1 - \ell_i^*\gamma_i}{\ell_i^*\|\mathbf{a}_i\|_2^2 - \gamma_i} & , \frac{1}{\gamma_i} > \ell_i^* > \frac{\gamma_i}{\|\mathbf{a}_i\|_2^2} \\ 0 & , \ell_i^* \geq \frac{1}{\gamma_i} \end{cases}. \tag{17}$$

If $\gamma_i \leq 0$ then,

$$x_i^* = \frac{1 - \ell_i^*\gamma_i}{\ell_i^*\|\mathbf{a}_i\|_2^2 - \gamma_i}. \tag{18}$$

Also,

$${\ell_{i+1}^*}^2 = \begin{cases} \frac{1}{\|\mathbf{a}_i\|_2^2} & , x_i^* = \infty \\ {\ell_i^*}^2 & , x_i^* = 0 \\ \ell_i^2 + \frac{(1 - \ell_i^*\gamma_i)^2}{\|\mathbf{a}_i\|_2^2 - \gamma_i^2} & , otherwise \end{cases} \tag{19}$$

where $\ell_1^* = 1/\|\mathbf{a}_0\|_2$.

Proof. We use Lemma 3. Let $r = \ell_i^*$, $p = s = 1$, $q = \|\mathbf{a}_i\|_2^2$ and $t = \gamma_i/\|\mathbf{a}_i\|_2^2$. Then $r - pt = \ell_i^* - \frac{\gamma_i}{\|\mathbf{a}_i\|_2^2}$, and $ps - rtq = 1 - \ell_i^*\gamma_i$. If $\gamma_i \leq 0$ then both $r - pt$ and $ps - rtq$ are positive, and, by Lemma 3 (case 3), we get that the optimal value is in $x_i^* = \frac{(ps-rtq)}{(rq-ptq)} = \frac{1-\ell_i^*\gamma_i}{\ell_i^*\|\mathbf{a}_i\|_2^2 - \gamma_i}$. Now, suppose $\gamma_i > 0$. Since $\gamma_i = y_i(\mathbf{w}_i^T\mathbf{a}_i)/\|\mathbf{w}_i\|_2 > 0$, then $\gamma_i/\|\mathbf{a}_i\|_2 = y_i(\mathbf{w}_i^T\mathbf{a}_i)/\|\mathbf{w}_i\|_2\|\mathbf{a}_i\|_2 \leq 1$ and hence, we get $\gamma_i < \|\mathbf{a}_i\|_2$. Since $\gamma_i < \|\mathbf{a}_i\|_2$, we have $\gamma_i/\|\mathbf{a}_i\|_2^2 < 1/\|\mathbf{a}_i\|_2 < 1/\gamma_i$. Therefore, we have the following three cases for ℓ_i^*:

Case I. $\ell_i^* \leq \gamma_i/\|\mathbf{a}_i\|_2^2$. Then, $r - pt = \ell_i^* - \gamma_i/\|\mathbf{a}_i\|_2^2 \leq 0$ and, $ps - rtq = 1 - \ell_i^*\gamma_i > 0$. By cases 1 and 5 in Lemma 3, we get $x_i^* = \infty$ and $\ell_{i+1}^* = 1/\|\mathbf{a}_i\|_2$.
Case II. $1/\gamma_i > \ell_i^* > \gamma_i/\|\mathbf{a}_i\|_2^2$. Then, $r - pt = \ell_i^* - \gamma_i/\|\mathbf{a}_i\|_2^2 > 0$ and, $ps - rtq = 1 - \ell_i^*\gamma_i > 0$. Hence, by case 3 in Lemma 3 we get $x_i^* = \frac{1-\ell_i^*\gamma_i}{\ell_i^*\|\mathbf{a}_i\|_2^2\gamma_i}$ and

$$\ell_{i+1}^* = \ell_{i+1}(x_i^*, \ell_i^*) = \frac{\ell_i^* + x_i^*}{\sqrt{1 + \|\mathbf{a}_i\|_2^2 x_i^{*2} + 2\gamma_i x_i^*}}. \tag{20}$$

By applying the value of x_i^* into (20) the result follows.
Case III. $1/\gamma_i \leq \ell_i^*$. Since $1/\gamma_i > \gamma_i/\|\mathbf{a}_i\|_2^2$ then, $r - pt = \ell_i^* - \gamma_i/\|\mathbf{a}_i\|_2^2 > 0$ and $ps - rtq = 1 - \ell_i^*\gamma_i \leq 0$. Therefore, using case 4 in Lemma 3 we get the result.

□

Recall the definition of x_i from (2). The implication of taking x_i to ∞ in the above update is to change the orientation of $\mathbf{w}_i$ to be the same as $\mathbf{a}_i$. That is, in case of $x_i^* = \infty$ we get, $\mathbf{w}_{i+1} = y_i\mathbf{a}_i$ and $\ell_{i+1}^* = 1/\|\mathbf{a}_i\|_2$. We have just proved,

Lemma 6. *Let $\{(\mathbf{a}_0, y_0), \cdots, (\mathbf{a}_T, y_T)\}$ be a sequence of examples where $\mathbf{a}_i \in \mathbb{R}^n$ and $y_i \in \{-1, +1\}$ for all $0 \leq i \leq T$. Let $\gamma > 0$ and $\mathbf{w} \in \mathbb{R}^n$ with $\|\mathbf{w}\|_2 = 1$, such that for all $0 \leq i \leq T$, $|\mathbf{w}^T\mathbf{a}_i| \geq \gamma$. Let $\mathbf{w}_1 = y_0\mathbf{a}_0$, $\ell_1^* = 1/\|\mathbf{a}_0\|_2$ and*

$$\mathbf{w}_{i+1} = \begin{cases} \mathbf{w}_i & , \gamma_i \geq \frac{1}{\ell_i^*}, \\ y_i\mathbf{a}_i & , \frac{1}{\ell_i^*} > \gamma_i \geq \|\mathbf{a}_i\|_2^2\ell_i^* \\ \mathbf{w}_i + \frac{(1-\ell_i^*\gamma_i)\|\mathbf{w}_i\|_2}{\ell_i^*\|\mathbf{a}_i\|_2^2 - \gamma_i}(y_i\mathbf{a}_i) & , \gamma_i < \min\{\ell_i^*\|\mathbf{a}_i\|_2^2, \frac{1}{\ell_i^*}\}. \end{cases} \tag{21}$$

and

$$\ell_{i+1}^{*2} = \begin{cases} \ell_i^{*2} & , \gamma_i \geq \frac{1}{\ell_i^*}, \\ \frac{1}{\|\mathbf{a}_i\|_2^2} & , \frac{1}{\ell_i^*} > \gamma_i \geq \|\mathbf{a}_i\|_2^2\ell_i^* \\ \ell_i^{*2} + \frac{(\ell_i^*\gamma_i - 1)^2}{\|\mathbf{a}_i\|_2^2 - \gamma_i^2} & , \gamma_i < \min\{\ell_i^*\|\mathbf{a}_i\|_2^2, \frac{1}{\ell_i^*}\}. \end{cases} \tag{22}$$

and γ_i is as in (7). Then after the ith update the cosine of the angle between $\mathbf{w}_i$ and $\mathbf{w}$ is at least

$$\cos(\theta_i) = \gamma\delta_i \geq \gamma\ell_i^*.$$

Figure 1 summarizes the NAROMMA algorithm.

The NAROMMA algorithm

1. Get $(\mathbf{a}_0, y_0)$; $\mathbf{w}_1 \leftarrow (y_0 \mathbf{a}_0)$; $i \leftarrow 1$; $\ell_1^* \leftarrow \frac{1}{\|\mathbf{a}_0\|_2}$.
2. Get $(\mathbf{x}, y)$.
3. $\gamma_i \leftarrow y(\mathbf{w}_i^T \mathbf{x})/\|\mathbf{w}_i\|_2$.
4. If $|(\mathbf{w}_i^T \mathbf{x})|/(\|\mathbf{w}_i\|_2 \|\mathbf{x}\|_2) = 1$, go to 8.
5. If $\gamma_i \geq 1/\ell_i^*$, then,
    5.1. $\mathbf{a}_i \leftarrow \mathbf{x}$; $y_i \leftarrow y$; $\mathbf{w}_{i+1} \leftarrow \mathbf{w}_i$; $\ell_{i+1}^* \leftarrow \ell_i^*$.
6. If $\frac{1}{\ell_i^*} > \gamma_i \geq \|\mathbf{a}_i\|_2^2 \ell_i^*$, then,
    6.1. $\mathbf{a}_i \leftarrow \mathbf{x}$; $y_i \leftarrow y$; $\mathbf{w}_{i+1} \leftarrow y_i \mathbf{a}_i$; $\ell_{i+1}^* \leftarrow 1/\|\mathbf{a}_i\|_2$.
7. If $\gamma_i < \min\{\ell_i^* \|\mathbf{a}_i\|_2^2, \frac{1}{\ell_i^*}\}$, then,
    7.1. $\mathbf{a}_i \leftarrow \mathbf{x}$; $y_i \leftarrow y$;
    7.2. $\mathbf{w}_{i+1} \leftarrow \mathbf{w}_i + \frac{(1-\ell_i^* \gamma_i)\|\mathbf{w}_i\|_2}{\ell_i^* \|\mathbf{a}_i\|_2^2 - \gamma_i}(y_i \mathbf{a}_i)$; $\ell_{i+1}^* \leftarrow \sqrt{\ell_i^{*2} + \frac{(\ell_i^* \gamma_i - 1)^2}{\|\mathbf{a}_i\|_2^2 - \gamma_i^2}}$.
8. $i \leftarrow i + 1$; Go to 2.

Fig. 1. The NAROMMA algorithm for binary classification.

3.2   Equivalence to Aggressive ROMMA

To prove that NAROMMA is equivalent to Aggressive ROMMA, it is enough to show that the algorithms' vectors have the same orientation after each update.

Theorem 1. *Let $\mathbf{u}_i$ denote the algorithm hypothesis used by algorithm Aggressive ROMMA after the ith update. Let $\mathbf{u}_1 = (y_0/\|\mathbf{a}_0\|_2^2)\mathbf{a}_0$. Let $\mathbf{v}_i$ be the hypothesis of the NAROMMA algorithm, and let $\mathbf{v}_1 = y_0 \mathbf{a}_0$, $\ell_1 = 1/\|\mathbf{a}_0\|_2$. Then for all i,*

1. $\|\mathbf{u}_i\|_2 = \ell_i^*$.
2. There exists some $\tau_i > 0$ such that, $\mathbf{u}_i = \tau_i \mathbf{v}_i$.

Proof. We prove by induction on i. For $i = 1$, the theorem trivially holds. Assume that the theorem holds for $i = t$, that is,

$$\|\mathbf{u}_t\|_2 = \ell_t^* \tag{23}$$

and,

$$\mathbf{u}_t = \tau_t \mathbf{v}_t \tag{24}$$

for some $\tau_t > 0$. To ease our discussion, we rewrite Aggressive ROMMA update as proposed in [5] as detailed in Table 1,
where

$$c_i = \frac{\|\mathbf{a}_i\|_2^2 \|\mathbf{u}_i\|_2^2 - y_i(\mathbf{u}_i^T \mathbf{a}_i)}{\|\mathbf{a}_i\|_2^2 \|\mathbf{u}_i\|_2^2 - (\mathbf{u}_i^T \mathbf{a}_i)^2} \tag{25}$$

and

$$d_i = \frac{\|\mathbf{u}_i\|_2^2 (y_i - (\mathbf{u}_i^T \mathbf{a}_i))}{\|\mathbf{a}_i\|_2^2 \|\mathbf{u}_i\|_2^2 - (\mathbf{u}_i^T \mathbf{a}_i)^2}. \tag{26}$$

In the same fashion, Table 2 summarizes the NAROMMA update rule.
Let $i = t + 1$. Let $\hat{\mathbf{u}}_t$ and $\hat{\mathbf{v}}_t$ denote the unit vectors in the directions of $\mathbf{u}_t$ and
$\mathbf{v}_t$, respectively. According to (23) and (24), we get that

$$\mathbf{u}_t = \|\mathbf{u}_t\|_2 \hat{\mathbf{u}}_t = \|\mathbf{u}_t\|_2 \hat{\mathbf{v}}_t = \ell_t^* \hat{\mathbf{v}}_t.$$

Hence, since $\ell_t^* > 0$ and using (7) we get,

$$y_t(\mathbf{u}_t^T \mathbf{a}_t) = y_t \ell_t^* (\hat{\mathbf{v}}_t^T \mathbf{a}_t) = y_t \ell_t^* \frac{1}{\|\mathbf{v}_t\|_2} (\mathbf{v}_t^T \mathbf{a}_t) = \ell_t^* \gamma_t. \tag{27}$$

We divide our discussion into three cases according to the update types.

Case I. Assume that Aggressive ROMMA does not perform any update on the
example $\mathbf{a}_t$ (update of type I). Then, according to Table 1, $y_t(\mathbf{u}_t^T \mathbf{a}_t) \geq 1$.
Using (27) and the fact that $\ell_t^* > 0$ we can conclude,

$$y_t(\mathbf{u}_t^T \mathbf{a}_t) \geq 1 \Leftrightarrow \gamma_t \geq 1/\ell_t^*. \tag{28}$$

Equation (28) implies that NAROMMA performs an update of type *I* if
and only if Aggressive ROMMA performs an update of type *I*. Hence, the
theorem holds for this case.

Case II. If Aggressive ROMMA performs an update of type *II*, then by Table 1
we get that $1 > y_t(\mathbf{u}_t^T \mathbf{a}_t) \geq \|\mathbf{a}_t\|_2^2 \|\mathbf{u}_t\|_2^2$. Therefore, since $\ell_t^* > 0$ and by (23),
(24) and (27) we get,

$$1 > y_t(\mathbf{u}_t^T \mathbf{a}_t) \geq \|\mathbf{a}_t\|_2^2 \|\mathbf{u}_t\|_2^2 \Leftrightarrow 1 > \ell_t^* \gamma_t \geq \|\mathbf{a}_t\|_2^2 \ell_t^{*2} \Leftrightarrow 1/\ell_t^* > \gamma_t \geq \|\mathbf{a}_t\|_2^2 \ell_t^*. \tag{29}$$

That is, as in the first case, Aggressive ROMMA and NAROMMA will make
their type *II* update simultaneously. Hence, we get that $\mathbf{v}_{t+1} = y_t \mathbf{a}_t$, $\ell_{t+1}^* =$
$1/\|\mathbf{a}_t\|_2$ and $\mathbf{u}_{t+1} = (y_t/\|\mathbf{a}_t\|_2^2)\mathbf{a}_t$. Therefore, the theorem trivially follows.

Case III. From the discussion in the previous two cases, it follows that Aggressive ROMMA makes an update of type *III* if and only if NAROMMA performs a type *III* update. By Table 1 we get for this case,

$$\|\mathbf{u}_{t+1}\|_2^2 = c_t^2 \|\mathbf{u}_t\|_2^2 + d_t^2 \|\mathbf{a}_t\|_2^2 + 2 c_t d_t (\mathbf{u}_t^T \mathbf{a}_t). \tag{30}$$

Since $y_t \in \{-1, 1\}$ and using (23), (24), (25), (26) and (27) we get that,

$$\|\mathbf{u}_{t+1}\|_2^2 = \ell_t^{*2} + \frac{(\ell_t^* \gamma_t - 1)^2}{\|\mathbf{a}_t\|_2^2 - \gamma_t^2} = \ell_{t+1}^{*2}. \tag{31}$$

Moreover, by choosing $\tau_{t+1} = (c_t \|\mathbf{u}_t\|_2)/\|\mathbf{v}_t\|_2$, and using the induction
assumption we can easily conclude that $\tau_{t+1} \mathbf{v}_{t+1} = \mathbf{u}_{t+1}$. Hence, the theorem holds for this case too.

$\square$

Table 1. Aggressive ROMMA update summary

Type	Condition	$\mathbf{u}_{i+1}$
I	$y_i(\mathbf{u}_i^T\mathbf{a}_i) \geq 1$	$\mathbf{u}_i$
II	$1 > y_i(\mathbf{u}_i^T\mathbf{a}_i) \geq \|\mathbf{a}_i\|_2^2\|\mathbf{u}_i\|_2^2$	$\frac{y_i}{\|\mathbf{a}_i\|_2^2}\mathbf{a}_i$
III	$y_i(\mathbf{u}_i^T\mathbf{a}_i) < \min\{\|\mathbf{a}_i\|_2^2\|\mathbf{u}_i\|_2^2, 1\}$	$c_i\mathbf{u}_i + d_i\mathbf{a}_i$

Table 2. NAROMMA update summary

Type	Condition	$\mathbf{v}_{i+1}$	ℓ_{i+1}^*
I	$\gamma_i \geq 1/\ell_i^*$;	$\mathbf{v}_i$;	ℓ_i^*;
II	$1/\ell_i^* > \gamma_i \geq \|\mathbf{a}_i\|_2^2\ell_i^*$;	$y_i\mathbf{a}_i$;	$\frac{1}{\|\mathbf{a}_i\|_2}$;
III	$\gamma_i < \min\{\ell_i^*\|\mathbf{a}_i\|_2^2, 1/\ell_i^*\}$;	$\mathbf{v}_i + \frac{\|\mathbf{v}_i\|_2(1-\ell_i^*\gamma_i)}{\ell_i^*\|\mathbf{a}_i\|_2^2-\gamma_i}y_i\mathbf{a}_i$;	$\ell_i^{*2} + \frac{(\ell_i^*\gamma_i-1)^2}{\|\mathbf{a}_i\|_2^2-\gamma_i^2}$;

4   The Maximum Cosine Perceptron Algorithm

We will use the MCF to generate the update rule of the MCP algorithm. The MCP algorithm is non-conservative, i.e. it updates its hypothesis on margin violation. We start our discussion by formulating a conservative algorithm, the *Conservative Maximum Cosine Perceptron* (CMCP) algorithm. By Lemma 1 we have,

$$\alpha_{i+1} \geq \frac{\alpha_i + \gamma x_i}{\sqrt{1 + \|\mathbf{a}_i\|_2^2 x_i^2 + 2\frac{y_i(\mathbf{w}_i^T\mathbf{a}_i)}{\|\mathbf{w}_i\|_2}x_i}}, \tag{32}$$

where $x_i = \lambda_i/\|\mathbf{w}_i\|_2$. The CMCP is a conservative algorithm hence, it makes an update when $y_i(\mathbf{w}_i^T\mathbf{a}_i) \leq 0$. Then, from (32) we can write,

$$\alpha_{i+1} \geq \frac{\alpha_i + \gamma x_i}{\sqrt{1 + \|\mathbf{a}_i\|_2^2 x_i^2}}. \tag{33}$$

Assuming that $\alpha_i \geq \gamma\ell_i$ for some ℓ_i we have,

$$\alpha_{i+1} \geq \gamma\frac{\ell_i + x_i}{\sqrt{1 + \|\mathbf{a}_i\|_2^2 x_i^2}}. \tag{34}$$

Inequality (34) formulates a local cosine bound for the CMCP algorithm. It holds for all $x_i \geq 0$, hence, by maximizing its right hand side we get an optimal lower bound for α_i. By Lemma 2, optimality is obtained in $x_i = 1/(\ell_i\|\mathbf{a}_i\|_2^2)$ and then,

$$\alpha_{i+1} \geq \gamma\sqrt{\ell_i^2 + \frac{1}{\|\mathbf{a}_i\|_2^2}}. \tag{35}$$

Let $\mathbf{w}_1 = y_0\mathbf{a}_0$ and $\ell_1 = 1/\|\mathbf{a}_0\|_2$, we get that, $\alpha_1 = \frac{\mathbf{w}^T\mathbf{w}_1}{\|\mathbf{w}_1\|_2} = \frac{|\mathbf{w}^T\mathbf{a}_0|}{\|\mathbf{a}_0\|_2} \geq \gamma\frac{1}{\|\mathbf{a}_0\|_2} = \gamma\ell_1$. And hence, by choosing

$$\ell_{i+1}^2 = \ell_i^2 + \frac{1}{\|\mathbf{a}_i\|_2^2} = \sum_{j=0}^{i} \frac{1}{\|\mathbf{a}_j\|_2^2} \tag{36}$$

we can conclude,

$$\alpha_i \geq \gamma\sqrt{\sum_{j=0}^{i-1} \frac{1}{\|\mathbf{a}_j\|_2^2}} = \gamma\ell_i. \tag{37}$$

We have just proved,

Lemma 7. *Let $\{(\mathbf{a}_0, y_0), \cdots, (\mathbf{a}_T, y_T)\}$ be a sequence of examples where $\mathbf{a}_i \in \mathbb{R}^n$ and $y_i \in \{-1, +1\}$ for all $0 \leq i \leq T$. Let $\gamma > 0$ and $\mathbf{w} \in \mathbb{R}^n$ with $\|\mathbf{w}\|_2 = 1$, such that for all $0 \leq i \leq T$, $|\mathbf{w}^T\mathbf{a}_i| \geq \gamma$. Let $\mathbf{w}_1 = y_0\mathbf{a}_0$, $\ell_1 = 1/\|\mathbf{a}_0\|_2$ and*

$$\mathbf{w}_{i+1} = \begin{cases} \mathbf{w}_i + \frac{\|\mathbf{w}_i\|_2}{\ell_i\|\mathbf{a}_i\|_2^2}(y_i\mathbf{a}_i) \, , & y_i(\mathbf{w}_i^T\mathbf{a}_i) \leq 0 \\ \mathbf{w}_i & , \ otherwise \end{cases} \tag{38}$$

be the update we use in the ith example, and

$$\ell_{i+1}^2 = \ell_i^2 + \frac{\mu_i}{\|\mathbf{a}_i\|_2^2}, \tag{39}$$

$$\mu_i = \begin{cases} 1 \, , & y_i(\mathbf{w}_i^T\mathbf{a}_i) \leq 0 \\ 0 \, , & otherwise. \end{cases} \tag{40}$$

Then, after the ith example, α_i is at least $\gamma\ell_i$.

To convert the CMCP algorithm to a non-conservative one, we use the same update rule not only on mistakes but also when the example is close to the algorithm's hyperplane, according to the following update rule,

$$\mathbf{w}_{i+1} = \begin{cases} \mathbf{w}_i + \frac{\|\mathbf{w}_i\|_2}{\ell_i\|\mathbf{a}_i\|_2^2}(y_i\mathbf{a}_i) \, , & y_i(\mathbf{w}_i^T\mathbf{a}_i) \leq \frac{\|\mathbf{w}_i\|_2}{2\ell_i} \\ \mathbf{w}_i & , \ otherwise. \end{cases} \tag{41}$$

Moreover, the value of ℓ_i is updated as follows,

$$\ell_{i+1}^2 = \begin{cases} \ell_i^2 + \frac{1-2\eta_i}{\|\mathbf{a}_i\|_2^2} \, , & y_i(\mathbf{w}_i^T\mathbf{a}_i) \leq \frac{\|\mathbf{w}_i\|_2}{2\ell_i} \\ \ell_i^2 & , \ otherwise \end{cases} \tag{42}$$

where for $y_i(\mathbf{w}_i^T\mathbf{a}_i) \leq \frac{\|\mathbf{w}_i\|_2}{2\ell_i}$,

$$\eta_i = \begin{cases} 0 & , \ y_i(\mathbf{w}_i^T\mathbf{a}_i) \leq 0 \\ \frac{y_i(\mathbf{w}_i^T\mathbf{a}_i)\ell_i}{\|\mathbf{w}_i\|_2} \, , & 0 < y_i(\mathbf{w}_i^T\mathbf{a}_i) \leq \frac{\|\mathbf{w}_i\|_2}{2\ell_i}. \end{cases} \tag{43}$$

Notice that when $\mathbf{a}_i$ is a counterexample, it contributes $1/\|\mathbf{a}_i\|_2^2$ to ℓ_i^2. Otherwise, it contributes $1 - 2\eta_i/\|\mathbf{a}_i\|_2^2$, which is always positive because of the update rule condition in (41). Figure 2 summarizes the algorithm.

The MCP algorithm

1. Get $(\mathbf{a}_0, y_0)$; $\mathbf{w}_1 \leftarrow (y_0 \mathbf{a}_0)$; $i \leftarrow 1$; $\ell_1 \leftarrow \frac{1}{\|\mathbf{a}_0\|_2}$.
2. Get $(\mathbf{x}, y)$.
3. If $y(\mathbf{w}_i^T \mathbf{x}) \leq \frac{\|\mathbf{w}_i\|_2}{2\ell_i}$, then,
    3.1. $\mathbf{a}_i \leftarrow \mathbf{x}$; $y_i \leftarrow y$; $\mathbf{w}_{i+1} \leftarrow \mathbf{w}_i + \frac{\|\mathbf{w}_i\|_2}{\ell_i \|\mathbf{a}_i\|_2^2}(y_i \mathbf{a}_i)$.
    3.2. If $y_i(\mathbf{w}_i^T \mathbf{a}_i) \leq 0$, Then, $\eta_i \leftarrow 0$.
    3.3. Else, $\eta_i \leftarrow \frac{y_i(\mathbf{w}_i^T \mathbf{a}_i)\ell_i}{\|\mathbf{w}_i\|_2}$.
    3.4. $\ell_{i+1} = \sqrt{\ell_i^2 + \frac{1-2\eta_i}{\|\mathbf{a}_i\|_2^2}}$.
4. Else, $\mathbf{w}_{i+1} \leftarrow \mathbf{w}_i$; $\ell_{i+1} \leftarrow \ell_i$.
5. $i \leftarrow i + 1$; Go to 2.

Fig. 2. The MCP algorithm for binary classification.

Lemma 8. *Let* $\{(\mathbf{a}_0, y_0), \cdots, (\mathbf{a}_T, y_T)\}$ *be a sequence of examples where* $\mathbf{a}_i \in \mathbb{R}^n$ *and* $y_i \in \{-1, +1\}$ *for all* $0 \leq i \leq T$. *Let* $\gamma > 0$ *and* $\mathbf{w} \in \mathbb{R}^n$ *with* $\|\mathbf{w}\|_2 = 1$, *such that for all* $0 \leq i \leq T$, $|\mathbf{w}^T \mathbf{a}_i| \geq \gamma$. *Let* $\mathbf{w}_1 = y_0 \mathbf{a}_0$ *and* $\ell_1 = 1/\|\mathbf{a}_0\|_2$. *Let* $\mathbf{w}_{i+1}$ *and* ℓ_{i+1} *be updated as defined in* (41) *and* (42) *respectively, after each example* $\mathbf{a}_i$. *Then after the ith example,* α_i *is at least* $\gamma \ell_i$.

Proof. We prove by induction on i, the index of the current example. For $i = 1$, $\alpha_1 = \frac{\mathbf{w}^T \mathbf{w}_1}{\|\mathbf{w}_1\|_2} = \frac{|\mathbf{w}^T \mathbf{a}_0|}{\|\mathbf{a}_0\|_2} \geq \gamma/\|\mathbf{a}_0\|_2 = \gamma \ell_1$. We assume the lemma holds for $i = k$, that is, $\alpha_k \geq \gamma \ell_k$, and prove for $i = k + 1$. We consider three cases,

Case I. if $y_k(\mathbf{w}_k^T \mathbf{a}_k) > \|\mathbf{w}_k\|_2/(2\ell_k)$, then, by (41) and (42) no update happens for $\mathbf{w}_k$ or ℓ_k and hence, the lemma is true for this case.

Case II. If $y_k(\mathbf{w}_k^T \mathbf{a}_k) \leq 0$, then, by Lemma 1, the induction assumption and (41), (42), (43) and using $x_k = 1/(\ell_k \|\mathbf{a}_k\|_2^2)$ we can write,

$$\alpha_{k+1} \geq \frac{\alpha_k + \gamma x_k}{\sqrt{1 + \|\mathbf{a}_k\|_2^2 x_k^2 + 2\frac{y_k(\mathbf{w}_k^T \mathbf{a}_k)}{\|\mathbf{w}_k\|_2} x_k}} \geq \frac{\gamma\left(\ell_k + \frac{1}{\ell_k \|\mathbf{a}_k\|_2^2}\right)}{\sqrt{1 + \frac{1}{\ell_k^2 \|\mathbf{a}_k\|_2^2}}} \geq$$

$$\geq \gamma \sqrt{\ell_k^2 + \frac{1}{\|\mathbf{a}_k\|_2^2}} \geq \gamma \ell_{k+1}.$$

Case III. If $0 < y_k(\mathbf{w}_k^T \mathbf{a}_k) \leq \|\mathbf{w}_k\|_2/(2\ell_k)$, then, by Lemma 1, the induction assumption, (41), (42) and (43) we can write,

$$\alpha_{k+1} \geq \frac{\alpha_k + \gamma x_k}{\sqrt{1 + \|\mathbf{a}_k\|_2^2 x_k^2 + 2\frac{y_k(\mathbf{w}_k^T \mathbf{a}_k)}{\|\mathbf{w}_k\|_2} x_k}} \geq$$

$$\geq \frac{\gamma \ell_k \left(1 + \frac{1}{\ell_k^2 \|\mathbf{a}_k\|_2^2}\right)}{\sqrt{1 + \frac{1}{\ell_k^2 \|\mathbf{a}_k\|_2^2} + \frac{2}{\ell_k^2 \|\mathbf{a}_k\|_2^2} \left(\frac{y_k (\mathbf{w}_k^T \mathbf{a}_k) \ell_k}{\|\mathbf{w}_k\|_2}\right)}} \geq$$

$$\geq \gamma \ell_k \frac{1 + \frac{1}{\ell_k^2 \|\mathbf{a}_k\|_2^2}}{\sqrt{1 + \frac{1 + 2\eta_k}{\ell_k^2 \|\mathbf{a}_k\|_2^2}}} \geq \gamma \ell_k \sqrt{1 + \frac{1 - 2\eta_k}{\ell_k^2 \|\mathbf{a}_k\|_2^2}} = \gamma \ell_{k+1}.$$

$\square$

Lemma 8 motivates the choice of the update condition of the MCP algorithm. Let $\mathbf{a}_i$ be the current example examined by the algorithm, and let us assume without loss of generality that $y_i = +1$. When $y_i(\mathbf{w}_i^T \mathbf{a}_i) > \|\mathbf{w}_i\|_2/(2\ell_i)$, by (41) no update happens. Let θ_i be the angle between the target hyperplane and the algorithm hyperplane. Let γ_i be as defined in (7). By Lemma 8 we get that, $\gamma_i > \frac{1}{\ell_i} > \frac{\gamma}{2\alpha_i} = \frac{1}{2} \frac{\gamma}{\cos \theta_i} > \frac{\gamma}{2}$ which implies that the example has the right label and is at distance of at least $\gamma/2$ from $\mathbf{w}_i$ and therefore no update occurs.

Theorem 2. *Let $S = (\mathbf{a}_0, y_0), \cdots, (\mathbf{a}_k, y_k)$ be a sequence of examples where $\mathbf{a}_i \in \mathbb{R}^n$, $y_i \in \{-1, +1\}$ and $\|\mathbf{a}_i\|_2 \leq R$ for all $0 \leq i \leq k$. Let $\mathbf{w}$ be some separating hyperplane for S, that is, there exists some $\gamma > 0$ such that for all i such that $0 \leq i \leq k$, $|\mathbf{w}^T \mathbf{a}_i| \geq \gamma$. And let $\|\mathbf{w}\|_2 = 1$. Let t be the number of mistakes the MCP algorithm makes on S, then, $t \leq (R/\gamma)^2$.*

Proof. Let $M \subseteq S$ denote the set of examples on which the MCP algorithm made a mistake. Similarly, let N be the set of examples on which the MCP algorithm made an update. Clearly, $M \subseteq N$. By (42) and (43) we get that $1 - 2\eta_i > 0$ for all i. Hence, we can conclude,

$$\ell_k^2 = \sum_{i: \mathbf{a}_i \in N} \frac{1 - 2\eta_i}{\|\mathbf{a}_i\|_2^2} \geq \sum_{i: \mathbf{a}_i \in M} \frac{1}{\|\mathbf{a}_i\|_2^2} \geq \sum_{i: \mathbf{a} \in M} \frac{1}{R^2} = \frac{t}{R^2}. \tag{44}$$

From Lemma (8) we get that,

$$1 \geq \alpha_k \geq \gamma \ell_k \geq \gamma \sqrt{\frac{t}{R^2}}. \tag{45}$$

By combining (44) and (45) we get the result. $\square$

5   Experiments

In this section we present experimental results that demonstrate the performance of the MCP algorithm vs. the well known algorithms PA and Aggressive ROMMA on the MNIST OCR database[1]. Every example in the MNIST database has two parts, the first is 28×28 matrix which represents the image of the

---

[1] See http://yann.lecun.com/exdb/mnist/ for information on obtaining this dataset.

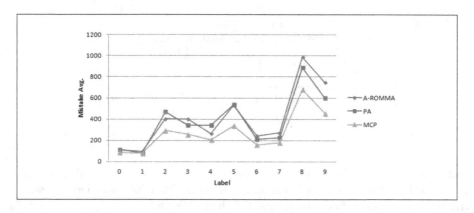

Fig. 3. Single label classifier mistake rates of MCP, PA, A-ROMMA on MNIST dataset.

corresponding digit. Each entry in the matrix takes values from $\{0, \cdots, 255\}$. The second part is a label taking values from $\{0, \cdots, 9\}$. The dataset consists of 60000 training examples and 10000 test examples. In the experiments we trained the algorithms for single preselected label l. When training on this, we replaced each labeled instance $(\mathbf{a}_i, y_i)$ by the binary-labeled instance $(\mathbf{a}_i, y_i^*)$, where $y_i^* = 1$ if $y_i = l$ and $y_i^* = -1$ otherwise. We have divided the training set to 60 buckets of examples each containing 1000 examples. For each label, we first chose a random permutation of the examples buckets, then we trained the algorithm via single pass over the training dataset according to the selected permutation. Then, we tested it on the test dataset. We repeated that for 20 random permutations for each label. At the end of the process, to calculate the mistake rates of each classifier, we took the average of the mistakes over the 20 rounds. Figure 3 summarizes the number of mistakes made by the three algorithms for all the ten labels on the test data. Actually it shows that MCP practically performs better than the other two algorithms under the restrictions of single dataset pass and hypothesis size that is linear in the number of the relevant features.

A   Proofs of Sect. 2

Proof. (Lemma 1).

$$\alpha_{i+1} = \frac{\mathbf{w}^T \mathbf{w}_{i+1}}{\|\mathbf{w}_{i+1}\|_2} = \frac{\|\mathbf{w}_i\|_2 \alpha_i + \|\mathbf{w}_i\|_2 x_i |\mathbf{w}^T \mathbf{a}_i|}{\sqrt{\|\mathbf{w}_i\|_2^2 + \lambda_i^2 \|\mathbf{a}_i\|_2^2 + 2\lambda_i y_i (\mathbf{w}_i^T \mathbf{a}_i)}} \geq$$

$$\geq \frac{\alpha_i + \gamma x_i}{\sqrt{1 + x_i^2 \|\mathbf{a}_i\|_2^2 + 2x_i \frac{y_i (\mathbf{w}_i^T \mathbf{a}_i)}{\|\mathbf{w}_i\|_2}}}.$$

This implies the result.   □

Proof. (Lemma 2). From solving $\partial\Phi(x)/\partial x = 0$ and checking the sign of $\frac{\partial\Phi(x)}{\partial x}$ we get the result.   □

Proof. (Lemma 3). We write $\Phi(x)$ in the following manner

$$\Phi(x) = \frac{(r - pt) + p(x + t)}{\sqrt{(s - qt^2) + q(x + t)^2}} \equiv \frac{r' + p'x'}{\sqrt{s' + q'x'^2}}, \qquad (46)$$

where $r' = r - pt$, $p' = p$, $x' = x + t$, $s' = s - qt^2$ and $q' = q$. Since $s + q \cdot x^2 + 2tq \cdot x > 0$ for all x, we have $q > 0$ and $\Delta = 4t^2q^2 - 4sq < 0$ which implies $q' > 0$ and $s' > 0$. Now by Lemma 2, if $r' = r - pt \neq 0$, the optimal value of $\Phi(x)$ is in

$$x_0 = x'_0 - t = \frac{p's'}{r'q'} - t = \frac{p(s - qt^2)}{q(r - pt)} - t = \frac{ps - rtp}{q(r - pt)}, \qquad (47)$$

and is equal to $\Phi(x_0) = sign(r')\sqrt{\frac{r'^2}{s'} + \frac{p'^2}{q'}} = sign(r - pt)\sqrt{\frac{(r-pt)^2}{s-qt^2} + \frac{p^2}{q}}$. This point is minimal if $r - pt < 0$ and maximal if $r - pt > 0$. If $r' = r - pt = 0$ then the function $\Phi(x)$ is monotone increasing if $p' = p > 0$ and monotone decreasing if $p' = p < 0$. Now we have six cases:

Case 1. If $r' = r - pt = 0$ and $p > 0$ then the function is monotone increasing, and therefore, $\Phi^* = \Phi(\infty) = p/\sqrt{q}$.

Case 2. If $r' = r - pt = 0$ and $p < 0$ then the function is monotone decreasing, and therefore, $\Phi^* = \Phi(0) = r/\sqrt{s}$.

Case 3. If $r' = r - pt > 0$ and $ps - rtq \geq 0$ then by (47) we get $x_0 = (ps - rtq)/q(r - pt) > 0$ and is a maximal point. Therefore, $\Phi^* = \Phi(x_0)$.

Case 4. If $r' = r - pt > 0$ and $ps - rtq < 0$ then by (47) we get $x_0 < 0$ and is a maximal point. Therefore, the function is monotone decreasing for $x > 0$; hence, $\Phi^* = \Phi(x_0)$.

Case 5. If $r' = r - pt < 0$ and $ps - rtq \geq 0$ then by (47) we get $x_0 \leq 0$ and is a minimal point. Therefore, $\Phi(x)$ is monotone increasing for $x > 0$ and $\Phi^* = \Phi(\infty) = p/\sqrt{q}$.

Case 6. If $r' = r - pt < 0$ and $ps - rtq < 0$ then by (47) we get $x_0 > 0$ and is a minimal point. Therefore, $\Phi^* = max(\Phi(\infty), \Phi(0)) = max\left(\frac{p}{\sqrt{q}}, \frac{r}{\sqrt{s}}\right)$.   □

References

1. Crammer, K., Dekel, O., Keshet, J., Shalev-Shwartz, S., Singer, Y.: Online passive-aggressive algorithms. J. Mach. Learn. Res. **7**, 551–585 (2006)
2. Cristianni, N., Shawe-Taylor, J.: Support Vector Machines and Other Kernel-based Learning Methods. Cambridge University Press, Cambridge (2000)
3. Gentile, C.: A new approximate maximal margin classification algorithm. J. Mach. Learn. Res. **2**, 213–242 (2001)

4. Kivinen, J., Smola, A.J., Williamson, R.C.: Online learning with kernels. IEEE Trans. Signal Process. **52**(8), 2165–2176 (2002)
5. Li, Y., Long, P.M.: The relaxed online maximum margin algorithm. Mach. Learn. **46**(1–3), 361–387 (2002)
6. Rosenblatt, F.: The perceptron: a probabilistic model for information storage and organization in the brain. Psychol. Rev. **65**, 386–407 (1958). (Reprinted in Neurocomputing (MIT Press, 1988))

Structural Online Learning

Mehryar Mohri[1,2] and Scott Yang[2(✉)]

[1] Google Research, 111 8th Avenue, New York, NY 10011, USA
mohri@cs.nyu.edu
[2] Courant Institute, 251 Mercer Street, New York, NY 10012, USA
yangs@cims.nyu.edu

Abstract. We study the problem of learning ensembles in the online setting, when the hypotheses are selected out of a base family that may be a union of possibly very complex sub-families. We prove new theoretical guarantees for the online learning of such ensembles in terms of the sequential Rademacher complexities of these sub-families. We also describe an algorithm that benefits from such guarantees. We further extend our framework by proving new structural estimation error guarantees for ensembles in the batch setting through a new data-dependent online-to-batch conversion technique, thereby also devising an effective algorithm for the batch setting which does not require the estimation of the Rademacher complexities of base sub-families.

1 Introduction

Ensemble methods are powerful techniques in machine learning for combining several predictors to define a more accurate one. They include notable methods such as bagging and boosting [4,11], and they have been successfully applied to a variety of scenarios including classification and regression.

Standard ensemble methods such as AdaBoost and Random Forests select base predictors from some hypothesis set $\mathcal{H}$, which may be the family of boosting stumps or that of decision trees with some limited depth. More complex base hypothesis sets may be needed to tackle some difficult modern tasks. At the same time, learning bounds for standard ensemble methods suggest a risk of overfitting when using very rich hypothesis sets, which has been further observed empirically [10,17].

Recent work in the batch setting has shown, however, that learning with such complex base hypothesis sets is possible using the *structure* of $\mathcal{H}$, that is its decomposition into subsets $\mathcal{H}_k$, $k = 1, \ldots, p$, of varying complexity. In particular, in [8], we introduced a new ensemble algorithm, *DeepBoost*, which we proved benefits from finer learning guarantees when using rich families as base classifier sets. In DeepBoost, the decisions in each iteration of which classifier to add to the ensemble and which weight to assign to that classifier depend on the complexity of the sub-family $\mathcal{H}_k$ to which the classifier belongs. This can be viewed as integrating the principle of structural risk minimization to each iteration of boosting.

© Springer International Publishing Switzerland 2016
R. Ortner et al. (Eds.): ALT 2016, LNAI 9925, pp. 223–237, 2016.
DOI: 10.1007/978-3-319-46379-7_15

This paper extends the *structural learning* idea of incorporating model selection in ensemble methods to the online learning setting. Specifically, we address the question: can one design ensemble algorithms for the online setting that admit strong guarantees even when using a complex $\mathcal{H}$? In Sect. 3, we first present a theoretical result guaranteeing the existence of a randomized algorithm that can compete against the best ensemble in $\mathcal{H}$ efficiently when this ensemble does not rely *too heavily* on complex base hypotheses. Motivated by this theory, we then design an online algorithm that benefits from such guarantees, for a wide family of hypotheses sets (Sect. 4). Finally, in Sect. 5, we further extend our framework by proving new structural estimation error guarantees for ensembles in the batch setting through a new data-dependent online-to-batch conversion technique. This also provides an effective algorithm for the batch setting which does not require the estimation of the Rademacher complexities of base hypothesis sets $\mathcal{H}_k$.

2   Notation and Preliminaries

Let $\mathcal{X}$ denote the input space and $\mathcal{Y}$ the output space. Let $L_t \colon \mathcal{Y} \to \mathbb{R}_+$ be a loss function. The online learning framework that we study is a sequential prediction setting that can be described as follows. At each time $t \in [1, T]$, the learner (or algorithm $\mathcal{A}$) receives an input instance x_t which he uses to select a hypothesis $h_t \in \mathcal{H} \subseteq \mathcal{Y}^{\mathcal{X}}$ and make a prediction $h_t(x_t)$. The learner then incurs the loss $L_t(h_t(x_t))$ based on the loss function L_t chosen by an adversary. The objective of the learner is to minimize his regret over T rounds, that is the difference of his cumulative loss $\sum_{t=1}^{T} L_t(h_t(x_t))$ and that of the best function in some benchmark hypothesis set $\mathcal{F} \subset \mathcal{Y}^{\mathcal{X}}$:

$$\mathrm{Reg}_T(\mathcal{A}) = \sum_{t=1}^{T} L_t(h_t(x_t)) - \min_{h \in \mathcal{F}} \sum_{t=1}^{T} L_t(h(x_t)).$$

In what follows, $\mathcal{F}$ will be assumed to coincide with $\mathcal{H}$, unless explicitly stated otherwise. The learner's algorithm may be randomized, in which case, at each round t, the learner draws hypothesis h_t from the distribution π_t he has defined at that round. The regret is then the difference between the expected cumulative loss and the expected cumulative loss of the best-in-class hypothesis: $\mathrm{Reg}_T(\mathcal{A}) = \sum_{t=1}^{T} \mathbb{E}[L_t(h_t(x_t))] - \min_{h \in \mathcal{H}} \sum_{t=1}^{T} \mathbb{E}[L_t(h(x_t))]$.

Clearly, the difficulty of the learner's regret minimization task depends on the richness of the competitor class $\mathcal{H}$. The more complex $\mathcal{H}$ is, the smaller the loss of the best function in $\mathcal{H}$ and thus the harder the learner's benchmark. This complexity can be captured by the notion of *sequential Rademacher complexity* introduced by [16]. Let $\mathcal{H}$ be a set of functions from $\mathcal{X}$ to $\mathbb{R}$. The sequential Rademacher complexity of a hypothesis $\mathcal{H}$ is denoted by $\mathfrak{R}_T^{\mathrm{seq}}(\mathcal{H})$ and defined by

$$\mathfrak{R}_T^{\mathrm{seq}}(\mathcal{H}) = \frac{1}{T} \sup_{\mathbf{x}} \mathbb{E}\left[\sup_{h \in \mathcal{H}} \sum_{t=1}^{T} \sigma_t h(x_t(\boldsymbol{\sigma})) \right], \tag{1}$$

where the supremum is taken over all $\mathcal{X}$-*valued complete binary trees* of depth T and where $\boldsymbol{\sigma} = (\sigma_1, \ldots, \sigma_T)$ is a sequence of i.i.d. Rademacher variables, each taking values in $\{\pm 1\}$ with probability $\frac{1}{2}$. Here, an $\mathcal{X}$-valued complete binary tree $\mathbf{x}$ is defined as a sequence $(x_1, \ldots, x_T)$ of mappings where $x_t \colon \{\pm 1\}^{t-1} \to \mathcal{X}$. The root x_1 can be thought of as some constant in $\mathcal{X}$. The left child of the root is $x_2(-1)$ and the right child is $x_2(1)$. A path in the tree is $\boldsymbol{\sigma} = (\sigma_1, \ldots, \sigma_{T-1})$. To simplify the notation, we write $x_t(\boldsymbol{\sigma})$ instead of $x_t(\sigma_1, \ldots, \sigma_{t-1})$. The sequential Rademacher complexity can be interpreted as the online counterpart of the standard Rademacher complexity widely used in the analysis of batch learning [1,13]. It has been used by [16] and [15] both to derive attainability results for some regret bounds and to guide the design of new online algorithms.

3   Theoretical Guarantees for Structural Online Learning

In this section, we present learning guarantees for structural online learning in binary classification. Hence, for any $t \in [1, T]$, the loss incurred at each time t by hypothesis h is $L_t(h(x_t)) = 1_{\{y_t h(x_t) < 0\}}$, with $y_t \in \mathcal{Y} = \{\pm 1\}$.

A *randomized player strategy* $\boldsymbol{\pi} = (\pi_1, \ldots, \pi_T)$ for a sequence of length T is a sequence of mappings $\pi_t \colon (\mathcal{X} \times \mathcal{Y})^{t-1} \to \mathcal{P}_{\mathcal{H}}$, $t \in [T]$, where $\mathcal{P}_{\mathcal{H}}$ is the family of distributions over $\mathcal{H}$. Thus, $\pi_t((x_1, y_1), \ldots, (x_{t-1}, y_{t-1}))$ is the distribution according to which the player selects a hypothesis $h \in \mathcal{H}$ at time t and which also depends on the past sequence $(x_1, y_1), \ldots, (x_{t-1}, y_{t-1})$ played against the adversary.

The following shows the existence of a randomized strategy that benefits from a margin-based regret guarantee in the online setting. This can be viewed as the counterpart of the classical margin-based learning bounds in the batch setting given by [13].

Theorem 1 (Proposition 25 [16]). *For any function class $\mathcal{H} \subset \mathbb{R}^{\mathcal{X}}$ of functions bounded by one, there exists a randomized player strategy given by $\boldsymbol{\pi}$ such that for any sequence $z_1, \ldots, z_T$ played by the adversary, $z_t = (x_t, y_t) \in \mathcal{X} \times \{\pm 1\}$, the following inequality holds:*

$$
\mathbb{E}\left[\frac{1}{T} \sum_{t=1}^{T} \mathop{\mathbb{E}}_{h_t \sim \pi_t} \left[1_{\{y_t h_t(x_t) < 0\}} \right] \right]
$$

$$
\leq \inf_{\gamma > 0} \left\{ \inf_{h \in \mathcal{H}} \frac{1}{T} \sum_{t=1}^{T} 1_{\{y_t h(x_t) < \gamma\}} + \frac{4}{\gamma} \mathcal{R}_T^{\mathrm{seq}}(\mathcal{H}) + \frac{1}{\sqrt{T}} \left(3 + \log \log \frac{1}{\gamma} \right) \right\}.
$$

We are not explicitly indicating the dependency of π_t on $(x_1, y_1), \ldots, (x_{t-1}, y_{t-1})$ to alleviate the notation. The theorem gives a guarantee for the expected error of a randomized strategy in terms of the empirical margin loss and the sequential Rademacher complexity of $\mathcal{H}$ scaled by γ for the best choice of $h \in \mathcal{H}$ and the best confidence margin γ. As with standard margin bounds, this is subject to a trade-off: for a larger γ the empirical margin loss is larger, while a smaller value of γ increases the complexity term. The result gives a very favorable guarantee when there exists a relatively large γ for which the empirical margin loss of the best h is relatively small.

While this result is remarkable for characterizing learnability against the best-in-class hypothesis, it does not identify and take advantage of any structure in the hypothesis set. The structural margin bound that we prove next specifically provides a guarantee that exploits the scenario where the hypothesis set admits a decomposition $\mathcal{H} = \cup_{k=1}^{p} \mathcal{H}_k$. For any $q \in \mathbb{N}$, we will denote by Δ_q the probability simplex in $\mathbb{R}^q$ and by $\mathrm{conv}(\mathcal{H})$ the convex hull of $\mathcal{H}$.

Theorem 2. *Let $\mathcal{H} \subset [-1,1]^{\mathcal{X}}$ be a family of functions admitting a decomposition $\mathcal{H} = \bigcup_{k=1}^{p} \mathcal{H}_k$. Then, there exists a randomized player strategy given by π on $\mathrm{conv}(\mathcal{H})$ such that for any sequence $((x_t, y_t))_{t \in [T]}$ in $\mathcal{X} \times \{\pm 1\}$, the following inequality holds:*

$$
\mathbb{E}\left[\frac{1}{T}\sum_{t=1}^{T} \underset{f_t \sim \pi_t}{\mathbb{E}}\left[1_{\{y_t f_t(x_t) < 0\}}\right]\right] \leq \inf_{\gamma > 0}\left\{ \inf_{\substack{f = \sum_{i=1}^{q} \alpha_i h_i \in \mathrm{conv}(\mathcal{H}) \\ \alpha \in \Delta_q, h_i \in \mathcal{H}_{k(h_i)}}} \frac{1}{T}\sum_{t=1}^{T} 1_{\{y_t f(x_t) < \gamma\}} \right.
$$

$$
\left. + \frac{6}{\gamma}\sum_{i=1}^{q} \alpha_i \mathfrak{R}_T^{\mathrm{seq}}(\mathcal{H}_{k(h_i)}) + \widetilde{\mathcal{O}}\left(\frac{1}{\gamma}\sqrt{\frac{\log p}{T}}\right) \right\},
$$

where $k(h_i) \in [1, p]$ is defined to be the smallest index k such that $h_i \in \mathcal{H}_k$ and $q \in \mathbb{N}$ so that f is an arbitrary element in the convex hull.

The theorem extends the margin bound of Theorem 1 given for a single hypothesis set to a guarantee for the convex hull of p hypothesis sets. Observe that, remarkably, the complexity term depends on the mixture weights α_i and hypotheses h_i defining the the best-in-class hypothesis $f = \sum_{i=1}^{q} \alpha_i h_i$. The complexity term is an $\boldsymbol{\alpha}$-average of the sequential Rademacher complexities. Thus, the theorem shows the existence of a randomized strategy π that achieves a favorable guarantee so long as the best-in-class hypothesis $f \in \mathrm{conv}(\mathcal{H})$ admits a decomposition for which the complexity term is relatively small, which directly depends on the amount of mixture weight assigned to more complex $\mathcal{H}_k$s versus less complex ones in the decomposition of f.

From a proof standpoint, it is enticing to use the fact that the sequential Rademacher complexity of a hypothesis set does not increase upon taking the convex hull. While this property yields an interesting result itself, it is not sufficiently fine for deriving the result of Theorem 2: in short, the resulting guarantee is then in terms of the maximum of the sequential Rademacher complexities instead of their $\boldsymbol{\alpha}$-average.

Proof. Fix $n \geq 1$. For any p-tuple of non-negative integers $\mathbf{N} = (N_1, \ldots, N_p) \in \mathbb{N}^p$ with $|\mathbf{N}| = \sum_{k=1}^{p} N_k = n$, consider the following family of functions:

$$
G_{\mathcal{H}, \mathbf{N}} = \left\{ \frac{1}{n}\sum_{k=1}^{p}\sum_{j=1}^{N_k} h_{k,j} \,\middle|\, \forall (k, j) \in [1, p] \times [N_k], h_{k,j} \in H_k \right\}.
$$

By the sub-additivity of the supremum operator, the sequential Rademacher complexity of $\mathcal{H}$ can be upper bounded as follows:

$$\mathfrak{R}_T^{seq}(G_{\mathcal{H},\mathbf{N}}) = \frac{1}{T}\sup_{\mathbf{x}}\mathbb{E}_{\sigma}\left[\sup_{h\in G_{\mathcal{H},N}}\sum_{t=1}^{T}\frac{1}{n}\sum_{k=1}^{p}\sum_{j=1}^{N_k}h_{k,j}(x_t(\sigma))\sigma_t\right]$$

$$\leq \frac{1}{T}\frac{1}{n}\sum_{k=1}^{p}\sum_{j=1}^{N_k}\sup_{\mathbf{x}}\mathbb{E}_{\sigma}\left[\sup_{h_{k,j}\in\mathcal{H}_k}\sum_{t=1}^{T}h_{k,j}(x_t(\sigma))\sigma_t\right] = \frac{1}{n}\sum_{k=1}^{p}N_k\mathfrak{R}_T^{seq}(\mathcal{H}_k).$$

In view of this inequality and the margin bound of Proposition 1, for any $G_{\mathcal{H},\mathbf{N}}$, there exists a player strategy $\pi^{\mathbf{N}}$ such that

$$\frac{1}{T}\mathbb{E}\left[\sum_{t=1}^{T}\mathbb{E}_{f_t\sim\pi_t^{\mathbf{N}}}[1_{\{y_tf_t(x_t)<0\}}]\right]$$

$$\leq \inf_{\gamma>0}\left\{\inf_{g\in G_{\mathcal{H},\mathbf{N}}}\frac{1}{T}\sum_{t=1}^{T}1_{\{y_tg(x_t)<\gamma\}} + \frac{4}{\gamma}\frac{1}{n}\sum_{k=1}^{p}N_k\mathfrak{R}_T^{seq}(\mathcal{H}_k) + \frac{3+\log\log\frac{1}{\gamma}}{\sqrt{T}}\right\}.$$

Now, let π^{exp} denote a randomized weighted majority strategy π^{exp} with the $\pi^{\mathbf{N}}$ strategies serving as experts [14]. Since there are at most p^n p-tuples $\mathbf{N}$ with $|\mathbf{N}| = n$, the regret of this randomized weighted majority strategy is bounded by $2\sqrt{T\log(p^n)}$ (see [6]). Thus, the following guarantee holds for the strategy π^{exp}:

$$\mathbb{E}\left[\sum_{t=1}^{T}\mathbb{E}_{f_t\sim\pi_t^{exp}}[1_{\{y_tf_t(x_t)<0\}}]\right] \leq \inf_{|\mathbf{N}|=n}\mathbb{E}\left[\sum_{t=1}^{T}\mathbb{E}_{f_t\sim\pi_t^{\mathbf{N}}}[1_{\{y_tf_t(x_t)<0\}}]\right] + \sqrt{4Tn\log p}.$$

In view of that, we can write

$$\frac{1}{T}\mathbb{E}\left[\sum_{t=1}^{T}\mathbb{E}_{f_t\sim\pi_t^{exp}}[1_{\{y_tf_t(x_t)<0\}}]\right]$$

$$\leq \inf_{\gamma>0}\left\{\inf_{\substack{g\in G_{\mathcal{H},\mathbf{N}}\\|\mathbf{N}|=n}}\frac{1}{T}\sum_{t=1}^{T}1_{\{y_tg(x_t)<\gamma\}} + \frac{4}{\gamma}\frac{1}{n}\sum_{k=1}^{p}N_k\mathfrak{R}_T^{seq}(\mathcal{H}_k)\right\} \tag{2}$$

$$+ \frac{3+\log\log\frac{1}{\gamma}}{\sqrt{T}} + 2\sqrt{\frac{n\log p}{T}}$$

$$= \inf_{\gamma>0}\left\{\inf_{\substack{g=\frac{1}{n}\sum_{i=1}^{q}n_ih_i\\h_i\in\mathcal{H}_{k(h_i)}}}\frac{1}{T}\sum_{t=1}^{T}1_{\{y_tg(x_t)<\gamma\}} + \frac{4}{\gamma}\frac{1}{n}\sum_{i=1}^{q}n_i\mathfrak{R}_T^{seq}(\mathcal{H}_{k(h_i)})\right\}$$

$$+ \frac{3+\log\log\frac{1}{\gamma}}{\sqrt{T}} + 2\sqrt{\frac{n\log p}{T}}. \tag{3}$$

Now, fix $(h_1, \ldots, h_q)$. Any $\alpha \in \Delta_q$ defines a distribution over $h_1, \ldots, h_q$. Sampling according to α and averaging leads to functions g of the form $g = \frac{1}{n} \sum_{i=1}^{q} n_i h_i$ for some q-tuple $\mathbf{n} = (n_1, \ldots, n_q)$ with $|\mathbf{n}| = n$. Let $f = \sum_{i=1}^{q} \alpha_i h_i$ for some $\alpha \in \Delta_q$. By the union bound, we can write, for any $\gamma > 0$ and (x_t, y_t),

$$
\begin{aligned}
\mathbb{E}_{\mathbf{n} \sim \alpha} \left[1_{y_t g(x_t) < \gamma} \right] &= \Pr_{\mathbf{n} \sim \alpha} \left[y_t g(x_t) < \gamma \right] = \Pr_{\mathbf{n} \sim \alpha} \left[y_t g(x_t) - y_t f(x_t) + y_t f(x_t) < \gamma \right] \\
&\leq \Pr_{\mathbf{n} \sim \alpha} \left[y_t g(x_t) - y_t f(x_t) < -\tfrac{\gamma}{2} \right] + \Pr_{\mathbf{n} \sim \alpha} \left[y_t f(x_t) < \tfrac{3\gamma}{2} \right] \\
&= \Pr_{\mathbf{n} \sim \alpha} \left[y_t g(x_t) - y_t f(x_t) < -\tfrac{\gamma}{2} \right] + 1_{y_t f(x_t) < \frac{3\gamma}{2}}.
\end{aligned}
$$

For any $\gamma > 0$ and (x_t, y_t), by Hoeffding's inequality, the following holds:

$$
\Pr_{\mathbf{n} \sim \alpha} \left[y_t g(x_t) - y_t f(x_t) < -\tfrac{\gamma}{2} \right] \leq e^{\frac{-n\gamma^2}{8}}.
$$

Plugging this inequality back into the previous one gives:

$$
\mathbb{E}_{\mathbf{n} \sim \alpha} \left[1_{y_t g(x_t) < \gamma} \right] \leq e^{\frac{-n\gamma^2}{8}} + 1_{y_t f(x_t) < \frac{3\gamma}{2}}. \tag{4}
$$

Fix $\gamma > 0$ and $h_1 \in \mathcal{H}_{k(h_1)}, \ldots, h_q \in \mathcal{H}_{k(h_q)}$ in inequality 2. Then, taking the expectation over α of both sides of the inequality and using inequality 4 combined with $\mathbb{E}[\frac{n_i}{n}] = \alpha_i$, we obtain that for any $\gamma > 0$, any $h_1 \in \mathcal{H}_{k(h_1)}, \ldots, h_q \in \mathcal{H}_{k(h_q)}$, and any $\alpha \in \Delta_q$, the following holds for $f = \sum_{i=1}^{q} \alpha_i h_i$:

$$
\begin{aligned}
\frac{1}{T} \mathbb{E} \left[\sum_{t=1}^{T} \mathbb{E}_{f_t \sim \pi_t^{\exp}} [1_{\{y_t f_t(x_t) < 0\}}] \right] &\leq \frac{1}{T} \sum_{t=1}^{T} 1_{y_t f(x_t) < \frac{3\gamma}{2}} + \frac{4}{\gamma} \sum_{i=1}^{q} \alpha_i \mathfrak{R}_T^{\text{seq}}(\mathcal{H}_{k(h_i)}) \\
&\quad + e^{\frac{-n\gamma^2}{8}} + \frac{3 + \log\log\frac{1}{\gamma}}{\sqrt{T}} + 2\sqrt{\frac{n \log p}{T}}.
\end{aligned}
$$

Thus, this inequality holds for any $\gamma > 0$, any $n \geq 1$, and any $f = \sum_{i=1}^{q} \alpha_i h_i \in \text{conv}(\mathcal{H})$. Choosing $n = \left\lceil \frac{4}{\gamma^2} \log \frac{\gamma^2 T}{16 \log p} \right\rceil$ and replacing $\frac{3\gamma}{2}$ by γ yields

$$
\begin{aligned}
&\frac{1}{T} \mathbb{E} \left[\sum_{t=1}^{T} \mathbb{E}_{f_t \sim \pi_t^{\exp}} [1_{\{y_t f_t(x_t) < 0\}}] \right] \\
&\leq \inf_{\gamma > 0} \left\{ \inf_{\substack{f = \sum_{i=1}^{q} \alpha_i h_i \in \text{conv}(\mathcal{H}) \\ \alpha \in \Delta_q, h_i \in \mathcal{H}_{k(h_i)}}} \frac{1}{T} \sum_{t=1}^{T} 1_{y_t f(x_t) < \gamma} + \frac{6}{\gamma} \sum_{i=1}^{q} \alpha_i \mathfrak{R}_T^{\text{seq}}(\mathcal{H}_{k(h_i)}) \right. \\
&\quad \left. + 6\sqrt{\frac{\log p}{\gamma^2 T}} + 6\sqrt{\left\lceil \frac{1}{\gamma^2} \log \left\lceil \frac{\gamma^2 T}{36 \log p} \right\rceil \right\rceil \frac{\log p}{T}} + \frac{3 + \log\log\frac{3}{2\gamma}}{\sqrt{T}} \right\},
\end{aligned}
$$

which completes the proof. $\qquad \square$

4   Algorithms for Structural Online Learning

While Theorem 2 proves the existence of a randomized strategy with favorable structural learning guarantees, it does not explicitly define one. In this section, we give a general algorithm that benefits from the structural online learning bound above.

In what follows, we will fix an arbitrary decomposition of the function class: $\mathcal{H} = \cup_{k=1}^{p} \mathcal{H}_k$. Moreover, we will assume that this decomposition $(\mathcal{H}_k)_{k=1}^{p}$ is *structurally online linear-learnable* in the sense that for any subset $\mathcal{H}_k \subset \mathcal{H}$, $k \in [1, p]$, there exists an online learning algorithm $\mathcal{A}_k$ such that for any time horizon T and every sequence $(x_t, L_t)_{t=1}^{T}$ where L_t is a linear loss function bounded by 1, $\mathcal{A}_k$ selects a sequence of functions $(h_t)_{t=1}^{T}$ satisfying $\sum_{t=1}^{T} L_t(h_t(x_t)) - \min_{h \in \mathcal{H}} \sum_{t=1}^{T} L_t(h(x_t)) = \mathrm{Reg}_{\mathcal{H}_k, T}(\mathcal{A}_k) = o(T)$. For instance, if $\mathcal{H}$ is finite, every decomposition is structurally online linear-learnable, which can be seen by applying a potential-based algorithm [6]. Note that the notion of structural online linear-learnability is a slight generalization of the concept introduced by [2].

We will also follow in this section the standard method of using a convex surrogate for the zero-one loss function. The will enable us to design algorithms that are both deterministic and can be used to achieve new structural PAC guarantees in the batch setting (the latter will be seen in Sect. 5). Let $\Phi \colon \mathbb{R} \to \mathbb{R}$ be any convex loss function upper bounding the zero-one loss. We further assume that Φ is G-Lipschitz. One standard example is the hinge loss, $\Phi(x) = (1 - x)1_{\{x \leq 1\}}$, which is 1-Lipschitz. Our goal is to design algorithms $\mathcal{A}$ that guarantee structural upper bounds on the following regret term: $\mathrm{Reg}_T(\mathcal{A}) = \max_{h \in \mathrm{conv}(\mathcal{H})} \sum_{t=1}^{T} \Phi(y_t h_t(x_t)) - \Phi(y_t h(x_t))$. For any x, we will denote by $\Phi'(x)$ an arbitrary element of the subgradient of Φ at x.

4.1   SOL.Boost Algorithm

At first glance, the structural learning bound of Theorem 2 seems unwieldy since the convex combination of the best-in-class hypothesis is not necessarily well-ordered with respect to the decomposition of the hypothesis set. Moreover, the proof of Theorem 2 is based on the existence of online learning algorithms for different subclasses of functions to which a meta-algorithm for learning with *experts* is applied. This is instructive, but it is also computationally infeasible because there are exponentially many experts in the proof.

Addressing the well-ordering issue and the computational problem will be essential to our algorithmic design. Towards the first point, we can "re-organize" the best-in-class hypothesis by writing:

$$\sum_{i=1}^{q} \alpha_i^* \mathfrak{R}_T(\mathcal{H}_{k(i^*)}) = \sum_{i=1}^{q} \sum_{k=1}^{p} 1_{\{k(i^*)=k\}} \alpha_i^* \mathfrak{R}_T(\mathcal{H}_{k(i^*)})$$

$$= \sum_{k=1}^{p} \sum_{i=1}^{q} 1_{\{k(i^*)=k\}} \alpha_i^* \mathfrak{R}_T(\mathcal{H}_k) = \sum_{k=1}^{p} \gamma_k^* \mathfrak{R}_T(\mathcal{H}_k).$$

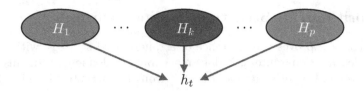

Fig. 1. Illustration of SOL.Boost Algorithm. The algorithm incorporates a meta-algorithm that measures the progress of each base algorithm. Note from the pseudo-code that the base algorithms are not assigned their true losses, but instead new hallucinated losses.

This suggests that learning against the convex hull of a hypothesis class with a structural decomposition can be equivalently cast as learning against each of its individual substructures along with some new set of convex weights. We will use this observation by applying an (efficient) experts-type algorithm to learn these new weights instead of the original weights from the best convex combination.

However, learning against each of the individual substructures proves to be a challenge in and of itself, since typical online learning algorithms, such as the weighted majority algorithm [14], are able to provide guarantees against only the single "best-in-class" hypothesis. Direct application of the experts algorithm on top of typical online learning algorithms for each $\mathcal{H}_k$ will only produce a regret guarantee against comparators of the form $\sum_{k=1}^{p} \alpha_k^* h_k^*$, $h_k^* \in \mathcal{H}_k$, $\boldsymbol{\alpha}^* \in \Delta_p$. On the other hand, Theorem 2 guarantees the existence of an algorithm that can attain a structural regret bound against arbitrary convex combinations in $\mathcal{H}$, including those that contain multiple base hypotheses from a single substructure $\mathcal{H}_k$. To attain this type of guarantees, we will linearize the loss and *hallucinate* different losses for each of the base online linear learning algorithms so that they learn well against the convex hull of each subclass.

Our algorithm, SOL.Boost, incorporates these two ideas to produce a guarantee in the form of the one given in Theorem 2. Figure 1 presents an illustration of the algorithm.

Theorem 3. *Let $\mathcal{H} \subset [-1,1]^{\mathcal{X}}$ be a hypothesis set admitting the decomposition $\mathcal{H} = \cup_{k=1}^{p} \mathcal{H}_k$ that is structurally online linear-learnable. For each $k \in [1,p]$, let $\mathcal{A}_k$ be an online algorithm that can minimize the regret of linear loss functions against $\mathcal{H}_k$, with regret $\mathrm{Reg}_{\mathcal{H}_k,T}(\mathcal{A}_k)$ over T rounds. Let $\Phi \colon \mathbb{R} \to \mathbb{R}$ be a G-Lipschitz convex upper bound on the zero-one loss. Then, SOL.Boost, initialized with $\eta < \sqrt{\frac{\log p}{T}}$, outputs a sequence of hypotheses $(h_t)_{t=1}^{T}$ that satisfies the following regret bound for any $(x_t, y_t)_{t=1}^{T}$:*

$$\sum_{t=1}^{T} \Phi(y_t h_t(x_t)) \leq \inf_{\substack{f = \sum_{i=1}^{q} \alpha_i h_i \in \mathrm{conv}(\mathcal{H}) \\ \boldsymbol{\alpha} \in \Delta_q, h_i \in \mathcal{H}_{k(h_i)}}} \left\{ \sum_{t=1}^{T} \Phi(y_t f(x_t)) \right.$$

$$\left. + \sum_{i=1}^{q} \alpha_i \mathrm{Reg}_{\mathcal{H}_{k(h_i)},T}(\mathcal{A}_{k(h_i)}) + \sqrt{G^2 T \log(p)} \right\}.$$

Algorithm 1. SOL.Boost

1: **Input:** Online linear learning algorithms $(\mathcal{A}_k)_{k=1}^p$ for $(\mathcal{H}_k)_{k=1}^p$, $(N_k)_{k=1}^p$ boosting stages, $\eta > 0$ learning rate, G Lipschitz constant for Φ.
2: **Initialize:** $w_{1,k} = 1, \forall k \in [1,p]$.
3: **for** $t = 1, \ldots, T$: **do**
4:     **Receive:** feature x_t.
5:     **for** $k = 1, \ldots, p$ **do**
6:        **Query:** algorithm $\mathcal{A}_k$ for hypothesis $h_{t,k}$ and prediction $h_{t,k}(x_t)$.
7:        **Set:** $\gamma_{t,k} = \frac{w_{t,k}}{\sum_{j=1}^p w_{t,j}}$.
8:     **end for**
9:     **Set:** predictor $h_t = \sum_{k=1}^p \gamma_{t,k} h_{t,k}$ and predict $h_t(x_t)$.
10:     **Receive:** label y_t.
11:     **for** $k = 1, \ldots, p$ **do**
12:        **Attribute:** loss $l_t(y_t h_{t,k}(x_t))$ to each $\mathcal{A}_k$, where l_t is the linear function $l_t : z \mapsto \Phi'\big(y_t h_t(x_t)\big) y_t z$.
13:     **end for**
14:     **for** $k = 1, \ldots, p$ **do**
15:        **Update:** weight $w_{t+1,k} = w_{t,k}\big(1 - \eta\, l_t\big(y_t h_{t,k}(x_t)\big)\big)$
16:     **end for**
17: **end for**

Proof. Let $\boldsymbol{\alpha}^* \in \Delta_q$ and $(h_i^*)_{i=1}^q \subset \mathcal{H}$ be such that $\sum_{i=1}^q \alpha_i^* h_i^* \in \text{conv}(\mathcal{H})$. For any $i \in [q]$ and $k \in [1,p]$, define $h_{k,i}^* = 1_{\{k(h_i^*)=k\}} h_i^*$. Then, it follows that

$$\sum_{i=1}^q \alpha_i^* h_i^* = \sum_{i=1}^q \sum_{k=1}^p 1_{\{k(h_i^*)=k\}} \alpha_i^* h_i^* = \sum_{k=1}^p \sum_{i=1}^q 1_{\{k(h_i^*)=k\}} \alpha_i^* 1_{\{k(h_i^*)=k\}} h_i^*$$

$$= \sum_{k=1}^p \left(\sum_{j=1}^q 1_{\{k(h_j^*)=k\}} \right) \left[\sum_{i=1}^q \frac{1_{\{k(h_i^*)=k\}} \alpha_i^*}{\sum_{j=1}^q 1_{\{k(h_j^*)=k\}}} h_{k,i}^* \right] = \sum_{k=1}^p \gamma_k^* \sum_{j=1}^q \beta_{k,j}^* h_{k,i}^*,$$

where $h_{k,i}^* = 1_{\{k(h_i^*)=k\}} h_i^*$, $\gamma_k^* = \sum_{j=1}^q 1_{\{k(h_j^*)=k\}}$, $\beta_{k,j}^* = \sum_{i=1}^q \frac{1_{\{k(h_i^*)=k\}} \alpha_i^*}{\sum_{j=1}^q 1_{\{k(h_j^*)=k\}}}$.
By convexity of the loss function Φ, we can write

$$\sum_{t=1}^T \Phi\left(y_t h_t(x_t)\right) - \Phi\left(y_t \sum_{k=1}^p \gamma_k^* \sum_{j=1}^q \beta_{k,j}^* h_{k,j}^*(x_t) \right)$$

$$\leq \sum_{t=1}^T \sum_{k=1}^p \Phi'\left(y_t h_t(x_t)\right) y_t h_{t,k}(x_t) \left[\gamma_{t,k} - \gamma_k^* \right]$$

$$+ \sum_{k=1}^p \gamma_k^* \sum_{t=1}^T \Phi'\left(y_t h_t(x_t)\right) y_t \left[h_{t,k}(x_t) - \sum_{j=1}^q \beta_{k,j}^* h_{k,j}^*(x_t) \right].$$

The first term in the last expression is bounded because the algorithm applies the Prod(η) algorithm [7] to the *hallucinated losses* above. Specifically, we can use the potential function $P(w_t) = \log\left(\sum_{k=1}^{p} w_{t,k}\right)$ to track the algorithm's progress against these surrogate losses and compute:

$$
\log\left(\frac{\sum_{k=1}^{p} w_{t+1,k}}{\sum_{j=1}^{p} w_{1,j}}\right) = \log\left(\prod_{s=1}^{t} \frac{\sum_{k=1}^{p} w_{s+1,k}}{\sum_{j=1}^{p} w_{s,j}}\right) = \sum_{s=1}^{t} \log\left(\frac{\sum_{k=1}^{p} w_{s+1,k}}{\sum_{j=1}^{p} w_{s,j}}\right)
$$

$$
= \sum_{s=1}^{t} \log\left(1 - \eta \sum_{k=1}^{p} \gamma_{s,k}\Phi'\left(y_t h_t(x_t)\right) y_t h_{t,k}(x_t)\right)
$$

$$
\leq \sum_{s=1}^{t} -\eta \sum_{k=1}^{p} \gamma_{s,k}\Phi'\left(y_t h_t(x_t)\right) y_t h_{t,k}(x_t),
$$

using the inequality $\log(1 + x) \leq x$ for $x \geq -\frac{1}{2}$. We can also write, for any $k \in [1, p]$,

$$
\log\left(\frac{\sum_{i=1}^{p} w_{t+1,i}}{\sum_{j=1}^{p} w_{1,j}}\right) \geq \log\left(\frac{w_{t+1,k}}{\sum_{j=1}^{p} w_{1,j}}\right) = -\log\left(\sum_{j=1}^{p} w_{1,j}\right) + \log\left(w_{t+1,k}\right)
$$

$$
\geq -\log\left(\sum_{j=1}^{p} w_{1,j}\right) - \sum_{s=1}^{t} \eta\Phi'\left(y_s h_s(x_s)\right) y_s h_{s,k}(x_s)
$$

$$
- \sum_{s=1}^{t} \left(\eta\Phi'\left(y_s h_s(x_s)\right) y_s h_{s,k}(x_s)\right)^2,
$$

in view of $\log(1 + x) \geq x - x^2$ for all $x \geq -\frac{1}{2}$, and the constraint on η. By concavity of the logarithm, this implies that for the $\gamma^* \in \Delta_p$ chosen above,

$$
\log\left(\frac{\sum_{i=1}^{p} w_{t+1,i}}{\sum_{j=1}^{p} w_{1,j}}\right) \geq \log\left(\frac{\sum_{k=1}^{p} \gamma_k^* w_{t+1,k}}{\sum_{j=1}^{p} w_{1,j}}\right) \geq \sum_{k=1}^{p} \gamma_k^* \log\left(\frac{w_{t+1,k}}{\sum_{j=1}^{p} w_{1,j}}\right)
$$

$$
\geq \sum_{k=1}^{p} \gamma_k^* \left[-\log\left(\sum_{j=1}^{p} w_{1,j}\right) + \sum_{s=1}^{t} -\eta\Phi'\left(y_s h_s(x_s)\right) y_s h_{s,k}(x_s)\right.
$$

$$
\left. - \left(\eta\Phi'\left(y_s h_s(x_s)\right) y_s h_{s,k}(x_s)\right)^2\right].
$$

Combining these calculations yields the inequality:

$$
\sum_{t=1}^{T} \sum_{k=1}^{p} \Phi'\left(y_t h_t(x_t)\right) y_s h_{t,k}(x_t)(\gamma_{t,k} - \gamma_k^*)
$$

$$
\leq \sum_{t=1}^{T} \sum_{k=1}^{p} \gamma_k^* \eta\left(\Phi'\left(y_t h_t(x_t)\right) y_s h_{t,k}(x_t)\right)^2 + \frac{1}{\eta} \log(p),
$$

since $w_{1,k} = 1 \; \forall k \in [1, p]$.

For the second term, notice that if for each $k \in [1, p]$ we attribute the loss $l_t(y_t h_{t,k}(x_t))$ to $\mathcal{A}_k$, where l_t is the linear function $l_t \colon z \mapsto \Phi'(y_t h_t(x_t))y_t z$, then the fact that l_t is linear implies that $\mathcal{A}_k$ attains some sublinear regret $\mathrm{Reg}_{\mathcal{H}_k,T}(\mathcal{A}_k)$ against $\mathcal{H}_k$: $\max_{h_k^* \in \mathcal{H}_k} \sum_{t=1}^{T} l_t(h_{t,k}(x_t)) - \sum_{t=1}^{T} l_t(h_k^*(x_t)) \leq \mathrm{Reg}_{\mathcal{H}_k,T}(\mathcal{A}_k)$. Since l_t is a linear loss, this directly implies that the regret guarantee $\mathrm{Reg}_{\mathcal{H}_k,T}(\mathcal{A}_k)$ extends to the convex hull of $\mathcal{H}_k$: $\max_{h_k^* \in \mathrm{conv}(\mathcal{H}_k)} \sum_{t=1}^{T} l_t(h_{t,k}(x_t)) - \sum_{t=1}^{T} l_t(h_k^*(x_t)) \leq \mathrm{Reg}_{\mathcal{H}_k,T}(\mathcal{A}_k)$. It now follows that:

$$\sum_{t=1}^{T} \Phi\Big(y_t h_t(x_t)\Big) - \Phi\Big(y_t \sum_{k=1}^{p} \gamma_k^* \sum_{j=1}^{q} \beta_{k,j}^* h_{k,j}^*(x_t)\Big)$$

$$\leq \sum_{k=1}^{p} \gamma_k^* \eta \sum_{t=1}^{T} \Big(\Phi'\Big(y_t h_t(x_t)\Big)y_s h_{t,k}(x_t)\Big)^2 + \frac{1}{\eta}\log(p) + \sum_{k=1}^{p} \gamma_k^* \mathrm{Reg}_{\mathcal{H}_k,T}(\mathcal{A}_k).$$

Moreover, we can rewrite the convex combination of the regret quantities in terms of the original convex combination weights of the comparator hypothesis:

$$\sum_{k=1}^{p} \gamma_k^* \mathrm{Reg}_{\mathcal{H}_k,T}(\mathcal{A}_k) = \sum_{k=1}^{p} \gamma_k^* \sum_{i=1}^{q} \beta_{k,i}^* \mathrm{Reg}_{\mathcal{H}_k,T}(\mathcal{A}_k)$$

$$= \sum_{k=1}^{p}\Big(\sum_{j=1}^{q} 1_{\{k(h_j^*)=k\}}\Big)\Big[\sum_{i=1}^{q} \frac{1_{\{k(h_i^*)=k\}}\alpha_i^*}{\sum_{j=1}^{q} 1_{\{k(h_j^*)=k\}}} \mathrm{Reg}_{\mathcal{H}_k,T}(\mathcal{A}_k)\Big]$$

$$= \sum_{k=1}^{p}\Big(\sum_{j=1}^{q} 1_{\{k(h_j^*)=k\}}\Big)\Big[\sum_{i=1}^{q} \frac{1_{\{k(h_i^*)=k\}}\alpha_i^*}{\sum_{j=1}^{q} 1_{\{k(h_j^*)=k\}}} \mathrm{Reg}_{\mathcal{H}_{k(h_i^*)},T}(\mathcal{A}_{k(h_i^*)})\Big]$$

$$= \sum_{k=1}^{p}\sum_{i=1}^{q} 1_{\{k(h_i^*)=k\}}\alpha_i^* \mathrm{Reg}_{\mathcal{H}_{k(h_i^*)},T}(\mathcal{A}_{k(h_i^*)}) = \sum_{i=1}^{q} \alpha_i^* \mathrm{Reg}_{\mathcal{H}_{k(h_i^*)},T}(\mathcal{A}_{k(h_i^*)}).$$

Choosing $\eta = \sqrt{\frac{\log(p)}{GT}}$ satisfies the conditions and yields the desired result. $\square$

One remarkable aspect of SOL.Boost is that it does not require knowledge of $\mathfrak{R}_T^{\mathrm{seq}}(\mathcal{H}_k)$ for any $\mathcal{H}_k$. As can be seen in Theorem 3, these complexity terms are replaced by the regret of each algorithm and are attained automatically. This is a significant advantage over the structural ensemble algorithms in the batch setting (e.g. DeepBoost [8]), which require the learner to either compute or estimate these quantities.

Moreover, the bound accompanied by SOL.Boost can vastly improve upon bounds that ignore the structural decomposition. The former realizes an average of all the regrets, and the latter is based upon the maximum regret among all base algorithms.

SOL.Boost updates all p base algorithms at each step. To improve the per-round computational cost, we can sample and query only a single base algorithm

---

Algorithm 2. Structural OTB

1: **Input:** Online algorithms $(\mathcal{A}_k)_{k=1}^p$ for decomposition $\mathcal{H} = \cup_{k=1}^p \mathcal{H}_k$, $\mathcal{J}$ family of contiguous intervals in $[1, T]$
2: **for** $t = 1, \ldots, T$: **do**
3:      **Receive:** feature x_t
4:      **Query:** algorithm $\mathcal{A}_k$ for hypothesis $h_{t,k}$
5:      **Receive:** label y_t and losses $L(h_{t,k}(x_t), y_t)$ for $k \in [1, p]$
6:      **Set:** $k_t = \operatorname{argmin}_{k \in [1,p]} L(h_{t,k}(x_t), y_t)$
7: **end for**
8: **Set:** $J_{\text{out}} = \operatorname{argmin}_{J \in \mathcal{J}} \frac{1}{|J|} \sum_{t \in J} L(h_{t,k_t}(x_t), y_t) + \sqrt{\frac{2C^2 \log(|\mathcal{J}|/\delta)}{|J|^{1/2}}}$
9: **Output:** $h_{J_{\text{out}}} = \frac{1}{|J_{\text{out}}|} \sum_{t \in J_{\text{out}}} h_{t,k_t}$

---

using techniques from the bandit literature (see e.g. [6]). This will come at the price of an extra $\sqrt{p}$ factor in the last term on the right-hand side of Theorem 3.

While it might be tempting to compare SOL.Boost to the work of [3], the algorithm presented here actually solves a regression problem, and it is more proper to compare it to the work of [2]. In fact, this is why we build upon the concept of online linear-learnability introduced in the latter paper.

The work of [12] may also seem related to the problem we consider, but it is in fact quite different since the hypothesis sets for online multiple kernel learning and online ensemble learning are distinct. Moreover, the guarantees procured there do not admit arbitrary structural decompositions as in Theorem 3 and are only in terms of the best base algorithm.

5   Online-to-Batch Conversion

In this section, we design an effective online-to-batch conversion technique for structural learning. Here, we assume that the learner receives a sample $((x_1, y_1), \ldots, (x_T, y_T))_{t=1}^T$ in $\mathcal{X} \times \mathcal{Y}$ drawn i.i.d. according to a distribution $\mathcal{D}$. The objective is to determine a hypothesis h based on a sequence of hypotheses $h_1, \ldots, h_T$ output by an online learning algorithm that admits a favorable generalization error $R(h) = \mathbb{E}_{(x,y) \sim \mathcal{D}} [L(x_t, y_t)]$.

We show that one can design algorithms whose generalization bounds account for the structure of the hypothesis set. Moreover, these are the first known structural generalization bounds in the batch setting that are in terms of the best-in-class hypothesis and can be viewed as an estimation error extension of the theoretical bounds in [8].

We consider a single static loss function $L: \mathcal{Y} \times \mathcal{Y} \to \mathbb{R}$ and assume that the difference between rounds is simply the label: $L_t(h_t(x_t)) = L(h_t(x_t), y_t)$. For convenience, we also assume that L is convex in its first argument. Note that any convex surrogate used in Sect. 4 will satisfy this. The online-to-batch conversion technique that we present is data-dependent and can be viewed as a structural extension of the method of [9]. At a high level, finding the subpath with the smallest cumulative loss will ensure small empirical estimation error, and

Fig. 2. Illustration of Structural OTB. The algorithm chooses the best a posteriori sub-path of hypotheses among all algorithms.

penalizing shorter paths will guarantee that the output will generalize well. This is not immediately obvious nor necessarily intuitive, since greedy optimization of empirical error without regularization does not lead to good generalization in many cases.

Theorem 4 presents the guarantee of our method. Figure 2 illustrates our algorithm, which benefits from the following guarantee.

Theorem 4. *Let $((x_1, y_1), \ldots, (x_T, y_T))$ be an i.i.d. sample drawn from a distribution $\mathcal{D}$ over $\mathcal{X} \times \mathcal{Y}$. Assume that the loss function $L: \mathcal{Y} \times \mathcal{Y} \to \mathbb{R}$ is bounded by a constant C and that each base online algorithm in the input of Structural OTB admits the following regret guarantee: for any $k \in [1, p]$, $h_k^* \in \mathcal{H}_k$, and contiguous subset $J \subset [1, T]$: $\sum_{t \in J} L(h_{t,k}(x_t), y_t) - L(h^*(x_t), y_t) \le \mathrm{Reg}_J(\mathcal{A}_k)$. Then, with probability at least $1 - \delta$, each of the following guarantees holds for Structural OTB:*

$$\mathop{\mathbb{E}}_{(x,y) \sim \mathcal{D}} [L(h_{J_{out}}(x), y)] \le \min_{J^* \in \mathcal{J}, \alpha^* \in \Delta_p, h_k^* \in \mathcal{H}_k} \left\{ \sum_{k=1}^{p} \alpha_k^* \frac{1}{|J^*|} \sum_{t \in J^*} L(h_k^*(x_t), y_t) \right.$$

$$\left. + \sum_{k=1}^{p} \alpha_k^* \mathrm{Reg}_{J^*}(\mathcal{A}_k) + \sqrt{\frac{2C^2 \log(|\mathcal{J}|/\delta)}{|J^*|}} \right\},$$

$$\mathop{\mathbb{E}}_{(x,y) \sim \mathcal{D}} [L(h_{J_{out}}(x), y)] \le \min_{J^* \in \mathcal{J}, \alpha^* \in \Delta_p, h_k^* \in \mathcal{H}_k} \left\{ \sum_{k=1}^{p} \alpha_k^* \mathop{\mathbb{E}}_{(x,y) \sim \mathcal{D}} [L(h_k^*(x), y)] \right.$$

$$\left. + \sum_{k=1}^{p} \alpha_k^* \mathrm{Reg}_{J^*}(\mathcal{A}_k) + \sqrt{\frac{2C^2 \log(p/\delta)}{T}} + \sqrt{\frac{2C^2 \log(|\mathcal{J}|/\delta)}{|J^*|}} \right\}.$$

Proof. Let $J \subset [1, T]$ be any subset, and denote $h_J = \frac{1}{|J|} \sum_{t \in J} h_{t,k_t}$. Notice that by convexity of L in the first coordinate, $\mathbb{E}_{(x,y) \sim \mathcal{D}} [L(h_J(x), y)] \le \frac{1}{|J|} \sum_{t \in J} \mathbb{E}_{(x,y) \sim \mathcal{D}} [L(h_{t,k_t}(x), y)]$.

Now for any $t \in J$, define $M_t = \frac{1}{|J|} \left(L(h_{t,k_t}(\cdot), \cdot) - \mathbb{E}_{(x,y) \sim \mathcal{D}} [L(h_{t,k_t}(x), y)] \right)$. By design, $(M_t)_{t \in J}$ is a sequence of martingale differences over J, such that if we reindex $J = [1, T_J]$, then $\mathbb{E}[M_s | M_1, \ldots, M_{s-1}] = 0$ for every $s \in [1, T_J]$.

Furthermore, by Azuma's inequality, we can guarantee that with probability at least $1 - \delta$, $\mathbb{E}_{(x,y) \sim \mathcal{D}} [L(h_{t,k_t}(x), y)] \le \frac{1}{|J|} \sum_{t \in J} L(h_{t,k_t}(x_t), y_t) + \sqrt{\frac{2C^2 \log(1/\delta)}{|J|}}$. By applying a union bound over all $J \in \mathcal{J}$, then with probability at least $1 - \delta$, the following bound holds for every $J \in \mathcal{J}$:

$$\mathop{\mathbb{E}}_{(x,y) \sim \mathcal{D}} [L(h_{t,k_t}(x), y)] \le \frac{1}{|J|} \sum_{t \in J} L(h_{t,k_t}(x_t), y_t) + \sqrt{\frac{2C^2 \log(|\mathcal{J}|/\delta)}{|J|}}.$$

Thus, it follows that

$$\mathop{\mathbb{E}}_{(x,y)\sim\mathcal{D}}[L(h_{J_{\mathrm{out}}}(x),y)] \leq \min_{J^*\in\mathcal{I}} \frac{1}{|J^*|} \sum_{t\in J^*} L(h_{t,k_t}(x_t),y_t) + \sqrt{\frac{2C^2\log(|\mathcal{I}|/\delta)}{|J^*|}}.$$

By the choice of k_t, we can further say that $L(h_{t,k_t}(x_t),y_t) \leq L(h_{t,k}(x_t),y_t)\ \forall k \in [1,p]$. In particular, this means that for any $J \in \mathcal{I}$,

$$\frac{1}{|J|}\sum_{t\in J} L(h_{t,k_t}(x_t),y_t) \leq \min_{\alpha^*\in\Delta_p} \frac{1}{|J|}\sum_{t\in J}\sum_{k=1}^{p}\alpha_k^* L(h_{t,k}(x_t),y_t)$$

$$\leq \min_{\alpha^*\in\Delta_p, h_k^*\in\mathcal{H}_k} \sum_{k=1}^{p}\alpha_k^*\left(\frac{1}{|J|}\sum_{t\in J} L(h_k^*(x_t),y_t) + \mathrm{Reg}_J(\mathcal{A}_k)\right).$$

Combining the above two inequalities yields the first result.

Furthermore, we can use Hoeffding's inequality over the best-in-class classifier's guarantee for each of the p subclasses and apply a union bound to say that

$$\mathop{\mathbb{E}}_{(x,y)\sim\mathcal{D}}[L(h_{J_{\mathrm{out}}}(x),y)] \leq \min_{J^*\in\mathcal{I},\alpha^*\in\Delta_p, h_k^*\in\mathcal{H}_k} \left\{ \sum_{k=1}^{p}\alpha_k^*\mathop{\mathbb{E}}_{(x,y)\sim\mathcal{D}}[L(h_k^*(x),y)] \right.$$

$$\left. + \sum_{k=1}^{p}\alpha_k^*\mathrm{Reg}_{J^*}(\mathcal{A}_k) + \sqrt{\frac{2C^2\log(p/\delta)}{T}} + \sqrt{\frac{2C^2\log(|\mathcal{I}|/\delta)}{|J^*|}} \right\}.$$

$\square$

Note that one natural choice of $\mathcal{I}$, as discussed by [9], is the set of all suffixes of $[1,T]$: $\{[1,T],[2,T],\ldots,[T,T]\}$. This was shown empirically to outperform the "data-independent" online-to-batch conversion methods of [5]. With this specific choice of $\mathcal{I}$, $|\mathcal{I}| = T$, and the logarithmic dependence on $|\mathcal{I}|$ is mild.

6   Conclusion

We presented a series of theoretical and algorithmic results for structural online learning. Our theory and algorithms can be further extended to cover other learning settings, including multi-class classification, regression and general online learning. In contrast with the batch algorithms for structural learning, our algorithms do not require the estimation of the Rademacher complexities in the decomposition of the hypothesis set. Moreover, our online-to-batch conversion algorithm provides an efficient alternative to the current structural ensemble methods used in the batch setting.

Acknowledgements. This work was partly funded by the NSF awards IIS-1117591 and CCF-1535987 and was also supported by the National Science Foundation Graduate Research Fellowship under Grant No. DGE 1342536.

References

1. Bartlett, P.L., Mendelson, S.: Rademacher and Gaussian complexities: risk bounds and structural results. J. Mach. Learn. Res. **3**, 463–482 (2002)
2. Beygelzimer, A., Hazan, E., Kale, S., Luo, H.: Online gradient boosting. In: Proceedings of NIPS, pp. 2449–2457 (2015)
3. Beygelzimer, A., Kale, S., Luo, H.: Optimal and adaptive algorithms for online boosting. In: ICML, volume 37 of JMLR Proceedings (2015)
4. Breiman, L.: Bagging predictors. Mach. Learn. **24**(2), 123–140 (1996)
5. Cesa-Bianchi, N., Conconi, A., Gentile, C.: On the generalization ability of on-line learning algorithms. IEEE Trans. Inf. Theor. **50**(9), 2050–2057 (2004)
6. Cesa-Bianchi, N., Lugosi, G.: Prediction, Learning, and Games. Cambridge University Press, New York (2006)
7. Cesa-Bianchi, N., Mansour, Y., Stoltz, G.: Improved second-order bounds for prediction with expert advice. Mach. Learn. **66**(2–3), 321–352 (2007)
8. Cortes, C., Mohri, M., Syed, U.: Deep boosting. In: Proceedings of ICML (2014)
9. Dekel, O., Singer, Y.: Data-driven online to batch conversions. In: NIPS, pp. 267–274 (2005)
10. Dietterich, T.G.: An experimental comparison of three methods for constructing ensembles of decision trees: bagging, boosting, and randomization. Mach. Learn. **40**(2), 139–157 (2000)
11. Freund, Y., Schapire, R.E.: A decision-theoretic generalization of on-line learning and an application to boosting. J. Comput. Syst. Sci. **55**(1), 119–139 (1997)
12. Jin, R., Hoi, S.C.H., Yang, T.: Online multiple kernel learning: algorithms and mistake bounds. In: Hutter, M., Stephan, F., Vovk, V., Zeugmann, T. (eds.) ALT 2010. LNCS, vol. 6331, pp. 390–404. Springer, Heidelberg (2010)
13. Koltchinskii, V., Panchenko, D.: Empirical margin distributions and bounding the generalization error of combined classifiers. Ann. Stat. **30**, 1–50 (2002)
14. Littlestone, N., Warmuth, M.K.: The weighted majority algorithm. Inf. Comput. **108**(2), 212–261 (1994)
15. Rakhlin, A., Shamir, O., Sridharan, K.: Relax, randomize: from value to algorithms. In: NIPS, pp. 2150–2158 (2012)
16. Rakhlin, A., Sridharan, K., Tewari, A.: Online learning: random averages, combinatorial parameters, and learnability. In: Proceedings of NIPS, pp. 1984–1992 (2010)
17. Rätsch, G., Onoda, T., Müller, K.-R.: Soft margins for adaboost. Mach. Learn. **42**(3), 287–320 (2001)

An Upper Bound for Aggregating Algorithm for Regression with Changing Dependencies

Yuri Kalnishkan[✉]

Computer Learning Research Centre and Department of Computer Science,
Royal Holloway, University of London, Egham, Surrey TW20 0EX, UK
Yuri.Kalnishkan@rhul.ac.uk

Abstract. The paper presents a competitive prediction-style upper bound on the square loss of the Aggregating Algorithm for Regression with Changing Dependencies in the linear case. The algorithm is able to compete with a sequence of linear predictors provided the sum of squared Euclidean norms of differences of regression coefficient vectors grows at a sublinear rate.

1 Introduction

We consider the on-line learning scenario with signals. The following events are repeated for $t = 1, 2, \ldots$. The learner sequentially reads a signal $x_t \in \mathbb{R}^n$, makes a prediction $\gamma_t \in \mathbb{R}$ on the basis of the signal and past observations, and sees the true outcome $y_t \in [-Y, Y]$. The quality of the learner's predictions are assessed using a loss function $\lambda(\gamma, y)$, which is $(\gamma - y)^2$ in this paper.

We want to develop strategies for the learner making sure it suffers low cumulative loss $\mathrm{Loss}(T) = \sum_{t=1}^{T} \lambda(\gamma_t, y_t)$ over T steps. We approach this task within the competitive on-line prediction framework. According to this framework, no mechanism (probabilistic or other) generating the signals and outcomes is postulated. Instead we take a pool of prediction strategies and aim to build one that suffers loss not much worse than any strategy from the pool on every possible sequence of signals and outcomes.

In [Vov01, AW01] a prediction strategy is built that competes against the pool of all linear predictors outputting $\gamma_t = \theta' x_t$ for a fixed $\theta \in \mathbb{R}^n$. (Unless otherwise stated, all vectors in this paper are column vectors and M' is the transpose of a matrix or vector M.) The strategy called Aggregating Algorithm for Regression (AAR; also known as Vovk-Azoury-Warmuth predictor) suffers loss satisfying

$$\mathrm{Loss}_{\mathrm{AAR}}(T) \le \inf_{\theta \in \mathbb{R}^n} \left((\theta' x_t - y_t)^2 + a\|\theta\|^2 \right) + nY^2 \ln \left(\frac{TB^2}{an} + 1 \right) \qquad (1)$$

on every sequence $(x_1, y_1), (x_2, y_2), \ldots, (x_T, y_T)$, where $B = \max_{t=1,2,\ldots,T} \|x_t\|$ and $Y = \max_{t=1,2,\ldots,T} |y_t|$, $T = 1, 2, \ldots$, and the number $a > 0$ is the parameter of the strategy. (In this paper, $\|x\|$ denotes the Euclidean norm.) AAR does not need to know either B, Y, or the time horizon T from the start.

© Springer International Publishing Switzerland 2016
R. Ortner et al. (Eds.): ALT 2016, LNAI 9925, pp. 238–252, 2016.
DOI: 10.1007/978-3-319-46379-7_16

Intuitively, AAR covers the situation when we need to learn the 'right' θ on the fly, while making predictions. The extra term $nY^2 \ln\left(\frac{TB^2}{an} + 1\right)$ grows logarithmically in T, which is a very small price to pay for not knowing the 'right' θ from the start. One may want to generalise the result to the situation when θ changes with time. Consider a prediction strategy using a sequence $\theta_1, \theta_2, \ldots$ to predict in the following way. On step t it predicts $\gamma_t = u'_t x_t$, where $u_t = \sum_{i=1}^{t} \theta_i$. Clearly, aiming to do as well as *any* such sequence is hopeless. To every sequence $(x_1, y_1), (x_2, y_2), \ldots$ one can fit a sequence $u_1, u_2, \ldots$ suffering zero loss, provided $x_t \neq 0$. However, one can hope to compete with a sequence of slowly changing u_t. If $\sum_{t=1}^{T} \|\theta_t\| = \|\theta_1\|^2 + \sum_{t=2}^{T} \|u_t - u_{t-1}\|^2$ grows slowly, can we have a strategy with an upper bound on the loss similar to (1)?

This problem has been approached using a variety of techniques. In [HW01] an algorithm based on Bregman divergence and gradient descent-type methods was proposed. The bounds obtained in [HW01] have multiplicative constants in front of the competitors' losses. In [BK07a] an algorithm called Aggregating Algorithm for Regression with Changing Dependencies (AARCh) based on Vovk's Aggregating Algorithm and extending the construction of AAR from [Vov01] was proposed. The bounds form [BK07a] have no multiplicative constant, but the final result is rather week. A recent paper [MVC15] has proposed a strategy LASER based on the last-step min-max approach of [For99]. The strategy takes a function $v(T) = O(T)$ and $a > 0$ as parameters and suffers loss satisfying

$$\text{Loss}_{\text{LASER}}(T) \leq \inf_{\substack{u_1, u_2, \ldots, u_T \in \mathbb{R}^n: \\ \sum_{t=2}^{T} \|u_t - u_{t-1}\|^2 \leq v(T)}} \left(\sum_{t=1}^{T} (u'_t x_t - y_t)^2 + a\|u_1\|^2 \right)$$

$$+ nY^2 \ln\left(\frac{TB^2}{an} + 1\right) + O((v(T))^{1/3} T^{2/3}). \quad (2)$$

The bound is far superior to that from [BK07a].

In this paper we improve the upper bound for AARCh from [BK07a] to achieve an extra term $O((v(T))^{1/3} T^{2/3})$ matching that of (2). The multiplicative constant in the extra term exhibits better dependency on the dimension, $n^{1/3}$ instead of $n^{2/3}$ in [MVC15].

As with LASER, in order to achieve this, AARCh should be optimised from the start using the prior knowledge of the time horizon T, the value of $v(T)$, B, and Y. Applying the Aggregating Algorithm allows one to dispense with some prior knowledge (with a notable exception of Y) but complicates the strategy.

One may note that the problem of competing with a sequence of us can be thought of as an extension to the regression framework of the problem of competing against switching experts in prediction with expert advice; see [AKCV12] for a comparison of bounds given by different approaches.

2 Preliminaries

2.1 Games and Prediction Strategies

A *game* $\mathfrak{G}$ is a triple of an *outcome space* Ω, prediction space Γ and a *loss function* $\lambda : \Gamma \times \Omega \to [0, +\infty]$.

A *prediction strategy* $\mathcal{S}$ for a game $\mathfrak{G}$ working with signals from a *signal space* X is a mapping $\mathcal{S} : (X \times \Omega)^* \times X \to \Gamma$. Intuitively, $\mathcal{S}$ supplies predictions for the learner acting according to this protocol:

Protocol 1
(1) FOR $t = 1, 2, \ldots$
(2)   the learner reads signal $x_t \in X$
(3)   the learner produces $\gamma_t \in \Gamma$
(4)   the learner sees $y_t \in \Omega$
(5) END FOR

On a sequence $(x_1, y_1), (x_2, y_2), \ldots, (x_T, y_T)$ the learner using the strategy $\mathcal{S}$ suffers cumulative loss

$$\mathrm{Loss}_{\mathcal{S}}(T) = \sum_{t=1}^{T} \lambda(\gamma_t, y_t) = \sum_{t=1}^{T} \lambda(\mathcal{S}(x_1, y_1, \ldots, x_{t-1}, y_{t-1}, x_t), y_t).$$

The index $\mathcal{S}$ will be dropped if it is clear from the context.

We will be considering square-loss games with $\Omega \subseteq \mathbb{R}$, $\Gamma = \mathbb{R}$ and $\lambda(\gamma, y) = (\gamma - y)^2$. For Ω we take different subsets of $\mathbb{R}$. Strictly speaking the theory of the Aggregating Algorithm (see [Vov01]) applies to the bounded game with $\Omega = [-Y, Y]$. However it often happens that the algorithm does not need to know Y in advance and Y only appears in the bound. Then we can say the algorithm applies to the case $\Omega = \mathbb{R}$.

2.2 Aggregating Algorithm for Regression with Changing Dependencies

The Aggregating Algorithm for Regression with Changing Dependencies (AARCh) was introduced in [BK07a] (see also [BK07b] for numerical experiments).

AARCh is a prediction strategy for a game with real outcomes and predictions and signals from $\mathbb{R}^n$. It takes as parameters a sequence $a_1, a_2, \ldots > 0$ and on step T predicts $\gamma_T = \tilde{y}'(\bar{K} + I)^{-1}\bar{k}$, where

$$\tilde{y} = \begin{pmatrix} y_1 \\ y_2 \\ \vdots \\ y_{T-1} \\ 0 \end{pmatrix}, \qquad \bar{k} = \begin{pmatrix} \frac{1}{a_1}x_1'x_T \\ \left(\frac{1}{a_1} + \frac{1}{a_2}\right)x_2'x_T \\ \vdots \\ \left(\frac{1}{a_1} + \frac{1}{a_2} + \cdots + \frac{1}{a_T}\right)x_T'x_T \end{pmatrix},$$

and

$$
\bar{K} = \begin{pmatrix}
\frac{1}{a_1}x_1'x_1 & \frac{1}{a_1}x_1'x_2 & \cdots & \frac{1}{a_1}x_1'x_T \\
\frac{1}{a_1}x_2'x_1 & \left(\frac{1}{a_1}+\frac{1}{a_2}\right)x_2'x_2 & \cdots & \left(\frac{1}{a_1}+\frac{1}{a_2}\right)x_2'x_T \\
\vdots & \vdots & \ddots & \vdots \\
\frac{1}{a_1}x_T'x_1 & \left(\frac{1}{a_1}+\frac{1}{a_2}\right)x_T'x_2 & \cdots & \left(\frac{1}{a_1}+\cdots+\frac{1}{a_T}\right)x_T'x_T
\end{pmatrix}
$$

(this is the dual form given in Sect. 3.3 of [BK07a]).

The algorithm is obtained by applying the Aggregating Algorithm in the bounded square loss game to a particular set of experts.

3   Main Result

In this section we formulate and discuss upper bounds on the cumulative square loss of AARCh.

Theorem 1. *For every sequence* $(x_1, y_1), (x_2, y_2), \ldots, (x_T, y_T) \in \mathbb{R}^n \times \mathbb{R}$, *the square loss of the learner using AARCh with positive parameters* $a_1, a, \ldots, a$ *satisfies*

$$
\mathrm{Loss}(T) \leq \inf_{u_1,\ldots,u_T \in \mathbb{R}^n} \left(\sum_{t=1}^{T}(u_t'x_t - y_t)^2 + a_1\|u_1\|^2 + a\sum_{t=2}^{T}\|u_t - u_{t-1}\|^2 \right) +
$$
$$
nY^2 \ln\left(1 + \frac{TB^2}{a_1 n}\right) + Y^2 B\left(T - \frac{1}{2}\right)\sqrt{\frac{n}{a}} - nY^2\ln 2 + \alpha(T, a), \quad (3)
$$

where $Y = \max_{t=1,\ldots,T}|y_t|$, $B = \max_{t=1,\ldots,T}\|x_t\|$, *and*

$$
\alpha(T, a) = nY^2\left(1 + \frac{B^2}{2an} - \sqrt{\frac{B^4}{4a^2n^2} + \frac{B^2}{an}}\right)^{2(T-1)} \leq nY^2. \quad (4)
$$

Clearly, the bound only makes sense if the terms on the right, apart from $\sum_{t=1}^{T}(u_t'x_t - y_t)^2$, are not too large. If the outcomes y_t are bounded, $|y_t| \leq Y$, then it is too easy to get loss not exceeding $Y^2 T$ by predicting 0 consistently. Thus an extra term growing faster than $O(T)$ makes little sense and $O(T)$ can only be useful if the constant is small. On the other hand, competing with sequences of u_t such that $\|u_t - u_{t-1}\|$ is large is futile: as pointed out in [MVC15], the sequence $u_t = x_t y_t / \|x_t\|^2$ leads to zero loss as long as $x_t \neq 0$. Thus one may want to obtain an extra term of the order $O(T)$ and, if possible, $o(T)$, by restricting the variability of u_t.

Let us find a optimising the sum $a \cdot \sum_{t=2}^{T}\|u_t - u_{t-1}\|^2 + Y^2 B(T - 1/2)\sqrt{\frac{n}{a}}$. If Y, B, T, and the order $v(T)$ of the growth of the sum $\sum_{t=2}^{T}\|u_t - u_{t-1}\|^2$ (cf. $V^{(2)}$ in [MVC15]) are known in advance, we can find the optimal a as follows.

Lemma 1. *For all positive v and c the minimum* $\min_{a>0}\left(av + \frac{c}{\sqrt{a}}\right)$ *is achieved at* $a = \left(\frac{c}{2v}\right)^{2/3}$ *and equals* $\frac{3}{2^{2/3}}c^{2/3}v^{1/3}$.

Proof. As $a \to 0$ or $a \to +\infty$, the expression tends to $+\infty$. We get the minimum by equating to zero the derivative

$$\frac{\partial}{\partial a}\left(av + \frac{c}{\sqrt{a}}\right) = v - \frac{c}{2a^{3/2}}.$$

$\square$

Corollary 1. *For every function* $v : \{2, 3, \ldots\} \to (0, +\infty)$ *and every sequence* $(x_1, y_1), (x_2, y_2), \ldots, (x_T, y_T) \in \mathbb{R}^n \times \mathbb{R}$ *such that* $\|x_t\| \le B$ *and* $|y_t| \le Y$ *for* $t = 1, 2, \ldots, T$, *the square loss of the learner using AARCh with parameters* $a_1, a, \ldots, a$, *where* $a_1 > 0$ *and*

$$a = a(T) = \frac{Y^{4/3}B^{2/3}n^{1/3}}{2^{2/3}} \cdot \frac{(T - 1/2)^{2/3}}{(v(T))^{2/3}}, \tag{5}$$

satisfies

$$\mathrm{Loss}(T) \le \inf_{\substack{u_1, \ldots, u_T \in \mathbb{R}^n \\ \sum_{t=2}^{T} \|u_t - u_{t-1}\|^2 \le v(T)}} \left(\sum_{t=1}^{T}(u_t' x_t - y_t)^2 + a_1\|u_1\|^2\right)$$

$$+ nY^2 \ln\left(\frac{TB^2}{a_1 n} + 1\right) + \frac{3}{2^{2/3}}Y^{4/3}B^{2/3}n^{1/3}\left(T - \frac{1}{2}\right)^{2/3}(v(T))^{1/3}$$

$$- nY^2 \ln 2 + \alpha(T, a(T)), \tag{6}$$

where $\alpha(T, a(T)) \le nY^2$ *is given by* (4).

If, moreover, $v(t) = o(T)$ *and* $1/v(T) = o(T^2)$ *as* $T \to +\infty$, *then*

$$\alpha(T, a(T)) \le nY^2 e^{-2\frac{B(T-1)}{\sqrt{a(T)n}}\left(1 - \frac{B}{2\sqrt{a(T)n}}\right)} \to 0 \tag{7}$$

as $T \to +\infty$.

Proof. It is easy to see that

$$0 < 1 + \frac{b}{2} - \sqrt{\frac{b^2}{4} + b} \le 1 + \frac{b}{2} - \sqrt{b}$$

for all $b \ge 0$. Applying the inequality $\ln(1 + x) \le x$ yields upper bound (7).

Since $v(t) = o(T)$, we get $(T - 1/2)/v(T) \to +\infty$ and thus $a(T) \to +\infty$ as $T \to +\infty$. The condition $1/v(T) = o(T^2)$ implies $T/\sqrt{a} \to +\infty$. Therefore the power in the term on the right-hand side tends to $-\infty$ and the term itself tends to 0 as $T \to +\infty$. $\square$

The main component of the extra term in the bound has the same order of growth in T, namely, $T^{2/3}(v(T))^{1/3}$, as in the bound for LASER in Corollary 12 of [MVC15]. If $v(T) = o(T)$ as $T \to +\infty$, the order of growth is sublinear.

However, the multiplicative coefficient differs and we get $\frac{3}{2^{2/3}}Y^{4/3}B^{2/3}n^{1/3}$ instead of $3 \cdot 2^{1/3}Y^{4/3}B^{2/3}n^{2/3}$. Our term is smaller by the factor of $2n^{1/3}$. See Remark 2 below for a discussion of the power[1] of n.

Having to know the time horizon T in advance to choose a is annoying. This problem can be eliminated by applying the Aggregating Algorithm. Suppose we know Y, B, and $v(T)$. Then we can apply the Aggregating Algorithm to a countable number of instances of AARCh, each using a from (5), $T = 2, 3, \ldots$ Let us assign to the instance corresponding to T a prior $p_0(T) = \frac{6}{\pi^2(T-1)^2}$, $T = 2, 3, \ldots$ and apply the AA. Bound (10) with $\eta = 1/(2Y^2)$ and $C(\eta) = 1$ give us the following corollary.

Corollary 2. *For $Y > 0$, $B > 0$, $a_1 > 0$ and a function $v : \{2, 3, \ldots\} \to (0, +\infty)$ there is a prediction strategy S that on every sequence $(x_1, y_1), \ldots, (x_T, y_T) \in \mathbb{R}^n \times \mathbb{R}$ such that $\|x_t\| \leq B$ and $|y_t| \leq Y$ for all $t = 1, 2, \ldots, T$ suffers square loss*

$$\mathrm{Loss}_S(T) \leq \inf_{\substack{u_1, \ldots, u_T \in \mathbb{R}^n \\ \sum_{t=2}^T \|u_t - u_{t-1}\|^2 \leq v(T)}} \left(\sum_{t=1}^T (u_t' x_t - y_t)^2 + a_1 \|u_1\|^2 \right) +$$

$$nY^2 \ln \left(\frac{TB^2}{a_1 n} + 1 \right) + \frac{3}{2^{2/3}} Y^{4/3} B^{2/3} n^{1/3} \left(T - \frac{1}{2} \right)^{2/3} (v(T))^{1/3} +$$

$$2Y^2 \ln T + 2Y^2 \ln \frac{\pi^2}{6} - nY^2 \ln 2 + \alpha_{Y,B,v}(T), \quad (8)$$

where $\alpha_{Y,B,v}(T) \leq nY^2$ and tends to zero as $T \to +\infty$ provided $v(T) = o(T)$ and $1/v(T) = o(T^2)$.

While the Aggregating Algorithm provides a way of computing S, the procedure is complicated. Arguing in a similar way, we can eliminate the dependency on B and reduce the dependency on the order of growth of $v(t)$ at a price of making the strategy even more complicated. The dependency on Y cannot be overcome this way though as the Aggregating Algorithm assumes Y is finite and known. (As Y grows to infinity, the maximum value $\eta = 1/(2Y^2)$ such that the game is mixable vanishes and renders bound (10) useless.)

In the rest of the paper we prove Theorem 1. Section 4 covers the steps done in [BK07a], Sect. 5 presents the original material, and Sect. 6 contains some remarks on the proof.

4   Deriving the Upper Bound on AARCh

In this section we review the derivation of the upper bound on AARCh from [BK07a] starting with the basics of prediction with expert advice and Vovk's Aggregating Algorithm after [Vov98, Vov01].

---

[1] The fact that the powers of n and T sum to 1 makes the straightforward kernelisation of the bound based on the representer theorem useless. This observation may potentially lead to a lower bound.

4.1   Prediction with Expert Advice

The goal of prediction with expert advice is constructing prediction strategies competitive with other strategies from a pool can be addressed within the framework of prediction with expert advice.

Suppose we have a *pool of experts* Θ. Predictions output by experts at any moment in time can be described by a function $\Theta \to \Gamma$. Let $\mathcal{E} \subseteq \Gamma^\Theta$ be a set of such functions that we allow (e.g., measurable functions). Prediction with expert advice is concerned with building merging strategies $\mathcal{M} : (\mathcal{E} \times \Omega)^* \times \mathcal{E} \to \Gamma$. Intuitively, $\mathcal{M}$ supplies predictions for the learner acting according to this protocol:

Protocol 2
(1) FOR $t = 1, 2, \ldots$
(2)   the learner reads experts' predictions γ_t^θ, $\theta \in \Theta$
(3)   the learner produces $\gamma_t \in \Gamma$
(4)   the learner sees $y_t \in \Omega$
(5) END FOR

Over T steps expert θ suffers loss $\mathrm{Loss}_\theta(T) = \sum_{t=1}^T \lambda(\gamma_t^\theta, y_t)$. Prediction with expert advice looks for merging strategies making sure that the cumulative loss of the learner is not much greater than the loss of every expert $\theta \in \Theta$.

4.2   Aggregating Algorithm

The Aggregating Algorithm (AA) was proposed in [Vov90, Vov98]. It is a rather general merging strategy.

The Aggregating Algorithm takes as parameters $\eta > 0$, a (prior) distribution P_0 on Θ, and a substitution rule, which will be defined later. On step t it forms the *generalised prediction*, which is a function $g_t : \Omega \to [0, +\infty]$ given by

$$g_t(y) = -\frac{1}{\eta} \ln \frac{\int_\Theta e^{-\eta\lambda(\gamma_t^\theta, y)} e^{-\eta\,\mathrm{Loss}_\theta(t-1)} P_0(d\theta)}{\int_\Theta e^{-\eta\,\mathrm{Loss}_\theta(t-1)} P_0(d\theta)}.$$

The generalised prediction is then converted to a prediction γ_t such that $\lambda(\gamma_t, y) \le C(\eta) g_t(y)$ for all $y \in \Omega$. Here $C(\eta)$ is the minimum constant permitted for the game. It is shown in Sect. 2.4 of [Vov01] that for the bounded square-loss game with $\Omega = [-Y, Y]$ we can take $C(\eta) = 1$ for $\eta \le 1/(2Y^2)$ (as can be seen from (10) below, in such situations one wants to maximise η, so $\eta = 1/(2Y^2)$ is used). A *substitution rule* maps generalised predictions into predictions. A convenient substitution rule leads to simple algorithms.

The Aggregating Algorithm ensures that the learner's loss satisfies

$$\mathrm{Loss}_{AA}(T) \le -\frac{C(\eta)}{\eta} \ln \int_\Theta e^{-\eta\,\mathrm{Loss}_\theta(T)} P_0(d\theta) \tag{9}$$

(this can be checked by induction). This inequality holds for all possible sequences of outcomes. If the pool is finite or countable, the integral reduces to the sum and by dropping from the sum all terms except for one we obtain the inequality

$$\text{Loss}_{\text{AA}}(T) \leq C(\eta)\,\text{Loss}_{\theta}(T) + \frac{C(\eta)}{\eta}\ln\frac{1}{P_0(\theta)} \tag{10}$$

for every expert θ. If the pool is not countable, as it is below, this general trick does not apply and we need to upper bound (9) for the particular case.

4.3   Constructing the Bound for AARCh

AARCh is obtained by applying AA in the context of a bounded square-loss game with the outcome space $\Omega = [-Y, Y]$ and the signal space $X = \mathbb{R}^n$ to the following experts. Fix a positive integer T and let $\Theta = (\mathbb{R}^n)^T$. We can consider elements of Θ as vectors of nT real components or sequences of T vectors from $\mathbb{R}^n$, $\theta = (\theta_1, \theta_2, \ldots, \theta_T)$. On step t expert θ predicts $\gamma_t^\theta = (\sum_{i=1}^t \theta_i)' x_t$.

Take $\eta = 1/(2Y^2)$; as mentioned above, we get $C(\eta) = 1$ for the bounded square-loss game. On Θ we consider the Gaussian prior with the density

$$p_0(\theta) = \prod_{t=1}^T \left[\left(\frac{\eta a_t}{\pi}\right)^{n/2} e^{-\eta a_t \|\theta_t\|^2}\right] = \left(\prod_{t=1}^T a_t^{n/2}\right)\left(\frac{\eta}{\pi}\right)^{Tn/2} e^{-\eta \sum_{t=1}^T a_t \|\theta_t\|^2},$$

where $a_1, a_2, \ldots, a_T > 0$ are the parameters of AARCh.

We will omit the derivation of the formulas for AARCh given in Sect. 2.2, but give the derivation of the upper bound. Bound (9) ensures that

$$\text{Loss}_{\text{AARCh}}(T) \leq -\frac{1}{\eta}\ln\int_{\mathbb{R}^{nT}} e^{-\eta\,\text{Loss}_{\theta}(T)} p_0(\theta) d\theta. \tag{11}$$

The loss of expert θ equals

$$\text{Loss}_{\theta}(T) = \sum_{t=1}^T \left(\left(\sum_{i=1}^t \theta_i\right)' x_t - y_t\right)^2 = \sum_{t=1}^T (\theta' w_t - y_t)^2,$$

where θ is interpreted as a column vector and

$$w_t' = (\underbrace{x_t', \ldots, x_t'}_{t \text{ times}}, \underbrace{0, \ldots, 0}_{(T-t)n \text{ zeros}})'.$$

This is a quadratic form in θ. Multiplying $e^{-\eta\,\text{Loss}_{\theta}(T)}$ by $p_0(\theta)$ adds a quadratic term to the power. The integral can be evaluated using the following proposition.

Proposition 1. *For a quadratic form $Q(\theta)$, $\theta \in \mathbb{R}^m$, with the quadratic part $\theta' A \theta$, where A is a symmetric positive definite $(m \times m)$-matrix, we get*

$$\int_{\mathbb{R}^m} e^{-Q(\theta)} = e^{-Q_0}\frac{\pi^{m/2}}{\sqrt{\det A}},$$

where $Q_0 = \min_{\theta \in \mathbb{R}^m} Q(\theta)$.

The proof of the proposition is essentially by completing the square and integration by substitution.

The matrix of the quadratic part of the negation of the form in the power in (11) is

$$\eta A = \eta \sum_{t=1}^{T} w_t w_t' + \eta \begin{pmatrix} a_1 I & & 0 \\ & \ddots & \\ 0 & & a_T I \end{pmatrix}.$$

It is easy to see that A is positive definite.

Proposition 2

$$\text{Loss}_{\text{AARCh}}(T) \leq \inf_{\theta_1, \dots, \theta_T \in \mathbb{R}^n} \left(\text{Loss}_{\theta_1, \dots, \theta_T}(T) + \sum_{t=1}^{T} a_t \|\theta_t\|^2 \right) + Y^2 \ln \frac{\det A}{\prod_{t=1}^{T} a_t^n},$$

where

$$A = \begin{pmatrix} \sum_{t=1}^{T} x_t x_t' + a_1 I & \sum_{t=2}^{T} x_t x_t' & \sum_{t=3}^{T} x_t x_t' & \vdots & x_T x_T' \\ \sum_{t=2}^{T} x_t x_t' & \sum_{t=2}^{T} x_t x_t' + a_2 I & \sum_{t=3}^{T} x_t x_t' & \vdots & x_T x_T' \\ \sum_{t=3}^{T} x_t x_t' & \sum_{t=3}^{T} x_t x_t' & \sum_{t=3}^{T} x_t x_t' + a_3 I & \vdots & x_T x_T' \\ \cdots & \cdots & \cdots & \ddots & \vdots \\ x_T x_T' & x_T x_T' & x_T x_T' & \cdots & x_T x_T' + a_T I \end{pmatrix}.$$

It remains to upper bound the determinant of A.

5 Upper Bounding the Determinant

By Theorem 7 of Sect. 2.10, [BB61], the determinant of a positive definite matrix does not exceed the product of determinants of the minors. Hence

$$\det A \leq \det \left(\sum_{t=1}^{T} x_t x_t' + a_1 I \right) \det A_2, \tag{12}$$

where

$$A_2 = \begin{pmatrix} \sum_{t=2}^{T} x_t x_t' + a_2 I & \sum_{t=3}^{T} x_t x_t' & \vdots & x_T x_T' \\ \sum_{t=3}^{T} x_t x_t' & \sum_{t=3}^{T} x_t x_t' + a_3 I & \vdots & x_T x_T' \\ \cdots & \cdots & \ddots & \vdots \\ x_T x_T' & x_T x_T' & \cdots & x_T x_T' + a_T I \end{pmatrix}.$$

Proposition 3 [CBCG05]. *For every positive integer* T, *all vectors* $x_1, x_2, \dots, x_T \in \mathbb{R}^n$ *such that* $\|x_t\| \leq B$, $t = 1, 2, \dots, T$, *and all* $a > 0$ *we have*

$$\frac{1}{a^n} \det \left(\sum_{t=1}^{T} x_t x_t' + a I \right) \leq \left(\frac{TB^2}{an} + 1 \right)^n.$$

The proof is by Proposition 5 given below.

We will now simplify the structure of A_2. From every block row, except for the last, we subtract the next row. We start from the first row and do this from top to bottom. Then from every block column, except for the last, we subtract the next block column, going right to left. This results in a block tridiagonal matrix $\widetilde{A}_2$ given by

$$
\begin{pmatrix}
x_2 x_2' + (a_2 + a_3)I & -a_3 I & & & 0 \\
-a_3 I & x_3 x_3' + (a_3 + a_4)I & -a_4 I & & \\
& \ddots & \ddots & & \ddots \\
& -a_{T-1}I & x_{T-1}x_{T-1}' + (a_{T-1} + a_T)I & -a_T I \\
& 0 & -a_T I & x_T x_T' + a_T I
\end{pmatrix}
$$

Subtracting block row j from block row i amounts to multiplication on the left by a block elementary matrix L_{ij} with determinant 1. Subtracting block column j from block row i amounts to multiplication on the right by $L'_{i,j}$. Thus

$$
\widetilde{A}_2 = L_{T-1,T} L_{T-2,T-1} \cdots L_{1,2} A_2 L'_{1,2} \cdots L'_{T-2,T-1} L'_{T-1,T}
$$

and therefore $\widetilde{A}_2$ is still symmetric and positive definite and $\det \widetilde{A}_2 = \det A_2$.

We now set $a_2 = a_3 = \ldots = a_T = a$ and let

$$
\overline{A}_2 = \frac{1}{a}\widetilde{A}_2 =
\begin{pmatrix}
\frac{x_2 x_2'}{a} + 2I & -I & & & 0 \\
-I & \frac{x_3 x_3'}{a} + 2I & -I & & \\
& \ddots & \ddots & & \ddots \\
& -I & \frac{x_{T-1}x_{T-1}'}{a} + 2I & -I \\
& 0 & -I & \frac{x_T x_T'}{a} + I
\end{pmatrix}
$$

The determinant of a block tridiagonal matrix can be calculated as follows.

Proposition 4 [Sal06]. *The determinant of a block tridiagonal matrix*

$$
M =
\begin{pmatrix}
G_1 & E_2 & 0 & & \\
F_2 & G_2 & E_3 & & \\
& \ddots & \ddots & \ddots & \\
& & F_{m-1} & G_{m-1} & E_m \\
& & 0 & F_m & G_m
\end{pmatrix}
$$

is $\det M = \prod_{k=1}^{m} \det \Lambda_k$, *where* $\Lambda_1 = G_1$ *and* $\Lambda_k = G_K - F_k \Lambda_{k-1}^{-1} E_k$, $k = 2, 3, \ldots, m$, *provided all required inversions can be performed.*

Proof. The proof is by reducing the matrix to the block upper triangular form and taking the product of determinants of the diagonal blocks. By subtracting from the second block row the first block row multiplied on the left by $F_2 G_1^{-1}$, we eliminate F_2 and get Λ_2 in place of G_2. The rest is by induction. $\qquad \square$

We get

$$\det \overline{A}_2 = \prod_{t=2}^{T} \det \Lambda_t, \tag{13}$$

where $\Lambda_2 = x_2 x_2'/a + 2I$, $\Lambda_t = x_t x_t'/a + 2I - \Lambda_{t-1}^{-1}$ for $t = 3, \ldots, T-1$, and $\Lambda_T = x_T x_T'/a + I - \Lambda_{T-1}^{-1}$.

Lemma 2. *All Λ_t, $t = 2, \ldots, T$, are well-defined symmetric positive definite matrices.*

Proof. Let us prove by induction that, for $t = 2, 3, \ldots, T-1$, Λ_t is symmetric positive definite and all its eigenvalues are greater than or equal to 1. The eigenvalues of Λ_2 are $2 + \|x_t\|^2/a, 2, \ldots, 2$ so the base of the induction holds.

If Λ_{t-1} satisfies the induction hypothesis, then it is invertible, Λ_{t-1}^{-1} is symmetric positive definite and all its eigenvalues are less than or equal to 1. The eigenvalues of $x_t x_t'/a + 2I$ are greater than or equal to 2. By the Courant-Fischer min-max theorem ([HJ13], Theorem 4.2.6) the eigenvalues of $x_t x_t'/a + 2I - \Lambda_{t-1}^{-1}$ are greater than or equal to 1.

A similar argument implies that eigenvalues of Λ_T are non-negative. However, if it is singular, then (13) implies that $\det \overline{A}_2 = 0$. Since $\overline{A}_2$ is positive definite, Λ_T is non-singular.   □

The matrix recursive formulas for Λ_t are difficult to analyse. We will use the following proposition to reduce them to scalar formulas.

Proposition 5. *If M is a symmetric positive semidefinite $(m \times m)$-matrix, then*

$$\det M \le \left(\frac{\operatorname{tr} M}{m} \right)^m;$$

if M is positive definite, then

$$\operatorname{tr} M^{-1} \ge \frac{m^2}{\operatorname{tr} M}.$$

(Notation $\operatorname{tr} M$ is used for the *trace* of a matrix M.)

Proof. Let $\lambda_1, \lambda_2, \ldots, \lambda_m$ be the eigenvalues of M, counting multiplicities. The inequalities for the arithmetic, geometric, and harmonic means

$$(\lambda_1 \lambda_2 \ldots \lambda_m)^{1/m} \le \frac{\lambda_1 + \lambda_2 + \ldots + \lambda_m}{m},$$

$$\frac{m}{\frac{1}{\lambda_1} + \frac{1}{\lambda_2} + \ldots + \frac{1}{\lambda_m}} \le \frac{\lambda_1 + \lambda_2 + \ldots + \lambda_m}{m}$$

(see Sect. 1.16 of [BB61]) imply the proposition.   □

Corollary 3. *The determinant of $\overline{A}_2$ satisfies $\det \overline{A}_2 \le (r_2 r_3 \ldots r_T)^n$, where the sequence r_t, $t = 2, 3, \ldots, T$, is defined by $r_2 = b + 2$, $r_t = b + 2 - 1/r_{t-1}$ for $t = 3, \ldots, T-1$, and $r_T = b + 1 - 1/r_{T-1}$ with $b = \frac{B^2}{an}$.*

Proof. It follows by induction that tr $\Lambda_t/n \leq r_t$. Indeed,

$$\frac{\text{tr } \Lambda_2}{n} \leq \frac{B^2}{an} + 2 = r_2,$$

$$\frac{\text{tr } \Lambda_t}{n} \leq \frac{B^2}{an} + 2 - \frac{\text{tr } \Lambda_{t-1}^{-1}}{n} \leq \frac{B^2}{na} + 2 - \frac{n}{\text{tr } \Lambda_{t-1}}$$

$$\leq b + 2 - \frac{1}{r_{t-1}} = r_t, \quad t = 3, \ldots, T-1,$$

and

$$\frac{\text{tr } \Lambda_T}{n} \leq \frac{B^2}{an} + 1 - \frac{\text{tr } \Lambda_{T-1}^{-1}}{n} \leq b + 1 - \frac{1}{r_{T-1}} = r_T.$$

We get $\det \Lambda_t \leq r_t^n$ and the corollary follows by (13). $\qquad\square$

The products $r_2 r_3 \ldots r_t$ form a recurrent sequence, which is easy to analyse.

Lemma 3. *The determinant of $\overline{A}_2$ satisfies $\det \left(\overline{A}_2 \right) \leq (d_T - d_{T-1})^n$, where the sequence d_t, $t = 0, 1, 2, \ldots$, is defined by $d_0 = 0$, $d_1 = 1$, and $d_t = (b+2)d_{t-1} - d_{t-2}$ for $t = 2, 3, \ldots$ with $b = \frac{B^2}{an}$.*

Proof. By induction we get $d_t = r_2 \ldots r_t$ for $t = 2, 3, \ldots, T-1$ and $d_T = r_2 \ldots r_{T-1}(r_T + 1) = r_2 \ldots r_{T-1} r_T + d_{T-1}$. $\qquad\square$

We need to study the behaviour of d_t.

Lemma 4. *For every $b > 0$ the sequence d_t from Lemma 3 satisfies*

$$d_T - d_{T-1} = \frac{1}{2} \left(\lambda_1^{T-1} \left(1 + \frac{b}{\sqrt{b^2 + 4b}} \right) + \lambda_2^{T-1} \left(1 - \frac{b}{\sqrt{b^2 + 4b}} \right) \right),$$

where $\lambda_1 = 1 + \frac{b}{2} + \frac{1}{2}\sqrt{b^2 + 4b}$, $\lambda_2 = 1 + \frac{b}{2} - \frac{1}{2}\sqrt{b^2 + 4b}$, and $T = 1, 2, \ldots$

Proof (Sketch). The recurrent formula for d_t can be written in the matrix form as

$$\begin{pmatrix} d_t \\ d_{t-1} \end{pmatrix} = R \begin{pmatrix} d_{t-1} \\ d_{t-2} \end{pmatrix}, \quad \text{where } R = \begin{pmatrix} b+2 & -1 \\ 1 & 0 \end{pmatrix},$$

and thus

$$\begin{pmatrix} d_T \\ d_{T-1} \end{pmatrix} = R^{T-1} \begin{pmatrix} d_1 \\ d_0 \end{pmatrix} = R^{T-1} \begin{pmatrix} 1 \\ 0 \end{pmatrix}.$$

In order to calculate R^{T-1}, we need to represent R in a convenient form. One can check that λ_1 and λ_2 are the eigenvalues of R and the corresponding eigenvectors can be chosen as

$$v_1 = (-\sqrt{b} - \sqrt{b+4}, \sqrt{b} - \sqrt{b+4})',$$
$$v_2 = (-\sqrt{b} + \sqrt{b+4}, \sqrt{b} + \sqrt{b+4})'.$$

We get $R = V\Lambda V^{-1}$, where Λ is the diagonal matrix with diagonal elements λ_1 and λ_2 and the columns of V are v_1 and v_2. Raising to power $T-1$ can be done as $R^{T-1} = V\Lambda^{T-1}V^{-1}$. The lemma follows by direct calculation. $\qquad\square$

The following simple facts will be used to upper bound $d_T - d_{T-1}$.

Lemma 5.   *For every $b > 0$ we get*

$$\frac{b}{\sqrt{b^2 + 4b}} \leq \frac{\sqrt{b}}{2}.$$

For every $b \geq 0$ we get

$$\frac{\lambda_2}{\lambda_1} = \lambda_2^2 = \left(1 + \frac{b}{2} - \frac{1}{2}\sqrt{b^2 + 4b}\right)^2 \leq 1,$$

and

$$\ln \lambda_1 \leq \sqrt{b},$$

where λ_1 and λ_2 are from Lemma 4.

Proof (Sketch). The first inequality follows from

$$\frac{b}{\sqrt{b^2 + 4b}} = \frac{\sqrt{b}}{\sqrt{b + 4}} \leq \frac{\sqrt{b}}{2}.$$

The equality involving lambdas can be checked by direct calculation. The inequality follows from

$$\frac{b}{2} \leq \sqrt{\frac{b^2}{4} + b} < 1 + \frac{b}{2}.$$

The last inequality follows by differentiation:

$$\frac{d}{db} \ln \lambda_1 = \frac{1}{\sqrt{b^2 + 4b}} \leq \frac{1}{\sqrt{4b}} = \frac{1}{2\sqrt{b}} = \frac{d}{db}\sqrt{b},$$

while for $b = 0$ we get $\ln \lambda_1 = \sqrt{b} = 0$.   □

We can now upper bound the extra term in Proposition 2 as

$$Y^2 \ln \frac{\det A}{a_1^n a^{n \cdot (T-1)}} \leq nY^2 \ln \left(\frac{TB^2}{a_1 n} + 1\right) + Y^2 \ln \det \overline{A}_2,$$

where $Y^2 \ln \det \overline{A}_2 \leq nY^2 \ln(d_T - d_{T-1})$ and

$$\ln(d_T - d_{T-1}) \leq \ln \frac{1}{2}\left(\lambda_1^{T-1}\left(1 + \frac{\sqrt{b}}{2}\right) + \lambda_2^{T-1}\right) =$$

$$- \ln 2 + (T-1)\ln \lambda_1 + \ln\left(1 + \frac{\sqrt{b}}{2}\right) + \ln\left(1 + \frac{1}{1 + \frac{\sqrt{b}}{2}}\left(\frac{\lambda_2}{\lambda_1}\right)^{T-1}\right) \leq$$

$$- \ln 2 + (T-1)\sqrt{b} + \frac{\sqrt{b}}{2} + \left(\frac{\lambda_2}{\lambda_1}\right)^{T-1},$$

where the last term is expanded in Lemma 5. Theorem 1 follows by substituting $b = \frac{B^2}{an}$.

6   Comments on the Proof

In this section we make some remarks about the proof.

Remark 1. Inequality (12) can be iterated, but that method would not lead to a good upper bound. For equal as, by using Stirling's formula we get

$$\ln \frac{\det A}{a^{nT}} \leq \ln \prod_{t=1}^{T} \left(\frac{tB^2}{an} + 1 \right)^n \approx n \ln T! + Tn \ln \frac{B^2}{an} \approx Tn \ln T - Tn \ln a \frac{n}{B^2}.$$

In order to get an extra term of the order $o(T)$, we must take $a(T)$ growing at about the same rate as T and thus ruin the growth of $a \cdot \sum_{t=2}^{T} \|u_t - u_{t-1}\|^2$.

Remark 2. A recurrent formula upper bounding the determinant of $\overline{A}_2$ can be obtained in a simpler way not involving Proposition 5 at a price of a small loss of quality.

If the diagonal blocks $x_t x_t'/a + cI$ in $\overline{A}_2$ are replaced by $\left(\frac{B^2}{a} + c \right) I$, the eigenvalues and the determinant may only increase. Indeed, each matrix $\frac{B^2}{a} I - x_t x_t'/a$ is positive semidefinite and adding the positive semidefinite block diagonal matrix will not increase the eigenvalues by the Courant-Fischer min-max theorem ([HJ13], Theorem 4.2.6). The resulting matrix turns out to be the Kronecker (tensor) product of I and the tridiagonal $(T \times T)$-matrix

$$\check{A}_2 = \begin{pmatrix} \frac{B^2}{a} + 2 & -1 & & & 0 \\ -1 & \frac{B^2}{a} + 2 & -1 & & \\ & \ddots & \ddots & \ddots & \\ & & -1 & \frac{B^2}{a} + 2 & -1 \\ 0 & & & -1 & \frac{B^2}{a} + 1 \end{pmatrix}.$$

Theorem 4.2.12 from [HJ94] on eigenvalues of the Kronecker product implies

$$\det \overline{A}_2 \leq \det(I \otimes \check{A}_2) = (\det I)^T (\det \check{A}_2)^n = (\det \check{A}_2)^n.$$

The determinant of $\check{A}_2$ can be calculated using the recurrence from [HJ13], Section 0.9.10 (this is effectively a non-block special case of Proposition 4). We get an upper bound on $\det \overline{A}_2$ similar to Lemma 3 but with $b = B^2/a$.

Then using Lemmas 4 and 5 we get an analogue of Theorem 1 with a slightly different α (which is not important) and $Y^2 B(T - 1/2) \frac{n}{\sqrt{a}}$ instead of $Y^2 B(T - 1/2)\sqrt{\frac{n}{a}}$. Applying Lemma 1 we get a counterpart of Corollary 1 but with the main extra term $\frac{3}{2^{2/3}} Y^{4/3} B^{2/3} n^{2/3} (T - 1/2)^{2/3} (v(T))^{1/3}$.

Acknowledgement. The author has been supported by the Leverhulme Trust through the grant RPG-2013-047 'Online self-tuning learning algorithms for handling historical information'. The author would like to thank Vladimir Vovk, Dmitry Adamskiy, and Vladimir V'yugin for useful discussions. Special thanks to Alexey Chernov, who helped to simplify the statement of the main result.

References

[AKCV12] Adamskiy, D., Koolen, W.M., Chernov, A., Vovk, V.: A closer look at adaptive regret. In: Bshouty, N.H., Stoltz, G., Vayatis, N., Zeugmann, T. (eds.) ALT 2012. LNCS, vol. 7568, pp. 290–304. Springer, Heidelberg (2012)

[AW01] Azoury, K.S., Warmuth, M.K.: Relative loss bounds for on-line density estimation with the exponential family of distributions. Mach. Learn. **43**, 211–246 (2001)

[BB61] Beckenbach, E.F., Bellman, R.E.: Inequalities. Springer, Heidelberg (1961)

[BK07a] Busuttil, S., Kalnishkan, Y.: Online regression competitive with changing predictors. In: Hutter, M., Servedio, R.A., Takimoto, E. (eds.) ALT 2007. LNCS (LNAI), vol. 4754, pp. 181–195. Springer, Heidelberg (2007)

[BK07b] Busuttil, S., Kalnishkan, Y.: Weighted kernel regression for predicting changing dependencies. In: Kok, J.N., Koronacki, J., Mantaras, R.L., Matwin, S., Mladenič, D., Skowron, A. (eds.) ECML 2007. LNAI, vol. 4701, pp. 535–542. Springer, Heidelberg (2007)

[CBCG05] Cesa-Bianchi, N., Conconi, A., Gentile, C.: A second-order perceptron algorithm. SIAM J. Comput. **34**(3), 640–668 (2005)

[For99] Forster, J.: On relative loss bounds in generalized linear regression. In: Ciobanu, G., Păun, G. (eds.) FCT 1999. LNCS, vol. 1684, pp. 269–280. Springer, Heidelberg (1999)

[HJ94] Horn, R.A., Johnson, C.R.: Topics in Matrix Analysis. Cambridge University Press, Cambridge (1994)

[HJ13] Horn, R.A., Johnson, C.R.: Matrix Analysis, 2nd edn. Cambridge University Press, Cambridge (2013)

[HW01] Herbster, M., Warmuth, M.K.: Tracking the best linear predictor. J. Mach. Learn. Res. **1**, 281–309 (2001)

[MVC15] Moroshko, E., Vaits, N., Crammer, K.: Second-order non-stationary on-line learning for regression. J. Mach. Learn. Res. **16**, 1481–1517 (2015)

[Sal06] Salkuyeh, D.K.: Comments on "A note on a three-term recurrence for a tridiagonal matrix". Appl. Math. Comput. **176**(2), 442–444 (2006)

[Vov90] Vovk, V.: Aggregating strategies. In: Proceedings of the 3rd Annual Workshop on Computational Learning Theory, pp. 371–383. Morgan Kaufmann, San Mateo (1990)

[Vov98] Vovk, V.: A game of prediction with expert advice. J. Comput. Syst. Sci. **56**, 153–173 (1998)

[Vov01] Vovk, V.: Competitive on-line statistics. Int. Stat. Rev. **69**(2), 213–248 (2001)

Things Bayes Can't Do

Daniil Ryabko[(✉)]

Inria, Villeneuve-d'Ascq, France
daniil@ryabko.net

Abstract. The problem of forecasting conditional probabilities of the next event given the past is considered in a general probabilistic setting. Given an arbitrary (large, uncountable) set C of predictors, we would like to construct a single predictor that performs asymptotically as well as the best predictor in C, on any data. Here we show that there are sets C for which such predictors exist, but none of them is a Bayesian predictor with a prior concentrated on C. In other words, there is a predictor with sublinear regret, but every Bayesian predictor must have a linear regret. This negative finding is in sharp contrast with previous results that establish the opposite for the case when one of the predictors in C achieves asymptotically vanishing error. In such a case, if there is a predictor that achieves asymptotically vanishing error for any measure in C, then there is a Bayesian predictor that also has this property, and whose prior is concentrated on (a countable subset of) C.

1 Introduction

The problem is probability forecasting in the most general setting. A sequence $x_1, \ldots, x_t, \ldots$ is generated by an unknown and arbitrary measure ν over the space of all infinite sequences. Here for simplicity we consider x_i coming from a finite set $\mathcal{X}$ (since we are after a negative result, this is not a limitation), but no other assumptions are made; in particular, x_i may be dependent and the dependence may be arbitrary. At each time step t a predictor ρ is required to give the conditional probabilities $\rho(x_{t+1}|x_1, \ldots, x_t)$ of the next outcome x_{t+1} given the observed past, before x_{t+1} is revealed and the process continues. We would like the predicted ρ probabilities to be as close as possible to the unknown ν probabilities $\nu(x_{t+1}|x_1, \ldots, x_t)$. The difference is measured with respect to some loss function L, which in this work we take to be the ν-expected average log loss (see the definitions below); however, it is clear that the main result applies more generally as well. Since ρ is required to give conditional probabilities given every possible sequence of past outcomes, ρ itself defines a probability measure over $\mathcal{X}^\infty$, and thus predictors and environments (mechanisms generating the data) are objects of the same kind.

To assist in the prediction task, we are given a set of predictors C. The performance of our predictor ρ is compared to that of the predictors in C, on sequences $x_1, \ldots, x_t \ldots$ generated by an arbitrary and unknown measure ν. Thus, we are interested in *regret* of using the predictor ρ as opposed to using the best (for this ν) predictor from C. The question we pose is whether this can be achieved

© Springer International Publishing Switzerland 2016
R. Ortner et al. (Eds.): ALT 2016, LNAI 9925, pp. 253–260, 2016.
DOI: 10.1007/978-3-319-46379-7_17

by some kind of combination of predictors in C, or whether it may be necessary to look elsewhere — outside of C (and its convex hull). More specifically, we are asking the question of *whether there exists a prior over C, such that the Bayesian predictor with this prior has the smallest possible regret (at least, asymptotically) with respect to the best measure in C.* The answer we obtain is negative: there are some classes C such that any Bayesian predictor has linear regret, while the best possible predictor has sublinear regret. Note that an example of such a set C is *necessarily uncountable*, since for a countable set C any prior with non-zero weights results in a Bayesian predictor with at most constant (in time) regret with respect to each predictor in C, and thus zero asymptotic average regret.

It is worth noting that the result is not about Bayesian versus non-Bayesian inference; in fact, in the last section of the paper it is argued that the negative finding applies not only to Bayesian predictors. Thus, the result means that in some cases, given a set of predictors, to construct a predictor that performs as close as possible to the best of them, one has to look elsewhere — somewhere completely outside of C (and its convex hull).

Prior Work. This is somewhat disturbing, since it contradicts both the intuition acquired from the literature on less general cases, and the positive results in related general settings. Specifically, the question has been studied extensively for specific families C of predictors, as well as in the non-probabilistic setting of prediction with expert advice. For specific families, the question dates back to Laplace who considered it for the case when C is the set of all Bernoulli i.i.d. measures, and, moreover, it is assumed that the measure ν to be predicted belongs to C. The latter assumption means that the problem is in the *realizable case*. The predictor suggested by Laplace is in fact a Bayesian predictor with the uniform prior over the parameter space. Moreover, a Bayesian predictor (with a different, Dirichlet, prior) is known to achieve optimal cumulative log loss in the realizable case of this problem, and, more generally for the case when C is the set of Markov processes of order k [4]. Bayesian predictors for a variety of other families are widely used, and their optimality can be often established even outside the Bayesian setting, including the settings where the measures to be predicted are outside the predictor's prior. For example, a Bayes mixture over all finite-memory processes predicts also all stationary processes [5].

In the setting of prediction with expert advice, one is given a finite set C of experts, and the predictor that competes with them is constructed that has a small regret (see, e.g., [1] for an extensive overview). A typical construction for such a combination of experts is obtained by attaching a weight to each expert's prediction, where the weight decreases exponentially with the loss accumulated — a construction that is clearly reminiscent of Bayesian updating.

In either setting, one is typically concerned with finite or countable classes, or with some specific parametric families of experts. The general case of the prediction problem has been formulated in [6], where it is shown that, if we are only interested in the realizable case, that is, the measure ν to be predicted belongs to C, then one can always do with a Bayes predictor. More precisely, if there is a predictor ρ whose error asymptotically vanishes with t on every $\nu \in C$, then there is a Bayesian predictor (with a prior over a measurable subset of C) that also has this property. Moreover, the prior can be always taken over a countable subset of C. This is shown without any assumptions on C whatsoever; in particular, C is not required to be measurable. The work [7] unifies the formulations of the realizable and the non-realizable (expert advice) problems, and also formulates the following semi-realizable problem to which the result of [6] is generalized: now ν is allowed to be any measure such that there is a measure μ in C whose error asymptotically vanishes on ν. Here, again, if anything works then there is a prior such that a Bayesian predictor with this prior works as well. The present work completes the picture (and answers an open question from [7]), showing that, unlike the realizable and semi-realizable case, the fully non-realizable case of the problem cannot always be solved by a Bayesian predictor.

The result of this work along with those cited above can be also put into the perspective of classical results on the consistency of Bayesian inference. Thus, in [2] it is shown that, roughly speaking, there may exist a prior with which a Bayesian predictor is inconsistent. In the context of the realizable case of the prediction problem, that is, if there is a consistent predictor, [6] shows that there always exists a prior with which a Bayesian predictor is consistent. Here we show that, in the nonrealizable case, there are cases where every Bayesian predictor with *every possible prior* is far from being as close as possible to being consistent.

2   Preliminaries

Let $\mathcal{X}$ be a finite set (the alphabet). Denote $\mathcal{X}^* := \cup_{k \in \mathbb{N}} \mathcal{X}^k$. The notation $x_{1..T}$ is used for $x_1, \dots, x_T$. We consider stochastic processes (probability measures) on $(\mathcal{X}^\infty, \mathcal{B})$ where $\mathcal{B}$ is the usual Borel sigma-field. We use E_μ for expectation with respect to a measure μ.

The loss we use in this paper is the expected log loss, which can be defined as the expected cumulative Kullback-Leibler divergence (KL divergence):

$$L_T(\nu, \rho) := E_\nu \sum_{t=1}^{T} \sum_{a \in \mathcal{X}} \nu(x_t = a | x_{1..t-1}) \log \frac{\nu(x_t = a | x_{1..t-1})}{\rho(x_t = a | x_{1..t-1})},$$

where ν, ρ are any measures over $\mathcal{X}^\infty$. In words, we take the expected (over data) cumulative (over time) KL divergence between ν- and ρ-conditional (on the past data) probability distributions of the next outcome. The expected log loss is easy to study because of the following identity

$$L_T(\nu, \rho) = -E_\nu \log \frac{\rho(x_{1..T})}{\nu(x_{1..T})}, \tag{1}$$

where on the right-hand side we have simply the KL divergence between measures μ and ρ restricted to the first T observations.

If we have two predictors μ and ρ, we can define the *regret* up to time T of (using the predictor) ρ as opposed to (using the predictor) μ on the measure ν (that is, ν generates the sequence to predict) as

$$R_T^\nu(\mu, \rho) := L_T(\nu, \rho) - L_T(\nu, \mu).$$

For a set of measures C one can also define the regret up to time T of ρ with respect to C on ν as $R_T^\nu(C, \rho) := \sup_{\mu \in C} R_T^\nu(\mu, \rho)$. For the case of a finite or compact C one often seeks to minimize $R_T^\nu(C, \rho)$. However, already for countably infinite sets C it may not be possible to bound $R_T^\nu(\mu, \rho)$ uniformly over C. This is why we will not make much use of $R_T^\nu(C, \rho)$, but rather work with its asymptotic version, defined as follows.

Define the asymptotic average regret as

$$\bar{R}^\nu(\mu, \rho) := \limsup_{T \to \infty} \frac{1}{T} R_T^\nu(\mu, \rho),$$

and

$$\bar{R}^\nu(C, \rho) := \sup_{\mu \in C} \bar{R}^\nu(\mu, \rho).$$

Note that, since we are after a negative result, working with asymptotic quantities only is not a limitation.

3   Main Result

Theorem 1. *There exist a set C of measures and a predictor ρ such that for every measure ν we have $\bar{R}^\nu(C, \rho) = 0$, yet for every Bayesian predictor φ with a prior concentrated on C there exists a measure ν such that $\bar{R}^\nu(C, \varphi) \geq c > 0$ where c is a (possibly large) constant. In other words, any Bayesian predictor must have a linear regret, while there exists a predictor with a sublinear regret.*

Remark 1 (Countable C). Note that any set C satisfying the theorem must necessarily be uncountable. Indeed, for any countable set $C = (\mu_k)_{k \in \mathbb{N}}$, take the Bayesian predictor $\varphi := \sum_{k \in \mathbb{N}} w_k \mu_k$, where w_k can be, for example, $\frac{1}{k(k+1)}$. Then, for any ν and any T, from (1) we obtain

$$L_T(\nu, \varphi) \leq -\log w_k + L_T(\nu, \mu_k).$$

That is to say, the regret of φ with respect to any μ_k is a constant independent of T (though it does depend on k), and thus for every ν we have $\bar{R}^\nu(C, \varphi) = 0$. It is worth nothing that the origins of the use of such countable mixtures for prediction trace back to [8,9].

Before passing to the proof of the theorem, we present here an informal exposition of the counterexample used in the proof and the idea why it works.

The example of the proof starts with taking a Bernoulli i.i.d. biased coin-toss measure, say, the one with the parameter $p = 1/3$, denoted β_p. Take then the set S of sequences typical for this measure, that is, all sequences for which the frequency of 1 is asymptotically $1/3$. We are interested in a predictor that predicts all measures concentrated on a single sequence from S, and we will ignore all other possible ν. The set of measures C is constructed as follows. Take any sequence $\mathbf{x}$ in S and define the measure $\mu_\mathbf{x}$ as the one that behaves exactly as Bernoulli $1/3$ on this sequence $\mathbf{x}$, and on all other sequences it behaves as some fixed (deterministic) measure. In other words, we have taken a Bernoulli 1/3 measure and split it into all its typical sequences, continuing it with a fixed arbitrary sequence everywhere else. Denote C the resulting set of measures. Note that the original measure β_p can be recovered with a Bayesian predictor from the set S. Indeed, it is enough to take β_p itself as a prior over S. Such a Baysian predictor will then be as good as β_p on any measure. Observe that for every $x_{1..T}$ it puts the weight of about $2^{-h_p T}$ on the set of sequences from S that start with $x_{1..T}$ (where h_p is the binary entropy for $p = 1/3$ of the example). The loss it achieves on measures from S is thus $h_p T$ and this is, in fact, also the minimax loss one can achieve on S. However, it is not possible to achieve the same loss (and to recover β_p) with a Bayesian predictor whose prior is concentrated on the set C. The trouble is that each measure μ in C attaches already too little weight to the sequence from S that it is based on. To be precise, the weight it attaches is the same $2^{-h_p T}$ that the Bayesian predictor gives to the corresponding deterministic sequence. Whatever extra prior weight a Bayesian predictor gives will only go towards regret; it cannot give a constant weight to each measure because there are uncountably many of them. In fact, the best it can do is give another $2^{-h_p T}$, which means that the resulting loss is going to be double the best possible one can obtain on measures from S with the best possible predictor, and, again, double of what one can obtain taking for each $\nu \in S$ the best $\mu \in C$. This results in linear regret, which is, as we show, is at least h_p in asymptotic average.

Proof. Let the alphabet $\mathcal{X}$ be ternary $\mathcal{X} = \{0, 1, 2\}$. For $\alpha \in (0, 1)$ denote $h(\alpha)$ the binary entropy $h(\alpha) := -\alpha \log \alpha - (1 - \alpha) \log(1 - \alpha)$. Fix an arbitrary $p \in (0, 1/2)$ and let β_p be the Bernoulli i.i.d. measure (produces only 0s and 1s) with parameter p. Let S be the set of sequences in $\mathcal{X}^\infty$ that have no 2s and such that the frequency of 1 is close to p:

$$S := \{\mathbf{x} \in \mathcal{X}^\infty : x_i \neq 2 \forall i, \text{ and}$$

$$\left| \frac{1}{t} |\{i = 1..t : x_i = 1\}| - p \right| \leq f(t) \text{ from some } t \text{ on}\},$$

where $f(t) = \log t / \sqrt{t}$. Clearly, $\beta_p(S) = 1$.

Define the set D_S as the set of all Dirac measures concentrated on a sequence from S, that is $D_S := \{\nu_{\mathbf{x}} : \nu_{\mathbf{x}}(\mathbf{x}) = 1, \mathbf{x} \in S\}$. Moreover, for each $\mathbf{x} \in S$ define the measure $\mu_{\mathbf{x}}$ as follows: $\mu_{\mathbf{x}}(X_{T+1}|X_{1..T}) = p$ coincides with β_p (that is, 1 w.p. p and 0 w.p. $1 - p$) if $X_{1..T} = x_{1..T}$, and outputs 2 w.p. 1 otherwise: $\mu_{\mathbf{x}}(2|X_{1..T}) = 1$ if $X_{1..T} \neq x_{1..T}$. That is, $\mu_{\mathbf{x}}$ behaves as β_p only on the sequence $\mathbf{x}$, and on all other sequences it just outputs 2 deterministically. This means, in particular, that many sequences have probability 0, and some probabilities above are defined conditionally on zero-probability events, but this is not a problem; see the remark in the end of the proof.

Finally, let $C := \{\mu_{\mathbf{x}} : \mathbf{x} \in S\}$. Next we will define the predictor ρ that predicts well all measures in C. First, introduce the measure δ that is going to take care of all the measures that output 2 w.p.1 from some time on. For each $a \in \mathcal{X}^*$ let δ_a be the measure that is concentrated on the sequence that starts with a and then consists of all 2s. Define $\delta := \sum_{a \in \mathcal{X}^*} w_a \delta_a$, where w_a are arbitrary positive numbers that sum to 1. Let also the measure β' be i.i.d. uniform over $\mathcal{X}$. Finally, define

$$\rho := 1/3(\beta_p + \beta' + \delta). \tag{2}$$

Next, let us show that, for every ν, the measure ρ predicts ν as well as any measure in C: its loss is an additive constant factor. In fact, it is enough to see this for all $\nu \in D_S$, and for all measures that output all 2s w.p.1 from some n on. For each ν in the latter set, from (2) the loss of ρ is upper-bounded by $\log 3 - \log w_a$, where w_a is the corresponding weight. This is a constant (does not depend on T). For the former set, again from the definition (2) for every $\nu_\mathbf{x} \in D_S$ we have (see also Remark 1)

$$L_T(\nu_\mathbf{x}, \rho) \leq \log 3 + L_T(\nu_\mathbf{x}, \beta_p) = Th_p + o(T),$$

while

$$\inf_{\mu \in C} L_T(\nu_\mathbf{x}, \mu) = L_T(\nu_\mathbf{x}, \mu_\mathbf{x}) = Th_p + o(T).$$

Therefore, for all ν we have

$$R_T^\nu(C, \rho) = o(T) \text{ and } \bar{R}^\nu(C, \rho) = 0.$$

Thus, we have shown that for every $\nu \in S$ there is a reasonably good predictor in C (here "reasonably good" means that its loss is linearly far from that of random guessing), and, moreover, there is a predictor ρ whose asymptotic regret is zero with respect to C.

Next we need to show that any Bayes predictor has $2Th_p + o(T)$ loss on at least some measure, which is double that of ρ, and which can be as bad as random guessing (or worse; depending on p). We will show something stronger: any Bayes predictor has asymptotic average loss of $2Th_p$ *on average* over all measures in S. So there will be many measures on which it is bad, not just one.

Let φ be any Bayesian predictor with its prior concentrated on C. Since C is parametrized by S, for any $x_{1..T} \in \mathcal{X}^T, T \in \mathbb{N}$ we can write $\varphi(x_{1..T}) = \int_S \mu_\mathbf{y}(x_{1..T})dW(\mathbf{y})$ where W is some measure over S (the prior). Moreover, using the notation $W(x_{1..k})$ for the W-measure of all sequences in S that start with $x_{1..k}$, from the definition of the measures $\mu_\mathbf{x}$, for every $\mathbf{x} \in S$ we have

$$\int_S \mu_\mathbf{y}(x_{1..T})dW(\mathbf{y}) = \int_{\mathbf{y} \in S: y_{1..T}=x_{1..T}} \beta_p(x_{1..T})dW(\mathbf{y}) = \beta_p(x_{1..T})W(x_{1..T}).$$

$$\tag{3}$$

We will consider the average

$$E_U \limsup \frac{1}{T} L_T(\nu_x, \varphi)dU(\mathbf{x}),$$

where the expectation is taken with respect to the measure U defined as the measure β_p restricted to S; in other words, U is approximately uniform over this set. Fix any $\nu_{\mathbf{x}} \in S$. Observe that $L_T(\nu_{\mathbf{x}}, \varphi) = -\log \varphi(x_{1..T})$. For the asymptotic regret, we can assume w.l.o.g. that the loss $L_T(\nu_{\mathbf{x}}, \varphi)$ is upper-bounded, say, by $T \log |\mathcal{X}|$ at least from some T on (for otherwise the statement already holds for φ). This allows us to use Fatou's lemma to bound

$$E_U \limsup \frac{1}{T} L_T(\nu_{\mathbf{x}}, \varphi)$$

$$\geq \limsup \frac{1}{T} E_U L_T(\nu_{\mathbf{x}}, \varphi) = \limsup -\frac{1}{T} E_U \log \varphi(\mathbf{x})$$

$$= \limsup -\frac{1}{T} E_U \log \beta_p(x_{1..T}) W(x_{1..T}), \quad (4)$$

where in the last equality we used (3). Moreover,

$$- E_U \log \beta_p(x_{1..T}) W(x_{1..T})$$

$$= - E_U \log \beta_p(x_{1..T}) + E_U \log \frac{U(x_{1..T})}{W(x_{1..T})} - E_U \log U(x_{1..T}) \geq 2h_p T + o(T),$$

$$(5)$$

where in the inequality we have used the fact that KL divergence is non-negative and the definition of U (that is, that $U = \beta_p|_S$). From this and (4) we obtain the statement of the theorem.

Finally, we remark that all the considered measures can be made non-zero everywhere by simply combining them with the uniform i.i.d. over $\mathcal{X}$ measure β', that is, taking for each measure ν the combination $\frac{1}{2}(\nu + \beta')$. This way all losses up to time T become bounded by $T \log |\mathcal{X}| + 1$, but the result still holds with a different constant. □

4 Discussion

We have shown that there are sets of predictors whose performance cannot be combined using any Bayesian predictor. While the result is stated for Bayesian predictors and for log loss, the example used to establish it seems to apply more generally. Indeed, it is clear that changing the loss won't change the result, only making the analysis slightly more cumbersome. More generally, the reason why any Bayesian predictor does not work in this example is that, since the set C considered is large, the predictor has to attach a quickly decreasing weight to each element in C, whereas each measure in C already attaches too little weight to the part of the event space of interest. In other words, the likelihood of the observations w.r.t. each predictor in C is too small to allow for any penalty. To combine predictors in C one has to *boost* the likelihood, rather than attach a penalty. Doing something like this would of course break a predictor on other sets C. This applies not only to Bayes. In fact, whatever general prediction principle one could consider, for example, the MDL principle (see, e.g., [3]), it appears to fail on the example presented. The same concerns expert-advice-style predictors. The problem, therefore, seems to be generic: to combine the predictive power of the predictors in the set, it is not enough to consider combinations of these predictors; rather, one has to look somewhere completely outside of C.

This suggests a more general question of how one can characterize those sets C of predictors for which it is enough to consider only the predictors inside C in order to effectively compete with them, as well as what can one do when this is not the case.

As far as prediction in the realizable case is concerned (i.e., $\nu \in C$, or, more generally, zero regret is possible), the following question remains open. It is shown in [6] that in this case, if any predictor works then there is a prior such that a Bayesian predictor with this prior works as well. However, this result is asymptotic. One can ask the question of whether the speed of convergence of the loss (to 0 in this case) can be matched by some Baysian predictor, if any of the measures in C is chosen to generate the data.

Another generalization is to the case when the best achievable regret is linear, either in the realizable case or in the non-realizable one. Thus, the set C of predictors may be so large that no predictor can have a sublinear regret. We still would like to have as small regret as possible with respect to this set. Since the set C is larger, the realizable case becomes more interesting. Can the smallest regret still be achieved with a Bayesian predictor?

Acknowledgements. The research presented in this paper was supported by CPER Nord-Pas de Calais/ FEDER DATA Advanced data science and technologies 2015–2020, by French Ministry of Higher Education and Research, Nord-Pas-de-Calais Regional Council.

References

1. Cesa-Bianchi, N., Lugosi, G.: Prediction, Learning, and Games. Cambridge University Press, Cambridge (2006)
2. Diaconis, P., Freedman, D.: On the consistency of Bayes estimates. Ann. Stat. **14**(1), 1–26 (1986)
3. Grünwald, P.: The Minimum Description Length Principle. MIT Press, Cambridge (2007)
4. Krichevsky, R.: Universal Compression and Retrival. Kluwer Academic Publications, Dordrecht (1993)
5. Ryabko, B.: Prediction of random sequences and universal coding. Probl. Inf. Transm. **24**, 87–96 (1988)
6. Ryabko, D.: On finding predictors for arbitrary families of processes. J. Mach. Learn. Res. **11**, 581–602 (2010)
7. Ryabko, D.: On the relation between realizable and non-realizable cases of the sequence prediction problem. J. Mach. Learn. Res. **12**, 2161–2180 (2011)
8. Solomonoff, R.J.: Complexity-based induction systems: comparisons and convergence theorems. IEEE Trans. Inf. Theor. IT **24**, 422–432 (1978)
9. Zvonkin, A.K., Levin, L.A.: The complexity of finite objects and the development of the concepts of information and randomness by means of the theory of algorithms. Russ. Math. Surv. **25**(6), 83–124 (1970)

On Minimaxity of Follow the Leader Strategy in the Stochastic Setting

Wojciech Kotłowski[(✉)]

Poznań University of Technology, Poznań, Poland
wkotlowski@cs.put.poznan.pl

Abstract. We consider the setting of prediction with expert advice with an additional assumption that each expert generates its losses i.i.d. according to some distribution. We first identify a class of "admissible" strategies, which we call permutation invariant, and show that every strategy outside this class will perform not better than some permutation invariant strategy. We then show that when the losses are binary, a simple Follow the Leader (FL) algorithm is the minimax strategy for this game, where minimaxity is simultaneously achieved for the expected regret, the expected redundancy, and the excess risk. Furthermore, FL has also the smallest regret, redundancy, and excess risk over all permutation invariant prediction strategies, simultaneously for all distributions over binary losses. When the losses are continuous in $[0, 1]$, FL remains minimax only when an additional trick called "loss binarization" is applied.

1 Introduction

In the game of prediction with expert advice [4,5], the learner sequentially decides on one of K experts to follow, and suffers loss associated with the chosen expert. The difference between the learner's cumulative loss and the cumulative loss of the best expert is called *regret*. The goal is to minimize the regret in the worst case over all possible loss sequences. A prediction strategy which achieves this goal (i.e., minimizes the worst-case regret) is called *minimax*. While there is no known solution to this problem in the general setting, it is possible to derive minimax algorithms for some special variants of this game: for 0/1 losses on the binary labels [4,5], for unit losses with fixed loss budget [2], and when $K = 2$ [9]. Interestingly, all these algorithms share a similar strategy of playing against a maximin adversary which assigns losses uniformly at random. They also have the *equalization* property: all data sequences lead to the same value of the regret. While this property makes them robust against the worst-case sequence, it also makes them over-conservative, preventing them from exploiting the case, when the actual data is not adversarially generated[1].

W. Kotłowski—This research was supported by the Polish National Science Centre under grant no. 2013/11/D/ST6/03050.

[1] There are various algorithms which combine almost optimal worst-case performance with good performance on "easy" sequences [6,10–13]; these algorithms, however, are not motivated from the minimax principle.

R. Ortner et al. (Eds.): ALT 2016, LNAI 9925, pp. 261–275, 2016.
DOI: 10.1007/978-3-319-46379-7_18

In this paper, we drop the analysis of worst-case performance entirely, and explore the minimax principle in a more constrained setting, in which the adversary is assumed to be *stochastic*. In particular, we associate with each expert k a fixed distribution P_k over loss values, and assume the observed losses of expert k are generated independently from P_k. The motivation behind our assumption is the practical usefulness of the stochastic setting: the data encountered in practice are rarely adversarial and can often be modeled as generated from a fixed (yet unknown) distribution. That is why we believe it is interesting to determine the minimax algorithm under this assumption. We immediately face two difficulties here. First, due to stochastic nature of the adversary, it is no longer possible to follow standard approaches of minimax analysis, such as backward induction [4,5] or sequential minimax duality [1,9], and we need to resort to a different technique. We define the notion of *permutation invariance* of prediction strategies. This let us identify a class of "admissible" strategies (which we call permutation invariant), and show that every strategy outside this class will perform not better than some permutation invariant strategy. Secondly, while the regret is a single, commonly used performance metric in the worst-case setting, the situation is different in the stochastic case. We know at least three potentially useful metrics in the stochastic setting: the *expected regret*, the *expected redundancy*, and the *excess risk* [8], and it is not clear, which of them should be used to define the minimax strategy.

Fortunately, it turns out that there exists a single strategy which is minimax with respect to all three metrics simultaneously. In the case of *binary* losses, which take out values from $\{0,1\}$, this strategy turns out to be the *Follow the Leader* (FL) algorithm, which chooses an expert with the smallest cumulative loss at a given trial (with ties broken randomly). Interestingly, FL is known to perform poorly in the worst-case, as its worst-case regret will grow linearly with T [5]. On the contrary, in the stochastic setting with binary losses, FL has the smallest regret, redundancy, and excess risk over all permutation invariant prediction strategies, *simultaneously for all distributions over binary losses!* In a more general case of continuous losses in the range $[0, 1]$, FL is provably suboptimal. However, by applying *binarization trick* to the losses [6], i.e. randomly setting them to $\{0,1\}$ such that the expectation matches the actual loss, and using FL on the binarized sequence (which results in the *binarized FL* strategy), we obtain the minimax strategy in the continuous case.

We note that when the excess risk is used as a performance metric, our setup falls into the framework of statistical decision theory [3,7], and the question we pose can be reduced to the problem of finding the minimax decision rule for a properly constructed loss function, which matches the excess risk on expectation. In principle, one could try to solve our problem by using the complete class theorem and search for the minimax rule within the class of (generalized) Bayesian decision rules. We initially followed this approach, but it turned out to be futile, as the class of distributions we are considering are all distributions in the range $[0, 1]$, and exploring prior distributions over such classes becomes very difficult. On the other hand, the analysis presented in this paper is relatively simple, and works not only for the excess risk, but also for the expected regret and the expected redundancy. To the best of our knowledge, both the results and the analysis presented here are novel.

The paper is organized as follows. In Sect. 2 we formally define the problem. The binary case is solved in Sect. 3, while Sect. 4 concerns continuous case. Section 5, concludes the paper and discusses an open problem.

2 Problem Setting

2.1 Prediction with Expert Advice in the Stochastic Setting

In the game of prediction with expert advice, at each trial $t = 1, \ldots, T$, the learner predicts with a distribution $\boldsymbol{w}_t = (w_{t,1}, \ldots, w_{t,K})$ over K experts. Then, the loss vector $\boldsymbol{\ell}_t = (\ell_{t,1}, \ldots, \ell_{t,K}) \in \mathcal{X}^K$ is revealed (where $\mathcal{X}$ is either $\{0, 1\}$ or $[0, 1]$), and the learner suffers loss:

$$\boldsymbol{w}_t \cdot \boldsymbol{\ell}_t = \sum_{k=1}^{K} w_{t,k} \ell_{t,k},$$

which can be interpreted as the expected loss the learner suffers by following one of the experts chosen randomly according to $\boldsymbol{w}_t$. Let $L_{t,k}$ denote the cumulative loss of expert k at the end of iteration t, $L_{t,k} = \sum_{q \leq t} \ell_{q,k}$. Let $\boldsymbol{\ell}^t$ abbreviate the sequence of losses $\boldsymbol{\ell}_1, \ldots, \boldsymbol{\ell}_t$. We will also use $\boldsymbol{\omega} = (\boldsymbol{w}_1, \ldots, \boldsymbol{w}_T)$ to denote the whole prediction strategy of the learner, having in mind that each distribution $\boldsymbol{w}_t$ is a function of the past $t-1$ outcomes $\boldsymbol{\ell}^{t-1}$. The performance of the strategy is measured by means of *regret*:

$$\sum_{t=1}^{T} \boldsymbol{w}_t \cdot \boldsymbol{\ell}_t - \min_k L_{T,k},$$

which is a difference between the algorithm's cumulative loss and the cumulative loss of the best expert. In the worst-case (adversarial) formulation of the problem, no assumption is made on the way the sequence of losses is generated, and hence the goal is then to find an algorithm which minimizes the worst-case regret over all possible sequences $\boldsymbol{\ell}^T$.

In this paper, we drop the analysis of worst-case performance and explore the minimax principle in the *stochastic* setting, defined as follows. We assume there are K distributions $\mathcal{P} = (P_1, \ldots, P_K)$ over $\mathcal{X}$, such that for each k, the losses $\ell_{t,k}$, $t = 1, \ldots, T$, are generated i.i.d. from P_k. Note that this implies that $\ell_{t,k}$ is independent from $\ell_{t',k'}$ whenever $t' \neq t$ or $k \neq k'$. The prediction strategy is then evaluated by means of *expected regret*:

$$R_{\text{eg}}(\boldsymbol{\omega}, \mathcal{P}) = \mathbb{E}\left[\sum_{t=1}^{T} \boldsymbol{w}_t(\boldsymbol{\ell}^{t-1}) \cdot \boldsymbol{\ell}_t - \min_k L_{T,k}\right],$$

where the expectation over the loss sequences ℓ^T with respect to distribution $\mathcal{P} = (P_1, \ldots, P_k)$, and we explicitly indicate the dependency of w_t on ℓ^{t-1}. However, the expected regret is not the only performance metric one can use in the stochastic setting. Instead of comparing the algorithm's loss to the loss of the best expert on the actual outcomes, one can choose the *best expected* expert as a comparator, which leads to a metric:

$$R_{\mathrm{ed}}(\boldsymbol{\omega}, \mathcal{P}) = \mathbb{E}\left[\sum_{t=1}^{T} \boldsymbol{w}_t(\boldsymbol{\ell}^{t-1}) \cdot \boldsymbol{\ell}_t\right] - \min_k \mathbb{E}\left[L_{T,k}\right],$$

which we call the *expected redundancy*, as it closely resembles a measure used in information theory to quantify the excess codelength of a prequential code [8]. Note that from Jensen's inequality it holds that $R_{\mathrm{ed}}(\boldsymbol{\omega}, \mathcal{P}) \geq R_{\mathrm{eg}}(\boldsymbol{\omega}, \mathcal{P})$ for any $\boldsymbol{\omega}$ and any $\mathcal{P}$, and the difference $R_{\mathrm{ed}}(\boldsymbol{\omega}, \mathcal{P}) - R_{\mathrm{eg}}(\boldsymbol{\omega}, \mathcal{P})$ is independent of $\boldsymbol{\omega}$ given fixed $\mathcal{P}$. This does not, however, imply that these metrics are equivalent in the minimax analysis, as the set of distributions $\mathcal{P}$ is chosen by the adversary *against* strategy $\boldsymbol{\omega}$ played by learner, and this choice will in general be different for the expected regret and the expected redundancy. Finally, the stochastic setting permits us to evaluate the prediction strategy by means of the *individual* rather than cumulative losses. Thus, it is reasonable to define the *excess risk* of the prediction strategy at time T:

$$R_{\mathrm{isk}}(\boldsymbol{\omega}, \mathcal{P}) = \mathbb{E}\left[\boldsymbol{w}_T(\boldsymbol{\ell}^{T-1}) \cdot \boldsymbol{\ell}_T\right] - \min_k \mathbb{E}\left[\ell_{T,k}\right],$$

a metric traditionally used in statistics to measure the accuracy of statistical procedures. Contrary to the expected regret and redundancy defined by means of cumulative losses of the prediction strategy, the excess risk concerns only a single prediction at a given trial; hence, without loss of generality, we can choose the last trial T in the definition. For the sake of clarity, we summarize the three measures in Table 1.

Table 1. Performance measures.

Expected regret:	$R_{\mathrm{eg}}(\boldsymbol{\omega}, \mathcal{P}) = \mathbb{E}\left[\sum_{t=1}^{T} \boldsymbol{w}_t(\boldsymbol{\ell}^{t-1}) \cdot \boldsymbol{\ell}_t - \min_k L_{T,k}\right]$
Expected redundancy:	$R_{\mathrm{ed}}(\boldsymbol{\omega}, \mathcal{P}) = \mathbb{E}\left[\sum_{t=1}^{T} \boldsymbol{w}_t(\boldsymbol{\ell}^{t-1}) \cdot \boldsymbol{\ell}_t\right] - \min_k \mathbb{E}\left[L_{T,k}\right]$
Excess risk:	$R_{\mathrm{isk}}(\boldsymbol{\omega}, \mathcal{P}) = \mathbb{E}\left[\boldsymbol{w}_T(\boldsymbol{\ell}^{T-1}) \cdot \boldsymbol{\ell}_T\right] - \min_k \mathbb{E}\left[\ell_{T,k}\right]$

Given performance measure R, we say that a strategy $\boldsymbol{\omega}^*$ is *minimax* with respect to R, if:

$$\sup_{\mathcal{P}} R(\boldsymbol{\omega}^*, \mathcal{P}) = \inf_{\boldsymbol{\omega}} \sup_{\mathcal{P}} R(\boldsymbol{\omega}, \mathcal{P}),$$

where the supremum is over all K-sets of distributions $(P_1, \ldots, P_K)$ on $\mathcal{X}$, and the infimum is over all prediction strategies.

2.2   Permutation Invariance

In this section, we identify a class of "admissible" prediction strategies, which we call permutation invariant. The name comes from the fact that the performance of these strategies remains invariant under any permutation of the distributions $\mathcal{P} = (P_1, \ldots, P_K)$. We show that for every prediction strategy, there exists a corresponding permutation invariant strategy with not worse expected regret, redundancy and excess risk in the worst-case with respect to all permutations of $\mathcal{P}$.

We say that a strategy $\boldsymbol{\omega}$ is *permutation invariant* if for any $t = 1, \ldots, T$, and any permutation $\sigma \in S_K$, where S_K denotes the group of permutations over $\{1, \ldots, K\}$, $\boldsymbol{w}_t(\sigma(\boldsymbol{\ell}^{t-1})) = \sigma(\boldsymbol{w}_t(\boldsymbol{\ell}^{t-1}))$, where for any vector $\boldsymbol{v} = (v_1, \ldots, v_K)$, we denote $\sigma(\boldsymbol{v}) = (v_{\sigma(1)}, \ldots, v_{\sigma(K)})$ and $\sigma(\boldsymbol{\ell}^{t-1}) = \sigma(\boldsymbol{\ell}_1), \ldots, \sigma(\boldsymbol{\ell}_{t-1})$. In words, if we σ-permute the indices of all past loss vectors, the resulting weight vector will be the σ-permutation of the original weight vector. Permutation invariant strategies are natural, as they only rely on the observed outcomes, not on the expert indices. The performance of these strategies remains invariant under any permutation of the distributions from $\mathcal{P}$:

Lemma 1. *Let $\boldsymbol{\omega}$ be permutation invariant. Then, for any permutation $\sigma \in S_K$, $\mathbb{E}_{\sigma(\mathcal{P})} \left[\boldsymbol{w}_t(\boldsymbol{\ell}^{t-1}) \cdot \boldsymbol{\ell}_t \right] = \mathbb{E}_{\mathcal{P}} \left[\boldsymbol{w}_t(\boldsymbol{\ell}^{t-1}) \cdot \boldsymbol{\ell}_t \right]$, and moreover $R(\boldsymbol{\omega}, \sigma(\mathcal{P})) = R(\boldsymbol{\omega}, \mathcal{P})$, where R is the expected regret, expected redundancy, or excess risk, and $\sigma(\mathcal{P}) = (P_{\sigma(1)}, \ldots, P_{\sigma(K)})$.*

Proof. We first show that the expected loss of the algorithm at any iteration $t = 1, \ldots, T$, is the same for both $\sigma(\mathcal{P})$ and $\mathcal{P}$:

$$
\begin{aligned}
\mathbb{E}_{\sigma(\mathcal{P})} \left[\boldsymbol{w}_t(\boldsymbol{\ell}^{t-1}) \cdot \boldsymbol{\ell}_t \right] &= \mathbb{E}_{\mathcal{P}} \left[\boldsymbol{w}_t(\sigma(\boldsymbol{\ell}^{t-1})) \cdot \sigma(\boldsymbol{\ell}_t) \right] = \mathbb{E}_{\mathcal{P}} \left[\sigma(\boldsymbol{w}_t(\boldsymbol{\ell}^{t-1})) \cdot \sigma(\boldsymbol{\ell}_t) \right] \\
&= \mathbb{E}_{\mathcal{P}} \left[\boldsymbol{w}_t(\boldsymbol{\ell}^{t-1}) \cdot \boldsymbol{\ell}_t \right],
\end{aligned}
$$

where the first equality is due to the fact, that permuting the distributions is equivalent to permuting the coordinates of the losses (which are random variables with respect to these distributions), the second equality exploits the permutation invariance of $\boldsymbol{\omega}$, while the third inequality uses a simple fact that the dot product is invariant under permuting both arguments. Therefore, the "loss of the algorithm" part of any of the three measures (regret, redundancy, risk) remains the same. To show that the "loss of the best expert" part of each measure is the same, note that for any $t = 1, \ldots, T$, $k = 1, \ldots, K$, $\mathbb{E}_{\sigma(\mathcal{P})} \left[\ell_{t,k} \right] = \mathbb{E}_{\mathcal{P}} \left[\ell_{t,\sigma(k)} \right]$, which implies:

$$
\min_k \mathbb{E}_{\sigma(\mathcal{P})} \left[\ell_{T,k} \right] = \min_k \mathbb{E}_{\mathcal{P}} \left[\ell_{T,\sigma(k)} \right] = \min_k \mathbb{E}_{\mathcal{P}} \left[\ell_{T,k} \right],
$$

$$
\min_k \mathbb{E}_{\sigma(\mathcal{P})} \left[L_{T,k} \right] = \min_k \mathbb{E}_{\mathcal{P}} \left[L_{T,\sigma(k)} \right] = \min_k \mathbb{E}_{\mathcal{P}} \left[L_{T,k} \right],
$$

$$
\mathbb{E}_{\sigma(\mathcal{P})} \left[\min_k L_{T,k} \right] = \mathbb{E}_{\mathcal{P}} \left[\min_k L_{T,\sigma(k)} \right] = \mathbb{E}_{\mathcal{P}} \left[\min_k L_{T,k} \right],
$$

so that the "loss of the best expert" parts of all measures are also the same for both $\sigma(\mathcal{P})$ and $\mathcal{P}$. □

We now show that permutation invariant strategies are "admissible" in the following sense:

Theorem 1. *For any strategy ω, there exists permutation invariant strategy $\widetilde{\omega}$, such that for any set of distributions $\mathcal{P}$,*

$$R(\widetilde{\omega}, \mathcal{P}) = \max_{\sigma \in S_K} R(\widetilde{\omega}, \sigma(\mathcal{P})) \leq \max_{\sigma \in S_K} R(\omega, \sigma(\mathcal{P})),$$

where R is either the expected regret, the expected redundancy or the excess risk. In particular, this implies that: $\sup_{\mathcal{P}} R(\widetilde{\omega}, \mathcal{P}) \leq \sup_{\mathcal{P}} R(\omega, \mathcal{P})$.

Proof. This first equality in the theorem immediately follows from Lemma 1. Define $\widetilde{\omega} = (\widetilde{w}_1, \ldots, \widetilde{w}_T)$ as:

$$\widetilde{w}_t(\boldsymbol{\ell}^{t-1}) = \frac{1}{K!} \sum_{\tau \in S_K} \tau^{-1}\Big(w_t(\tau(\boldsymbol{\ell}^{t-1}))\Big).$$

Note that $\widetilde{\omega}$ is a valid prediction strategy, since $\widetilde{w}_t$ is a function of $\boldsymbol{\ell}^{t-1}$ and a distribution over K experts ($\widetilde{w}_t$ is a convex combination of $K!$ distributions, so it is a distribution itself). Moreover, $\widetilde{\omega}$ is permutation invariant:

$$
\begin{aligned}
\widetilde{w}_t(\sigma(\boldsymbol{\ell}^{t-1})) &= \frac{1}{K!} \sum_{\tau \in S_K} \tau^{-1}\Big(w_t(\tau\sigma(\boldsymbol{\ell}^{t-1}))\Big) \\
&= \frac{1}{K!} \sum_{\tau \in S_K} (\tau\sigma^{-1})^{-1}\Big(w_t(\tau(\boldsymbol{\ell}^{t-1}))\Big) \\
&= \frac{1}{K!} \sum_{\tau \in S_K} \sigma\tau^{-1}\Big(w_t(\tau(\boldsymbol{\ell}^{t-1}))\Big) = \sigma(\widetilde{w}_t(\boldsymbol{\ell}^{t-1})),
\end{aligned}
$$

where the second equality is from replacing the summation index $\tau \mapsto \tau\sigma$. Now, note that the expected loss of $\widetilde{w}_t$ is:

$$
\begin{aligned}
\mathbb{E}_{\mathcal{P}}\left[\widetilde{w}_t(\boldsymbol{\ell}^{t-1}) \cdot \boldsymbol{\ell}_t\right] &= \frac{1}{K!} \sum_{\tau \in S_K} \mathbb{E}_{\mathcal{P}}\left[\tau^{-1}\left(w_t(\tau(\boldsymbol{\ell}^{t-1}))\right) \cdot \boldsymbol{\ell}_t\right] \\
&= \frac{1}{K!} \sum_{\tau \in S_K} \mathbb{E}_{\mathcal{P}}\left[w_t(\tau(\boldsymbol{\ell}^{t-1})) \cdot \tau(\boldsymbol{\ell}_t)\right] \\
&= \frac{1}{K!} \sum_{\tau \in S_K} \mathbb{E}_{\tau^{-1}(\mathcal{P})}\left[w_t(\boldsymbol{\ell}^{t-1}) \cdot \boldsymbol{\ell}_t\right] \\
&= \frac{1}{K!} \sum_{\sigma \in S_K} \mathbb{E}_{\sigma(\mathcal{P})}\left[w_t(\boldsymbol{\ell}^{t-1}) \cdot \boldsymbol{\ell}_t\right].
\end{aligned}
$$

Since the "loss of the best expert" parts of all three measures are invariant under any permutation of $\mathcal{P}$ (see the proof of Lemma 1), we have:

$$R(\widetilde{\omega}, \mathcal{P}) = \frac{1}{K!} \sum_{\sigma \in S_K} R(\omega, \sigma(\mathcal{P})) \leq \max_{\sigma \in S_K} R(\omega, \sigma(\mathcal{P})). \tag{1}$$

This implies that:

$$\sup_{\mathcal{P}} R(\widetilde{\omega}, \mathcal{P}) \leq \sup_{\mathcal{P}} \max_{\sigma \in S_K} R(\omega, \sigma(\mathcal{P})) = \sup_{\mathcal{P}} R(\omega, \mathcal{P}).$$

□

Theorem 1 states that strategies which are not permutation-invariant do not give any advantage over permutation-invariant strategies even when the set of distributions $\mathcal{P}$ is fixed (and even possibly known to the learner), but the adversary can permute the distributions to make the learner incur the most loss. We also note that one can easily show a slightly stronger version of Theorem 1: if strategy ω is not permutation invariant, and it holds that $R(\omega, \mathcal{P}) \neq R(\omega, \tau(\mathcal{P}))$ for some set of distributions and permutation τ, then $R(\widetilde{\omega}, \mathcal{P}) < \max_{\sigma \in S_K} R(\omega, \sigma(\mathcal{P}))$. This follows from the fact that the inequality in (1) becomes sharp.

2.3   Follow the Leader Strategy

Given loss sequence $\boldsymbol{\ell}^{t-1}$, let $N = |\argmin_{j=1,\ldots,K} L_{t-1,j}|$ be the size of the leader set at the beginning of trial t. We define the *Follow the Leader* (FL) strategy $\boldsymbol{w}_t^{\mathrm{fl}}$ such that $w_{t,k}^{\mathrm{fl}} = \frac{1}{N}$ if $k \in \argmin_j L_{t-1,j}$ and $w_{t,k}^{\mathrm{fl}} = 0$ otherwise. In other words, FL predicts with the current leader, breaking ties uniformly at random. It is straightforward to show that such defined FL strategy is permutation invariant.

3   Binary Losses

In this section, we set $\mathcal{X} = \{0, 1\}$, so that all losses are binary. In this case, each P_k is a Bernoulli distribution. Take any permutation invariant strategy ω. It follows from Lemma 1 that for any $\mathcal{P}$, and any permutation $\sigma \in S_K$, $\mathbb{E}_{\mathcal{P}}\left[\boldsymbol{w}_t(\boldsymbol{\ell}^{t-1}) \cdot \boldsymbol{\ell}_t\right] = \mathbb{E}_{\sigma(\mathcal{P})}\left[\boldsymbol{w}_t(\boldsymbol{\ell}^{t-1}) \cdot \boldsymbol{\ell}_t\right]$. Averaging this equality over all permutations $\sigma \in S_K$ gives:

$$\mathbb{E}_{\mathcal{P}}\left[\boldsymbol{w}_t(\boldsymbol{\ell}^{t-1}) \cdot \boldsymbol{\ell}_t\right] = \underbrace{\frac{1}{K!}\sum_{\sigma}\mathbb{E}_{\sigma(\mathcal{P})}\left[\boldsymbol{w}_t(\boldsymbol{\ell}^{t-1}) \cdot \boldsymbol{\ell}_t\right]}_{=:\ \overline{\mathrm{loss}}_t(\boldsymbol{w}_t, \mathcal{P})}, \tag{2}$$

where we defined $\overline{\mathrm{loss}}_t(\boldsymbol{w}_t, \mathcal{P})$ to be permutation-averaged expected loss at trial t. We now show the main result of this paper, a surprisingly strong property of FL strategy, which states that FL minimizes $\overline{\mathrm{loss}}_t(\boldsymbol{w}_t, \mathcal{P})$ *simultaneously* over all K-sets of distributions. Hence, FL is not only optimal in the worst case, but is actually optimal for permutation-averaged expected loss *for any* $\mathcal{P}$, even if $\mathcal{P}$ is known to the learner! The consequence of this fact (by (2)) is that *FL has the smallest expected loss among all permutation invariant strategies for any* $\mathcal{P}$ (again, even if $\mathcal{P}$ is known to the learner).

Theorem 2. *Let* $\omega^{\text{fl}} = (w_1^{\text{fl}}, \ldots, w_T^{\text{fl}})$ *be the FL strategy. Then, for any K-set of distributions* $\mathcal{P} = (P_1, \ldots, P_K)$ *over binary losses, for any strategy* $\omega = (w_1, \ldots, w_T)$, *and any* $t = 1, \ldots, T$:

$$\overline{\text{loss}}_t(w_t^{\text{fl}}, \mathcal{P}) \leq \overline{\text{loss}}_t(w_t, \mathcal{P}).$$

Proof. For any distribution P_k over binary losses, let $p_k := P_k(\ell_{t,k} = 1) = \mathbb{E}_{P_k}[\ell_{t,k}]$. We have:

$$\overline{\text{loss}}_t(w_t, \mathcal{P}) = \frac{1}{K!} \sum_\sigma \mathbb{E}_{\sigma(\mathcal{P})} \left[w_t(\ell^{t-1}) \cdot \ell_t \right] \tag{3}$$

$$= \frac{1}{K!} \sum_\sigma \mathbb{E}_{\sigma(\mathcal{P})} \left[w_t(\ell^{t-1}) \right] \cdot \mathbb{E}_{\sigma(\mathcal{P})} [\ell_t]$$

$$= \frac{1}{K!} \sum_\sigma \sum_{\ell^{t-1}} \left(\prod_{k=1}^K p_{\sigma(k)}^{L_{t-1,k}} (1-p_{\sigma(k)})^{t-1-L_{t-1,k}} \right) \left(\sum_{k=1}^K w_{t,k}(\ell^{t-1}) p_{\sigma(k)} \right)$$

$$= \frac{1}{K!} \sum_{\ell^{t-1}} \sum_{k=1}^K w_{t,k}(\ell^{t-1}) \underbrace{\left(\sum_\sigma \prod_{j=1}^K p_{\sigma(j)}^{L_{t-1,j}} (1-p_{\sigma(j)})^{t-1-L_{t-1,j}} p_{\sigma(k)} \right)}_{=: \ \overline{\text{loss}}_t(w_t, \mathcal{P}|\ell^{t-1})},$$

where in the second equality we used the fact that w_t depends on ℓ^{t-1} and does not depend on ℓ_t. Fix ℓ^{t-1} and consider the term $\overline{\text{loss}}_t(w_t, \mathcal{P}|\ell^{t-1})$. This term is linear in w_t, hence it is minimized by $w_t = e_k$ for some $k = 1, \ldots, K$, where e_k is the k-th standard basis vector with 1 on the k-th coordinate, and zeros on the remaining coordinates. We will drop the trial index and use a shorthand notation $L_j = L_{t-1,j}$, for $j = 1, \ldots, K$, and $L = (L_1, \ldots, L_K)$. In this notation, we rewrite $\overline{\text{loss}}_t(w_t, \mathcal{P}|\ell^{t-1})$ as:

$$\overline{\text{loss}}_t(w_t, \mathcal{P}|\ell^{t-1}) = \sum_{k=1}^K w_{t,k}(\ell^{t-1}) \left(\sum_\sigma \prod_{j=1}^K p_{\sigma(j)}^{L_j} (1 - p_{\sigma(j)})^{t-1-L_j} p_{\sigma(k)} \right). \tag{4}$$

We will show that for any $\mathcal{P}$, and any ℓ^{t-1} (and hence, any L), $\overline{\text{loss}}_t(w_t, \mathcal{P}|\ell^{t-1})$ is minimized by setting $w_t = e_{k^*}$ for any $k^* \in \arg\min_j L_j$. In other words, we will show that for any $\mathcal{P}$, L, any $k^* \in \arg\min_j L_j$, and any $k = 1, \ldots, K$,

$$\overline{\text{loss}}_t(e_{k^*}, \mathcal{P}|\ell^{t-1}) \leq \overline{\text{loss}}_t(e_k, \mathcal{P}|\ell^{t-1}).$$

or equivalently, using (4), that for any $\mathcal{P}$, L, $k^* \in \arg\min_j L_j$, and $k = 1, \ldots, K$,

$$\sum_\sigma \prod_{j=1}^K p_{\sigma(j)}^{L_j} (1 - p_{\sigma(j)})^{t-1-L_j} p_{\sigma(k^*)} \leq \sum_\sigma \prod_{j=1}^K p_{\sigma(j)}^{L_j} (1 - p_{\sigma(j)})^{t-1-L_j} p_{\sigma(k)}. \tag{5}$$

We proceed by induction on K. Take $K = 2$ and note that when $k^* = k$, there is nothing to prove, as both sides of (5) are identical. Therefore, without loss of generality, assume $k^* = 1$ and $k = 2$, which implies $L_1 \leq L_2$. Then, (5) reduces to:

$$p_1^{L_1} p_2^{L_2} (1 - p_1)^{t-1-L_1} (1 - p_2)^{t-1-L_2} p_1$$
$$+ p_2^{L_1} p_1^{L_2} (1 - p_2)^{t-1-L_1} (1 - p_1)^{t-1-L_2} p_2$$
$$\leq p_1^{L_1} p_2^{L_2} (1 - p_1)^{t-1-L_1} (1 - p_2)^{t-1-L_2} p_2$$
$$+ p_2^{L_1} p_1^{L_2} (1 - p_2)^{t-1-L_1} (1 - p_1)^{t-1-L_2} p_1,$$

After rearranging the terms, it amounts to show that:

$$(p_1 p_2)^{L_1} \left((1 - p_1)(1 - p_2) \right)^{t-1-L_2} (p_1 - p_2)$$
$$\times \left((p_2(1 - p_1))^{L_2 - L_1} - (p_1(1 - p_2))^{L_2 - L_1} \right) \leq 0.$$

But this will hold if:

$$(p_1 - p_2)\left((p_2(1 - p_1))^{L_2 - L_1} - (p_1(1 - p_2))^{L_2 - L_1} \right) \leq 0. \tag{6}$$

If $L_1 = L_2$, (6) clearly holds; therefore assume $L_1 < L_2$. We prove the validity of (6) by noticing that:

$$p_2(1 - p_1) > p_1(1 - p_2) \quad \Longleftrightarrow \quad p_2 > p_1,$$

which means that the two factors of the product on the left-hand side of (6) have the opposite sign (when $p_1 \neq p_2$) or are zero at the same time (when $p_1 = p_2$). Hence, we proved (6), which implies (5) when $k^* = 1$ and $k = 2$. The opposite case $k^* = 2, k = 1$ with $L_2 \leq L_1$ can be shown with exactly the same line of arguments by simply exchanging the indices 1 and 2.

Now, we assume (5) holds for $K - 1 \geq 2$ experts and any $\mathcal{P} = (P_1, \ldots, P_{K-1})$, any $L = (L_1, \ldots, L_{K-1})$, any $k^* \in \operatorname{argmin}_{j=1,\ldots,K-1} L_j$, and any $k = 1, \ldots, K - 1$, and we show that it also holds for K experts. Take any $k^* \in \operatorname{argmin}_{j=1,\ldots,K} L_j$, and any $k = 1, \ldots, K$. Without loss of generality, assume that $k^* \neq 1$ and $k \neq 1$ (it is always possible find expert different than k^* and k, because there are $K \geq 3$ experts). We expand the sum over permutations on the left-hand side of (5) with respect to the value of $\sigma(1)$:

$$\sum_{s=1}^{K} p_s^{L_1} (1 - p_s)^{t-1-L_1} \sum_{\sigma: \sigma(1)=s} \prod_{j=2}^{K} p_{\sigma(j)}^{L_j} (1 - p_{\sigma(j)})^{t-1-L_j} p_{\sigma(k^*)},$$

and we also expand the sum on the right-hand side of (5) in the same way. To prove (5), it suffices to show that every term in the sum over s on the left-hand side is not greater than the corresponding term in the sum on the right-hand side, i.e. to show that for any $s = 1, \ldots, K$,

$$\sum_{\sigma:\,\sigma(1)=s}\prod_{j=2}^{K}p_{\sigma(j)}^{L_j}(1-p_{\sigma(j)})^{t-1-L_j}p_{\sigma(k^*)} \leq \sum_{\sigma:\,\sigma(1)=s}\prod_{j=2}^{K}p_{\sigma(j)}^{L_j}(1-p_{\sigma(j)})^{t-1-L_j}p_{\sigma(k)}.$$

$$(7)$$

We now argue that this inequality follows directly from the inductive assumption by dropping L_1 and P_s, and applying (5) to such a $(K-1)$-expert case. More precisely, note that the sum on both sides of (7) goes over all permutations on indices $(1,\ldots,s-1,s+1,\ldots,K)$ and since $k,k^* \neq 1$, $k^* \in \operatorname{argmin}_{j=2,\ldots,K} L_j$ and $k \geq 2$. Hence, applying (5) to $K-1$ expert case with $K-1$ distributions $(P_1,P_2,\ldots,P_{s-1},P_{s+1},\ldots,P_K)$ (or any permutation thereof), and $K-1$ integers $(L_2,\ldots,L_K)$ immediately implies (7).

Thus, we proved (5) which states that $\overline{\text{loss}}_t(w_t,\mathcal{P}|\ell^{t-1})$ is minimized by any leader $k^* \in \operatorname{argmin}_j L_j$, where $L_j = \sum_{q=1}^{t-1}\ell_{q,j}$. This means $\overline{\text{loss}}_t(w_t,\mathcal{P}|\ell^{t-1})$ is also minimized by the FL strategy w_t^{fl}, which distributes its mass uniformly over all leaders. Since FL minimizes $\overline{\text{loss}}_t(w_t,\mathcal{P}|\ell^{t-1})$ for any ℓ^{t-1}, by (3) it also minimizes $\overline{\text{loss}}_t(w_t,\mathcal{P})$.   □

Note that the proof did not require uniform tie breaking over leaders, as any distribution over leaders would work as well. Uniform distribution, however, makes the FL strategy permutation invariant.

The consequence of Theorem 2 is the following corollary which states the minimaxity of FL strategy for binary losses:

Corollary 1. *Let* $\omega^{\text{fl}} = (w_1^{\text{fl}},\ldots,w_T^{\text{fl}})$ *be the FL strategy. Then, for any* $\mathcal{P}$ *over binary losses, and any permutation invariant strategy* ω:

$$R(\omega^{\text{fl}},\mathcal{P}) \leq R(\omega,\mathcal{P}).$$

where R *is the expected regret, expected redundancy, or excess risk. This implies:*

$$\sup_{\mathcal{P}} R(\omega^{\text{fl}},\mathcal{P}) = \inf_{\omega}\sup_{\mathcal{P}} R(\omega,\mathcal{P}),$$

where the supremum is over all distributions on binary losses, and the infimum over all (not necessarily permutation invariant) strategies.

Proof. The second statement immediately follows from the first statement and Theorem 1. For the first statement, note that the "loss of the best expert" part of each measure only depends on $\mathcal{P}$. Hence, we only need to show that for any $t = 1,\ldots,T$,

$$\mathbb{E}_{\mathcal{P}}\left[w_t^{\text{fl}} \cdot \ell_t\right] \leq \mathbb{E}_{\mathcal{P}}\left[w_t \cdot \ell_t\right].$$

Since w_t^{fl} and w_t are permutation invariant, Lemma 1 shows that $\mathbb{E}_{\mathcal{P}}\left[w_t^{\text{fl}} \cdot \ell_t\right] = \overline{\text{loss}}_t(w_t^{\text{fl}},\mathcal{P})$, and similarly, $\mathbb{E}_{\mathcal{P}}\left[w_t \cdot \ell_t\right] = \overline{\text{loss}}_t(w_t,\mathcal{P})$. Application of Theorem 2 finishes the proof.

4 Continuous Losses

In this section, we consider the general case $\mathcal{X} = [0, 1]$ of continuous loss vectors. We give a modification of FL and prove its minimaxity. We later justify the modification by arguing that the plain FL strategy is not minimax for continuous losses.

4.1 Binarized FL

The modification of FL is based on the procedure we call *binarization*. A similar trick has already been used in [6] to deal with non-integer losses in a different context. We define a binarization of any loss value $\ell_{t,k} \in [0,1]$ as a Bernoulli random variable $b_{t,k}$ which takes out value 1 with probability $\ell_{t,k}$ and value 0 with probability $1 - \ell_{t,k}$. In other words, we replace each non-binary loss $\ell_{t,k}$ by a random binary outcome $b_{t,k}$, such that $\mathbb{E}[b_{t,k}] = \ell_{t,k}$. Note that if $\ell_{t,k} \in \{0,1\}$, then $b_{t,k} = \ell_{t,k}$, i.e. binarization has no effect on losses which are already binary. Let us also define $\boldsymbol{b}_t = (b_{t,1}, \ldots, b_{t,K})$, where all K Bernoulli random variables $b_{t,k}$ are independent. Similarly, $\boldsymbol{b}^t$ will denote a binary loss sequence $\boldsymbol{b}_1, \ldots, \boldsymbol{b}_t$, where the binarization procedure was applied independently (with a new set of Bernoulli variables) for each trial t. Now, given the loss sequence $\boldsymbol{\ell}^{t-1}$, we define the *binarized FL* strategy $\boldsymbol{w}^{\mathrm{bfl}}$ by:

$$\boldsymbol{w}_t^{\mathrm{bfl}}(\boldsymbol{\ell}^{t-1}) = \mathbb{E}_{\boldsymbol{b}^{t-1}}\left[\boldsymbol{w}_t^{\mathrm{fl}}(\boldsymbol{b}^{t-1})\right],$$

where $\boldsymbol{w}_t^{\mathrm{fl}}(\boldsymbol{b}^{t-1})$ is the standard FL strategy applied to binarized losses $\boldsymbol{b}^{t-1}$, and the expectation is over internal randomization of the algorithm (binarization variables).

Note that if the set of distributions $\mathcal{P}$ has support only on $\{0, 1\}$, then $\boldsymbol{w}_t^{\mathrm{bfl}} \equiv \boldsymbol{w}_t^{\mathrm{fl}}$. On the other hand, these two strategies may differ significantly for non-binary losses. However, we will show that for any K-set of distributions $\mathcal{P}$ (with support in $[0, 1]$), $\boldsymbol{w}_t^{\mathrm{bfl}}$ will behave in the same way as $\boldsymbol{w}_t^{\mathrm{fl}}$ would behave on some particular K-set of distributions over binary losses. To this end, we introduce *binarization of a K-set of distributions* $\mathcal{P}$, defined as $\mathcal{P}^{\mathrm{bin}} = (P_1^{\mathrm{bin}}, \ldots, P_K^{\mathrm{bin}})$, where P_k^{bin} is a distribution with support $\{0, 1\}$ such that:

$$\mathbb{E}_{P_k^{\mathrm{bin}}}[\ell_{t,k}] = P_k^{\mathrm{bin}}(\ell_{t,k} = 1) = \mathbb{E}_{P_k}[\ell_{t,k}].$$

In other words, P_k^{bin} is a Bernoulli distribution which has the same expectation as the original distribution (over continuous losses) P_k. We now show the following results:

Lemma 2. *For any K-set of distributions $\mathcal{P} = (P_1, \ldots, P_K)$ with support on $\mathcal{X} = [0, 1]$,*

$$\mathbb{E}_{\boldsymbol{\ell}^t \sim \mathcal{P}}\left[\boldsymbol{w}_t^{\mathrm{bfl}}(\boldsymbol{\ell}^{t-1}) \cdot \boldsymbol{\ell}_t\right] = \mathbb{E}_{\boldsymbol{\ell}^t \sim \mathcal{P}^{\mathrm{bin}}}\left[\boldsymbol{w}_t^{\mathrm{fl}}(\boldsymbol{\ell}^{t-1}) \cdot \boldsymbol{\ell}_t\right].$$

Proof. Let p_k be the expectation of $\ell_{t,k}$ according to either P_k or P_k^{bin}, $p_k :=$ $\mathbb{E}_{P_k}[\ell_{t,k}] = \mathbb{E}_{P_k^{\text{bin}}}[\ell_{t,k}]$. Since for any prediction strategy $\boldsymbol{\omega}$, $\boldsymbol{w}_t$ depends on $\boldsymbol{\ell}^{t-1}$ and does not depend on $\boldsymbol{\ell}_t$, we have:

$$\mathbb{E}_{\mathcal{P}}\left[\boldsymbol{w}_t^{\text{bfl}} \cdot \boldsymbol{\ell}_t\right] = \mathbb{E}_{\mathcal{P}}\left[\boldsymbol{w}_t^{\text{bfl}}\right] \cdot \mathbb{E}_{\mathcal{P}}\left[\boldsymbol{\ell}_t\right] = \mathbb{E}_{\mathcal{P}}\left[\boldsymbol{w}_t^{\text{bfl}}\right] \cdot \boldsymbol{p},$$

where $\boldsymbol{p} = (p_1, \ldots, p_K)$. Similarly,

$$\mathbb{E}_{\mathcal{P}^{\text{bin}}}\left[\boldsymbol{w}_t^{\text{fl}} \cdot \boldsymbol{\ell}_t\right] = \mathbb{E}_{\mathcal{P}^{\text{bin}}}\left[\boldsymbol{w}_t^{\text{fl}}\right] \cdot \boldsymbol{p}.$$

Hence, we only need to show that $\mathbb{E}_{\mathcal{P}}\left[\boldsymbol{w}_t^{\text{bfl}}\right] = \mathbb{E}_{\mathcal{P}^{\text{bin}}}\left[\boldsymbol{w}_t^{\text{fl}}\right]$. This holds because $\boldsymbol{w}_t^{\text{bfl}}$ "sees" only the binary outcomes resulting from the joint distribution of $\mathcal{P}$ and the distribution of binarization variables:

$$\mathbb{E}_{\boldsymbol{\ell}^{t-1} \sim \mathcal{P}}\left[\boldsymbol{w}_t^{\text{bfl}}(\boldsymbol{\ell}^{t-1})\right] = \mathbb{E}_{\boldsymbol{\ell}^{t-1} \sim \mathcal{P}, \boldsymbol{b}^{t-1}}\left[\boldsymbol{w}_t^{\text{fl}}(\boldsymbol{b}^{t-1})\right],$$

and for any $b_{t,k}$, the probability (jointly over P_k and the binarization variables) of $b_{t,k} = 1$ is the same as probability of $\ell_{t,k} = 1$ over the distribution P_k^{bin}:

$$
\begin{aligned}
P(b_{t,k} = 1) &= \int_{[0,1]} P(b_{t,k} = 1 | \ell_{t,k}) P_k(\ell_{t,k}) \mathrm{d}\ell_{t,k} \\
&= \int_{[0,1]} \ell_{t,k} P_k(\ell_{t,k}) \mathrm{d}\ell_{t,k} = p_t = P^{\text{bin}}(\ell_{t,k} = 1). \quad (8)
\end{aligned}
$$

Hence,

$$\mathbb{E}_{\boldsymbol{\ell}^{t-1} \sim \mathcal{P}, \boldsymbol{b}^{t-1}}\left[\boldsymbol{w}_t^{\text{fl}}(\boldsymbol{b}^{t-1})\right] = \mathbb{E}_{\boldsymbol{\ell}^{t-1} \sim \mathcal{P}^{\text{bin}}}\left[\boldsymbol{w}_t^{\text{fl}}(\boldsymbol{\ell}^t)\right].$$

$\square$

Lemma 3. *For any K-set of distributions $\mathcal{P} = (P_1, \ldots, P_K)$ with support on $\mathcal{X} = [0,1]$,*

$$R(\boldsymbol{\omega}^{\text{bfl}}, \mathcal{P}) \le R(\boldsymbol{\omega}^{\text{fl}}, \mathcal{P}^{\text{bin}}),$$

where R is either the expected regret, the expected redundancy, or the excess risk.

Proof. Lemma 2 shows that the expected loss of $\boldsymbol{\omega}^{\text{bfl}}$ on $\mathcal{P}$ is the same as the expected loss of $\boldsymbol{\omega}^{\text{fl}}$ on $\mathcal{P}^{\text{bin}}$. Hence, to prove the inequality, we only need to consider the "loss of the best expert" part of each measure. For the expected redundancy, and the expected regret, it directly follows from the definition of $\mathcal{P}^{\text{bin}}$ that for any t, k, $\mathbb{E}_{\mathcal{P}}[\ell_{t,k}] = \mathbb{E}_{\mathcal{P}^{\text{bin}}}[\ell_{t,k}]$, hence $\min_k \mathbb{E}_{\mathcal{P}}[\ell_{T,k}] = \min_k \mathbb{E}_{\mathcal{P}^{\text{bin}}}[\ell_{T,k}]$, and similarly, $\min_k \mathbb{E}_{\mathcal{P}}[L_{T,k}] = \min_k \mathbb{E}_{\mathcal{P}^{\text{bin}}}[L_{T,k}]$. Thus, for the expected redundancy and the excess risk, the lemma actually holds with equality.

For the expected regret, we will show that $\mathbb{E}_{\mathcal{P}}[\min_k L_{T,k}] \ge \mathbb{E}_{\mathcal{P}^{\text{bin}}}[\min_k L_{T,k}]$, which will finish the proof. Denoting $B_{T,k} = \sum_{t=1}^{T} b_{t,k}$, we have:

$$
\begin{aligned}
\mathbb{E}_{\boldsymbol{\ell}^T \sim \mathcal{P}^{\text{bin}}}\left[\min_k L_{T,k}\right] &= \mathbb{E}_{\boldsymbol{\ell}^T \sim \mathcal{P}, \boldsymbol{b}^T}\left[\min_k B_{T,k}\right] \\
&\le \mathbb{E}_{\boldsymbol{\ell}^T \sim \mathcal{P}}\left[\min_k \mathbb{E}_{\boldsymbol{b}^T}[B_{T,k} | \boldsymbol{\ell}^T]\right] \\
&= \mathbb{E}_{\boldsymbol{\ell}^T \sim \mathcal{P}}[\min_k L_{T,k}],
\end{aligned}
$$

where the first equality is from the fact that for any $b_{t,k}$, the probability (jointly over P_k and the binarization variables) of $b_{t,k} = 1$ is the same as probability of $\ell_{t,k} = 1$ over the distribution P_k^{bin} (see (8) in the proof of Lemma 2), while the inequality follows from Jensen's inequality applied to the concave function $\min(\cdot)$. □

Theorem 3. *Let* $\boldsymbol{\omega}^{\text{bfl}} = (\boldsymbol{w}_1^{\text{bfl}}, \ldots, \boldsymbol{w}_T^{\text{bfl}})$ *be the binarized FL strategy. Then:*

$$\sup_{\mathcal{P}} R(\boldsymbol{\omega}^{\text{bfl}}, \mathcal{P}) = \inf_{\boldsymbol{\omega}} \sup_{\mathcal{P}} R(\boldsymbol{\omega}, \mathcal{P}),$$

where R is the expected regret, expected redundancy, or excess risk, the supremum is over all K-sets of distributions on $[0, 1]$, and the infimum is over all prediction strategies.

Proof. Lemma 3 states that for any K-set of distributions $\mathcal{P}$, $R(\boldsymbol{\omega}^{\text{bfl}}, \mathcal{P}) \leq R(\boldsymbol{\omega}^{\text{fl}}, \mathcal{P}^{\text{bin}})$. Furthermore, since $\boldsymbol{\omega}^{\text{bfl}}$ is the same as $\boldsymbol{\omega}^{\text{fl}}$ when all the losses are binary, $R(\boldsymbol{\omega}^{\text{bfl}}, \mathcal{P}^{\text{bin}}) = R(\boldsymbol{\omega}^{\text{fl}}, \mathcal{P}^{\text{bin}})$, and hence $R(\boldsymbol{\omega}^{\text{bfl}}, \mathcal{P}) \leq R(\boldsymbol{\omega}^{\text{bfl}}, \mathcal{P}^{\text{bin}})$, i.e. for every $\mathcal{P}$ over continuous losses, there is a corresponding $\mathcal{P}^{\text{bin}}$ over binary losses which incurs at least the same regret/redundancy/risk to $\boldsymbol{\omega}^{\text{bfl}}$. Therefore,

$$\sup_{\mathcal{P} \text{ on } [0,1]} R(\boldsymbol{\omega}^{\text{bfl}}, \mathcal{P}) = \sup_{\mathcal{P} \text{ on } \{0,1\}} R(\boldsymbol{\omega}^{\text{bfl}}, \mathcal{P}) = \sup_{\mathcal{P} \text{ on } \{0,1\}} R(\boldsymbol{\omega}^{\text{fl}}, \mathcal{P}).$$

By the second part of Corollary 1, for any prediction strategy $\boldsymbol{\omega}$:

$$\sup_{\mathcal{P} \text{ on } \{0,1\}} R(\boldsymbol{\omega}^{\text{fl}}, \mathcal{P}) \leq \sup_{\mathcal{P} \text{ on } \{0,1\}} R(\boldsymbol{\omega}, \mathcal{P}) \leq \sup_{\mathcal{P} \text{ on } [0,1]} R(\boldsymbol{\omega}, \mathcal{P}),$$

which finishes the proof. □

Theorem 3 states that the binarized FL strategy is the minimax prediction strategy when the losses are continuous on $[0, 1]$. Note that the same arguments would hold for any other loss range $[a, b]$, where the binarization on losses would convert continuous losses to the binary losses with values in $\{a, b\}$.

4.2 Vanilla FL is Not Minimax for Continuous Losses

We introduced the binarization procedure to show that the resulting binarized FL strategy is minimax for continuous losses. So far, however, we did not exclude the possibility that the plain FL strategy (without binarization) could also be minimax in the continuous setup. In this section, we prove (by counterexample) that this is not the case, so that the binarization procedure is justified. We will only consider excess risk for simplicity, but one can use similar arguments to show a counterexample for the expected regret and the expected redundancy as well.

The counterexamples proceeds by choosing the simplest non-trivial setup of $K = 2$ experts and $T = 2$ trials. We will first consider the case of binary losses and determine the minimax excess risk. Take two distributions P_1, P_2 on binary losses and denote $p_1 = P_1(\ell_{t,1} = 1)$ and $p_2 = P_2(\ell_{t,2} = 1)$, assuming (without loss of generality) that $p_1 \leq p_2$. The excess risk of the FL strategy (its expected loss in the second trial minus the expected loss of the first expert) is given by:

$$P(\ell_{1,1} = 0, \ell_{1,2} = 1)p_1 + P(\ell_{1,2} = 0, \ell_{1,1} = 1)p_2 + P(\ell_{1,1} = \ell_{1,2})\frac{p_1 + p_2}{2} - p_1,$$

which can be rewritten as:

$$\underbrace{p_2(1 - p_1)p_1 + p_1(1 - p_2)p_2}_{=2p_1p_2 - p_1p_2(p_1 + p_2)} + \underbrace{(p_1p_2 + (1 - p_1)(1 - p_2))\frac{p_1 + p_2}{2}}_{=p_1p_2(p_1 + p_2) - (p_1 + p_2)^2 + \frac{p_1 + p_2}{2}} - p_1$$

$$= \frac{p_2 - p_1}{2} - \frac{(p_2 - p_1)^2}{2}.$$

Denoting $\delta = p_2 - p_1$, the excess risk can be concisely written as $\frac{\delta}{2} - \frac{\delta^2}{2}$. Maximizing over δ gives $\delta^* = \frac{1}{2}$ and hence the maximum risk of FL on binary losses is equal to $\frac{1}{8}$.

Now, the crucial point to note is that this is also the minimax risk on *continuous* losses. This follows because the binarized FL strategy (which is the minimax strategy on continuous losses) achieves the maximum risk on binary losses (for which it is equivalent to the FL strategy), as follows from the proof of Theorem 3. What remains to be shown is that there exist distributions P_1, P_2 on continuous losses which force FL to suffer more excess risk than $\frac{1}{8}$. We take P_1 with support on two points $\{\epsilon, 1\}$, where ϵ is a very small positive number, and $p_1 = P_1(\ell_{t,1} = 1)$. Note that $\mathbb{E}[\ell_{t,1}] = p_1 + \epsilon(1 - p_1)$. P_2 has support on $\{0, 1 - \epsilon\}$, and let $p_2 = P_2(\ell_{t,2} = 1 - \epsilon)$, which means that $\mathbb{E}[\ell_{t,2}] = p_2(1 - \epsilon)$. We also assume $\mathbb{E}[\ell_{t,1}] < \mathbb{E}[\ell_{t,2}]$ i.e. expert 1 is the "better" expert, which translates to $p_1 + \epsilon(1 - p_1) < p_2(1 - \epsilon)$. The main idea in this counterexample is that by using ϵ values, all "ties" are resolved in favor of expert 2, which makes the FL algorithm suffer more loss. More precisely, this risk of FL is now given by:

$$p_2(1 - p_1)p_1 + p_1(1 - p_2)p_2 + \underbrace{(p_1p_2 + (1 - p_1)(1 - p_2))p_2}_{\text{ties}} - p_1 + O(\epsilon).$$

Choosing, e.g. $p_1 = 0$ and $p_2 = 0.5$, gives $\frac{1}{4} + O(\epsilon)$ excess risk, which is more than $\frac{1}{8}$, given that we take ϵ sufficiently small.

5   Conclusions and Open Problem

In this paper, we determined the minimax strategy for the stochastic setting of prediction with expert advice in which each expert generates its losses i.i.d. according to some distribution. Interestingly, the minimaxity is achieved by a single strategy, simultaneously for three considered performance measures: the expected regret, the expected redundancy, and the excess risk. We showed that when the losses are binary, the Follow the Leader algorithm is the minimax strategy for this game, and furthermore, it also has the smallest expected regret, expected redundancy, and excess risk among all permutation invariant prediction strategies for *every* distribution over the binary losses simultaneously, even among (permutation invariant) strategies which know the distributions of the losses. When the losses are continuous in $[0, 1]$, FL remains minimax only when an additional trick called "loss binarization" is applied, which results in the binarized FL strategy.

Open Problem. The setting considered in this paper concerns distributions over loss vectors which are i.i.d. between trials and i.i.d. between experts. It would be interesting to determined the minimax strategy in a more general setting, when the adversary can choose any joint distribution over loss vectors (still i.i.d. between trials, but not necessarily i.i.d. between experts). We did some preliminary computational experiment, which showed that that FL is not minimax in this setting, even when the losses are restricted to be binary.

References

1. Abernethy, J., Agarwal, A., Bartlett, P.L., Rakhlin, A.: A stochastic view of optimal regret through minimax duality. In: COLT (2009)
2. Abernethy, J., Warmuth, M.K., Yellin, J.: When random play is optimal against an adversary. In: COLT, pp. 437–445, July 2008
3. Berger, J.O.: Statistical Decision Theory and Bayesian Analysis. Springer, Berlin (1985)
4. Cesa-Bianchi, N., Freund, Y., Haussler, D., Helmbold, D.P., Schapire, R.E., Warmuth, M.K.: How to use expert advice. J. ACM **44**(3), 427–485 (1997)
5. Cesa-Bianchi, N., Lugosi, G.: Prediction, Learning, and Games. Cambridge University Press, Cambridge (2006)
6. van Erven, T., Kotłowski, W., Warmuth, M.K.: Follow the leader with dropout perturbations. In: COLT, pp. 949–974 (2014)
7. Ferguson, T.: Mathematical Statistics: A Decision Theoretic Approach. Academic Press, London (1967)
8. Grünwald, P.D.: The Minimum Description Length Principle. MIT Press, Cambridge (2007)
9. Koolen, W.M.: Combining strategies efficiently: high-quality decisions from conflicting advice. Ph.D. thesis, ILLC, University of Amsterdam (2011)
10. Koolen, W.M., van Erven, T.: Second-order quantile methods for experts and combinatorial games. In: COLT, pp. 1155–1175 (2015)
11. Luo, H., Schapire, R.E.: Achieving all with no parameters: AdaNormalHedge. In: COLT, pp. 1286–1304 (2015)
12. de Rooij, S., van Erven, T., Grünwald, P.D., Koolen, W.M.: Follow the leader if you can, hedge if you must. J. Mach. Learn. Res. **15**(1), 1281–1316 (2014)
13. Sani, A., Neu, G., Lazaric, A.: Exploiting easy data in online optimization. In: NIPS, pp. 810–818 (2014)

A Combinatorial Metrical Task System Problem Under the Uniform Metric

Takumi Nakazono[1,3], Ken-ichiro Moridomi[1(✉)], Kohei Hatano[2], and Eiji Takimoto[1]

[1] Department of Informatics, Kyushu University, Fukuoka, Japan
{moridomi.kenichiro,eiji}@inf.kyushu-u.ac.jp
[2] Library, Kyushu University, Fukuoka, Japan
hatano@inf.kyushu-u.ac.jp
[3] Toshiba Solutions Corporation, Fukuoka, Japan

Abstract. We consider a variant of the metrical task system (MTS) problem under the uniform metric, where each decision corresponds to some combinatorial object in a fixed set (e.g., the set of all s-t paths of a fixed graph). Typical algorithms such as Marking algorithm are not known to solve this problem efficiently and straightforward implementations takes exponential time for many classes of combinatorial sets. We propose a modification of Marking algorithm, which we call Weighted Marking algorithm. We show that Weighted Marking algorithm still keeps $O(\log n)$ competitive ratio for the standard MTS problem with n states. On the other hand, combining with known sampling techniques for combinatorial sets, Weighted Marking algorithm works efficiently for various classes of combinatorial sets.

1 Introduction

The metrical task system is defined as a repeated game between the player and the adversary. Given a fixed set C of states a metric $\delta : C \times C \to \mathbb{R}_+$ and a initial state $c_0 \in C$, for each round $t = 1, \ldots, T$, (i) the adversary reveals a (processing) cost function $f_t : C \to \mathbb{R}_+$, (ii) the player chooses a state $c_t \in C$, and (iii) the player incurs the processing cost $f_t(c_t)$ and the moving cost $\delta(c_t, c_{t-1})$. The goal of the algorithm is minimizing the cumulative (processing and moving) cost. The performance of the algorithm is measured by the competitive ratio, that is, the ratio of the cumulative cost of the algorithm to the cumulative cost of the best fixed sequence of states in hindsight.

In the expert setting, i.e., where the decision set consists of n states, there are many existing works on the MTS [4,5,8,11,14]. In particular, for the uniform metric δ (which is defined as $\delta(i,j) = 1$ if $i \neq j$ and otherwise $\delta(i,j) = 0$), the MTS problem is well studied [1,4,8,14]. Borodin et al. show the lower bound of the competitive ratio of any randomized algorithm is H_n, where H_n is the n-th harmonic number [8]. Especially, Abernethy et al. provide an algorithm which uses the method of convex optimization, and shows the upper bound of the competitive ratio of the algorithm is $H_n + O(\log \log n)$ [1].

© Springer International Publishing Switzerland 2016
R. Ortner et al. (Eds.): ALT 2016, LNAI 9925, pp. 276–287, 2016.
DOI: 10.1007/978-3-319-46379-7_19

When we consider the situation where the decision set C is a combinatorial set from $\{0,1\}^d$ (e.g., the set of spanning trees or s-t paths of a graph), the computational issue arises. A natural example of a combinatorial MTS is a routing problem. For example, we consider a routing problem. Consider a fixed network $G = (V, E)$ where V is the set of routers (nodes) and $E \subseteq V \times V$ is the set of d edges between routers and V includes two routers, the source s and the sink t. The decision set C is the set of paths from s to t, whose size is exponential in d. In general, for typical combinatorial sets, the size could be exponential in the dimension size d as well and straightforward implementations of known algorithms for the MTS take exponential time as well since time complexity of these algorithms is proportional to the size n of the decision set.

In this paper, for the uniform metric, we propose a modification of the Marking algorithm [8], which we call the Weighted Marking algorithm. The weighted Marking algorithm employs an exponential weighting scheme and can be viewed as an analogue of the Hedge algorithm [12] for the MTS problem, whereas the Marking algorithm is an analogue of the classical Halving algorithm. We prove that the Weighted Marking algorithm retains $O(\log n)$ competitive ratio for the standard MTS problem with n states. The expected running time of the Weighted Marking algorithm at each round is the same as that of the original one.

On the other hand, combining with efficient sampling techniques w.r.t. exponential weights on combinatorial objects (k-sets, s-t paths [18], stars in a graph [10] permutation matrices [10,15], permutation vectors [2]), the Weighted Marking algorithm works efficiently for various classes of combinatorial sets.

1.1 Related Work

There are some existing works for combinatorial metrical task systems. Blum et al. provide algorithms for the list update problem [6]. For the k-server problem, which can be viewed as a combinatorial MTS problem, Koutsoupias et al. provide a deterministic algorithm [17]. Bansal et al. improve the results of Koutsoupias et al. by a randomization technique [3]. These algorithms are efficient and perform well for specific problems, i.e., the list update problem and the k-server problem. However, these algorithms are specialized for limited decision sets and we cannot use them for other problems.

Buchbinder et al. consider combinatorial MTS problems where the decision space is defined as a matroid [9]. The concept of matroid can express various classes of combinatorial objects such as spanning trees. They show a unified algorithm with a guaranteed competitive ratio. Their analysis is, however, applicable for a continuous "relaxed" space only. It is not known if there exists a rounding scheme that approximately preserves the moving cost over the relaxed space. Gupta et al. also consider combinatorial MTS problems over the basis of a matroid [13]. They give a rounding algorithm and prove the competitive ratio of a rounded solution, for a class of metrics including the Hamming distance but not the uniform metric.

2   Preliminaries

A metrical task system (MTS) is a pair (C, δ) where C is a set of states and $\delta : C \times C \to \mathbb{R}_+$ is a metric. In particular, we consider a combinatorial setting where C is a subset of $\{0,1\}^d$ for some dimension $d > 0$. We denote by n the size of C, that is, $n = |C|$. Typically, n is exponentially large in d. Moreover, we only consider the uniform metric δ, that is,

$$\delta(c_1, c_2) = \begin{cases} 1 & \text{if } c_1 \neq c_2, \\ 0 & \text{if } c_1 = c_2. \end{cases}$$

The combinatorial MTS problem for (C, δ) is defined as the following protocol between the algorithm and the adversary.

First the adversary chooses a task sequence $\sigma = (\ell_1, \ell_2, \ldots, \ell_T)$, where each $\ell_t \in [0, 1]^d$ is called a loss vector. In other words, we assume the oblivious setting. For a given initial state $c_0 \in C$, the protocol proceeds in rounds, where in each round $t = 1, 2, \ldots, T$,

1. the algorithm receives the loss vector $\ell_t \in [0, 1]^d$,
2. the algorithm chooses a state $c_t \in C$, and
3. the algorithm suffers a cost given by $c_t \cdot \ell_t + \delta(c_t, c_{t-1})$.

The first term $c_t \cdot \ell_t$ of the cost is called the processing cost at round t, and the second term $\delta(c_t, c_{t-1})$ is called the moving cost at round t.

For a task sequence σ, the cumulative cost of an algorithm A is defined as

$$\text{cost}_A(\sigma) = \sum_{t=1}^{T} (c_t \cdot \ell_t + \delta(c_t, c_{t-1})),$$

and the cumulative cost of the best offline solution is defined as

$$\text{cost}_{\text{OPT}}(\sigma) = \min_{(c_1^*, c_2^*, \ldots, c_T^*) \in C^T} \sum_{t=1}^{T} (c_t^* \cdot \ell_t + \delta(c_t^*, c_{t-1}^*)).$$

We measure the performance of algorithm A by its competitive ratio, which is defined as

$$\text{CR}(\sigma) = \frac{\mathbb{E}[\text{cost}_A(\sigma)]}{\text{cost}_{\text{OPT}}(\sigma)},$$

where the expectation is with respect to the internal randomness of A. The goal of the algorithm is to minimize the worst case competitive ratio $\max_\sigma \text{CR}(\sigma)$. Note that the usual (non-combinatorial) MTS problem is a special case where C consists of unit vectors.

We also require the algorithm A to produce a state c_t in time polynomial in d for each round t. Typically, the size n of C is exponential in d, and so we cannot directly maintain all states c in C. Therefore, we assume two oracles to access the state set C efficiently. The first one is the linear optimization oracle, which solves the following decision problem:

$$\underline{\mathrm{OPT}(C)}$$

Input: $L \in \mathbb{R}_+^d$

Output: $\begin{cases} 0 & \text{if } \min_{c \in C} c \cdot L < 1, \\ 1 & \text{otherwise.} \end{cases}$

The assumption of this oracle is natural since the linear optimization problem has a polynomial time algorithm for many useful state sets C.

The second one is a sampling oracle, which chooses a state c randomly according to a certain probability distribution over C, where the distribution is specified by a given parameter $L \in \mathbb{R}_+^d$. In particular, we consider two kinds of sampling oracles, which will be defined later.

3   The Marking Algorithm

Here we apply the Marking algorithm [8] to the combinatorial MTS problem. The Marking algorithm is a simple randomized algorithm whose competitive ratio is upper bounded by $2H_n \leq 2(\ln n + 1)$, where H_n is the n-th harmonic number.

Below we describe how the Marking algorithm works. For a naive implementation, it maintains the cumulative processing costs $l[c]$ for all states $c \in C$. For each round t,

1. Observe the loss vector ℓ_t and update $l[c] = l[c] + c \cdot \ell_t$ for all $c \in C$.
2. If $l[c_{t-1}] < 1$ then output $c_t = c_{t-1}$.
3. Else choose a state c_t uniformly at random from the set of states c with $l[c] < 1$, and output c_t.
4. If no such states exist, then reset $l[c] = 0$ for all $c \in C$ and choose a state c_t uniformly at random from C, and output c_t.

Note that Lines 2 and 3 intuitively mean that the Marking algorithm does not change states until $l[c_t] \geq 1$. As is well known as a folklore (See, e.g., [7]), we can assume without loss of generality that the loss vectors ℓ_t are small enough so that $l[c_t] \leq 1$ always holds. In the appendix we give more detailed discussion. In other words, the Marking algorithm changes states only when $l[c_t] = 1$.

Of course, the naive implementation of the Marking algorithm is not efficient because it maintains the cumulative processing cost $l[c]$ for all states $c \in C$. Instead, we can maintain the cumulative loss vector $L = \sum_t \ell_t$, which implicitly maintains $l[c]$ as $l[c] = c \cdot L$ for all c. Furthermore, the sampling problem at Line 3 can be restated as the following problem in terms of L, which we call Sampling 1.

$\underline{\text{Sampling 1}}$

Input: $L \in \mathbb{R}_+^d$,

Output: $c \in C_L = \{c \in C \mid c \cdot L < 1\}$ uniformly at random.

Note that the problem Sampling 1 is only defined when $C_L \neq \emptyset$, but we can check whether the condition holds by using the linear optimization oracle for OPT(C). Moreover, the uniform sampling at Line 4 is also restated as Sampling 1 with $L = 0$. So, if we assume a linear optimization oracle for OPT(C) and a sampling oracle for Sampling 1, then we can emulate the Marking algorithm in $O(d)$ time per round. We give this implementation of the Marking algorithm in Algorithm 1.

Algorithm 1. An implementation of the Marking algorithm

Input: A linear optimization oracle for OPT(C) and a sampling oracle for Sampling 1
Initialize: Let $L = 0$.
For each round $t = 1, 2, \ldots, T$,

1. Observe the loss vector ℓ_t and update $L = L + \ell_t$.
2. Let $c_t = c_{t-1}$ and output c_t.
3. If $c_t \cdot L \geq 1$, then
   (a) If $\min_{c \in C} c \cdot L \geq 1$, then reset $L = 0$.   // use the linear optimization oracle
   (b) Choose a state $c_t \in C_L$ uniformly at random.   // use the sampling oracle

---

The question that naturally arises is that for what state set C, the problem Sampling 1 is efficiently solved. Unfortunately, we do not know any non-trivial sets C that have polynomial time algorithm for Sampling 1. We could use MCMC sampling methods to design approximate sampling, but it seems hard to show theoretically guaranteed performance bounds for many natural state sets C.

4   The Weighted Marking Algorithm

The computational cost of the sampling problem Sampling 1 would be due to the fact that the support of the sampling distribution is restricted to the set C_L. So, we relax the distribution to a continuous distribution whose support is not restricted to C_L.

Specifically, we propose the following sampling problem, called Sampling 2.

Sampling 2
Input: $L \in \mathbb{R}_+^d$,

Output: $c \in C$ chosen with probability $\pi_L(c) = \dfrac{\exp(-\eta c \cdot L)}{\sum_{c \in C} \exp(-\eta c \cdot L)}$,

where $\eta > 0$ is a parameter.

In words, the new sampling distribution π_L is such that $\pi_L(c)$ is a monotone decreasing function with respect to its cumulative processing cost $l[c] = c \cdot L$. So, the probability that a state c with large $l[c]$ is chosen is very low, and thus we will see that the support of π_L is essentially restricted to a set $\{c \in C \mid c \cdot L < L\}$ for some $L > 1$.

Unlike Sampling 1, there are known efficient implementations of Sampling 2 for several combinatorial objects such as k-sets, s-t paths [18], permutation matrices [10,15], stars in a graph [10] and permutation vectors [2].

Now we modify the Marking algorithm by assuming the sampling oracle for Sampling 2, as well as assuming the linear optimization oracle for OPT(C). The modified version is called the Weighted Marking algorithm. The difference from the Marking algorithm is that (1) it does not change states until its cumulative processing cost reaches L instead of 1, and (2) it uses π_L as the sampling distribution instead of the uniform distribution over C_L. Note that the Weighted Marking algorithm resets the cumulative loss vector as $\boldsymbol{L} = \boldsymbol{0}$ when $\min_{c \in C} c \cdot \boldsymbol{L}$ reaches 1, which is the same condition as the Marking algorithm. So, unlike the Marking algorithm, resetting $\boldsymbol{L}$ may happen at some round where the cumulative processing cost of the current state does not reach L, since $L \neq 1$.

The detailed description of the Weighted Marking algorithm is given in Algorithm 2.

Algorithm 2. Weighted Marking algorithm

Input: A linear optimization oracle for OPT(C) and a sampling oracle for Sampling 2
Parameter: $\eta > 0$ and $L > 1$ such that $ne^{-\eta L} \leq e^{-\eta}/2$.
Initialize: Let $\boldsymbol{L} = \boldsymbol{0}$.
For each round $t = 1, 2, \ldots, T$,

1. Observe the loss vector $\boldsymbol{\ell}_t$ and update $\boldsymbol{L} = \boldsymbol{L} + \boldsymbol{\ell}_t$.
2. Let $c_t = c_{t-1}$ and output c_t.
3. If $\min_{c \in C} c \cdot \boldsymbol{L} \geq 1$ then   // use the linear optimization oracle
   (a) Reset $\boldsymbol{L} = \boldsymbol{0}$.
   (b) Choose a state $c_t \in C$ with probability $\pi_L(c)$ // use the sampling oracle
4. Else if $c_t \cdot \boldsymbol{L} \geq L$, then
   (a) Repeat
       Choose a state $c_t \in C$ with probability $\pi_L(c)$ // use the sampling oracle
       Until $c_t \cdot \boldsymbol{L} < L$.

For convenience, we define the notion of phases for analyzing the behavior of the Weighted Marking algorithm. A phase is an interval $\{t \mid t_b \leq t \leq t_e\}$ of rounds such that the resetting happens at round $t_b - 1$ and t_e but does not happen at every round $t_b \leq t < t_e$.

Again, as is well known as a folklore, we assume without loss of generality that the loss vectors $\boldsymbol{\ell}_t$ are small enough so that it always holds that $\min_{c \in C} c \cdot \boldsymbol{L} \leq 1$ at Line 3 and it always hold that $c_t \cdot \boldsymbol{L} \leq L$ at Line 4. In other words, a phase ends (resetting happens) only when $\min_{c \in C} c \cdot \boldsymbol{L} = 1$ and states c_t are changed only when $c_t \cdot \boldsymbol{L} = L$. These assumptions greatly simplifies the analysis.

More formally, the assumption is described as follows:

Assumption 1. *Whenever the previous state c_{t-1} satisfies $c_{t-1} \cdot \boldsymbol{L} < L$, where $\boldsymbol{L}$ is the cumulative loss vectors up to round $t - 1$ in the current phase, and the* *phase did not end at round* $t - 1$, *i.e.*, $\min_{c^* \in C} c^* \cdot L < L$, *then* ℓ_t *satisfies the two conditions:*

1. $c_{t-1} \cdot (L + \ell_t) \leq L$, *and*
2. $\min_{c^* \in C} c^* \cdot (L + \ell_t) \leq 1$.

We assume Assumption 1 holds throughout this section. In the appendix, we briefly explain why the assumption holds without loss of generality.

In the next theorem, we give an upper bound of the competitive ratio of the Weighted Marking algorithm.

Theorem 1. *Let* $\eta = \ln 2n$, *and* $L = 2$. *Then for any task sequence* $\sigma = (\ell_1, \ell_2, \ldots, \ell_T)$, *the competitive ratio of the Weighted Marking algorithm is upper bounded by*

$$\mathrm{CR}(\sigma) \leq 6e \ln n + 9.$$

Moreover, the expected running time per round is $O(d + T_{\mathrm{lin}} + T_{\mathrm{Samp2}})$, *where* T_{lin} *is the running time of the linear optimization oracle and* T_{Samp2} *is that of the sampling oracle for Sampling 2.*

To prove this theorem, we show that the cumulative moving cost in each phase is $O(\log n)$. So in the following, we fix a particular phase $I = \{t_b, \ldots, t_e\}$. For each round $t \in I$, L_t denotes the cumulative loss vector L at Line 1 at round t. Note by definition that $\min_{c^* \in C} c^* \cdot L_{t_e} = 1$.

Let $G = \{c \in C \mid c \cdot L_{t_e} < L\}$ be the goal set, meaning that if we choose a state in G at some round $t \in I$, i.e., $c_t \in G$, then the Weighted Marking algorithm never changes the state until the end of the phase. Note that $c^* \in G$ and so $G \neq \emptyset$. Let $c_1, c_2, \ldots, c_n$ be the members of C. (This is an abuse of notation. Do not confuse them with the states c_t the algorithm chooses at round t). For any $c_i \notin G$, we can define $t_i \in I$ such that $c_i \cdot L_{t_i} = L$. Then, without loss of generality, we assume $t_1 \leq t_2 \leq \cdots \leq t_{n-|G|}$ and $c_n = c^*$, i.e., $c_n \cdot L_{t_e} = 1$. Moreover, we assume $|G| = 1$ just for simplicity. Clearly, the algorithm changes states only at some rounds in $\{t_1, \ldots, t_{n-1}\}$. Let $t^{(k)}$ be the round where the algorithm makes the k-th change of states. For any state $c \in C$, we define the weight function $W_k(c)$ as

$$W_k(c) := \begin{cases} e^{-\eta c \cdot L_{t^{(k)}}} & \text{if } c \cdot L_{t^{(k)}} < L, \\ 0 & \text{if } c \cdot L_{t^{(k)}} \geq L. \end{cases}$$

Let $\overline{W}_k := \sum_{c \in C} W_k(c)$. Then $W_k(c)/\overline{W}_k$ is the probability of choosing state c at the k-th change of states. One can see that $W_k(c)$ is monotonically decreasing w.r.t. k because L_t is monotonically increasing vector w.r.t. t.

If the best offline solution changes its state in the phase, then its cumulative moving cost is at least 1, and otherwise its cumulative processing cost is at least 1 by the definition of the phase. This immediately implies the following lemma.

Lemma 2. *For any sequence of loss vectors* $(\ell_1, \ell_2, \cdots, \ell_T)$, *the best offline solution suffers cost at least 1 on each phase.*

On the other hand, whenever the Weighted Marking algorithm changes states (i.e., suffers the moving cost of 1) from $c_{t^{(k-1)}}$ to $c_{t^{(k)}}$, then its cumulative processing cost from $t^{(k-1)}$ to $t^{(k)}$ is at most L. This implies the following lemma.

Lemma 3. *For any sequence of loss vectors $(\ell_1, \ell_2, \cdots, \ell_T)$, the cumulative processing cost of the Weighted Marking algorithm is at most L times the cumulative moving cost on each phase.*

The following lemma provides the probability of ending a phase.

Lemma 4. *For any $\alpha \in (0,1)$ and for any k, if $\alpha \overline{W}_k \leq e^{-\eta}$ holds then the phase will end at the $k+1$-th change of the state with probability at least α.*

Proof. By the assumption $c_n \cdot L_{t_e} = 1$, if the algorithm choose c_n then the algorithm will change its state at the end of the phase t_e, i.e. if the state c_n is chosen then the phase rests only 1 change. By $c_n \cdot L \leq 1$, we get $W_k(c_n) \geq e^{-\eta}$ for any k. Using this and the condition of the lemma, we get

$$\alpha \leq \frac{e^{-\eta}}{\overline{W}_k} \leq \frac{W_k(c_n)}{\overline{W}_k}.$$

Here, the right hand side is the probability of the state c_n will be chosen by the Weighted Marking algorithm. $\square$

The following lemma guarantees the probability of choosing c_n becomes higher at each change of the state.

Lemma 5. *For any $\alpha \in (0,1)$, for any k, if $\alpha \overline{W}_k \geq e^{-\eta}$ holds then*

$$\Pr[\overline{W}_{k+1} \leq \alpha \overline{W}_k] > \alpha.$$

Proof. Summing up weights of states from $n, n-1, \cdots$ and consider when the sum gets greater than $\alpha \overline{W}_k$. E.g. consider i_k s.t. $\sum_{i=i_k+1}^{n} W_k(c_i) \leq \alpha \overline{W}_k$ and $\sum_{i=i_k}^{n} W_k(c_i) > \alpha \overline{W}_k$.

Assume that the Weighted Marking algorithm chooses the state c_s at the k-th change of the state. If $s \geq i_k$, the algorithm changes its state at $t^{(k+1)}$ and then $W_{k+1}(c_i) = 0$ for any $i \geq i_k$ by the definition of W and i_k. Thus,

$$\overline{W}_{k+1} = \sum_{i=1}^{n} W_{k+1}(c_i) = \sum_{i=i_k+1}^{n} W_{k+1}(c_i).$$

Because W_k is monotonically decreasing w.r.t. k, one can get

$$\sum_{i=i_k+1}^{n} W_{k+1}(c_i) \leq \sum_{i=i_k+1}^{n} W_k(c_i) \leq \alpha \overline{W}.$$

So we get if $s \geq i_k$ then $\overline{W}_{k+1} \leq \alpha \overline{W}$. The probability of the Weighted Marking algorithm choosing the state c_s such that $s \geq i_k$ satisfies

$$\Pr[s \geq i_k] = \frac{\sum_{i=i_k}^{n} W_k(c_i)}{\overline{W}_k} > \frac{\alpha \overline{W}_k}{\overline{W}_k} = \alpha.$$

$\square$

By Lemma 4, one can get the following immediately.

Lemma 6. *For any $\alpha \in (0,1)$ and round $t^{(k)}$, if $\alpha \overline{W}_k \le e^{-\eta}$ then the expected number of remaining changes of states in the phase is less than $\frac{1}{\alpha} + 1$.*

Because of W_k is monotonically decreasing w.r.t. k and Lemma 5, one can get the following lemma.

Lemma 7. *For any k, for any $\alpha \in (0,1)$, if $\alpha \overline{W}_k \le e^{-\eta}$ then the expectation of m such that $\overline{W}_{k+m} \le \alpha \overline{W}_k$ is $\mathbb{E}[m] < \frac{1}{\alpha}$.*

We say that a sequence $\overline{W} = \{\overline{W}_1, \overline{W}_2, \cdots, \overline{W}_K\}$ of weights is α-fast decreasing at the round $t^{(k+1)}$ if $\overline{W}_{k+1} \ge \alpha \overline{W}_k$ holds.

Proof (Proof of Theorem 1). Assume that the Weighted Marking algorithm changes its state at K times in a phase. By Lemma 6, if $\alpha \overline{W}_{k'} \le e^{-\eta}$ holds then we have

$$\mathbb{E}[K] \le k' + \frac{1}{\alpha} + 1.$$

Thus, we need to estimate k' s.t. $\alpha \overline{W}_{k'} \le e^{-\eta}$ to bound $\mathbb{E}[K]$.

Let $\alpha \overline{W}_{k'} \le e^{-\eta}$ holds after α-fast decreasing K' times, then

$$\alpha^{K'} \overline{W}_0 \le \overline{W}_{k'} \le \frac{e^{\eta}}{\alpha}.$$

By $\overline{W}_0 = n$, we get $\alpha^{K'} n \le \frac{e^{-\eta}}{\alpha}$ and rearranging, $K' \le \frac{1}{\ln \frac{1}{\alpha}}(\ln n + \eta) - 1$. Using Lemma 7,

$$\mathbb{E}[k'] \le \mathbb{E}[m]K' = \frac{1}{\alpha}K' = \frac{1}{\alpha \ln \frac{1}{\alpha}}(\ln n + \eta) - \frac{1}{\alpha}.$$

Thus, the number of changing of states at a phase is

$$\mathbb{E}[K] \le \mathbb{E}[k'] + \frac{1}{\alpha} + 1 = \frac{1}{\alpha \ln \frac{1}{\alpha}}(\ln n + \eta) + 1.$$

The bound of $\mathbb{E}[K]$ is minimized when $\alpha = 1/e$. So we get $\mathbb{E}[K] \le e(\ln n + \eta) + 1$. Setting $\eta = \ln 2n$, we get $\mathbb{E}[K] \le 2e \ln n + 3$. By Lemma 3,

$$\mathbb{E}[(\text{cumulative processing cost})] \le L \times \mathbb{E}[(\text{cumulative moving cost})].$$

At each phase, we have

$$\mathbb{E}[\text{Cumulative loss}]$$
$$= \mathbb{E}[\text{Cumulative processing cost}] + \mathbb{E}[\text{Cumulative moving cost}]$$
$$\le 3 \times \mathbb{E}[K]$$
$$\le 6e \ln n + 9.$$

By Lemma 2, at each phase the best offline solution has the cumulative processing cost at least 1. Thus we get the bound of the competitive ratio.   □

Next, we prove the running time of the Weighted Marking algorithm. The key point of analysis of the Weighted Marking algorithm is the number of calls to the oracle for Sampling 2 at Line 4-(a) of the pseudo code. The following lemma gives a theoretical bound of retrying.

Lemma 8. *The expected number of calls to the sampling oracle at Line 4-(a) is at most 2.*

Proof. For any state c such that $c \cdot L \geq L$, the probability that the sampling oracle chooses c is

$$\frac{\exp(-\eta c \cdot L)}{\sum_{c'} \exp(-\eta c' \cdot L)} \leq \frac{\exp(-\eta L)}{\exp(-\eta c_n \cdot L)}$$

$$\leq \frac{\exp(-\eta L)}{\exp(-\eta)}$$

since $c_n \cdot L < 1$. By the union bound, the probability that the sampling oracle chooses some c with $c \cdot L \geq L$ is at most

$$\frac{n \exp(-\eta L)}{\exp(-\eta)} = \frac{1}{2}$$

by our choice of η and L.   □

5   Conclusion and Future Work

In this paper, we proposed the Weighted Marking algorithm for combinatorial MTS problems under the uniform metric space, and proved its competitive ratio is at most $6e \ln n + 9 = O(\log n)$. We showed that, by combining with existing sampling techniques for exponential weights over combinatorial objects, the proposed algorithm runs efficiently for several combinatorial classes, e.g., s-t paths and k-sets.

There are several open problems to investigate. First one is to provide a lower bound of the competitive ratio of the combinatorial MTS. In particular, it still remains open to prove $\Omega(\log d)$ or $\Omega(\log n)$ lower bounds for some combinatorial class of the decision set.

Secondly, it is not known if FPL [16] is applicable for the combinatorial MTS problem. If so, the sampling oracle is no longer necessary and we could efficiently solve MTS problems for more classes of combinatorial objects.

Finally, the hardness of the Sampling 1(C) is not known, either. Our conjecture is, it is #P hard for a specific class.

Acknowledgments. We thank anonymous reviewers for useful comments. Hatano is grateful to the supports from JSPS KAKENHI Grant Number 16K00305. Takimoto is grateful to the supports from JSPS KAKENHI Grant Number 15H02667. In addition, the authors acknowledge the support from MEXT KAKENHI Grant Number 24106010 (the ELC project).

A   On Assumption 1

As is well known as a folklore, we can assume without loss of generality that the loss vectors ℓ_t are small enough, so that Assumption 1 is satisfied.

Assumption 1. *Whenever the previous state c_{t-1} satisfies $c_{t-1} \cdot L < L$, where L is the cumulative loss vectors up to round $t-1$ in the current phase, and the phase did not end at round $t-1$, i.e., $\min_{c^* \in C} c^* \cdot L < 1$, then ℓ_t satisfies the two conditions:*

1. $c_{t-1} \cdot (L + \ell_t) \leq L$, and
2. $\min_{c^* \in C} c^* \cdot (L + \ell_t) \leq 1$.

This is because, when ℓ_t violates the assumption, then we can replace ℓ_t by a sequence of non-negative loss vectors $\ell_{t_1}, \ell_{t_2}, \ldots, \ell_{t_k}$ so that $\ell_t = \ell_{t_1} + \cdots + \ell_{t_k}$ and the new sequence of loss vectors satisfy the assumption in the following way:

1. If the first condition is violated, i.e., $c_{t-1} \cdot (L + \ell_t) = a > L$, then we let

$$\alpha_1 = \frac{L - c_{t-1} \cdot L}{a - c_{t-1} \cdot L}.$$

Otherwise, we let $\alpha_1 = 1$. In the former case, we can easily verify that $0 < \alpha_1 < 1$ and $c_{t-1} \cdot (L + \alpha_1 \ell_t) = L$.
2. If the second condition is violated, i.e., $\min_{c^* \in C} c^* \cdot (L + \ell_t) > 1$, then we let $0 < \alpha_2 < 1$ be such that $\min_{c^* \in C} c^* \cdot (L + \alpha_2 \ell_t) = 1$. Otherwise, we let $\alpha_2 = 1$. Note that, in the former case, we can find such α_2 efficiently by binary search.
3. Let $\alpha = \min\{\alpha_1, \alpha_2\}$ and $\ell_{t_1} = \alpha \ell_t$ and $\ell_{t_2} = (1 - \alpha)\ell_t$. Then, clearly ℓ_{t_1} satisfies Assumption 1. If ℓ_{t_2} still violates the assumption, then repeat the same procedure for ℓ_{t_2} recursively.

References

1. Abernethy, J., Bartlett, P.L., Buchbinder, N., Stanton, I.: A regularization approach to metrical task systems. In: Hutter, M., Stephan, F., Vovk, V., Zeugmann, T. (eds.) ALT 2015. Lecture Notes in Artificial Intelligence (LNAI), vol. 6331, pp. 270–284. Springer, Heidelberg (2010). doi:10.1007/978-3-642-16108-7_23
2. Ailon, N., Hatano, K., Takimoto, E.: Bandit online optimization over the permutahedron. In: Auer, P., Clark, A., Zeugmann, T., Zilles, S. (eds.) ALT 2015. Lecture Notes in Artificial Intelligence (LNAI), vol. 8776, pp. 215–229. Springer, Heidelberg (2014). doi:10.1007/978-3-319-11662-4_16
3. Bansal, N., Buchbinder, N., Madry, A., Naor, J.S.: A polylogarithmic-competitive algorithm for the k-server problem. J. ACM **62**(5), 40:1–40:49 (2015)
4. Bartal, Y., Blum, A., Burch, C., Tomkins, A.: A polylog(n)-competitive algorithm for metrical task systems. In: Proceedings of the Twenty-ninth Annual ACM Symposium on Theory of Computing, STOC 1997, pp. 711–719. ACM, New York (1997)

5. Bartal, Y., Bollobas, B., Mendel, M.: A Ramsey-type theorem for metric spaces and its applications for metrical task systems and related problems. In: Proceedings of 42nd IEEE Symposium on Foundations of Computer Science, pp. 396–405, October 2001

6. Blum, A., Chawla, S., Kalai, A.: Static optimality and dynamic search-optimality in lists and trees. In: Proceedings of the Thirteenth Annual ACM-SIAM Symposium on Discrete Algorithms, SODA 2002, pp. 1–8. Society for Industrial and Applied Mathematics, Philadelphia (2002)

7. Borodin, A., El-Yaniv, R.: Online Computation and Competitive Analysis. Cambridge University Press, New York (1998)

8. Borodin, A., Linial, N., Saks, M.E.: An optimal on-line algorithm for metrical task system. J. ACM **39**(4), 745–763 (1992)

9. Buchbinder, N., Chen, S., Naor, J.S., Shamir, O.: Unified algorithms for online learning and competitive analysis. Math. Oper. Res. **41**(2), 612–625 (2016)

10. Cesa-Bianchi, N., Lugosi, G.: Combinaotrial bandits. J. Comput. Syst. Sci. **78**(5), 1404–1422 (2012)

11. Fiat, A., Mendel, M.: Better algorithms for unfair metrical task systems and applications. SIAM J. Comput. **32**(6), 1403–1422 (2003)

12. Freund, Y., Schapire, R.E.: A decision-theoretic generalization of on-line learning and an application to boosting. J. Comput. Syst. Sci. **55**(1), 119–139 (1997)

13. Gupta, A., Talwar, K., Wieder, U.: Changing bases: multistage optimization for matroids and matchings. In: Esparza, J., Fraigniaud, P., Husfeldt, T., Koutsoupias, E. (eds.) ICALP 2015. LNCS, vol. 8572, pp. 563–575. Springer, Heidelberg (2014). doi:10.1007/978-3-662-43948-7_47

14. Irani, S., Seiden, S.: Randomized algorithms for metrical task systems. Theor. Comput. Sci. **194**(12), 163–182 (1998)

15. Jerrum, M., Sinclair, A., Vigoda, E.: A polynomial-time approximation algorithm for the permanent of a matrix with nonnegative entries. J. ACM **51**, 671–697 (2004)

16. Kalai, A., Vempala, S.: Efficient algorithms for online decision problems. J. Comput. Syst. Sci. **71**(3), 291–307 (2005)

17. Koutsoupias, E., Papadimitriou, C.H.: On the k-server conjecture. J. ACM **42**(5), 971–983 (1995)

18. Takimoto, E., Warmuth, M.K.: Path kernels and multiplicative updates. J. Mach. Learn. Res. **4**(5), 773–818 (2004)

Competitive Portfolio Selection Using Stochastic Predictions

Tuğkan Batu and Pongphat Taptagaporn[(✉)]

Department of Mathematics, London School of Economics, London, UK
{t.batu,p.taptagaporn}@lse.ac.uk

Abstract. We study a portfolio selection problem where a player attempts to maximise a utility function that represents the growth rate of wealth. We show that, given some stochastic predictions of the asset prices in the next time step, a sublinear expected regret is attainable against an optimal greedy algorithm, subject to tradeoff against the "accuracy" of such predictions that learn (or improve) over time. We also study the effects of introducing transaction costs into the model.

1 Introduction

In the field of portfolio management, the problem of how to distribute wealth among a number of assets to maximise wealth gain (or some notion of utility, e.g., mean-variance tradeoff) has been the focus of much academic and industrial research. Most of the studies in this field were previously from the perspective of financial mathematics and economics, and would usually assume some underlying distribution for the price process, e.g., Brownian Motion.

In the 1990's, a new field emerged that uses online learning to design growth-optimal portfolio selection models, following Cover's original work [8]. This model was shown to be competitive to the best CRP: an investment strategy that maintains a fixed proportion of wealth in each of the m assets for each time step, performing any required rebalancing as to maintain these proportions as the asset prices change. In particular, Cover showed sublinear regret on all possible outcomes of price sequence

$$\max_{x^T} \left(\log S_T^* - \log \hat{S}_T \right) = O(\log T),$$

where S_T^* and $\hat{S}_T$ are the wealth obtained by the best CRP and Cover's universal portfolio over T time steps (for some price sequence x^T), respectively. Most interestingly, the sublinear regret implies that the (per time step) log-wealth growth achieved by Cover's model converges to that of the best CRP as $T \to \infty$, without making any assumption on the price process (that is, in a model-free sense).

© Springer International Publishing Switzerland 2016
R. Ortner et al. (Eds.): ALT 2016, LNAI 9925, pp. 288–302, 2016.
DOI: 10.1007/978-3-319-46379-7_20

1.1 Our Contributions

Our result goes beyond the restriction imposed by the CRP, and instead, we devise a model that is competitive with the best greedy portfolio in a stochastic setting: one that makes the optimal decision as if it knows the next time step's price. To do this, we suppose that our model has access to a price prediction $\tilde{x}_t$ (of the next time step, $t+1$) that follows some probability distribution $\tilde{x}_t \sim \mathcal{D}_t(x_t)$, where x_t is the later observed price change. In this model, we quantify the precise relationship between the expected regret and the accuracy of such predictions. Note that we allow the prediction accuracy to vary over time, as reflected by the dependence of $\mathcal{D}_t$ on the current time step t. We demonstrate that for certain probability distributions $\mathcal{D}_t$,

$$\mathbb{E}_{\tilde{x}_t \sim \mathcal{D}_t(x_t)}\left[\max_{x^T} \left(\log S_T^* - \log \hat{S}_T \right) \right] = o(T)$$

is attainable, subject to some restrictions on the accuracy of $\tilde{x}_t$'s: namely, that the integral of the tail probabilities (of misestimation) must converge to zero as t grows. Intuitively, this is equivalent to improving our predictions through learning from past outcomes, and the requirement is that the model must be learning at a rate fast enough as to satisfy a certain sufficient condition that we will later prove. We also show a bound on the variance of regret in these cases.

Note that we also consider transaction costs for transferring wealth between assets (similar to Blum and Kalai [6]), as there is usually costs associated with buying and selling financial assets in practice (spreads, brokerage charge, etc.). However, we will prove that sublinear expected regret (over all possible price paths) is not attainable in the case of non-zero transaction costs (unless we assume that the price increases in each time step are independently distributed), unlike in the case of zero transaction costs.

Lastly, we show that our portfolio selection model can be computed efficiently using linear programming.

1.2 Related Work

The first published work combining the studies of portfolio theory with regret minimisation was by Cover [8]. Since then, there has been much follow up work and extensions to Cover's original portfolio model. Of particular interest to us, Blum and Kalai [6] extended the original model to account for transaction costs. However, the transaction costs plays a minor role in the Blum and Kalai model as it does not affect the decision process beyond that the penalty reduces the wealth that was retained. In particular, there was no cost-versus-wealth tradeoff, to assess whether shifting the portfolio would be beneficial over the cost this would incur, due to the limitation of the CRP model. We introduced a counterpart to the above that balances the reward from rebalancing the portfolio (based on information received from a price prediction) against the transaction cost incurred, and find an optimal point in between as to maximise cost adjusted wealth again.

Transaction costs aside, we compare our model to a less restrictive benchmark than in [8] because the best greedy portfolio is at least as good as the best CRP (in terms of the wealth obtained). However, we instead proved a bound on expected regret (as a function of the distributions $\mathcal{D}_t$) rather than worst-case regret, as we assume that we have additional knowledge in the form of price predictions, bringing us from an adversarial setting to a stochastic one. Note that when considering non-zero transaction costs, neither the greedy portfolio nor the best CRP is strictly better than the other.

Some other works that introduced notions similar to predictions [3,9] used a concept called "side information". This is where the adversary reveals a side information (say, an integer between 1 and y) and the CRP restriction is applied on each state separately. In particular, there is now y different CRPs that may be used, depending on the side information in that particular time step. The benchmark in this case is the best set of y CRP's that achieves the best wealth, given the observed sequence of side information. However, the regret bound of this model assumes that y is finite and does not grow with T, meaning that sublinear regret does not hold if the benchmark model uses a different portfolio in every time step (i.e., the side information never repeats). We do not have such restriction in our model.

More recent efforts to incorporate predictions into online learning problems can be found in [7,16]; these works look at the more general case of convex loss functions, but their regret is still benchmarked against the best CRP (which is substantially weaker than the best greedy portfolio). Some other variants of universal portfolio models can be found in [1,2,4,10,11,13,15,18]. Most of these models are based on the idea of taking a weighted combination of CRPs over the set of all possible portfolio vectors.

Portfolio optimisation is a fundamental problem studied in mathematical finance literature [14,17], wherein models with stochastic price changes is the norm. For example, price changes distributed log-normally is analogous to Geometric Brownian Motion [5,12,14], a well-understood model used in that field. However, our study and model, motivated by a machine learning perspective to maximise growth-rate of wealth (as opposed to, say, mean-variance optimisation in modern portfolio theory) yields incomparable results.

2   Preliminaries

Consider the scenario where we have m assets available for trading over T time steps. Define $x_t = (x_t(1), \ldots, x_t(m)) \in \mathbb{R}_+^m$ as a real-valued vector of price relatives at time step t; the i-th element of this vector is the ratio of the respective true market prices of Asset i at time t and time $t - 1$. For convention, x_t is defined for $1 \leq t \leq T$, and we denote by x^T the vector $(x_1, \ldots, x_T)$. The space $\mathcal{B}$ of portfolio vectors is defined as

$$\mathcal{B} := \{b \in \mathbb{R}_+^m : \sum_{i=1}^{m} b(i) = 1\},$$

where $b(i)$ is the proportion of the portfolio b's total wealth allocated to Asset i. Typically, we may need to redistribute wealth between assets as to obtain the portfolio vector chosen for the next time step. We will call this process of redistributing wealth *rebalancing*. We denote by $\theta(b, b', x)$ the multiplicative factor of decrease in wealth due to rebalancing from portfolio b (after observing the price change x) to portfolio b', which we will define in more details in the next section. Then, we can define the wealth of a portfolio model $(b_1, \ldots, b_T)$ as[1]

$$S_T = \prod_{t=1}^{T} b_t x_t \theta(b_{t-1}, b_t, x_{t-1}).$$

As a convention, we assume that there are no transaction costs associated with the initial positioning before the first time step: that is, $b_0 := b_1$, $x_0 = (1, \ldots, 1)$, and, thus, $\theta(b_0, b_1, x_0) = 1$. Broadly speaking, S_T is the product of the wealth change across all time steps $t = 1, \ldots, T$, where, at each step, we first pay a factor of $\theta(b_{t-1}, b_t, x_{t-1})$ transaction cost for rebalancing b_{t-1} to b_t, and then experience a change $b_t x_t$ in wealth, once the price change is observed. Similarly, for the portfolio models denoted as $(\hat{b}_1, \ldots, \hat{b}_T)$ and $(b_1^*, \ldots, b_T^*)$, respectively, we will use $\hat{S}_T$ and S_T^*, respectively, to denote the wealth generated by the corresponding portfolio model.

Note that a CRP (from [8]) imposes the additional constraint that the portfolio vector is the same throughout every time step, that is, $b_1 = \ldots = b_T$.

Although the portfolio model investigated here has the restriction that all the wealth must be invested in one of the m assets, this can be extended to a portfolio of $m + 1$ assets where the first m asset is as before, and the last one represents cash. Therefore, the returns x_t now has $m + 1$ dimension where the last element could represent risk-free interest rate, analogous to much of the work in financial mathematics.

2.1 Transaction Costs

The concept of transaction costs was first introduced into the study of online portfolios selection by Blum and Kalai [6], wherein their model charge a fixed percentage of commission on the purchase, but not on the sale, of assets. This is equivalent to charging commission on the purchase and sale of assets equally, as the wealth from any asset we sold will have to be used to purchase another asset (by the constraints of the problem setting). We will use the same model here, though the choice of model doesn't significantly affect our results.

Given portfolio vectors $b_{t-1}, b_t \in \mathcal{B}$ and price-relatives vector x_{t-1}, we want to rebalance from the vector $b_{t-1}' := b_{t-1} \cdot x_{t-1} \in \mathbb{R}^m$ to $b_t \in \mathcal{B} \subset \mathbb{R}^m$. Given a transaction cost factor $c \in [0, 1]$ indicating the proportion of cost to be paid from the value of assets purchased, the proportion of wealth retained after rebalancing can be expressed recursively as

$$\theta := \theta(b_{t-1}, b_t, x_{t-1}) = 1 - c \sum_{i:\beta_i > 0} \beta_i,$$

---

[1] The notations $b_t x_t$ is used as a short-hand for vector dot product.

where $\beta_i = \theta b_t(i) - b_{t-1}(i) \cdot x_{t-1}(i) = \theta b_t(i) - b'_{t-1}(i)$ indicates the quantity of Asset i that needs to be sold or bought, depending on its sign. Intuitively, θ represents the proportion of the total wealth left after rebalancing. In the worst case, the market value of b' is at least $1 - c$ of the market value of b after rebalancing. In particular, rebalancing a portfolio will always retain at least $1 - c$ proportion of its wealth.

2.2   Problem Setting

At time $t \in [T]$, suppose our model has access to a prediction such that it follows some probability distribution with respect to the later observed price change: that is, $\tilde{x}_t \sim \mathcal{D}_t(x_t)$. Note that the distribution $\mathcal{D}_t$ may depend on the current time step t (hence, the subscript) and x_t, possibly hiding further dependencies on additional parameters such as variance. Based on this prediction, we can compute a portfolio vector as to optimise the wealth.

Definition 1 (Portfolio Model). *For each $t \in [T]$, given a predicted price-change $\tilde{x}_t$ of the observed price change x_t such that $\tilde{x}_t \sim \mathcal{D}_t(x_t)$ for some probability distribution $\mathcal{D}_t$, the portfolio vector at time t is specified by*

$$\hat{b}_t := \arg\max_{b \in \mathcal{B}} b\tilde{x}_t \theta(\hat{b}_{t-1}, b, x_{t-1}).$$

Our benchmark model, which we call the optimal greedy portfolio, is defined similarly as, for each time t,

$$b_t^* = \arg\max_{b \in \mathcal{B}} b x_t \theta(b_{t-1}^*, b, x_{t-1}).$$

Note that the above models considers the tradeoff between the transaction cost of shifting to a "better" portfolio against the expected benefit of doing such a rebalancing given the prediction or actual outcome, respectively. In the case where the optimisation yields multiple solutions, we canonically choose the one with the least transaction costs. This will be made more precise in Sect. 5.

3   Main Results

In this section, we present our technical contributions. In particular, we investigate how close the wealth of our portfolio model is to the benchmark model, in expectation over the random choices of $\tilde{x}_t \sim \mathcal{D}_t(x_t)$ and adversarially chosen x_t, for $t \in [T]$.

Firstly, we show the expected-regret bound of the portfolio model $\hat{b}$ against b^*, in terms of the distribution of the predicted price change $\tilde{x}_t$ relative to the later observed price change x_t. This will lead us to a sufficient condition to obtain a sublinear expected regret (and, additionally, sublinear variance of regret) in the case of zero transaction costs. Then, we show that sublinear expected regret is unattainable in general in the case of non-zero transaction costs, no matter how small $c > 0$ is.

3.1   Expected-Regret Bound

As a measure of performance, we consider the expected-regret $\mathbb{E}[R]$ of our portfolio model against the optimal greedy portfolio model: namely,

$$\mathbb{E}_{\tilde{x}_t \sim \mathcal{D}_t(x_t)} \left[\max_{x^T} \left(\log S_T^* - \log \hat{S}_T \right) \right].$$

This can be interpreted as enumerating through all possible price predictions $\tilde{x}^T$ and choosing the outcome of price sequence x^T that maximises regret for each choice of $\tilde{x}^T$. Each of these choices of $\tilde{x}^T$ occurs with some probability depending on x^T and $\mathcal{D}_t$ for $t \in [T]$, and we take the expectation over these probabilities.

We analyse the expected regret $\mathbb{E}[R]$, where the choices of portfolio vectors depend directly on the random choices of $\tilde{x}_t \sim \mathcal{D}_t(x_t)$ and x_t is chosen adversarially, for each $t \in [T]$. The theorem below gives an upper bound on the expected regret as a function of the distributions $\mathcal{D}_t$ of predictions in each time step.

Theorem 2. *The expected regret of our portfolio model from Definition 1 can be bounded from above as*

$$\mathbb{E}[R] \leq \gamma + 2 \sum_{t=1}^{T} \int_0^\infty \Pr_{\tilde{x}_t \sim \mathcal{D}_t(x_t)} [\tilde{x}_t \notin (e^{-z} x_t, e^z x_t)] \, dz \,,$$

where γ accounts for the regret arising from the positioning error of our portfolio and is defined as

$$\gamma = - \sum_{t=1}^{T} \mathbb{E} \left[\log \frac{\theta(\hat{b}_{t-1}, b_t^*, x_{t-1})}{\theta(b_{t-1}^*, b_t^*, x_{t-1})} \right].$$

Proof. We fix some time t and consider the ratio of the single-time-step wealth change of our portfolio to that of the benchmark at time t in order to bound the regret arising from that time step. The regret associated with the time step t has two sources: positioning error of the current portfolio that results in transaction costs and inaccurate price predictions. We define

$$\rho_t = \frac{\theta(\hat{b}_{t-1}, b_t^*, x_{t-1})}{\theta(b_{t-1}^*, b_t^*, x_{t-1})}$$

to capture the regret arising from the positioning error of the portfolio at time step t: for example, when b_{t-1}^* was in a better position than $\hat{b}_{t-1}$ to minimise transaction costs when rebalancing at time t.

Now, suppose that $(1 - \delta)x_t \preceq \tilde{x}_t \preceq (1 - \delta)^{-1} x_t$,[2] at time step t, for some δ such that $0 \leq \delta < 1$. Then, for any $\hat{b}_t, b_t^*, \hat{b}_{t-1}, b_{t-1}^* \in \mathcal{B}$, we have the following bound on the ratio of the single-time-step wealths:

---

[2] The notations $\preceq, \succeq, \prec$, and $\succ$ denote component-wise vector inequalities.

$$\frac{\hat{b}_t x_t \theta(\hat{b}_{t-1}, \hat{b}_t, x_{t-1})}{b_t^* x_t \theta(b_{t-1}^*, b_t^*, x_{t-1})} \geq (1-\delta)\frac{\hat{b}_t \tilde{x}_t \theta(\hat{b}_{t-1}, \hat{b}_t, x_{t-1})}{b_t^* x_t \theta(b_{t-1}^*, b_t^*, x_{t-1})} \tag{1}$$

$$\geq (1-\delta)^2 \frac{\hat{b}_t \tilde{x}_t \theta(\hat{b}_{t-1}, \hat{b}_t, x_{t-1})}{b_t^* \tilde{x}_t \theta(b_{t-1}^*, b_t^*, x_{t-1})} \tag{2}$$

$$\geq (1-\delta)^2 \rho_t. \tag{3}$$

In the above, (1) is due to $x_t \succeq (1-\delta)\tilde{x}_t$, (2) is due to $\tilde{x}_t \succeq (1-\delta)x_t$, and (3) is due to the fact that

$$\hat{b}_t \tilde{x}_t \theta(\hat{b}_{t-1}, \hat{b}_t, x_{t-1}) \geq b_t^* \tilde{x}_t \theta(\hat{b}_{t-1}, b_t^*, x_{t-1}) = \rho_t b_t^* \tilde{x}_t \theta(b_{t-1}^*, b_t^*, x_{t-1}),$$

as $\hat{b}_t$ was chosen to maximise its single-time-step wealth by Definition 1. For each time step $t \in [T]$, we define deviation δ_t of x_t and $\tilde{x}_t$ as

$$\delta_t := \min\{\delta \geq 0 \mid (1-\delta)x_t \succeq \tilde{x}_t \succeq (1-\delta)^{-1}x_t\}.$$

Intuitively, this is the deviation of the predicted price change from the observed price change. We can now calculate the expected regret as follows.

$$\mathbb{E}[R] = \mathbb{E}\left[\max_{x^T} \log\left(\frac{S_T^*}{\hat{S}_T}\right)\right]$$

$$= \mathbb{E}\left[\max_{x^T} \log\left(\prod_{t=1}^{T} \frac{b_t^* x_t \theta(b_{t-1}^*, b_t^*, x_{t-1})}{\hat{b}_t x_t \theta(\hat{b}_{t-1}, \hat{b}_t, x_{t-1})}\right)\right]$$

$$\leq \mathbb{E}\left[\log\left(\prod_{t=1}^{T} (1-\delta_t)^{-2} \rho_t^{-1}\right)\right] \tag{4}$$

$$\leq \sum_{t=1}^{T} 2\mathbb{E}\left[-\log(1-\delta_t)\right] - \mathbb{E}\left[\log \rho_t\right], \tag{5}$$

where (4) is by the inequality from (3), and (5) follows from linearity of expectation. We now will now use $\gamma = -\sum_{t=1}^{T} \mathbb{E}[\log \rho_t]$ to denote the "positioning error," and continue our analysis of the first term on the right hand side of the inequality.

$$\sum_{t=1}^{T} \mathbb{E}\left[-\log(1-\delta_t)\right] = \sum_{t=1}^{T} \int_0^\infty \Pr_{\tilde{x}_t}[-\log(1-\delta_t) \geq z]\, dz$$

$$= \sum_{t=1}^{T} \int_0^\infty \Pr_{\tilde{x}_t}[1-\delta_t \leq e^{-z}]\, dz,$$

$$= \sum_{t=1}^{T} \int_0^\infty 1 - \Pr_{\tilde{x}_t}[1-\delta_t > e^{-z}]\, dz,$$

$$= \sum_{t=1}^{T} \int_0^\infty 1 - \Pr_{\tilde{x}_t}[e^{-z}x_t \prec \tilde{x}_t \prec e^z x_t]\, dz,$$

where the last line above is obtained from applying the definition of δ_t, giving us the bound on expected regret. $\square$

Note that the quantity γ in Theorem 2 captures the positioning error of our model arising from transaction costs. Hence, in the absence of transaction costs (that is, when $c = 0$), we have that $\gamma = 0$. In fact, we later prove in Sect. 3.3 that, in general, $\gamma = \Omega(T)$ for non-zero transaction costs (that is, when $c > 0$), by showing that there exists a sequence x^T that yields an expected regret at least linear in T.

We also observe that $\gamma = 0$ in the weaker case when x_t is a random variable that is independent of x_{t-1} (hence, also independent of b_{t-1}^* and $\hat{b}_{t-1}$), for all time steps $t \in [T]$, whereas Theorem 2 is stronger as it makes no assumption on how x_t are chosen. This is because

$$\mathbb{E}[\log \theta(b_{t-1}^*, b_t^*, x_{t-1})] = \mathbb{E}[\log \theta(\hat{b}_{t-1}, \hat{b}_t, x_{t-1})],$$

intuitively meaning that the random choice of x_t and $\tilde{x}_t$ are just as likely be favourable to b_{t-1}^* as it is to $\hat{b}_{t-1}$. For example, suppose that we define $\tilde{x}_t = (1, ..., 1)$ and x_t is drawn from some log-normal distribution with mean $\tilde{x}_t$. Then, this is equivalent to assuming that the returns x_t follows a Geometric Brownian Motion and that the current price is the best prediction of the next time step's price; similar to the assumption surrounding much of the work in financial mathematics.

Finally, setting γ aside, the result above gives us a good intuition on what the expected regret looks like. Namely, in each time step the regret can be thought of to be no larger than the sum of an integral of the tail probabilities. Having a small expected regret then hinges on bounding these tail probabilities.

3.2   Variance-of-Regret Bound

We can now prove a bound on the variance of regret, using much of the ideas from the proof of the bound on expected regret in Theorem 2.

Theorem 3. *The variance of regret of our portfolio model from Definition 1 can be bounded from above as*

$$\mathrm{Var}[R] \leq \eta + 4 \sum_{t=1}^{T} \int_0^\infty \Pr_{\tilde{x}_t \sim \mathcal{D}_t(x_t)} [\tilde{x}_t \notin (e^{-\sqrt{z}} x_t, e^{\sqrt{z}} x_t)] \, dz \,,$$

where η accounts for the variance in the regret arising from the positioning error and the covariance of the single-time-step wealth ratios, defined as

$$\eta = - \sum_{t=1}^{T} \mathrm{Var}\left[\log \frac{\theta(\hat{b}_{t-1}, b_t^*, x_{t-1})}{\theta(b_{t-1}^*, b_t^*, x_{t-1})} \right]$$

$$+ \sum_{t=1}^{T} \sum_{j \neq t} \mathrm{cov}\left[\frac{b_t^* x_t \theta(b_{t-1}^*, b_t^*, x_{t-1})}{\hat{b}_t x_t \theta(\hat{b}_{t-1}, \hat{b}_t, x_{t-1})}, \frac{b_j^* x_j \theta(b_{j-1}^*, b_j^*, x_{j-1})}{\hat{b}_j x_j \theta(\hat{b}_{j-1}, \hat{b}_j, x_{j-1})} \right].$$

Proof

$$\mathrm{Var}[R] = \mathrm{Var}\Big[\max_{x^T}\log\Big(\frac{S_T^*}{\hat{S}_T}\Big)\Big]$$

$$= \mathrm{Var}\Big[\max_{x^T}\log\Big(\prod_{t=1}^{T}\frac{b_t^* x_t \theta(b_{t-1}^*, b_t^*, x_{t-1})}{\hat{b}_t x_t \theta(\hat{b}_{t-1}, \hat{b}_t, x_{t-1})}\Big)\Big]$$

$$\leq \mathrm{Var}\Big[\log\Big(\prod_{t=1}^{T}(1-\delta_t)^{-2}\rho_t^{-1}\Big)\Big]$$

$$\leq \eta + 4\sum_{t=1}^{T}\mathrm{Var}\Big[-\log(1-\delta_t)\Big],$$

where η is the term representing the positioning errors and covariance terms, as described in the theorem statement. We continue to simplify the remaining part of the equation, making use of the inequality $\mathrm{Var}[R] \leq \mathbb{E}[R^2]$. Thus, we get

$$\sum_{t=1}^{T}\mathrm{Var}\Big[-\log(1-\delta_t)\Big] \leq \sum_{t=1}^{T}\mathbb{E}\Big[(-\log(1-\delta_t))^2\Big]$$

$$= \sum_{t=1}^{T}\int_0^\infty \Pr_{\tilde{x}_t}[-\log(1-\delta_t) \geq \sqrt{z}]\, dz$$

$$= \sum_{t=1}^{T}\int_0^\infty \Pr_{\tilde{x}_t}[1-\delta_t \leq e^{-\sqrt{z}}]\, dz,$$

$$= \sum_{t=1}^{T}\int_0^\infty 1 - \Pr_{\tilde{x}_t}[1-\delta_t > e^{-\sqrt{z}}]\, dz,$$

$$= \sum_{t=1}^{T}\int_0^\infty 1 - \Pr_{\tilde{x}_t}[e^{-\sqrt{z}}x_t \prec \tilde{x}_t \prec e^{\sqrt{z}}x_t]\, dz,$$

where the last line above is obtained from applying the definition of δ_t (as defined in the proof of Theorem 2), giving us the desired result.   □

Similarly to the case for expected regret discussed in the previous section, we also have that $\eta = 0$ in the zero-transaction cost scenario (that is, $c = 0$) or x_t is independently distributed from x_{t-1} for $t \in [T]$.

3.3   Linear Expected Regret for Non-zero Transaction Costs

We will now show that for any class of non-trivial distributions $\mathcal{D}_t$, the expected-regret bound above will not be sublinear for non-zero transaction cost (in effect, showing that γ is not necessarily sublinear for any $c > 0$). This is because there exists a sequence of returns x_t for $t \in [T]$ that will favour b_t^* position, hence, yielding a large enough long-term regret. Here, we define a *non-trivial* *distribution* as one where the preimage of the cumulative distribution function is non-empty at some value inside a constant interval around $\frac{1}{2}$. Note that any class of continuous distributions satisfies this criteria.

Theorem 4. *Given non-trivial $\mathcal{D}_t$, for all $t \in [T]$, $\mathbb{E}[R] = \Omega(T)$ when transaction cost c is non-zero.*

Proof. To prove that the expected regret is not necessarily sublinear in the case of non-zero transaction cost, it is enough to come up with a sequence of x_t that breaks this sub-linearity. Therefore, we will give a way to construct such x_t for each $t \in [T]$ in the two-asset case ($m = 2$), where b_t^* and $\hat{b}_t$ will always take the values of either $(0, 1)$ or $(1, 0)$ by our construction of the re-balancing scheme from Sect. 5.

For time step t, assume that $\hat{b}_{t-1} = (0, 1)$, without loss of generality, with b_{t-1}^* is $(0, 1)$ or $(1, 0)$. We will calculate the single-time-step loss

$$\frac{b_t^* x_t \theta(b_{t-1}^*, b_t^*, x_{t-1})}{\hat{b}_t x_t \theta(\hat{b}_{t-1}, \hat{b}_t, x_{t-1})}$$

in these two cases separately.

State 1 (Different). $b_{t-1}^* = (1, 0)$

The adversary chooses $x_t = (1, 1 - c)$, resulting in a single-time-step loss of $\frac{1}{1-c}$, regardless of the choice $\tilde{x}_t \sim \mathcal{D}_t(x_t)$.

State 2 (Same). $b_{t-1}^* = (0, 1)$

The adversary chooses $x_t = (\xi_t, 1)$, where ξ_t is chosen such that

$$\Pr_{\tilde{x}_t \sim \mathcal{D}_t((\xi_t, 1))} \left[\frac{\tilde{x}_t(1)}{\tilde{x}_t(2)} > \frac{1}{1 - c} \right] = \frac{1}{2}.$$

Intuitively, this is the choice of price relative vector where the portfolio model (as represented by $\hat{b}_t$) has equal probabilities of shifting or staying put. This implies that $\Pr_{\tilde{x}_t \sim \mathcal{D}_t(x_t)}[\hat{b}_t = b_t^*] = \frac{1}{2}$, and the single-time-step loss may be as small as 1 in this case. Note that this choice of ξ_t exists if the preimage of the CDF of $\mathcal{D}_t$ at $\frac{1}{2}$ is non-empty. One can easily extend this proof to cases where the preimage of the CDF is non-empty at some value inside a constant interval around $\frac{1}{2}$.

With this information, we can model the dynamics of the portfolio as a Markov chain with these two states (Different and Same). The transition probability matrix of that Markov chain, assuming worst-case, i.e., the lowest probability of staying in "different" is

$$\begin{pmatrix} 0 & 1 \\ \frac{1}{2} & \frac{1}{2} \end{pmatrix},$$

which implies a limiting distribution $\pi = (\frac{1}{3}, \frac{2}{3})$. Using this, the expected regret (over all possible x_t) can be lower-bounded by the linear expected regret (over the particular choice of x_t, as described above).

$$\mathbb{E}[R] = \mathbb{E}\left[\max_{x^T} \log \left(\frac{S_T^*}{\hat{S}_T} \right) \right]$$

$$\geq \mathbb{E}\left[\log \left(\frac{S_T^*}{\hat{S}_T} \right) \right]$$

$$= \sum_{t=1}^{T} \mathbb{E}\left[\log \frac{b_t^* x_t \theta(b_{t-1}^*, b_t^*, x_{t-1})}{\hat{b}_t x_t \theta(\hat{b}_{t-1}, \hat{b}_t, x_{t-1})} \right]$$

$$= -\frac{1}{3} \sum_{t=1}^{T} \log(1 - c) = \Theta(T),$$

where the last line follows from the fact that the portfolio needs to shift all its wealth in one third of the steps in the long run (due to the limiting distribution of the Markov chain above), each of which incurs a loss factor of $1 - c$.   □

So now we have established that we cannot hope for sublinear expected regret in the presence of transaction costs, no matter the choice of $\mathcal{D}_t$ (as long as it is non-trivial). However, we will later show in Sect. 4 that a few sensible choices for $\mathcal{D}_t$ will indeed yield sublinear expected regret (and variance of regret) in the case $c = 0$.

4   Special Cases for the Distributions of Predictions

Given the above results are for a generically distributed $\tilde{x}_t \sim \mathcal{D}_t(x_t)$, we will now look at some particular cases for $\mathcal{D}_t$ and compute the required quality of prediction in order to achieve sublinear expected regret. Herein we will assume that $c = 0$, as Theorem 4 shows that we cannot hope for sublinear expected regret in the presence of transaction costs.

Firstly, we shall assume that $\mathcal{D}_t$ is parametrised by two variables μ_t (mean) and σ_t (standard deviation). We will look only at log-returns (rather than absolute returns); this is quite a standard notion in financial mathematics for a number of reasons [5, 12, 14]. In particular, we will say that the log-predicted returns $(\ln \tilde{x}_t)$ are distributed around the mean (defined as the log-observed returns, $\ln x_t$) with some standard deviation σ_t. Formally, $\ln \tilde{x}_t \sim \mathcal{D}_{\ln x_t, \sigma_t^2}$ for some distribution $\mathcal{D}$, or simply $\tilde{x}_t \sim \ln \mathcal{D}_{\ln x_t, \sigma_t^2}$ for short-hand. As our portfolio vector is multi-dimensional, we will use $\sigma_t = (\sigma_t, ..., \sigma_t) \in \mathbb{R}_+^m$, apply the logarithm and distribution element-wise: that is,

$$\ln x_t = \ln(x_t(1), ..., x_t(m)) = (\ln x_t(1), ..., \ln x_t(m)),$$

and, thus,

$$\ln \mathcal{D}_{\ln x_t, \sigma_t^2} = \ln \mathcal{D}_{\ln x_t(1), \sigma_t^2} \times ... \times \ln \mathcal{D}_{\ln x_t(m), \sigma_t^2}.$$

Note that Chebyshev's inequality is too loose to obtain a reasonable bound for a generalised distribution $\mathcal{D}$:

$$\mathbb{E}[R] \leq 2 \sum_{t=1}^{T} \int_{0}^{\infty} \Pr_{\tilde{x}_t \sim \mathcal{D}_t(x_t)} [\tilde{x}_t \notin (e^{-z} x_t, e^z x_t)] \, dz \leq 2 \sum_{t=1}^{T} \int_{0}^{\infty} \frac{\sigma_t^2}{z^2} \, dz,$$

where the last inequality is due to Chebyshev's, which states that

$$Pr(|x - \mu| \geq z) \leq \sigma_t^2 / z^2.$$

As a result, the last integral evaluates to $+\infty$. Therefore, the next three subsection looks at the required σ_t, for $t \in [T]$, to obtain sublinear expected regret for three particular cases of $\mathcal{D}$: uniform, linear, and normal.

4.1   Log-Uniformly Distributed Predictions

Suppose that $\tilde{x}_t \sim \ln \mathcal{U}_{\ln x_t, \sigma_t^2}$, where $\mathcal{U}$ is the uniform distribution on the log-returns between the range $[-\sigma_t, \sigma_t]$ with the following probability density function

$$f(y) = \begin{cases} \frac{1}{2\sigma_t} & \text{if } 0 \leq |y - \ln x_t| \leq \sigma_t, \\ 0 & \text{otherwise.} \end{cases}$$

In this case, applying Theorems 2 and 3 yields

$$\mathbb{E}[R] \leq 2 \sum_{t=1}^{T} \int_{0}^{\sigma_t} 1 - \frac{z}{\sigma_t} \, dz = \sum_{t=1}^{T} \sigma_t,$$

$$\text{Var}[R] \leq 4 \sum_{t=1}^{T} \int_{0}^{\sigma_t} 1 - \frac{\sqrt{z}}{\sigma_t} \, dz = 4 \sum_{t=1}^{T} \sigma_t - \frac{2}{3} \sqrt{\sigma_t}.$$

Thus, $\sigma_t \to 0$ at any speed will yield sublinear expected regret and variance of regret, hence, making no other restriction on the required rate of learning.

4.2   Log-Linearly Distributed Predictions

Suppose that $\tilde{x}_t \sim \ln \mathcal{L}_{\ln x_t, \sigma_t^2}$, where $\mathcal{L}$ is the linearly-decreasing distribution with largest density at the mean, $\ln x_t$. More precisely, it has the following probability density function

$$f(y) = \begin{cases} \frac{1}{\sigma_t} - \frac{|y - \ln x_t|}{\sigma_t^2} & \text{if } 0 \leq |y - \ln x_t| \leq \sigma_t, \\ 0 & \text{otherwise.} \end{cases}$$

In this case, applying Theorems 2 and 3 yields

$$\mathbb{E}[R] \leq 2 \sum_{t=1}^{T} \int_{0}^{\sigma_t} (1 - 2\frac{z}{\sigma_t} + \frac{z^2}{\sigma_t^2}) \, dz = 2 \sum_{t=1}^{T} \frac{\sigma_t}{3} = \frac{2}{3} \sum_{t=1}^{T} \sigma_t,$$

$$\text{Var}[R] \le 4 \sum_{t=1}^{T} \int_0^{\sigma_t} (1 - 2\frac{\sqrt{z}}{\sigma_t} + \frac{z}{\sigma_t^2})\, dz = 4 \sum_{t=1}^{T} \sigma_t - \frac{4}{3}\sqrt{\sigma_t} + \frac{1}{2} = \Theta(T).$$

so $\sigma_t \to 0$ at any speed will yield sublinear expected regret, but the bound on the variance of regret is linear in T.

4.3   Log-Normally Distributed Predictions

We will now look at the particular case when $\mathcal{D}_t$ is log-normally distributed (analogous to Geometric Brownian Motion). Suppose that $\tilde{x}_t \sim \ln \mathcal{N}_{\ln x_t, \sigma_t^2}$, then

$$\mathbb{E}[R] \le 4 \sum_{t=1}^{T} \int_0^{\infty} \Pr_{y \sim \mathcal{N}_{0,1}} [y > z/\sigma_t]\, dz.$$

To achieve a sublinear expected regret then depends on the ability to obtain an appropriate sequence of predictions with σ_t such that

$$\frac{1}{T} \sum_{t=1}^{T} \int_0^{\infty} \Pr_{y \sim \mathcal{N}_{0,1}} [y > z/\sigma_t]\, dz \to 0,$$

as $T \to \infty$. This has a very natural interpretation; the above condition can be viewed as an integral over the tail probabilities of the standard normal distribution, where the size of the tail is determined by σ_t.

Clearly, $\sigma_t = O(1)$ for all $t \in [T]$ is not a sufficient condition as the tail probabilities will not tend to zero for small values of z, so we must necessarily have that $\sigma_t \to 0$ as $t \to \infty$. However, it is unclear what rate of convergence would be required for this condition to hold. We suspect that $\sigma_t = O(1/\log t)$ suffices, but this remains to be shown and leaves an interesting open question. Similarly, the variance of regret in this case can be bounded as

$$\text{Var}[R] \le 8 \sum_{t=1}^{T} \int_0^{\infty} \Pr_{y \sim \mathcal{N}_{0,1}} [y > \sqrt{z}/\sigma_t]\, dz.$$

5   Portfolio Computation

The θ function can be viewed as a variant of the earth mover's distance, which, in turn, can be formulated as a transportation or flow problem and solved using a linear program. Here, we present an LP for computing $\hat{b}$ (and, hence, for similarly computing b^*) by first computing θ. The input to the computation is the original allocation vector $w = (w_1, \dots, w_m)$ (corresponding to $K\hat{b}$, where K is the total wealth before rebalancing and $b \in \mathcal{B}$) and the target portfolio vector given as $q = (q_1, \dots, q_m)$ (with $\sum_i q_i = 1$). The variables of the LP are the wealth W resulting after the rebalancing and f_{ij}, for $i, j \in [m]$, that corresponds to wealth that needs to be transferred from Asset i to Asset j.

$$\max W$$

subject to

$$\sum_{j \in [m]} f_{ij} \leq w_i \qquad\qquad \forall i = 1, \ldots, m \qquad\qquad (6)$$

$$f_{jj} + (1 - c) \cdot \sum_{\substack{i \in [m] \\ i \neq j}} f_{ij} \geq W \cdot q_j \qquad\qquad \forall j = 1, \ldots, m \qquad\qquad (7)$$

$$f_{ij} \geq 0 \qquad\qquad \forall i, j = 1, \ldots, m \qquad\qquad (8)$$

The constraints in (6) ensure that the wealth transferred out of each asset is bounded by the current wealth in that asset. The constraints in (7) ensure that the wealth that stays in each asset plus the wealth transferred into that asset, minus the incurred transaction costs, are sufficient to reach the target portfolio vector with a total wealth of W. Finally, the flow of wealth will always be positive by (8). Note that the sets of constraints in (6) and (7) will be satisfied tightly in an optimal solution. First of all, for any $i \in [m]$, total flow $\sum_{j \in [m]} f_{ij}$ out of Asset i will be equal to w_i, because any increase in the total flow $\sum_{i,j} f_{ij}$ can be distributed over the assets according to q, creating slack in each constraint in (7) and allowing a strictly larger value for W. Similarly, if the flow into any Asset j, given as $f_{jj} + (1-c) \cdot \sum_{i \in [m], i \neq j} f_{ij}$, was strictly larger than $W \cdot q_j$, then this excess flow can be shifted to other assets to create slack in each constraint in (7), which, in turn, allows W to be increased. The fact that the constraints in (6) and (7) are tight for an optimal solution shows that all the wealth in the previous time step is used during rebalancing and the resulting portfolio distribution adheres to q. Finally, by the maximisation of W, we get that the optimal solution to the LP gives the value of θ, and also $\hat{b}$ (by summing up all of the flow in/out of each asset f_{ij}). In the case where there are multiple optimal solutions, we choose the one with the lowest $\sum_{j \in [m]} f_{ij}$, for $i = 1, ..., m$ sequentially; that is, we break ties by minimising the outflow from the smallest to the largest i.

References

1. Agarwal, A., Hazan, E.: Efficient algorithms for online game playing and universal portfolio management. In: Electronic Colloquium on Computational Complexity, 13(033) (2006). http://eccc.hpi-web.de/eccc-reports/2006/TR06-033/index.html
2. Agarwal, A., Hazan, E., Kale, S., Schapire, R.E.: Algorithms for portfolio management based on the Newton method. In: Cohen, W.W., Moore, A. (eds.) Machine Learning, Proceedings of the Twenty-Third International Conference (ICML 2006). ACM International Conference Proceeding Series, vol. 148, pp. 9–16. ACM (2006). http://doi.acm.org/10.1145/1143844.1143846
3. Bean, A.J., Singer, A.C.: Universal switching and side information portfolios under transaction costs using factor graphs. In: Proceedings of the IEEE International Conference on Acoustics, Speech, and Signal Processing, ICASSP 2010, pp. 1986–1989. IEEE (2010). http://dx.doi.org/10.1109/ICASSP.2010.5495255

4. Bean, A.J., Singer, A.C.: Factor graph switching portfolios under transaction costs. In: Proceedings of the IEEE International Conference on Acoustics, Speech, and Signal Processing, ICASSP 2011, pp. 5748–5751. IEEE (2011). http://dx.doi.org/10.1109/ICASSP.2011.5947666

5. Black, F., Scholes, M.S.: The pricing of options and corporate liabilities. J. Polit. Econ. **81**(3), 637–654 (1973). https://ideas.repec.org/a/ucp/jpolec/v81y1973i3p637-54.html

6. Blum, A., Kalai, A.: Universal portfolios with and without transaction costs. Mach. Learn. **35**(3), 193–205 (1999). http://dx.doi.org/10.1023/A:1007530728748

7. Chiang, C.K., Yang, T., Lee, C.J., Mahdavi, M., Lu, C.J., Jin, R., Zhu, S.: Online optimization with gradual variations. In: COLT, p. 6-1 (2012)

8. Cover, T.M.: Universal portfolios. Math. Finan. **1**(1), 1–29 (1991). http://dx.doi.org/10.1111/j.1467-9965.1991.tb00002.x

9. Cover, T.M., Ordentlich, E.: Universal portfolios with side information. IEEE Trans. Inf. Theor. **42**(2), 348–363 (1996). http://dx.doi.org/10.1109/18.485708

10. Györfi, L., Walk, H.: Empirical portfolio selection strategies with proportional transaction costs. IEEE Trans. Inf. Theor. **58**(10), 6320–6331 (2012). http://dx.doi.org/10.1109/TIT.2012.2205131

11. Hazan, E., Agarwal, A., Kale, S.: Logarithmic regret algorithms for online convex optimization. Mach. Learn. **69**(2–3), 169–192 (2007). http://dx.doi.org/10.1007/s10994-007-5016-8

12. Karatzas, I., Shreve, S.E.: Brownian Motion and Stochastic Calculus. Graduatetexts in Mathematics. Springer, New York (1991). http://opac.inria.fr/record=b1079144, autres tirages corrigs: 1996, 1997, 1999, 2000, 2005

13. Kivinen, J., Warmuth, M.K.: Averaging expert predictions. In: Fischer, P., Simon, H.U. (eds.) EuroCOLT 1999. LNCS (LNAI), vol. 1572, pp. 153–167. Springer, Heidelberg (1999). http://dx.doi.org/10.1007/3-540-49097-3_13

14. Merton, R.C.: Optimum consumption and portfolio rules in a continuous-time model. J. Econ. Theor. **3**(4), 373–413 (1971). https://ideas.repec.org/a/eee/jetheo/v3y1971i4p373-413.html

15. Ordentlich, E., Cover, T.M.: On-line portfolio selection. In: Proceedings of the Ninth Annual Conference on Computational Learning Theory, COLT 1996, pp. 310–313. ACM, New York (1996). http://doi.acm.org/10.1145/238061.238161

16. Rakhlin, A., Sridharan, K.: Online learning with predictable sequences. In: COLT, pp. 993–1019 (2013)

17. Sharpe, W.F.: Capital asset prices: a theory of market equilibrium under conditions of risk. J. Finan. **19**(3), 425–442 (1964). https://ideas.repec.org/a/bla/jfinan/v19y1964i3p425-442.html

18. Stoltz, G., Lugosi, G.: Internal regret in on-line portfolio selection. Mach. Learn. **59**(1–2), 125–159 (2005). http://dx.doi.org/10.1007/s10994-005-0465-4

Bandits and Reinforcement Learning

Q(λ) with Off-Policy Corrections

Anna Harutyunyan[1]($\boxtimes$), Marc G. Bellemare[2], Tom Stepleton[2],
and Rémi Munos[2]

[1] VU Brussel, Brussels, Belgium
aharutyu@vub.ac.be
[2] Google DeepMind, London, UK
bellemare@google.com, stepleton@google.com, munos@google.com

Abstract. We propose and analyze an alternate approach to off-policy multi-step temporal difference learning, in which off-policy returns are corrected with the current Q-function in terms of rewards, rather than with the target policy in terms of transition probabilities. We prove that such approximate corrections are sufficient for off-policy convergence both in policy evaluation and control, provided certain conditions. These conditions relate the distance between the target and behavior policies, the eligibility trace parameter and the discount factor, and formalize an underlying tradeoff in off-policy TD(λ). We illustrate this theoretical relationship empirically on a continuous-state control task.

1 Introduction

In reinforcement learning (RL), learning is off-policy when samples generated by a *behavior* policy are used to learn about a distinct *target* policy. The usual approach to off-policy learning is to disregard, or altogether discard transitions whose target policy probabilities are low. For example, Watkins's Q(λ) [22] cuts the trajectory backup as soon as a non-greedy action is encountered. Similarly, in policy evaluation, importance sampling methods [9] weight the returns according to the mismatch in the target and behavior probabilities of the corresponding actions. This approach treats transitions conservatively, and hence may unnecessarily terminate backups, or introduce a large amount of variance.

Many off-policy methods, in particular of the Monte Carlo kind, have no other option than to judge off-policy actions in the probability sense. However, *temporal difference* methods [15] in RL maintain an approximation of the value function along the way, with *eligiblity traces* [23] providing a continuous link between one-step and Monte Carlo approaches. The value function assesses actions in terms of the following expected cumulative reward, and thus provides a way to directly correct immediate *rewards*, rather than transitions. We show in

---

A. Harutyunyan—This work was carried out during an internship at Google Deep-Mind.

R. Ortner et al. (Eds.): ALT 2016, LNAI 9925, pp. 305–320, 2016.
DOI: 10.1007/978-3-319-46379-7_21

this paper that such approximate corrections can be sufficient for off-policy convergence, subject to a tradeoff condition between the eligibility trace parameter and the distance between the target and behavior policies. The two extremes of this tradeoff are one-step Q-learning, and on-policy learning. Formalizing the continuum of the tradeoff is one of the main insights of this paper.

In particular, we propose an off-policy return operator that augments the return with a correction term, based on the current approximation of the Q-function. We then formalize three algorithms stemming from this operator: (1) off-policy $Q^\pi(\lambda)$, and its special case (2) *on*-policy $Q^\pi(\lambda)$, for policy evaluation, and (3) $Q^*(\lambda)$ for off-policy control.

In policy evaluation, both on- and off-policy $Q^\pi(\lambda)$ are novel, but closely related to several existing algorithms of the TD(λ) family. Section 7 discusses this in detail. We prove convergence of $Q^\pi(\lambda)$, subject to the $\lambda - \varepsilon$ tradeoff where $\varepsilon \stackrel{\text{def}}{=} \max_x \|\pi(\cdot|x) - \mu(\cdot|x)\|_1$ s a measure of dissimilarity between the behavior and target policies. More precisely, we prove that for any amount of "off-policy-ness" $\varepsilon \in [0, 2]$ there is an inherent maximum allowed backup length value $\lambda = \frac{1-\gamma}{\gamma\varepsilon}$, and taking λ below this value guarantees convergence to Q^π without involving policy probabilities. This is desirable due to the instabilities and variance introduced by the likelihood ratio products in the importance sampling approach [10].

In control, $Q^*(\lambda)$ is in fact identical to Watkins's Q(λ), except it does not cut the eligiblity trace at off-policy actions. Sutton and Barto [17] mention such a variation, which they call *naive* Q(λ). We analyze this algorithm for the first time and prove its convergence for small values of λ. Although we were not able to prove a $\lambda - \varepsilon$ tradeoff similar to the policy evaluation case, we provide empirical evidence for the existence of such a tradeoff, confirming the intuition that naive Q(λ) is "not as naive as one might at first suppose" [17].

We first give the technical background, and define our operators. We then specify the incremental versions of our algorithms based on these operators, and state their convergence. We follow by proving convergence: subject to the $\lambda - \varepsilon$ tradeoff in policy evaluation, and more conservatively, for small values of λ in control. We illustrate the tradeoff emerge empirically in the Bicycle domain in the control setting. Finally, we conclude by placing our algorithms in context within existing work in TD(λ).

2   Preliminaries

We consider an environment modelled by the usual discrete-time Markov Decision Process $(\mathcal{X}, \mathcal{A}, \gamma, P, r)$ composed of the finite state and action spaces $\mathcal{X}$ and $\mathcal{A}$, a discount factor γ, a transition function P mapping each $(x, a) \in (\mathcal{X}, \mathcal{A})$ to a distribution over $\mathcal{X}$, and a reward function $r : \mathcal{X} \times \mathcal{A} \to [-R_{\text{MAX}}, R_{\text{MAX}}]$. A *policy* π maps a state $x \in \mathcal{X}$ to a distribution over $\mathcal{A}$. A Q-function Q is a mapping $\mathcal{X} \times \mathcal{A} \to \mathbb{R}$. Given a policy π, we define the operator P^π over Q-functions:

$$(P^\pi Q)(x, a) \stackrel{\text{def}}{=} \sum_{x' \in \mathcal{X}} \sum_{a' \in \mathcal{A}} P(x' \,|\, x, a)\pi(a' \,|\, x')Q(x', a').$$

To each policy π corresponds a unique Q-function Q^π which describes the expected discounted sum of rewards achieved when following π:

$$Q^\pi \stackrel{\text{def}}{=} \sum_{t\geq 0} \gamma^t (P^\pi)^t r, \tag{1}$$

where for any operator X, $(X)^t$ denotes t successive applications of X, and where we commonly treat r as one particular Q-function. We write the *Bellman operator* $\mathcal{T}^\pi$, and the *Bellman equation* for Q^π:

$$\mathcal{T}^\pi Q \stackrel{\text{def}}{=} r + \gamma P^\pi Q,$$
$$\mathcal{T}^\pi Q^\pi = Q^\pi = (I - \gamma P^\pi)^{-1} r. \tag{2}$$

The *Bellman optimality operator* $\mathcal{T}$ is defined as $\mathcal{T}Q \stackrel{\text{def}}{=} r + \gamma \max_\pi P^\pi Q$, and it is well known e.g., [1,11] that the optimal Q-function $Q^* \stackrel{\text{def}}{=} \sup_\pi Q^\pi$ is the unique solution to the Bellman optimality equation

$$\mathcal{T}Q = Q. \tag{3}$$

We write $\text{GREEDY}(Q) \stackrel{\text{def}}{=} \{\pi | \pi(a|x) > 0 \Rightarrow Q(x,a) = \max_{a'} Q(x,a')\}$ to denote the set of greedy policies w.r.t. Q. Thus $\mathcal{T}Q = \mathcal{T}^\pi Q$ for any $\pi \in \text{GREEDY}(Q)$.

Temporal difference (TD) learning [15] rests on the fact that iterates of both operators $\mathcal{T}^\pi$ and $\mathcal{T}$ are guaranteed to converge to their respective fixed points Q^π and Q^*. Given a sample experience x, a, r, x', a', SARSA(0) [13] updates its Q-function estimate at k^{th} iteration as follows:

$$Q_{k+1}(x,a) \leftarrow Q_k(x,a) + \alpha_k \delta,$$
$$\delta = r + \gamma Q_k(x',a') - Q_k(x,a),$$

where δ is the *TD-error*, and $(\alpha_k)_{k\in\mathbb{N}}$ a sequence of nonnegative stepsizes. One need not only consider short experiences, but may sample trajectories $x_0, a_0, r_0, x_1, a_1, r_1, \ldots$, and accordingly apply $\mathcal{T}^\pi$ (or $\mathcal{T}$) repeatedly. A particularly flexible way of doing this is via a weighted sum A^λ of such *n-step* operators:

$$\mathcal{T}_\lambda^\pi Q \stackrel{\text{def}}{=} A^\lambda[(\mathcal{T}^\pi)^{n+1}Q]$$
$$= Q + (I - \lambda\gamma P^\pi)^{-1}(\mathcal{T}^\pi Q - Q),$$
$$A^\lambda[f(n)] \stackrel{\text{def}}{=} (1-\lambda)\sum_{n\geq 0} \lambda^n f(n).$$

Naturally, Q^π remains the fixed point of $\mathcal{T}_\lambda^\pi$. Taking $\lambda = 0$ yields the usual Bellman operator $\mathcal{T}^\pi$, and $\lambda = 1$ removes the recursion on the approximate Q-function, and restores Q^π in the *Monte Carlo* sense. It is well-known that λ trades off the bias from *bootstrapping* with an approximate Q-function, with the variance from using a sampled multi-step return [4], with intermediate values of λ usually performing best in practice [14,16]. The above λ-operator can be efficiently implemented in the online setting via a mechanism called *eligibility traces*.

As we will see in Sect. 7, it in fact corresponds to a number of online algorithms, each subtly different, of which SARSA(λ) [13] is the canonical instance.

Finally, we make an important distinction between the *target policy* π, which we wish to estimate, and the *behavior policy* μ, from which the actions have been generated. If $\mu = \pi$, the learning is said to be *on-policy*, otherwise it is *off-policy*. We will write $\mathbb{E}_\mu$ to denote expectations over sequences $x_0, a_0, r_0, x_1, a_1, r_1, \ldots$, $a_i \sim \mu(\cdot|x_i)$, $x_{i+1} \sim P(\cdot|x_i, a_i)$ and assume conditioning on $x_0 = x$ and $a_0 = a$ wherever appropriate. Throughout, we will write $\|\cdot\|$ for supremum norm.

3   Off-Policy Return Operators

We will now describe the Monte Carlo *off-policy corrected return operator* $\mathcal{R}^{\pi,\mu}$ that is at the heart of our contribution. Given a target π, and a return generated by the behavior μ, the operator $\mathcal{R}^{\pi,\mu}$ attempts to approximate a return that would have been generated by π, by utilizing a correction built from a current approximation Q of Q^π. Its application to Q at a state-action pair (x, a) is defined as follows:

$$(\mathcal{R}^{\pi,\mu}Q)(x,a) \stackrel{\text{def}}{=} r(x,a) + \mathbb{E}_\mu\Big[\sum_{t\geq 1}\gamma^t\big(r_t + \underbrace{\mathbb{E}_\pi Q(x_t,\cdot) - Q(x_t,a_t)}_{\text{off-policy correction}}\big)\Big], \quad (4)$$

where we use the shorthand $\mathbb{E}_\pi Q(x,\cdot) \equiv \sum_{a\in\mathcal{A}}\pi(a|x)Q(x,a)$.

That is, $\mathcal{R}^{\pi,\mu}$ gives the usual expected discounted sum of future rewards, but each reward in the trajectory is augmented with an *off-policy correction*, which we define as the difference between the *expected* (with respect to the target policy) Q-value and the Q-value for the taken action. Thus, how much a reward is corrected is determined by both the approximation Q, and the target policy probabilities. Notice that if actions are similarly valued, the correction will have little effect, and learning will be roughly on-policy, but if the Q-function has converged to the correct estimates Q^π, the correction takes the immediate reward r_t to the expected reward with respect to π exactly. Indeed, as we will see later, Q^π is the fixed point of $\mathcal{R}^{\pi,\mu}$ for any behavior policy μ.

We define the n-step and λ-versions of $\mathcal{R}^{\pi,\mu}$ in the usual way:

$$\mathcal{R}^{\pi,\mu}_\lambda Q \stackrel{\text{def}}{=} A^\lambda[\mathcal{R}^{\pi,\mu}_n], \quad (5)$$

$$(\mathcal{R}^{\pi,\mu}_n Q)(x,a) \stackrel{\text{def}}{=} r(x,a) + \mathbb{E}_\mu\Big[\sum_{t=1}^n\gamma^t\big(r_t + \mathbb{E}_\pi Q(x_t,\cdot) - Q(x_t,a_t)\big)$$
$$+ \gamma^{n+1}\mathbb{E}_\pi Q(x_{n+1},\cdot)\Big].$$

Note that the λ parameter here takes us from TD(0) to the Monte Carlo version of our operator $\mathcal{R}^{\pi,\mu}$, rather than the traditional Monte Carlo form (1).

4   Algorithm

We consider the problems of *off-policy policy evaluation* and *off-policy control*. In both problems we are given data generated by a sequence of behavior policies

---

Algorithm 1. Q(λ) with off-policy corrections

Given: Initial Q-function Q_0, stepsizes $(\alpha_k)_{k \in \mathbb{N}}$
  for $k = 1 \ldots$ **do**
    Sample a trajectory $x_0, a_0, r_0, \ldots, x_{T_k}$ from μ_k
    $Q_{k+1}(x, a) \leftarrow Q_k(x, a) \quad \forall x, a$
    $e(x, a) \leftarrow 0 \quad \forall x, a$
    for $t = 0 \ldots T_k - 1$ **do**
      $\delta_t^{\pi_k} \leftarrow r_t + \gamma \mathbb{E}_{\pi_k} Q_{k+1}(x_{t+1}, \cdot) - Q_{k+1}(x_t, a_t)$
      for all $x \in \mathcal{X}, a \in \mathcal{A}$ **do**
        $e(x, a) \leftarrow \lambda \gamma e(x, a) + \mathbb{I}\{(x_t, a_t) = (x, a)\}$
        $Q_{k+1}(x, a) \leftarrow Q_{k+1}(x, a) + \alpha_k \delta_t^{\pi_k} e(x, a)$
      end for
    end for
  end for

On-policy $\mathbf{Q}^\pi(\lambda)$: $\mu_k = \pi_k = \pi$.
Off-policy $\mathbf{Q}^\pi(\lambda)$: $\mu_k \neq \pi_k = \pi$.
$\mathbf{Q}^*(\lambda)$: $\pi_k \in \text{GREEDY}(Q_k)$.

$(\mu_k)_{k \in \mathbb{N}}$. In policy evaluation, we wish to estimate Q^π for a fixed target policy π. In control, we wish to estimate Q^*. Our algorithm constructs a sequence $(Q_k)_{k \in \mathbb{N}}$ of estimates of Q^{π_k} from trajectories sampled from μ_k, by applying the $\mathcal{R}_\lambda^{\pi_k, \mu_k}$-operator:

$$Q_{k+1} = \mathcal{R}_\lambda^{\pi_k, \mu_k} Q_k, \tag{6}$$

where π_k is the kth interim target policy. We distinguish between three algorithms:

Off-policy $\mathbf{Q}^\pi(\lambda)$ for policy evaluation: $\pi_k = \pi$ is the fixed target policy. We write the corresponding operator $\mathcal{R}_\lambda^\pi$.
On-policy $\mathbf{Q}^\pi(\lambda)$ for policy evaluation: for the special case of $\mu_k = \mu = \pi$.
$\mathbf{Q}^*(\lambda)$ for off-policy control: $(\pi_k)_{k \in \mathbb{N}}$ is a sequence of greedy policies with respect to Q_k. We write the corresponding operator $\mathcal{R}_\lambda^*$.

We wish to write the update (6) in terms of a simulated trajectory $x_0, a_0, r_0, \ldots, x_{T_k}$ drawn according to μ_k. First, notice that (5) can be rewritten:

$$\mathcal{R}_\lambda^{\pi, \mu} Q(x, a) = Q(x, a) + \mathbb{E}_\mu \Big[\sum_{t \geq 0} (\lambda \gamma)^t \delta_t^\pi \Big],$$

$$\delta_t^\pi \overset{\text{def}}{=} r_t + \gamma \mathbb{E}_\pi Q(x_{t+1}, \cdot) - Q(x_t, a_t),$$

where δ_t^π is the *expected* TD-error. The *offline* forward view[1] is then

$$Q_{k+1}(x, a) \leftarrow Q_k(x, a) + \alpha_k \sum_{t=0}^{T_k} (\gamma \lambda)^t \delta_t^{\pi_k}, \tag{7}$$

---

[1] The true online version can be derived as given by van Seijen and Sutton [20].

While (7) resembles many existing TD(λ) algorithms, it subtly differs from all of them, due to $\mathcal{R}_\lambda^{\pi,\mu}$ (rather than T_λ^π) being at its basis. Section 7 discusses the distinctions in detail. The practical *every-visit* [17] form of (7) is written

$$Q_{k+1}(x,a) \leftarrow Q_k(x,a) + \alpha_k \sum_{t=0}^{T} \delta_t^{\pi_k} \sum_{s=0}^{t} (\gamma\lambda)^{t-s} \mathbb{I}\{(x_s,a_s) = (x,a)\}, \qquad (8)$$

and the corresponding online backward view of all three algorithms is summarized in Algorithm 1.

The following theorem states that when μ and π are sufficiently close, the off-policy $Q^\pi(\lambda)$ algorithm converges to its fixed point Q^π.

Theorem 1. *Consider the sequence of Q-functions computed according to Algorithm 1 with fixed policies μ and π. Let $\varepsilon = \max_x \|\pi(\cdot|x) - \mu(\cdot|x)\|_1$. If $\lambda\varepsilon < \frac{1-\gamma}{\gamma}$, then under the same conditions required for the convergence of TD(λ) (1–3 in Sect. 5.3) we have, almost surely:*

$$\lim_{k\to\infty} Q_k(x,a) = Q^\pi(x,a).$$

We state a similar, albeit weaker result for $Q^*(\lambda)$.

Theorem 2. *Consider the sequence of Q-functions computed according to Algorithm 1 with π_k the greedy policy with respect to Q_k. If $\lambda < \frac{1-\gamma}{2\gamma}$, then under the same conditions required for the convergence of TD(λ) (1–3 in Sect. 5.3) we have, almost surely:*

$$\lim_{k\to\infty} Q_k(x,a) = Q^*(x,a).$$

The proofs of these theorems rely on showing that $\mathcal{R}_\lambda^\pi$ and $\mathcal{R}_\lambda^*$ are contractions (under the stated conditions), and invoking classical stochastic approximation convergence to their fixed point (such as Proposition 4.5 from [2]). We will focus on the contraction lemmas, which are the crux of the proofs, then outline the sketch of the online convergence argument.

Discussion. Theorem 1 states that for *any* $\lambda \in [0,1]$ there exists some degree of "off-policy-ness" $\varepsilon < \frac{1-\gamma}{\lambda\gamma}$ under which Q_k converges to Q^π. This is the $\lambda - \varepsilon$ tradeoff for the off-policy $Q^\pi(\lambda)$ learning algorithm for policy evaluation. In the control case, the result of Theorem 2 is weaker as it only holds for values of λ smaller than $\frac{1-\gamma}{2\gamma}$. Notice that this threshold corresponds to the policy evaluation case for $\varepsilon = 2$ (arbitrary off-policy-ness). We were not able to prove convergence to Q^* for any $\lambda \in [0,1]$ and some $\varepsilon > 0$. This is left as an open problem for now[2].

The main technical difficulty lies in the fact that in control, the greedy policy with respect to the current Q_k may change drastically from one step to the next,

---

[2] For a general convergence result (for any λ and any ε), we refer the reader to the follow-up work [7].

while Q_k itself changes incrementally (under small learning steps α_k). So the current Q_k may not offer a good off-policy correction to evaluate the new greedy policy. In order to circumvent this problem we may want to use slowly changing target policies π_k. For example we could keep π_k fixed for slowly increasing periods of time. This can be seen as a form of optimistic policy iteration [11] where policy improvement steps alternate with approximate policy evaluation steps (and when the policy is fixed, Theorem 1 guarantees convergence to the value function of that policy). Another option would be to define π_k as the empirical average $\pi_k \stackrel{\text{def}}{=} \frac{1}{k} \sum_{i=1}^{k} \pi'_i$ of the previous greedy policies π'_i. We conjecture that defining π_k such that (1) π_k changes slowly with k, and (2) π_k becomes increasingly greedy, then we could extend the $\lambda - \varepsilon$ tradeoff of Theorem 1 to the control case. This is left for future work.

5   Analysis

We begin by verifying that the fixed points of $\mathcal{R}_\lambda^{\pi,\mu}$ in the policy evaluation and control settings are Q^π and Q^*, respectively. We then prove the contractive properties of these operators: $\mathcal{R}_\lambda^\pi$ is always a contraction and will converge to its fixed point, $\mathcal{R}_\lambda^*$ is a contraction for particular choices of λ (given in terms of γ). The contraction coefficients depend on λ, γ, and ε: the distance between policies. Finally, we give a proof sketch for online convergence of Algorithm 1.

Before we begin, it will be convenient to rewrite (4) for all state-action pairs:

$$\mathcal{R}^{\pi,\mu}Q = r + \sum_{t \geq 1} \gamma^t (P^\mu)^{t-1}[P^\mu r + P^\pi Q - P^\mu Q].$$

We can then write $\mathcal{R}_\lambda^\pi$ and $\mathcal{R}_\lambda^*$ from (5) as follows:

$$\mathcal{R}_\lambda^\pi Q \stackrel{\text{def}}{=} Q + (I - \lambda\gamma P^\mu)^{-1}[T^\pi Q - Q], \tag{9}$$

$$\mathcal{R}_\lambda^* Q \stackrel{\text{def}}{=} Q + (I - \lambda\gamma P^\mu)^{-1}[TQ - Q]. \tag{10}$$

It is not surprising that the above along with the Bellman equations (2) and (3) directly yields that Q^π and Q^* are the fixed points of $\mathcal{R}_\lambda^\pi$ and $\mathcal{R}_\lambda^*$:

$$\mathcal{R}_\lambda^\pi Q^\pi = Q^\pi,$$
$$\mathcal{R}_\lambda^* Q^* = Q^*.$$

It then remains to analyze the behavior of $\mathcal{R}_\lambda^{\pi,\mu}$ as it gets iterated.

5.1 λ-Return for Policy Evaluation: $Q^\pi(\lambda)$

We first consider the case with a fixed arbitrary policy π. For simplicity, we take μ to be fixed as well, but the same will hold for any sequence $(\mu_k)_{k \in \mathbb{N}}$, as long as each μ_k satisfies the condition imposed on μ.

Lemma 1. *Consider the policy evaluation algorithm $Q_k = (\mathcal{R}_\lambda^\pi)^k Q$. Assume the behavior policy μ is ε-away from the target policy π, in the sense that $\max_x \|\pi(\cdot|x) - \mu(\cdot|x)\|_1 \le \varepsilon$. Then for $\varepsilon < \frac{1-\gamma}{\lambda\gamma}$, the sequence $(Q_k)_{k \ge 1}$ converges to Q^π exponentially fast: $\|Q_k - Q^\pi\| = O(\eta^k)$, where $\eta = \frac{\gamma}{1-\lambda\gamma}(1 - \lambda + \lambda\varepsilon) < 1$.*

Proof. First notice that

$$
\begin{aligned}
\|P^\pi - P^\mu\| &= \sup_{\|Q\| \le 1} \|(P^\pi - P^\mu)Q\| \\
&= \sup_{\|Q\| \le 1} \max_{x,a} \left| \sum_y P(y|x,a) \sum_b ((\pi(b|y) - \mu(b|y)) Q(y,b) \right| \\
&\le \max_{x,a} \sum_y P(y|x,a) \sum_b |\pi(b|y) - \mu(b|y)| \le \varepsilon.
\end{aligned}
$$

Let $B = (I - \lambda\gamma P^\mu)^{-1}$ be the resolvent matrix. From (9) we have

$$
\begin{aligned}
\mathcal{R}_\lambda^\pi Q - Q^\pi &= B\big[T^\pi Q - Q + (I - \lambda\gamma P^\mu)(Q - Q^\pi)\big] \\
&= B\big[r + \gamma P^\pi Q - Q^\pi - \lambda\gamma P^\mu(Q - Q^\pi)\big] \\
&= B\big[\gamma P^\pi(Q - Q^\pi) - \lambda\gamma P^\mu(Q - Q^\pi)\big] \\
&= \gamma B\big[(1 - \lambda)P^\pi + \lambda(P^\pi - P^\mu)\big](Q - Q^\pi).
\end{aligned}
$$

Taking the sup norm, since μ is ε-away from π:

$$
\|\mathcal{R}_\lambda^\pi Q - Q^\pi\| \le \eta \|Q - Q^\pi\|
$$

for $\eta = \frac{\gamma}{1-\lambda\gamma}(1 - \lambda + \lambda\varepsilon) < 1$. Thus $\|Q_k - Q^\pi\| = O(\eta^k)$.

5.2 λ-Return for Control: $Q^*(\lambda)$

We next consider the case where the kth target policy π_k is greedy with respect to the value estimate Q_k. The following Lemma states that is possible to select a small, but nonzero λ and still guarantee convergence.

Lemma 2. *Consider the off-policy control algorithm $Q_k = (\mathcal{R}_\lambda^*)^k Q$. Then*

$$
\|\mathcal{R}_\lambda^* Q_k - Q^*\| \le \frac{\gamma + \lambda\gamma}{1 - \lambda\gamma} \|Q_k - Q^*\|,
$$

and for $\lambda < \frac{1-\gamma}{2\gamma}$ the sequence $(Q_k)_{k \ge 1}$ converges to Q^* exponentially fast.

Proof. Fix μ and let $B = (I - \lambda\gamma P^\mu)^{-1}$. Using (10), we write

$$
\begin{aligned}
\mathcal{R}_\lambda^* Q - Q^* &= B\left[TQ - Q + (I - \lambda\gamma P^\mu)(Q - Q^*)\right] \\
&= B\left[TQ - Q^* - \lambda\gamma P^\mu(Q - Q^*)\right].
\end{aligned}
$$

Taking the sup-norm, since $\|TQ - Q^*\| \le \gamma\|Q - Q^*\|$, we deduce the result:

$$
\|\mathcal{R}_\lambda^* Q - Q^*\| \le \frac{\gamma + \lambda\gamma}{1 - \lambda\gamma} \|Q - Q^*\|.
$$

5.3   Online Convergence

We are now ready to prove the online convergence of Algorithm 1. Let the following hold for every sample trajectory τ_k and all $x \in \mathcal{X}, a \in \mathcal{A}$:

1. **Minimum visit frequency:** $\sum_{t \geq 0} \mathbb{P}\{x_t, a_t = x, a\} \geq D > 0$.
2. **Finite trajectories:** $\mathbb{E}_{\mu_k} T_k^2 < \infty$, where T_k is the length of τ_k.
3. **Bounded stepsizes:** $\sum_{k \geq 0} \alpha_k(x, a) = \infty$, $\sum_{k \geq 0} \alpha_k^2(x, a) < \infty$.

Assumption 2 requires trajectories to be finite w.p. 1, which is satisfied by *proper* behavior policies. Equivalently, we may require from the MDP that all trajectories eventually reach a zero-value absorbing state. The proof closely follows that of Proposition 5.2 from [2], and requires rewriting the update in the suitable form, and verifying Assumptions (a) through (d) from their Proposition 4.5.

Proof. (Sketch) Let $z_{k,t}(x, a) \stackrel{\text{def}}{=} \sum_{s=0}^{t} (\gamma \lambda)^{t-s} \mathbb{I}\{(x_s, a_s) = (x, a)\}$ denote the accumulating trace. It follows from Assumptions 1 and 2 that the total update at phase k is bounded, which allows us to write the online version of (8) as

$$Q_{k+1}^o(x, a) \leftarrow (1 - D_k \alpha_k) Q_k^o(x, a) + D_k \alpha_k \big(\mathcal{R}_\lambda^{\pi_k, \mu_k} Q_k^o(x, a) + w_k + u_k \big)$$

$$w_k \stackrel{\text{def}}{=} (D_k)^{-1} \Big[\sum_{t \geq 0} z_{k,t} \delta_t^{\pi_k} - \mathbb{E}_{\mu_k} \Big[\sum_{t \geq 0} z_{k,t} \delta_t^{\pi_k} \Big] \Big],$$

$$u_k \stackrel{\text{def}}{=} (D_k \alpha_k)^{-1} \big(Q_{k+1}^o(x, a) - Q_{k+1}(x, a) \big),$$

where $D_k(x, a) \stackrel{\text{def}}{=} \sum_{t \geq 0} \mathbb{P}\{x_t, a_t = x, a\}$, and we use the shorthand $y_k \equiv y_k(x, a)$ for α_k, D_k, w_k, u_k, and $z_{k,t}$. Combining Assumptions 1 and 2, we have $0 < D \leq D_k(x, a) < \infty$, which, combined in turn with Assumption 3, assures that the new stepsize sequence $\tilde{\alpha}_k(x, a) = (D_k \alpha_k)(x, a)$ satisfies Assumption (a) of Prop. 4.5. Assumptions (b) and (d) require the variance of the noise term $w_k(x, a)$ to be bounded, and the residual $u_k(x, a)$ to converge to zero, both of which can be shown identically to the corresponding results from [2], if Assumption 2 and Assumption (a) are satisfied. Finally, Assumption (c) is satisfied by Lemmas 1 and 2 for the policy evaluation and control cases, respectively.[3] We conclude that the sequence $(Q_k^o)_{k \in \mathbb{N}}$ converges to Q^π or Q^* in the respective settings, w.p. 1.

6   Experimental Results

Although we do not have a proof of the $\lambda - \varepsilon$ tradeoff (see Sect. 4) in the control case , we wished to investigate whether such a tradeoff can be observed experimentally. To this end, we applied $Q^*(\lambda)$ to the Bicycle domain [12]. Here, the agent must simultaneously balance a bicycle and drive it to a goal position. Six real-valued variables describe the state – angle, velocity, etc. – of the bicycle. The reward function is proportional to the angle to the goal, and gives -1 for

---

[3] Note that the control case goes through without modifications, for the values of λ prescribed by Lemma 2.

falling and +1 for reaching the goal. The discount factor is 0.99. The Q-function was approximated using multilinear interpolation over a uniform grid of size $10 \times \cdots \times 10$, and the stepsize was tuned to 0.1. We are chiefly interested in the interplay between the λ parameter in $Q^*(\lambda)$ and an ε-greedy exploration policy. Our main performance indicator is the frequency at which the goal is reached by the greedy policy after 500,000 episodes of training. We report three findings:

1. Higher values of λ lead to improved learning;
2. Very low values of ε exhibit lower performance; and
3. The Q-function diverges when λ is high relative to ε.

Together, these findings suggest that there is indeed a $\lambda-\varepsilon$ tradeoff in the control case as well, and lead us to conclude that with proper care it can be beneficial to do off-policy control with $Q^*(\lambda)$.

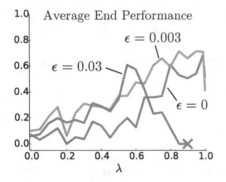 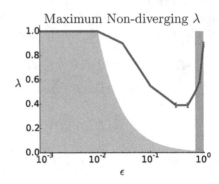

Fig. 1. Left. Performance of $Q^*(\lambda)$ on the Bicycle domain. Each configuration is an average of five trials. The 'X' marks the lowest value of λ for which $\varepsilon = 0.03$ causes divergence. **Right.** Maximum non-diverging λ in function of ε. The left-hand shaded region corresponds to our hypothesized bound. Parameter settings in the right-hand shaded region do not produce meaningful policies.

Learning Speed and Performance. Figure 1 (left) depicts the performance of $Q^*(\lambda)$, in terms of the goal-reaching frequency, for three values of ε. The agent performs best ($p < 0.05$) for $\varepsilon \in [0.003, 0.03]$ and high (w.r.t. ε) values of λ.[4]

Divergence. For each value of ε, we determined the highest *safe* choice of λ which did not result in divergence. As Fig. 1 (right) illustrates, there is a marked decrease in what is a safe value of λ as ε increases. Note the left-hand shaded region corresponding to the *policy evaluation* bound $\frac{1-\gamma}{\gamma\varepsilon}$. Supporting our hypothesis on the true bound on λ (Sect. 5), it appears clear that the maximum safe value of λ depends on ε. In particular, notice how $\lambda = 1$ stops diverging exactly where predicted by this bound.

---

[4] Recall that Randløv and Alstrøm's agent was trained using SARSA(λ) with $\lambda = 0.95$.

7   Related Work

In this section, we place the presented algorithms in context of the existing work in TD(λ) [17], focusing in particular on action-value methods. As usual, let $(x_t, a_t, r_t)_{t \geq 0}$ be a trajectory generated by following a behavior policy μ, i.e. $a_t \sim \mu(\cdot | x_t)$. At time s, SARSA(λ) [13] updates its Q-function as follows:

$$Q_{s+1}(x_s, a_s) \leftarrow Q_s(x_s, a_s) + \alpha_s(\underbrace{A^\lambda R_s^{(n)} - Q(x_s, a_s)}_{\Delta_s}), \tag{11}$$

$$R_s^{(n)} = \sum_{t=s}^{s+n} \gamma^{t-s} r_t + \gamma^{n+1} Q(x_{s+n+1}, a_{s+n+1}), \tag{12}$$

where Δ_s denotes the update made at time s, and can be rewritten in terms of one-step TD-errors:

$$\Delta_s = \sum_{l \geq s} (\lambda \gamma)^{t-s} \delta_t, \tag{13}$$

$$\delta_t = r_t + \gamma Q(x_{t+1}, a_{t+1}) - Q(x_t, a_t).$$

SARSA(λ) is an on-policy algorithm and converges to the value function Q^μ of the behavior policy. Different algorithms arise by instantiating $R_s^{(n)}$ or Δ_s from (11) differently. Table 1 provides the full details, while in text we will specify the most revealing components of the update.

7.1   Policy Evaluation

One can imagine considering *expectations* over action-values at the corresponding states $\mathbb{E}_\pi Q(x_t, \cdot)$, in place of the value of the sampled action $Q(x_t, a_t)$, i.e.:

$$\delta_t = r_t + \gamma \mathbb{E}_\pi Q(x_{t+1}, \cdot) - \mathbb{E}_\pi Q(x_t, \cdot). \tag{14}$$

This is the one-step update for *General Q-Learning* [19], which is a generalization of *Expected SARSA* [21] to arbitrary policies. We refer to the direct eligibility trace extensions of these algorithms formed via Equations (11)-(13) by General Q(λ) and Expected SARSA(λ) (first mentioned by Sutton et al. [18]) Unfortunately, in an off-policy setting, General Q(λ) will not converge to the value function Q^π of the target policy, as stated by the following proposition.

Proposition 1. *The stable point of General Q(λ) is $Q^{\mu,\pi} = (I - \lambda\gamma(P^\mu - P^\pi) - \gamma P^\pi)^{-1} r$ which is the fixed point of the operator $(1 - \lambda)T^\pi + \lambda T^\mu$.*

Proof. Writing the algorithm in operator form, we get

$$\mathcal{R}Q = (1 - \lambda) \sum_{n \geq 0} \lambda^n \left[\sum_{t=0}^{n} \gamma^t (P^\mu)^t r + \gamma^{n+1} (P^\mu)^n P^\pi Q \right]$$

$$= \sum_{t \geq 0} (\lambda\gamma)^t (P^\mu)^t \left[r + (1 - \lambda)\gamma P^\pi Q \right] = (I - \lambda\gamma P^\mu)^{-1} \left[r + (1 - \lambda)\gamma P^\pi Q \right].$$

Thus the fixed point $Q^{\mu,\pi}$ of $\mathcal{R}$ satisfies the following:

$$Q^{\mu,\pi} = (I - \lambda\gamma P^\mu)^{-1}\left[r + (1-\lambda)\gamma P^\pi Q^{\mu,\pi}\right] = (1-\lambda)T^\pi Q^{\mu,\pi} + \lambda T^\mu Q^{\mu,\pi}.$$

Solving for $Q^{\mu,\pi}$ yields the result.

Alternatively to replacing both terms with an expectation, one may only replace the value at the *next* state x_{t+1} by $\mathbb{E}_\pi Q(x_{t+1}, \cdot)$, obtaining:

$$\delta_t^\pi = r_t + \gamma\mathbb{E}_\pi Q(x_{t+1}, \cdot) - Q(x_t, a_t). \tag{15}$$

This is exactly our policy evaluation algorithm $Q^\pi(\lambda)$. Specifically, when $\pi = \mu$, we get the on-policy $Q^\pi(\lambda)$. The induced *on-policy* correction may serve as a variance reduction term for Expected SARSA(λ) (it may be helpful to refer to the n-step return in Table 1 to observe this), but we leave variance analysis of this algorithm for future work. When $\pi \neq \mu$, we recover off-policy $Q^\pi(\lambda)$, which (under the stated conditions) converges to Q^π.

Target Policy Probability Methods: The algorithms above directly descend from basic SARSA(λ), but often learning off-policy requires special treatment. For example, a typical off-policy technique is importance sampling (IS) [10]. It is a classical Monte Carlo method that allows one to sample from the available distribution, but obtain (unbiased or consistent) samples of the desired one, by reweighing the samples with their likelihood ratio according to the two distributions. That is, the updates for the ordinary *per-decision* IS algorithm for policy evaluation are made as follows:

$$\Delta_s = \sum_{t \geq s}(\lambda\gamma)^{t-s}\delta_t \prod_{i=s+1}^{t}\frac{\pi(a_i|x_i)}{\mu(a_i|x_i)}$$

$$\delta_t = r_t + \gamma\frac{\pi(a_{t+1}|s_{t+1})}{\mu(a_{t+1}|s_{t+1})}Q(x_{t+1}, a_{t+1}) - Q(x_t, a_t).$$

This family of algorithms converges to Q^π with probability 1, under any soft, stationary behavior μ [9]. There are several (recent) off-policy algorithms that reduce the variance of IS methods, at the cost of added bias [3,5,6].

However, off-policy $Q^\pi(\lambda)$ is perhaps related closest to the *Tree-Backup (TB) algorithm*, also discussed by Precup et al. [9]. Its one-step TD-error is the same as (15), the algorithms back up the same tree, and neither requires knowledge of the behavior policy μ. The important difference is in the weighting of the updates. As an off-policy precaution, TB(λ) weighs updates along a trajectory with the cumulative target probability of that trajectory up until that point:

$$\Delta_s = \sum_{t \geq s}(\lambda\gamma)^{t-s}\delta_t^\pi \prod_{i=s+1}^{t}\pi(a_i|x_i). \tag{16}$$

Table 1. Comparison of the update rules of several learning algorithms using the λ-return. We show both the n-step return and the resulting update rule for the λ-return from any state x_s when following a behavior policy $a_t \sim \mu(\cdot|x_t)$. **Top part, policy evaluation algorithms:** SARSA(λ), Expected SARSA(λ), General Q(λ), Per-Decision Importance Sampling (PDIS(λ)), TB(λ), and $Q^\pi(\lambda)$, in both on-policy (i.e. $\pi = \mu$) and off-policy settings (with a target policy $\pi \neq \mu$). Note the same $Q^\pi(\lambda)$ equation applies to both on- and off-policy settings. We abbreviate $\pi_i \equiv \pi(a_i|x_i)$, $\mu_i \equiv \mu(a_i|x_i)$, $\rho_i \equiv \pi_i/\mu_i$, and write $\mathbb{E}_\pi^{a\neq b} Q(x,\cdot) \equiv \sum_{a\in A\backslash b} \pi(a|x) Q(x,a)$. **Bottom part, control algorithms:** Watkins's Q(λ), Peng and Williams's Q(λ), and $Q^*(\lambda)$. The **FP** column denotes the stable point of these algorithms (i.e. the fixed point of the expected update), regardless of whether the algorithm converges to it. General Q(λ) may converge to $Q^{\mu,\pi}$ defined as the fixed point of the Bellman operator $(1-\lambda)T^\pi + \lambda T^\mu$. The fixed point of Watkins's Q(λ) is Q^* but the case $\lambda > 0$ may not be significantly better than $\lambda = 0$ (regular Q-learning) if the behavior policy is different from the greedy one. The fixed point $Q^{\mu,*}$ of Peng and Williams's Q(λ) is the fixed point of $(1-\lambda)T + \lambda T^\mu$, which is different from Q^* when $\mu \neq \pi$ (see Proposition 1). The algorithms analyzed in this paper are $Q^\pi(\lambda)$ and $Q^*(\lambda)$, for which convergence to respectively Q^π and Q^* occurs under some conditions (see Lemmas).

Algorithm	n-step return	Update rule for the λ-return	FP
TD(λ) (on-policy)	$\sum_{t=s}^{s+n} \gamma^{t-s} r_t + \gamma^{n+1} V(x_{s+n+1})$	$\sum_{t\geq s} (\gamma\lambda)^{t-s} \delta_t$ $\delta_t = r_t + \gamma V(x_{t+1}) - V(x_t)$	V^μ
SARSA(λ) (on-policy)	$\sum_{t=s}^{s+n} \gamma^{t-s} r_t + \gamma^{n+1} Q(x_{s+n+1}, a_{s+n+1})$	$\sum_{t\geq s} (\gamma\lambda)^{t-s} \delta_t$	Q^μ
$\mathbb{E}$ SARSA(λ) (on-policy)	$\sum_{t=s}^{s+n} \gamma^{t-s} r_t + \gamma^{n+1} \mathbb{E}_\mu Q(x_{s+n+1}, \cdot)$	$\sum_{t\geq s} (\gamma\lambda)^{t-s} \delta_t + \mathbb{E}_\mu Q(x_{s}, \cdot) - Q(x_s, a_s)$ $\delta_t = r_t + \gamma \mathbb{E}_\mu Q(x_{t+1}, \cdot) - \mathbb{E}_\mu Q(x_t, \cdot)$	Q^μ
General Q(λ) (off-policy)	$\sum_{t=s}^{s+n} \gamma^{t-s} r_t + \gamma^{n+1} \mathbb{E}_\pi Q(x_{s+n+1}, \cdot)$	$\sum_{t\geq s} (\gamma\lambda)^{t-s} \delta_t + \mathbb{E}_\pi Q(x_{s}, \cdot) - Q(x_s, a_s)$ $\delta_t = r_t + \gamma \mathbb{E}_\pi Q(x_{t+1}, \cdot) - \mathbb{E}_\pi Q(x_t, \cdot)$	$Q^{\mu,\pi}$
PDIS(λ) (off-policy)	$\sum_{t=s}^{s+n} \gamma^{t-s} r_t \prod_{i=s+1}^t \rho_i$ $+ \gamma^{n+1} Q(x_{s+n+1}, a_{s+n+1}) \prod_{i=s}^{s+n} \rho_i$	$\sum_{t\geq s} (\gamma\lambda)^{t-s} \delta_t \prod_{i=s+1}^t \rho_i$ $\delta_t = r_t + \gamma\rho_{t+1} Q(x_{t+1}, a_{t+1}) - Q(x_t, a_t)$	Q^π
TB(λ) (off-policy)	$\sum_{t=s}^{s+n} \gamma^{t-s} \prod_{i=s+1}^t \pi_i [r_t + \gamma \mathbb{E}_\pi^{a\neq a_{t+1}} Q(x_{t+1}, \cdot)]$ $+ \gamma^{n+1} \prod_{i=s+1}^{s+n-1} \pi_i Q(x_{s+n+1}, a_{s+n+1})$	$\sum_{t\geq s} (\gamma\lambda)^{t-s} \delta_t \prod_{i=s+1}^t \pi_i$ $\delta_t = r_t + \gamma \mathbb{E}_\pi Q(x_{t+1}, \cdot) - Q(x_t, a_t)$	Q^π
$Q^\pi(\lambda)$ (on/off-policy)	$\sum_{t=s}^{s+n} \gamma^{t-s} [r_t + \mathbb{E}_\pi Q(x_t, \cdot) - Q(x_t, a_t)]$ $+ \gamma^{n+1} \mathbb{E}_\pi Q(x_{s+n+1}, \cdot)$	$\sum_{t\geq s} (\gamma\lambda)^{t-s} \delta_t$ $\delta_t = r_t + \gamma \mathbb{E}_\pi Q(x_{t+1}, \cdot) - Q(x_t, a_t)$	Q^π
Q(λ) (Watkins's)	$\sum_{t=s}^{s+n} \gamma^{t-s} r_t + \gamma^{n+1} \max_a Q(x_{s+n+1}, a)$ (for any $n < \tau = \arg\min_{u\geq 1} \mathbb{1}[\pi_{s+u} \neq \mu_{s+u}]$)	$\sum_{t=s}^{s+\tau} (\gamma\lambda)^{t-s} \delta_t \prod_{i=s+1}^t$ $\delta_t = r_t + \gamma \max_a Q(x_{t+1}, a) - Q(x_t, a_t)$	Q^*
Q(λ) (P & W's)	$\sum_{t=s}^{s+n} \gamma^{t-s} r_t + \gamma^{n+1} \max_a Q(x_{s+n+1}, a)$	$\sum_{t\geq s} (\gamma\lambda)^{t-s} \delta_t + \max_a Q(x_s, a) - Q(x_s, a_s)$ $\delta_t = r_t + \gamma \max_a Q(x_{t+1}, a) - \max_a Q(x_t, a)$	$Q^{\mu,*}$
$Q^*(\lambda)$	$\sum_{t=s}^{s+n} \gamma^{t-s} [r_t + \max_a Q(x_t, a) - Q(x_t, a_t)]$ $+ \gamma^{n+1} \max_a Q(x_{s+n+1}, a)$	$\sum_{t\geq s} (\gamma\lambda)^{t-s} \delta_t$ $\delta_t = r_t + \gamma \max_a Q(x_{t+1}, a) - Q(x_t, a_t)$	Q^*

The weighting simplifies the convergence argument, allowing TB(λ) to converge to Q^{π} without further restrictions on the distance between μ and π [9]. The drawback of TB(λ) is that in the case of near on-policy-ness (when μ is close to π) the product of the probabilities cuts the traces unnecessarily (especially when the policies are stochastic). What we show in this paper, is that plain TD-learning *can* converge off-policy with no special treatment, subject to a tradeoff condition on λ and ε. Under that condition, $Q^{\pi}(\lambda)$ applies both on- and off-policy, without modifications. An ideal algorithm should be able to automatically cut the traces (like TB(λ)) in case of extreme off-policy-ness while reverting to $Q^{\pi}(\lambda)$ when being near on-policy.

7.2   Control

Perhaps the most popular version of Q(λ) is due to Watkins and Dayan [22]. Off-policy, it truncates the return and bootstraps as soon as the behavior policy takes a non-greedy action, as described by the following update:

$$\Delta_s = \sum_{t=s}^{s+\tau} (\lambda\gamma)^{t-s} \delta_t, \tag{17}$$

where $\tau = \min\{u \geq 1 : a_{s+u} \notin \arg\max_a Q(x_{s+u}, a)\}$. Note that this update is a special case of (16) for deterministic greedy policies, with $\prod_{i=s+1}^{t} \mathbb{I}\{a_i \in \arg\max_a Q(x_i, a)\}$ replacing the probability product. When the policies μ and π are not too similar, and λ is not too small, the truncation may greatly reduce the benefit of complex backups.

Q(λ) of Peng and Williams [8] is meant to remedy this, by being a hybrid between SARSA(λ) and Watkins's Q(λ). Its n-step return $\sum_{t=s}^{s+n} \gamma^{t-s} r_t + \gamma^{n+1} \max_a Q(x_{s+n+1}, a)$ requires the following form for the TD-error:

$$\delta_t = r(x_t, a_t) + \gamma \max_a Q(x_{t+1}, a) - \max_a Q(x_t, a).$$

This is, in fact, the same update rule as the General Q(λ) defined in (14), where π is the greedy policy. Following the same steps as in the proof of Proposition 1, the limit of this algorithm (if it converges) will be the fixed point of the operator $(1 - \lambda)\mathcal{T} + \lambda\mathcal{T}^{\mu}$ which is different from Q^* unless the behavior is always greedy.

Sutton and Barto [17] mention another, *naive* version of Watkins's Q(λ) that does not cut the trace on non-greedy actions. That is exactly the Q*(λ) algorithm described in this paper. Notice that despite the similarity to Watkins's Q(λ), the equivalence representation for Q*(λ) is different from the one that would be derived by setting $\tau = \infty$ in (17), since the n-step return uses the *corrected* immediate reward $r_t + \gamma \max_a Q(x_t, a) - Q(x_t, a_t)$ instead of the immediate reward alone. This correction is invisible in Watkins's Q(λ), since the behavior policy is assumed to be greedy, before the return is cut off.

8   Conclusion

We formulated new algorithms of the TD(λ) family for off-policy policy evaluation and control. Unlike traditional off-policy learning algorithms, these methods do not involve weighting returns by their policy probabilities, yet under the right conditions converge to the correct TD fixed points. In policy evaluation, convergence is subject to a tradeoff between the degree of bootstrapping λ, distance between policies ε, and the discount factor γ. In control, determining the existence of a non-trivial ε-dependent bound for λ remains an open problem. Supported by telling empirical results in the Bicycle domain, we hypothesize that such a bound exists, and closely resembles the $\frac{1-\gamma}{\gamma\varepsilon}$ bound from the policy evaluation case.

Acknowledgements. The authors thank Hado van Hasselt, André Barreto, Georg Ostrovski, Hubert Soyer, and others at Google DeepMind for their helpful input to the paper, as well as the anonymous reviewers for their thoughtful feedback.

References

1. Bellman, R.: Dynamic Programming. Princeton University Press, Princeton (1957)
2. Bertsekas, D.P., Tsitsiklis, J.N.: Neuro-Dynamic Programming. Athena Scientific, Belmont (1996)
3. Hallak, A., Tamar, A., Munos, R., Mannor, S.: Generalized emphatic temporal difference learning: bias-variance analysis (2015). arXiv:1509.05172
4. Kearns, M.J., Singh, S.P.: Bias-variance error bounds for temporal difference updates. In: Conference on Computational Learning Theory, pp. 142–147 (2000)
5. Mahmood, A.R., Sutton, R.S.: Off-policy learning based on weighted importance sampling with linear computational complexity. In: Conference on Uncertainty in Artificial Intelligence (2015)
6. Mahmood, A.R., Huizhen, Y., White, M., Sutton, R.S.: Emphatic temporal-difference learning. arXiv preprint arXiv:1507.01569 (2015)
7. Munos, R., Stepleton, T., Harutyunyan, A., Bellemare, M.G.: Safe and efficient off-policy reinforcement learning. In: Advances in Neural Information Processing Systems (2016)
8. Peng, J., Williams, R.J.: Incremental multi-step q-learning. Mach. Learn. **22**(1–3), 283–290 (1996)
9. Precup, D., Sutton, R.S., Singh, S.: Eligibility traces for off-policy policy evaluation. In: International Conference on Machine Learning (2000)
10. Precup, D., Sutton, R.S., Dasgupta, S.: Off-policy temporal-difference learning with function approximation. In: International Conference on Machine Learning (2001)
11. Puterman, M.L.: Markov Decision Processes: Discrete Stochastic Dynamic Programming, 1st edn. Wiley, New York (1994)
12. Randløv, J., Alstrøm, P.: Learning to drive a bicycle using reinforcement learning and shaping. In: International Conference on Machine Learning (1998)
13. Rummery, G.A., Niranjan, M.: On-line q-learning using connectionist systems. Technical report, Cambridge University Engineering Department (1994)

14. Singh, S., Dayan, P.: Analytical mean squared error curves for temporal difference learning. Mach. Learn. **32**(1), 5–40 (1998)
15. Sutton, R.S.: Learning to predict by the methods of temporal differences. Mach. learn. **3**(1), 9–44 (1988)
16. Sutton, R.S.: Generalization in reinforcement learning: successful examples using sparse coarse coding. In: Advances in Neural Information Processing Systems (1996)
17. Sutton, R.S., Barto, A.G.: Reinforcement Learning: An Introduction. Cambridge University Press, Cambridge (1998)
18. Sutton, R.S., Mahmood, A.R., Precup, D., van Hasselt, H.: A new q (λ) with interim forward view and monte carlo equivalence. In: International Conference on Machine Learning, pp. 568–576 (2014)
19. van Hasselt, H.P.: Insights in reinforcement learning: formal analysis and empirical evaluation of temporal-difference learning algorithms. Ph.D. thesis, Universiteit Utrecht, January 2011
20. van Seijen, H., Sutton, R.S.: True online TD(λ). In: International Conference on Machine Learning, pp. 692–700 (2014)
21. van Seijen, H., van Hasselt, H., Whiteson, S., Wiering, M.: A theoretical and empirical analysis of expected Sarsa. In: Adaptive Dynamic Programming and Reinforcement Learning, pp. 177–184. IEEE (2009)
22. Watkins, C.J.C.H., Dayan, P.: Q-learning. Mach. Learn. **8**, 272–292 (1992)
23. Watkins, C.J.C.H.: Learning from delayed rewards. Ph.D. thesis, King's College, Cambridge (1989)

On the Prior Sensitivity of Thompson Sampling

Che-Yu Liu[1] and Lihong Li[2(✉)]

[1] ORFE, Princeton University, Princeton, NJ 08544, USA
cheliu@princeton.edu
[2] Microsoft Research, One Microsoft Way, Redmond, WA 98052, USA
lihongli@microsoft.com

Abstract. The empirically successful Thompson Sampling algorithm for stochastic bandits has drawn much interest in understanding its theoretical properties. One important benefit of the algorithm is that it allows domain knowledge to be conveniently encoded as a prior distribution to balance exploration and exploitation more effectively. While it is generally believed that the algorithm's regret is low (high) when the prior is good (bad), little is known about the exact dependence. This paper is a first step towards answering this important question: focusing on a special yet representative case, we fully characterize the algorithm's worst-case dependence of regret on the choice of prior. As a corollary, these results also provide useful insights into the *general* sensitivity of the algorithm to the choice of priors, when no structural assumptions are made. In particular, with p being the prior probability mass of the true reward-generating model, we prove $O(\sqrt{T/p})$ and $O(\sqrt{(1-p)T})$ regret upper bounds for the poor- and good-prior cases, respectively, as well as *matching* lower bounds. Our proofs rely on a fundamental property of Thompson Sampling and make heavy use of martingale theory, both of which appear novel in the Thompson-Sampling literature and may be useful for studying other behavior of the algorithm.

1 Introduction

Thompson Sampling (TS), also known as *probability matching* and *posterior sampling*, is a popular strategy for solving stochastic bandit problems. An important benefit of this algorithm is that it allows domain knowledge to be conveniently encoded as a prior distribution to address the exploration-exploitation tradeoff more effectively. In this paper, we focus on the sensitivity of the algorithm to the prior it uses. In the rest of this section, we first define the bandit setting and notation, and describe Thompson Sampling; we will then discuss previous works that are most related to the present paper.

1.1 Thompson Sampling for Stochastic Bandits

In the multi-armed bandit problem, an agent is repeatedly faced with K possible actions. At each time step $t = 1, \ldots, T$, the agent chooses an action $I_t \in \mathcal{A} :=$

Most of this work was done when C.Y. Liu was an intern at Microsoft.

© Springer International Publishing Switzerland 2016
R. Ortner et al. (Eds.): ALT 2016, LNAI 9925, pp. 321–336, 2016.
DOI: 10.1007/978-3-319-46379-7_22

$\{1, \ldots, K\}$, then receives reward $X_{I_t,t} \in \mathbb{R}$. An *eligible* action-selection strategy chooses actions at step t based only on past observed rewards $\mathcal{H}_t = \{I_s, X_{I_s,s}; 1 \le s < t\}$ and potentially on an external source of randomness. More background on the bandit problem can be found in a recent survey [8].

We make the following stochastic assumption on the underlying reward-generating mechanism. Let Θ be a countable[1] set of possible reward-generating models. When $\theta \in \Theta$ is the true underlying model, the rewards $(X_{i,t})_{t \ge 1}$ are i.i.d. random variables taking values in $[0,1]$ drawn from some *known* distribution $\nu_i(\theta)$ with mean $\mu_i(\theta)$. Of course, the agent knows neither the true underlying model nor the optimal action that yields the highest expected reward. The performance of the agent is measured by the regret incurred for not always selecting the optimal action. More precisely, the *frequentist regret* (or *regret* for short) for an eligible action-selection strategy π under a certain reward-generating model θ is defined as

$$R_T(\theta, \pi) := \mathbb{E} \sum_{t=1}^{T} \left(\max_{i \in \mathcal{A}} \mu_i(\theta) - \mu_{I_t}(\theta) \right), \tag{1}$$

where the expectation is taken with respect to the rewards $(X_{i,t})_{i \in \mathcal{A}, t \ge 1}$, generated according to the model θ, and the potential external source of randomness.

If one imposes a prior distribution p over Θ, then it is natural to consider the following notion of average regret known as *Bayes regret*:

$$\bar{R}_T(\pi) := \mathbb{E}_{\theta \sim p} R_T(\theta, \pi) = \sum_{\theta \in \Theta} R_T(\theta, \pi) p(\theta). \tag{2}$$

The Thompson Sampling strategy was proposed in probably the very first paper on multi-armed bandits [29]. This strategy takes as input a prior distribution p_1 for $\theta \in \Theta$. At each time t, let p_t be the posterior distribution for θ given the prior p_1 and the history $\mathcal{H}_t = \{I_s, X_{I_s,s}; 1 \le s < t\}$. Thompson Sampling selects an action randomly according to its posterior probability of being the optimal action. Equivalently, Thompson Sampling first draws a model θ_t from p_t (independently from the past given p_t) and it pulls $I_t \in \arg\max_{i \in \mathcal{A}} \mu_i(\theta_t)$. For concreteness, we assume that the distributions $(\nu_i(\theta))_{i \in \mathcal{A}, \theta \in \Theta}$ are absolutely continuous with respect to some common measure ν on $[0,1]$ with likelihood functions $(\ell_i(\theta)(\cdot))_{i \in \mathcal{A}, \theta \in \Theta}$. The posterior distributions p_t can be computed recursively by Bayes rule as follows:

$$p_{t+1}(\theta) = \frac{p_t(\theta) \ell_{I_t}(\theta)(X_{I_t,t})}{\sum_{\eta \in \Theta} p_t(\eta) \ell_{I_t}(\eta)(X_{I_t,t})}.$$

We denote by $\mathrm{TS}(p_1)$ the Thompson Sampling strategy with prior p_1.

Two remarks are in order. First, the setup above is a *discretized* version of rather general bandit problems. For example, the K-armed bandit is a special

---

[1] Note that in this paper, we do not impose any continuity structure on the reward distributions $\nu(\theta)$ with respect to $\theta \in \Theta$. Therefore, it is easy to see that when Θ is uncountable, the (frequentist) regret of Thompson Sampling, as defined in Eq. 1, in the worst-case scenario is linear in time under most underlying models $\theta \in \Theta$.

case, where Θ is the Cartesian product of the sets of reward distributions of all arms. As another example, in linear bandits [1,12], Θ is a set of candidate coefficient vectors that determine the expected reward function. Discretization of Θ provides a convenient yet useful approximation that leads to simplicity in expositions and analysis. Such an abstract formulation is analogous to the expert setting widely studied in the online-learning literature [10]; also see a recent study of Thompson Sampling with 2 and 3 experts [15].

Second, although we assume reward are bounded, some results in the paper, especially Lemma 1 that may be of independent interest, still hold with unbounded rewards.

1.2 Related Work

Recently, Thompson Sampling has gained a lot of interest, largely due to its empirical successes [11,14,25,28]. Furthermore, this strategy is often easy to be combined with complex reward models and easy to implement [13,20,30]. While asymptotic, no-regret results are known [25], these empirical successes inspired finite-time analyses that deepen our understanding of this old strategy.

For the classic K-armed bandits, regret bounds comparable to the the more widely studied UCB algorithms are obtained [3,4,18,19], matching a well-known asymptotic lower bound [21]. For linear bandits of dimension d, an $\widetilde{O}(d\sqrt{TK})$ upper bound has been proved [5]. All these bounds, while providing interesting insights about the algorithm, assume *non-informative priors* (often *uniform priors*), and essentially show that Thompson Sampling has a comparable regret to other popular strategies, especially those based on upper confidence bounds. Unfortunately, the bounds do *not* show what role prior plays in the performance of the algorithm. In contrast, a *variant* of Thompson Sampling is proposed, with a bound that depends explicitly on the entropy of the prior [23]. However, their bound has an $O(T^{2/3})$ dependence on T that is likely sub-optimal.

Another line of work in the literature focuses on the *Bayes regret* with an *informative prior*. Previous work has shown that, for any prior in the two-armed case, TS is a 2-approximation to the optimal strategy that minimizes the "stochastic" (Bayes) regret [17]. It has also been shown that in the K-armed case, the Bayes regret of TS is always upper bounded by $O(\sqrt{KT})$ for any prior [9,26]. These results were later improved [27] to a prior-dependent bound $O(\sqrt{H(q)KT})$ where q is the prior distribution of the optimal action, defined as $q(i) = \mathbb{P}_{\theta \sim p_1}(i = \text{argmax}_{j \in \mathcal{A}} \mu_j(\theta))$, and $H(q) = -\sum_{i=1}^{K} q(i) \log q(i)$ is the entropy of q. While this bound elegantly quantifies, in terms of *averaged* regret, how Thompson Sampling exploits prior distributions, it does not tell how well Thompson Sampling works in *individual* problems. Indeed, in the analysis of Bayes regret, it is unclear what a "good" prior means from a theoretical perspective, as the definition of Bayes regret essentially assumes the prior is correctly specified. In the extreme case where prior p_1 is a point mass, $H(q) = 0$ and the Bayes regret is trivially 0.

To the best of our knowledge, *our work is the first to consider frequentist regret of Thompson Sampling with an informative prior*. Specifically, we focus on understanding TS's sensitivity to the choice of prior, making progress towards a better understanding of such a popular Bayesian algorithm. It is shown that, while a strong prior can lower the Bayes regret substantially [27], such a benefit comes with a cost: if the true model happens to be assigned a low prior (the poor-prior case), the frequentist regret will be very large, which is consistent with a recent result on Pareto regret frontier [22]. Our findings suggest Thompson Sampling can be *under*-exploring in general. Techniques like those in the "mini-monster" algorithm [2] may be necessary to modify Thompson Sampling to make it less prior-sensitive. It is an open question whether such modified Thompson Sampling algorithms can still take advantage of an informative prior to enjoy a small Bayes regret.

Finally, our analysis makes critical use of a certain martingale property of Thompson Sampling. Although martingales have been applied to hypothesis testing, for example, in analyzing the statistical behavior of likelihood ratios [7], our use of martingales to analyze the behavior of posteriors in TS is new, to the best of our knowledge. Moreover, a different martingale property was used by other authors to study the Bayesian multi-armed bandit problem, where the reward at the current "state" is the same as the expected reward over the distribution of the next state when a play is made in the current state [16,17]. Their martingale property is different from ours: their martingales apply to the reward at the current state, while ours refers to the inverse of the posterior probability mass of the true model (see Sect. 3 for details).

2   Main Results

Naturally, we expect the regret of Thompson Sampling to be small when the true reward-generating model is given a large prior probability mass, and vice versa. An interesting and important question is to understand the sensitivity of the algorithm's regret to the prior it takes as input. We take a minimalist approach, and investigate a special yet meaningful case. Our results fully characterize the worst-case dependence of TS's regret on the prior, which also provides important insights into a more general case as a corollary. Furthermore, our analysis appears novel to the best of our knowledge, making heavy use of martingale techniques to analyze the behavior of the posterior probability. Such techniques may be useful for studying other bandit algorithms.

Similar to the expert setting [10], we assume access to a set of candidate models, $\Theta = \{\theta_1, \theta_2, \ldots, \theta_N\}$ for $N \geq 2$. This setting is referred to as **K-Actions-And-N-Models**, where K is the cardinality of the action set. For simplicity, in this work, we restrict ourselves to the binary action case: $K = 2$. Finally, the special case with $N = 2$ and $K = 2$ is called 2-**Actions-And-2-Models**.

Two comments are in order. First, our goal in this work is *not* to solve these specialized bandit problems, but rather to understand prior sensitivity of TS. Such seemingly simplistic problems happen to be nontrivial enough to be useful in our constructive proof of matching lower bounds. Second, we aim to understand TS's prior sensitivity *without* making any structural assumptions about Θ. A natural next step of this work is to investigate, with a structural Θ (e.g., linear), how robust TS is to the prior.

Our upper-bound analysis requires the following smoothness assumption of the likelihood functions of models in Θ. Note that this assumption is needed only in the upper-bound analysis, but *not* in the lower-bound proofs.

Assumption 1 *(Smoothness).* *There exists constant $s > 1$ such that ν-almost surely, for $i \in \{1, 2\}$, $s^{-1} \cdot \ell_i(\theta_1) \leq \ell_i(\theta_2) \leq s \cdot \ell_i(\theta_1)$.*

Remark 1. While this assumption does not hold for all distributions, it holds for some important ones, such as Bernoulli distributions $Bern(p)$ with mean $p \in (0, 1)$. On one hand, the assumption essentially avoids situations where a single application of Bayes rule can change posteriors by too much, analogous to bounded gradients or rewards in most online-learning literature. On the other hand, a small s value in the assumption tends to create hard problems for Thompson Sampling, since models are less distinguishable. Therefore, the assumption does *not* trivialize the problem.

The first main result of this paper is the following upper bound; see Sect. 4 for more details:

Theorem 1. *Consider the 2-Actions-And-2-Models case and assume that Assumption 1 holds. Then, the regret of Thompson Sampling with prior p_1 satisfies $R_T(\theta_1, TS(p_1)) = O(s\sqrt{T/p_1(\theta_1)})$. Moreover, when $p_1(\theta_1) \geq 1 - \frac{1}{8s^2}$, we have $R_T(\theta_1, TS(p_1)) = O(s^4\sqrt{(1 - p_1(\theta_1))T})$.*

Remark 2. The above upper bounds have the same dependence on T and $p_1(\theta_1)$ as the lower bounds to be given in Theorems 2 and 3 below. Moreover, both bounds are increasing functions of the smoothness parameter s. Because problems with small s tend to be harder for Thompson Sampling, our upper bounds are tight up to a universal constant for a fairly general class of hard problems. We conjecture that the dependence on s is an artifact of our proof techniques and can be removed to get tighter upper bounds for all problem instances of the 2-Actions-And-2-Models case.

The next two theorems give *matching* lower bounds for the poor- and good-prior cases, respectively. More details are given in Sect. 5.

Theorem 2. *Consider the 2-Actions-And-2-Models case. Let p_1 be a prior distribution and $T \geq \frac{1}{p_1(\theta_1)}$. Consider the following specific problem instance: $\nu_1(\theta_1) = Bern\left(\frac{1}{2} + \Delta\right)$, $\nu_1(\theta_2) = Bern\left(\frac{1}{2} - \Delta\right)$, $\nu_2(\theta_1) = \nu_2(\theta_2) = Bern\left(\frac{1}{2}\right)$, where $\Delta = 1/\sqrt{8p_1(\theta_1)T}$. Then, the regret of Thompson Sampling with prior p_1 satisfies the following: if $p_1(\theta_1) \leq \frac{1}{2}$, then $R_T(\theta_1, TS(p_1)) \geq \frac{1}{168\sqrt{2}}\sqrt{\frac{T}{p_1(\theta_1)}}$.*

Theorem 3. *Consider the 2-Actions-And-2-Models case. Let p_1 be a prior distribution and $T \geq \frac{1}{1 - p_1(\theta_1)}$. Consider the following specific problem instance* with Bernoulli reward distributions: $\nu_1(\theta_1) = \nu_1(\theta_2) = Bern\left(\frac{1}{2}\right)$, $\nu_2(\theta_1) = Bern\left(\frac{1}{2} - \Delta\right)$, $\nu_2(\theta_2) = Bern\left(\frac{1}{2} + \Delta\right)$, where $\Delta = \sqrt{\frac{1}{8(1-p_1(\theta_1))T}}$. Then the regret of Thompson Sampling with prior p_1 satisfies $R_T(\theta_1, TS(p_1)) \geq \frac{1}{10\sqrt{2}}\sqrt{(1-p_1(\theta_1))T}$.

The lower bounds in the 2-Actions-And-2-Models case easily imply the lower bounds in the general case.

Corollary 1. *(General Lower Bounds) Consider the case with two actions and an arbitrary countable Θ. Let p_1 be a prior over Θ and $\theta^* \in \Theta$ be the true model. Then, there exist problem instances where the regrets of Thompson Sampling are $\Omega(\sqrt{\frac{T}{p_1(\theta^*)}})$ and $\Omega(\sqrt{(1-p_1(\theta^*))T})$ for small $p_1(\theta^*)$ and large $p_1(\theta^*)$, respectively.*

Remark 3. These lower bounds show that the performance of Thompson Sampling can be quite sensitive to the choice of input prior, especially when the prior is poorly chosen.

Due to space limit, we can only include the more important, novel or challenging parts of the analysis in the paper. A complete proof, together with simulation results corroborating our theoretical findings, are given in a full version [24].

2.1 Comparison to Previous Results

Note that an upper bound in the K-Actions-And-N-Models case can be derived from an earlier result [27], which upper-bounds the Bayes regret, $\bar{R}_T(TS(p_1))$:

$$R_T(\theta_1, TS(p_1)) \leq \frac{\bar{R}_T(TS(p_1))}{p_1(\theta_1)} = O\left(\frac{\sqrt{H(q)KT}}{p_1(\theta_1)}\right),$$

where $\theta_1 \in \Theta$ is the unknown, true model. On one hand, in the 2-Actions-And-2-Models case, the above upper bound becomes $O\left(\sqrt{\log\left(\frac{1}{p_1(\theta_1)}\right)\frac{T}{p_1(\theta_1)}}\right)$ for small $p_1(\theta_1)$, and $O\left(\sqrt{\log\left(\frac{1}{1-p_1(\theta_1)}\right)(1-p_1(\theta_1))T}\right)$ for large $p_1(\theta_1)$. Our upper bounds in Theorem 1 remove the extraneous logarithmic terms in these upper bounds. On the other hand, the above general upper bound can be further upper bounded by $O\left(\frac{\sqrt{T}}{p_1(\theta_1)}\right)$ for small $p_1(\theta_1)$ and $O\left(\sqrt{\log\left(\frac{1}{1-p_1(\theta_1)}\right)(1-p_1(\theta_1))T}\right)$ for large $p_1(\theta_1)$. We conjecture that these general upper bounds can be improved to match our lower bounds in Corollary 1, especially for small $p_1(\theta_1)$. But it remains open how to extend our proof techniques for the 2-Actions-And-2-Models case to get tight general upper bounds.

It is natural to compare Thompson Sampling to exponentially weighted algorithms, a well-known family of algorithms that can also take advantage of prior knowledge. If we see each model $\theta \in \Theta$ as an expert who recommends the optimal action based on distributions specified by θ, and use the prior p_1 as the initial weights assigned to the experts, then the EXP4 algorithm [6] has a regret of $O\left(KT\gamma + \frac{1}{\gamma}\log\frac{1}{p_1(\theta^*)}\right)$, with a parameter $\gamma \in (0,1)$. For the sake of simplicity, we only do the comparison in the 2-Actions-And-2-Models case. By trying to match or even beat the upper bounds in Theorem 1, we reach the choice that $\gamma = \sqrt{H(p_1)/T}$. Assuming that θ_1 is the true model, the bound becomes $O\left(\sqrt{\log\left(\frac{1}{p_1(\theta_1)}\right)\frac{T}{p_1(\theta_1)}}\right)$ for small $p_1(\theta_1)$, and $O\left(\sqrt{\log\left(\frac{1}{1-p_1(\theta_1)}\right)(1-p_1(\theta_1))T}\right)$ for large $p_1(\theta_1)$. Thus, although EXP4 is not a Bayesian algorithm, it has the same worst-case dependence on prior as Thompson Sampling, up to logarithmic factors. This is partly explained by the fact that such algorithms are designed to perform well in the worst-case (adaptive adversarial) scenario. On the contrary, by design, Thompson Sampling takes advantage of prior information more efficiently in most cases, especially when there is certain structure on the model space Θ [9]. Note that in this paper, we do not impose any structure on Θ, thus our lower bounds do not contradict existing results in the literature with non-informative priors (where $p(\theta^*)$ can be very small as Θ is typically large).

Finally, our proof techniques are new in the Thompson Sampling literature, to the best of our knowledge. The key observation is that the inverse of the posterior probability of the true underlying model is a martingale (Lemma 1). It allows us to use results and techniques from martingale theory to quantify the time and probability that the posterior distribution hits a certain threshold. Then, the regret of Thompson Sampling can be analyzed separately before and after hitting times.

3   Preliminaries

In this section, we study a fundamental martingale property of Thompson Sampling and its implications. The results are essential to proving our upper bounds in Sect. 4. Note that a similar property holds for posterior updates using Bayes rule, which however does not involve action selection.

Throughout this paper, for a random variable Y, we will use the shorthand $\mathbb{E}_t[Y]$ for the conditional expectation $\mathbb{E}[Y|\mathcal{H}_t]$. Moreover, we denote by $\mathbb{E}^\theta[Y]$ the expectation of Y when θ is the true underlying model, i.e., when $X_{i,t}$ has distribution $\nu_i(\theta)$. The notation $\mathbb{P}^\theta[\cdot]$ is similarly defined. Furthermore, we use the shorthand $a \wedge b$ for $\min\{a, b\}$.

Lemma 1. *(Martingale Property) Assume that Θ is countable and that $\theta^* \in \Theta$ is the true reward-generating model. Then, the stochastic process $(p_t(\theta^*)^{-1})_{t \geq 1}$ is a martingale with respect to the filtration $(\mathcal{H}_t)_{t \geq 1}$.*

Proof. First, recall that conditioned on $\mathcal{H}_t$, p_t is deterministic. Then one has

$$
\begin{aligned}
\mathbb{E}_t^{\theta^*}[p_{t+1}(\theta^*)^{-1}] &= \mathbb{E}_t^{\theta^*}\left[\frac{\sum_{\eta\in\Theta} p_t(\eta)\ell_{I_t}(\eta)(X_{I_t,t})}{p_t(\theta^*)\ell_{I_t}(\theta^*)(X_{I_t,t})}\right] \\
&= \sum_{i=1}^K \mathbb{P}_t^{\theta^*}(I_t = i)\mathbb{E}_t^{\theta^*}\left[\frac{\sum_{\eta\in\Theta} p_t(\eta)\ell_i(\eta)(X_{i,t})}{p_t(\theta^*)\ell_i(\theta^*)(X_{i,t})}\right] \\
&= \sum_{i=1}^K \mathbb{P}_t^{\theta^*}(I_t = i)\int \frac{\sum_{\eta\in\Theta} p_t(\eta)\ell_i(\eta)(x)}{p_t(\theta^*)\ell_i(\theta^*)(x)}\ell_i(\theta^*)(x)\,d\nu(x) \\
&= p_t(\theta^*)^{-1}\sum_{i=1}^K \mathbb{P}_t^{\theta^*}(I_t = i)\int \sum_{\eta\in\Theta} p_t(\eta)\ell_i(\eta)(x)\,d\nu(x) \\
&= p_t(\theta^*)^{-1}\sum_{i=1}^K \mathbb{P}_t^{\theta^*}(I_t = i) = p_t(\theta^*)^{-1},
\end{aligned}
$$

where the second last equality follows from the fact that $\int \ell_i(\eta)(x)\,d\nu(x) = 1$ for any $\eta \in \Theta$. $\qquad\square$

Consider the 2-Actions-And-2-Models case. Let $A, B \in (0,1)$ be two constants such that $A > p_1(\theta_1) > B$. We define the following *hitting times* and *hitting probabilities*: $\tau_A = \inf\{t \geq 1, p_t(\theta_1) \geq A\}$, $\tau_B = \inf\{t \geq 1, p_t(\theta_1) \leq B\}$, $q_{A,B} = \mathbb{P}^{\theta_1}(\tau_A < \tau_B)$, and $q_{B,A} = \mathbb{P}^{\theta_1}(\tau_A > \tau_B)$. The martingale property above implies the following results which will be used repeatedly in the proofs of our results.

Lemma 2. *Consider the 2-Actions-And-2-Models case with $\Delta > 0$, where Δ is as defined in Theorem 2. Then, we have $\tau_A < +\infty$ almost surely. Furthermore, assume that $\tau_B < \infty$ and that there exists constant $\gamma > 0$ so that $p_{\tau_B}(\theta_1) \geq \gamma$ almost surely, then*

$$
q_{A,B} = \frac{\mathbb{E}^{\theta_1}[p_{\tau_B}(\theta_1)^{-1}|\tau_A > \tau_B] - p_1(\theta_1)^{-1}}{\mathbb{E}^{\theta_1}[p_{\tau_B}(\theta_1)^{-1}|\tau_A > \tau_B] - \mathbb{E}^{\theta_1}[p_{\tau_A}(\theta_1)^{-1}|\tau_A < \tau_B]} \quad and
$$

$$
q_{B,A} = \frac{p_1(\theta_1)^{-1} - \mathbb{E}^{\theta_1}[p_{\tau_A}(\theta_1)^{-1}|\tau_A < \tau_B]}{\mathbb{E}^{\theta_1}[p_{\tau_B}(\theta_1)^{-1}|\tau_A > \tau_B] - \mathbb{E}^{\theta_1}[p_{\tau_A}(\theta_1)^{-1}|\tau_A < \tau_B]}.
$$

Finally, $q_{B,A} \leq \frac{B}{p_1(\theta_1)}$ and $q_{B,A} \leq \frac{1-p_1(\theta_1)}{A-B}$.

Proof. We first argue that $\tau_A < +\infty$ almost surely. Define the event $E = \{\tau_A = +\infty\}$. Under the event E, $p_t(\theta_1)$ is always upper bounded by A for any t. Thus

$$
R_T(\theta_1, TS(p_1)) = \Delta \cdot \mathbb{E}^{\theta_1}\sum_{t=1}^T p_t(\theta_2) \geq \mathbb{P}^{\theta_1}(E)\Delta(1 - A)T.
$$

It follows that

$$\bar{R}_T(TS(p_1)) \geq p_1(\theta_1)\, R_T(\theta_1, TS(p_1)) \geq p_1(\theta_1)\mathbb{P}^{\theta_1}(E)\Delta(1-A)T.$$

However, it was proven [9] that the Bayes risk $\bar{R}_T(TS(p_1))$ is always upper bounded by $O(\sqrt{T})$. Therefore we must have $\mathbb{P}^{\theta_1}(E) = 0$; that is $\tau_A < +\infty$ almost surely. This implies that $p_{\tau_A \wedge \tau_B}(\theta_1)$ is well defined and $q_{A,B} + q_{B,A} = 1$.

Now, by Lemma 1, $(p_t(\theta_1)^{-1})_{t \geq 1}$ is a martingale. It is easy to verify that τ_A and τ_B are both stopping times with respect to the filtration $(\mathcal{H}_t)_{t \geq 1}$. Then it follows from Doob's optional stopping theorem that for any t, $\mathbb{E}^{\theta_1}[p_{t \wedge \tau_A \wedge \tau_B}(\theta_1)^{-1}] = p_1(\theta_1)^{-1}$. Moreover, for any $t \geq 1$, $p_{t \wedge \tau_A \wedge \tau_B}(\theta_1)^{-1} \leq \gamma^{-1}$ (Note that by definition, $\gamma \leq B$). Hence, by Lebesgue's dominated convergence theorem, $\mathbb{E}^{\theta_1}[p_{t \wedge \tau_A \wedge \tau_B}(\theta_1)^{-1}] \longrightarrow \mathbb{E}^{\theta_1}[p_{\tau_A \wedge \tau_B}(\theta_1)^{-1}]$ as $t \to +\infty$. Thus,

$$
\begin{aligned}
p_1(\theta_1)^{-1} &= \mathbb{E}^{\theta_1}[p_{\tau_A \wedge \tau_B}(\theta_1)^{-1}] \\
&= q_{A,B}\mathbb{E}^{\theta_1}[p_{\tau_A}(\theta_1)^{-1}|\tau_A < \tau_B] + q_{B,A}\mathbb{E}^{\theta_1}[p_{\tau_B}(\theta_1)^{-1}|\tau_A > \tau_B].
\end{aligned}
$$

The above equality combined with $q_{A,B} + q_{B,A} = 1$ gives the desired expressions for $q_{A,B}$ and $q_{B,A}$. Finally, we have

$$
\begin{aligned}
q_{B,A} &= \frac{p_1(\theta_1)^{-1} - \mathbb{E}^{\theta_1}[p_{\tau_A}(\theta_1)^{-1}|\tau_A < \tau_B]}{\mathbb{E}^{\theta_1}[p_{\tau_B}(\theta_1)^{-1}|\tau_A > \tau_B] - \mathbb{E}^{\theta_1}[p_{\tau_A}(\theta_1)^{-1}|\tau_A < \tau_B]} \\
&\leq \frac{p_1(\theta_1)^{-1}}{\mathbb{E}^{\theta_1}[p_{\tau_B}(\theta_1)^{-1}|\tau_A < \tau_B]} \leq \frac{B}{p_1(\theta_1)}
\end{aligned}
$$

and

$$
\begin{aligned}
q_{B,A} &= \frac{p_1(\theta_1)^{-1} - \mathbb{E}^{\theta_1}[p_{\tau_A}(\theta_1)^{-1}|\tau_A < \tau_B]}{\mathbb{E}^{\theta_1}[p_{\tau_B}(\theta_1)^{-1}|\tau_A > \tau_B] - \mathbb{E}^{\theta_1}[p_{\tau_A}(\theta_1)^{-1}|\tau_A < \tau_B]} \\
&\leq \frac{p_1(\theta_1)^{-1} - 1}{B^{-1} - A^{-1}} = \frac{AB}{p_1(\theta_1)}\frac{1 - p_1(\theta_1)}{A - B} \leq \frac{1 - p_1(\theta_1)}{A - B}.
\end{aligned}
$$

$\square$

4   Upper Bounds

In this section, we focus on the 2-Actions-And-2-Models case. We present and prove our results on the upper bounds for the frequentist regret of Thompson Sampling. Due to space limitation, we only sketch the proof for the poor-prior case (first part of Theorem 1); complete proofs, including those for the good-prior case, will appear in a long version.

We start with a simple lemma that follows immediate from Assumption 1:

Lemma 3. *Under Assumption 1, regardless of either θ_1 or θ_2 being the true underlying model, for any $\theta \in \{\theta_1, \theta_2\}$, $s^{-1} \cdot p_t(\theta) \leq p_{t+1}(\theta) \leq s \cdot p_t(\theta)$ ν-almost surely.*

The next lemma describes how the posterior probability mass of the true model evolves over time. It can be proved by direct, although a bit tedious, calculations.

Lemma 4. *Consider the 2-Actions-And-2-Models case. We have the following inequalities concerning various functionals of the stochastic process* $(p_t(\theta_1))_{t\geq 1}$.

(a) *For* $t \geq 1$, $\mathbb{E}_t^{\theta_1}\left[\log(p_t(\theta_1)^{-1}) - \log(p_{t+1}(\theta_1)^{-1})\right]$
$\geq \frac{1}{2}\sum_{i\in\{1,2\}} p_t(\theta_i)p_t(\theta_2)^2|\mu_i(\theta_1) - \mu_i(\theta_2)|^2$.

(b) *For* $t \geq 1$, $\mathbb{E}^{\theta_1}[p_{t+1}(\theta_1)] \geq \mathbb{E}^{\theta_1}[p_t(\theta_1)]$ *and*
$\mathbb{E}_t^{\theta_1}\left[p_{t+1}(\theta_1) - p_t(\theta_1)\right] \leq \sum_{i\in\{1,2\}} p_t(\theta_i)p_t(\theta_1)p_t(\theta_2)\mathbb{E}^{\theta_1}\left[\frac{\ell_i(\theta_1)(X_{i,t})}{\ell_i(\theta_2)(X_{i,t})} - 1\right]$.

(c) *For* $t \geq 1$, $\mathbb{E}_t^{\theta_1}\left[(1 - p_{t+1}(\theta_1))^{-1} - (1 - p_t(\theta_1))^{-1}\right]$
$= \sum_{i\in\{1,2\}} p_t(\theta_i)\frac{p_t(\theta_1)}{p_t(\theta_2)}\mathbb{E}^{\theta_1}\left[\frac{\ell_i(\theta_1)(X_{i,t})}{\ell_i(\theta_2)(X_{i,t})} - 1\right]$
$\geq \frac{p_t(\theta_1)^2}{2p_t(\theta_2)}|\mu_1(\theta_1) - \mu_1(\theta_2)|^2 + \frac{p_t(\theta_1)}{2}|\mu_2(\theta_1) - \mu_2(\theta_2)|^2$.

(d) $R_T(\theta_1, TS(p_1)) \leq \Delta T(1 - p_1(\theta_1))$.

We now introduce some notation. Let $\Delta = \mu_1(\theta_1) - \mu_2(\theta_1)$, $\Delta_1 = |\mu_1(\theta_1) - \mu_1(\theta_2)|$ and $\Delta_2 = |\mu_2(\theta_1) - \mu_2(\theta_2)|$. Obviously, $\Delta \leq \Delta_1 + \Delta_2$. We assume $\Delta > 0$ to avoid the generated case. To simplify notation, define the regret function $R_T(\cdot)$ by $R_T(p_1(\theta_1)) = R_T(\theta_1, TS(p_1))$. Since the immediate regret of each step is at most Δ, we immediately have $R_T(p_1(\theta_1)) \leq \Delta T$. Furthermore, we have the following useful and intuitive monotone property, which can be proved by a dynamic-programming argument inspired by previous work [17, Section 3].

Lemma 5. R_T *is a decreasing function of* $p_1(\theta_1)$.

The proofs of the upper bounds rely on several propositions that reveal interesting recursions of Thompson Sampling's regret as a function of the prior. Although these propositions use similar analytic techniques, they differ in many important details. Due to space limitation, we only sketch the proof of Proposition 1.

Proposition 1. *Consider the 2-Actions-And-2-Models case and assume that Assumption 1 holds. Then for any* $T > 0$ *and* $p_1(\theta_1) \in (0, 1)$, *we have*

$$R_T(p_1(\theta_1)) \leq \left(96\log\frac{3s}{2} + 6\right)\sqrt{\frac{T}{p_1(\theta_1)}} + R_T\left(\frac{1}{3}\right).$$

Proof (Sketch). We recall that θ_1 is assumed to be the true reward-generating model in the proposition, and use the same notation as in Lemma 2. First, the desired inequality is trivial if $p_1(\theta_1) \geq \frac{1}{3}$ since $R_T(\cdot)$ is a decreasing function. Moreover, if $\Delta \leq 2\sqrt{\frac{1}{p_1(\theta_1)T}}$, then $R_T(p_1(\theta_1)) \leq \Delta T \leq 2\sqrt{\frac{T}{p_1(\theta_1)}}$, which completes the proof. Thus, we can assume that $p_1(\theta_1) \leq \frac{1}{3}$ and $\Delta > 2\sqrt{\frac{1}{p_1(\theta_1)T}}$. Let

$A = \frac{3}{2}p_1(\theta_1)$ and $B = \frac{1}{\Delta}\sqrt{\frac{p_1(\theta_1)}{T}}$. Then, it is easy to see that $B \leq \frac{1}{2}p_1(\theta_1) \leq \frac{1}{2} \leq 1 - A$.

Now, the first step is to upper bound $\mathbb{E}^{\theta_1}[\tau_A \wedge \tau_B - 1]$. By Lemma 4(a), we have for $t \leq \tau_A \wedge \tau_B - 1$ that,

$$\mathbb{E}_t^{\theta_1}\left[\log(p_t(\theta_1)^{-1}) - \log(p_{t+1}(\theta_1)^{-1})\right] \geq \frac{1}{2}p_t(\theta_1)p_t(\theta_2)^2\Delta_1^2 + \frac{1}{2}p_t(\theta_2)^3\Delta_2^2$$

$$\geq \frac{p_t(\theta_2)^2 B}{2}(\Delta_1^2 + \Delta_2^2) \geq \frac{B\Delta^2}{16}.$$

In other words, $\left(\log(p_t(\theta_1)^{-1}) + t\frac{B\Delta^2}{16}\right)_{t \leq \tau_A \wedge \tau_B}$ is a supermartingale. Applying Doob's optional stopping theorem to the stopping times $\sigma_1 = t \wedge \tau_A \wedge \tau_B$ and $\sigma_2 = 1$ and letting $t \to +\infty$ by using Lebesgue's dominated convergence theorem and the monotone convergence theorem, we have

$$\mathbb{E}^{\theta_1}[\tau_A \wedge \tau_B - 1] \leq \frac{16}{B\Delta^2}\mathbb{E}^{\theta_1}\left[\log\frac{p_{\tau_A \wedge \tau_B}(\theta_1)}{p_1(\theta_1)}\right]$$

$$\leq \frac{16}{B\Delta^2}\log\frac{sA}{p_1(\theta_1)} = \frac{16}{B\Delta^2}\log\frac{3s}{2},$$

where we have used Lemma 3 in the second last step.

Next, the regret of Thompson Sampling can be decomposed as follows

$$R_T(p_1(\theta_1))$$
$$= \Delta \cdot \mathbb{E}^{\theta_1}[\tau_A \wedge \tau_B - 1] + q_{B,A} \cdot \mathbb{E}^{\theta_1}[R_T(p_{\tau_B}(\theta_1))|\tau_A > \tau_B]$$
$$+ q_{A,B} \cdot \mathbb{E}^{\theta_1}[R_T(p_{\tau_A}(\theta_1))|\tau_A < \tau_B]$$
$$\leq \frac{16}{B\Delta}\log\frac{3s}{2} + \frac{B}{p_1(\theta_1)}\Delta T + R_T\left(\frac{3}{2}p_1(\theta_1)\right)$$
$$= \left(16\log\frac{3s}{2} + 1\right)\sqrt{\frac{T}{p_1(\theta_1)}} + R_T\left(\frac{3}{2}p_1(\theta_1)\right),$$

where in the second last step, we have used the facts that $q_{B,A} \leq \frac{B}{p_1(\theta_1)}$ (by Lemma 2), $p_{\tau_A}(\theta_1) \geq A = \frac{3}{2}p_1(\theta_1)$, and $R_T(\cdot)$ is a decreasing function (by Lemma 5). Because the above recurrence inequality holds for all $p_1(\theta_1) \leq \frac{1}{3}$, simple calculations lead to the desired inequality. $\qquad\square$

Using similar proof techniques, one can prove the following recursion:

Proposition 2. *Consider the 2-Actions-And-2-Models case and assume that Assumption 1 holds. Then, for any $T > 0$ and $p_1(\theta_1) \leq \frac{1}{2}$, we have*

$$R_T(p_1(\theta_1)) \leq \left(\frac{16s}{p_1(\theta_1)^2} + 1\right)\sqrt{T} + \frac{1}{2}R_T\left(\frac{1}{2s}p_1(\theta_1)\right).$$

With the technical lemmas and propositions developed so far, we are now ready to prove the first upper bound of Theorem 1, for p small. The second bound for large p can be proved in a similar fashion, although the details are quite different [24].

Proof (of the first part in Theorem 1). For convenience, define $\beta = 96 \log \frac{3s}{2} + 6$. By Propositions 1 and 2,

$$R_T\left(\frac{1}{3}\right) \leq (144s + 1)\sqrt{T} + \frac{1}{2}R_T\left(\frac{1}{6s}\right)$$

$$\leq (144s + 1)\sqrt{T} + \frac{1}{2}\beta\sqrt{6sT} + \frac{1}{2}R_T\left(\frac{1}{3}\right).$$

Therefore,

$$R_T\left(\frac{1}{3}\right) \leq \left(288s + \beta\sqrt{6s} + 2\right)\sqrt{T}.$$

Using again Proposition 1, one has for any $p_1(\theta_1) \in (0, 1)$,

$$R_T(p_1(\theta_1)) \leq \beta\sqrt{\frac{T}{p_1(\theta_1)}} + R_T\left(\frac{1}{3}\right)$$

$$\leq \beta\sqrt{\frac{T}{p_1(\theta_1)}} + \left(288s + \beta\sqrt{6s} + 2\right)\sqrt{T}$$

$$\leq \beta\sqrt{\frac{T}{p_1(\theta_1)}} + \left(288s + \beta\sqrt{6s} + 2\right)\sqrt{\frac{T}{p_1(\theta_1)}}$$

$$\leq \left(288s + \beta(\sqrt{6s} + 1) + 2\right)\sqrt{\frac{T}{p_1(\theta_1)}} \leq 1490s\sqrt{\frac{T}{p_1(\theta_1)}},$$

where the last step follows from the inequalities $\beta = 96 \log \frac{3s}{2} + 6 \leq 300\sqrt{s}$ and $\sqrt{6s} + 1 \leq 4\sqrt{s}$ for $s > 1$. □

5   Lower Bounds

In this section, we give a proof for the lower bound when the prior is poor (Theorem 2); the other case (Theorem 3) is left in the long version [24]. The following technical lemma is needed, which can be proved by direct calculations:

Lemma 6. *Let* $-\sqrt{\frac{1}{8}} \leq \Delta \leq \sqrt{\frac{1}{8}}$. *Let* ℓ_1 *and* ℓ_2 *be the density functions of the Bernoulli distributions* $Bern\left(\frac{1}{2} + \Delta\right)$ *and* $Bern\left(\frac{1}{2} - \Delta\right)$ *with respect to the counting measure on* $[0, 1]$. *Then* $\mathbb{E}_{X \sim Bern\left(\frac{1}{2}+\Delta\right)}\left[\frac{\ell_1(X)}{\ell_2(X)} - 1\right] \leq 32\Delta^2$.

Proof (of Theorem 2). Let $A = \frac{3}{2}p_1(\theta_1)$. Clearly, $A \leq \frac{3}{4}$. Recall that $\tau_A = \inf\{t \geq 1, p_t(\theta_1) \geq A\}$. Using Lemmas 4(b) and 6, one has for $t \leq \tau_A - 1$,

$$\mathbb{E}_t^{\theta_1}\left[p_{t+1}(\theta_1) - p_t(\theta_1)\right]$$

$$\leq \sum_{i \in \{1,2\}} p_t(\theta_i)p_t(\theta_1)p_t(\theta_2)\mathbb{E}^{\theta_1}\left[\frac{\ell_i(\theta_1)(X_{i,t})}{\ell_i(\theta_2)(X_{i,t})} - 1\right]$$

$$= p_t(\theta_1)^2 p_t(\theta_2)\mathbb{E}^{\theta_1}\left[\frac{\ell_1(\theta_1)(X_{1,t})}{\ell_1(\theta_2)(X_{1,t})} - 1\right] \leq 32A^2\Delta^2 = 72p_1(\theta_1)^2\Delta^2.$$

Therefore, $\left(p_t(\theta_1) - 72p_1(\theta_1)^2\Delta^2 t\right)_{t \leq \tau_A}$ is a supermartingale. Now, using Doob's optional stopping theorem, one has $\mathbb{E}^{\theta_1}\left[p_{t \wedge \tau_A \wedge T}(\theta_1) - (t \wedge \tau_A \wedge T)72p_1(\theta_1)^2\Delta^2\right] \leq p_1(\theta_1) - 72p_1(\theta_1)^2\Delta^2$ for any $t \geq 1$.

Moreover, using Lebesgue's dominated convergence theorem and the monotone convergence theorem,

$$\mathbb{E}^{\theta_1}\left[p_{t \wedge \tau_A \wedge T}(\theta_1) - (t \wedge \tau_A \wedge T)72p_1(\theta_1)^2\Delta^2\right]$$
$$\longrightarrow \mathbb{E}^{\theta_1}\left[p_{\tau_A \wedge T}(\theta_1) - (\tau_A \wedge T)72p_1(\theta_1)^2\Delta^2\right]$$

as $t \to +\infty$. Hence,

$$\mathbb{E}^{\theta_1}[\tau_A \wedge T - 1] \geq \frac{1}{72p_1(\theta_1)^2\Delta^2}\mathbb{E}^{\theta_1}\left[p_{\tau_A \wedge T}(\theta_1) - p_1(\theta_1)\right].$$

One one side, if $\mathbb{P}^{\theta_1}(\tau_A \wedge T = T) \geq \frac{1}{21}$, then $\mathbb{E}^{\theta_1}[\tau_A \wedge T] \geq \mathbb{P}^{\theta_1}(\tau_A \wedge T = T)T \geq \frac{T}{21}$. On the other side, if $\mathbb{P}^{\theta_1}(\tau_A \wedge T = \tau_A) \geq \frac{20}{21}$, then $\mathbb{E}^{\theta_1}\left[p_{\tau_A \wedge T}(\theta_1)\right] \geq \mathbb{P}^{\theta_1}(\tau_A \wedge T = \tau_A)A \geq \frac{10}{7}p_1(\theta_1)$ and thus

$$\mathbb{E}^{\theta_1}[\tau_A \wedge T - 1] \geq \frac{1}{72p_1(\theta_1)^2\Delta^2}\left(\frac{10}{7}p_1(\theta_1) - p_1(\theta_1)\right) = \frac{T}{21}.$$

In both cases, we have $\mathbb{E}^{\theta_1}[\tau_A \wedge T - 1] \geq \frac{T}{21}$.
Finally, one has

$$R_T(\theta_1, TS(p_1)) = \Delta\mathbb{E}^{\theta_1}\left[\sum_{t=1}^{T}(1 - p_t(\theta_1))\right] \geq \Delta\mathbb{E}^{\theta_1}\left[\sum_{t=1}^{\tau_A \wedge T - 1}(1 - p_t(\theta_1))\right]$$

$$\geq \Delta(1 - A)\mathbb{E}^{\theta_1}[\tau_A \wedge T - 1] \geq \frac{\Delta T}{84} = \frac{1}{168\sqrt{2}}\sqrt{\frac{T}{p_1(\theta_1)}},$$

where we have used the fact that $1 - A \geq \frac{1}{4}$. $\square$

Proof (of Theorem 3). Using Lemmas 4(c) and 6, one has

$$\mathbb{E}_t^{\theta_1}\left[p_{t+1}(\theta_2)^{-1} - p_t(\theta_2)^{-1}\right] = \sum_{i \in \{1,2\}} p_t(\theta_i)\frac{p_t(\theta_1)}{p_t(\theta_2)}\mathbb{E}^{\theta_1}\left[\frac{\ell_i(\theta_1)(X_{i,t})}{\ell_i(\theta_2)(X_{i,t})} - 1\right]$$

$$= p_t(\theta_1)\mathbb{E}^{\theta_1}\left[\frac{\ell_2(\theta_1)(X_{2,t})}{\ell_2(\theta_2)(X_{2,t})} - 1\right] \leq 32\Delta^2.$$

Then for any $t \leq T$,

$$\mathbb{E}^{\theta_1}\left[p_t(\theta_2)^{-1}\right] \leq \frac{1}{1-p_1(\theta_1)} + 32(t-1)\Delta^2 = \frac{1+4(t-1)/T}{1-p_1(\theta_1)} \leq \frac{5}{1-p_1(\theta_1)}.$$

By Jensen's inequality, we have for any $t \leq T$, $\mathbb{E}^{\theta_1}\left[p_t(\theta_2)\right] \geq \left(\mathbb{E}^{\theta_1}\left[p_t(\theta_2)^{-1}\right]\right)^{-1} \geq \frac{1-p_1(\theta_1)}{5}$. Hence,

$$R_T(\theta_1, TS(p_1)) = \Delta \cdot \mathbb{E}^{\theta_1}\sum_{t=1}^{T} p_t(\theta_2) \geq \Delta T \frac{1-p_1(\theta_1)}{5} \geq \frac{1}{10\sqrt{2}}\sqrt{(1-p_1(\theta_1))T}.$$

$\square$

6 Conclusions

In this work, we studied an important aspect of the popular Thompson Sampling strategy for stochastic bandits — its sensitivity to the prior. Focusing on a special yet nontrivial problem, we fully characterized its worst-case dependence of regret on prior, both for the good- and bad-prior cases, with matching upper and lower bounds. The lower bounds are also extended to a more general case as a corollary, quantifying inherent sensitivity of the algorithm when the prior is poor and when no structural assumptions are made.

These results suggest a few interesting directions for future work, only four of which are outlined here. One is to close the gap between upper and lower bounds for the general, multiple-model case. We conjecture that a tighter upper bound is likely to match the lower bound in Corollary 1. The second is to consider prior sensitivity for structured stochastic bandits, where models in Θ are related in certain ways. For example, in the discretized version of the multi-armed bandit problem [4], the prior probability mass of the true model is exponentially small when a uniform prior is used, but strong frequentist regret bound is still possible. Sensitivity analysis for such problems can provide useful insights and guidance for applications of Thompson Sampling. Thrid, it remains open whether there exists an algorithm whose worst-case regret bounds are better than those of Thompson Sampling for any range of $p_1(\theta^*)$, with θ^* being the true underlying model. This question is related to the recent study of Pareto regret front [22]. We conjecture that the answer is negative, especially in the 2-Actions-And-2-Models case. Finally, it is interesting to consider problem-dependent regret bounds that often scale logarithmically with T.

Acknowledgments. We thank Sébastien Bubeck and the anonymous reviewers for helpful advice that improves the presentation of the paper.

References

1. Abbasi-Yadkori, Y., Pál, D., Szepesvári, C.: Improved algorithms for linear stochastic bandits. In: NIPS, pp. 2312–2320 (2011)

2. Agarwal, A., Hsu, D., Kale, S., Langford, J., Li, L., Schapire, R.E.: Taming the monster: a fast and simple algorithm for contextual bandits. In: ICML, pp. 1638–1646 (2014)
3. Agrawal, S., Goyal, N.: Analysis of Thompson sampling for the multi-armed bandit problem. In: COLT, pp. 39.1–39.26 (2012)
4. Agrawal, S., Goyal, N.: Further optimal regret bounds for Thompson sampling. In: AISTATS, pp. 99–107 (2013)
5. Agrawal, S., Goyal, N.: Thompson sampling for contextual bandits with linear payoffs. In: ICML, pp. 127–135 (2013)
6. Auer, P., Cesa-Bianchi, N., Freund, Y., Schapire, R.: The non-stochastic multi-armed bandit problem. SIAM J. Comput. **32**(1), 48–77 (2002)
7. Bartroff, J., Lai, T.L., Shih, M.-C.: Sequential Experimentation in Clinical Trials: Design and Analysis, vol. 298. Springer, Heildelberg (2013)
8. Bubeck, S., Cesa-Bianchi, N.: Regret analysis of stochastic and nonstochastic multi-armed bandit problems. Found. Trends Mach. Learn. **5**(1), 1–122 (2012)
9. Bubeck, S., Liu, C.Y.: Prior-free and prior-dependent regret bounds for Thompson sampling. In: NIPS, pp. 638–646 (2013)
10. Cesa-Bianchi, N., Lugosi, G.: Prediction, Learning, and Games. Cambridge University Press, Cambridge (2006)
11. Chapelle, O., Li, L.: An empirical evaluation of Thompson sampling. In: NIPS, pp. 2249–2257 (2011)
12. Chu, W., Li, L., Reyzin, L., Schapire, R.E.: Contextual bandits with linear payoff functions. In: AISTATS, pp. 208–214 (2011)
13. Gopalan, A., Mannor, S., Mansour, Y.: Thompson sampling for complex online problems. In: ICML, pp. 100–108 (2014)
14. Graepel, T., Candela, J.Q., Borchert, T., Herbrich, R.: Web-scale Bayesian click-through rate prediction for sponsored search advertising in Microsoft's Bing search engine. In: ICML, pp. 13–20 (2010)
15. Gravin, N., Peres, Y., Sivan, B.: Towards optimal algorithms for prediction with expert advice. In: SODA, pp. 528–547 (2016)
16. Guha, S., Munagala, K.: Approximation algorithms for Bayesian multi-armed bandit problems. arXiv preprint arXiv: 1306.3525v2 (2013)
17. Guha, S., Munagala, K.: Stochastic regret minimization via Thompson sampling. In: COLT, pp. 317–338 (2014)
18. Honda, J., Takemura, A.: Optimality of Thompson sampling for Gaussian bandits depends on priors. In: AISTATS, pp. 375–383 (2014)
19. Kaufmann, E., Korda, N., Munos, R.: Thompson sampling: an asymptotically optimal finite-time analysis. In: Bshouty, N.H., Stoltz, G., Vayatis, N., Zeugmann, T. (eds.) ALT 2012. LNCS, vol. 7568, pp. 199–213. Springer, Heidelberg (2012)
20. Komiyama, J., Honda, J., Nakagawa, H.: Optimal regret analysis of Thompson sampling in stochastic multi-armed bandit problem with multiple plays. In: ICML, pp. 1152–1161 (2015)
21. Lai, T.L., Robbins, H.: Asymptotically efficient adaptive allocation rules. Adv. Appl. Math. **6**, 4–22 (1985)
22. Lattimore, T.: The pareto regret frontier for bandits. In: NIPS, pp. 208–216 (2015)
23. Li, L.: Generalized Thompson sampling for contextual bandits. Technical report MSR-TR-2013-136, Microsoft Research (2013)
24. Liu, C.Y., Li, L.: On the prior sensitivity of Thompson sampling (2015). arXiv:1506.03378
25. May, B.C., Korda, N., Lee, A., Leslie, D.S.: Optimistic Bayesian sampling in contextual-bandit problems. J. Mach. Learn. Res. **13**, 2069–2106 (2012)

26. Russo, D., Van Roy, B.: Learning to optimize via posterior sampling. Math. Oper. Res. **39**(4), 1221–1243 (2014)
27. Russo, D., Van Roy, B.: An information-theoretic analysis of Thompson sampling. J. Mach. Learn. Res. **17**(68), 1–30 (2016)
28. Scott, S.L.: A modern Bayesian look at the multi-armed bandit. Appl. Stoch. Models Bus. Ind. **26**, 639–658 (2010)
29. Thompson, W.: On the likelihood that one unknown probability exceeds another in view of the evidence of two samples. Bull. Am. Math. Soc. **25**, 285–294 (1933)
30. Xia, Y., Li, H., Qin, T., Yu, N., Liu, T.-Y.: Thompson sampling for budgeted multi-armed bandits. In: IJCAI, pp. 3960–3966 (2015)

Clustering

Finding Meaningful Cluster Structure Amidst Background Noise

Shrinu Kushagra[1(✉)], Samira Samadi[2], and Shai Ben-David[1]

[1] University of Waterloo, Waterloo, Canada
skushagr@uwaterloo.ca, shai@cs.uwaterloo.ca
[2] Georgia Institute of Technology, Atlanta, USA
ssamadi6@gatech.edu

Abstract. We consider efficient clustering algorithm under data cluster-ability assumptions with added noise. In contrast with most literature on this topic that considers either the adversarial noise setting or some noise generative model, we examine a realistically motivated setting in which the only restriction about the noisy part of the data is that it does not create significantly large "clusters". Another aspect in which our model deviates from common approaches is that we stipulate the goals of clustering as discovering meaningful cluster structure in the data, rather than optimizing some objective (clustering cost).

We introduce efficient algorithms that discover and cluster every subset of the data with meaningful structure and lack of structure on its complement (under some formal definition of such "structure"). Notably, the success of our algorithms does not depend on any upper bound on the fraction of noisy data.

We complement our results by showing that when either the notions of structure or the noise requirements are relaxed, no such results are possible.

1 Introduction

Clustering is an umbrella term for a wide variety of unsupervised data processing techniques. Being widely applied in practice, it comes in many variations that are hard to encompass in a precise single definition. A relatively comprehensive description is that clustering aims to group together data instances that are similar, while separating dissimilar objects. Most of the common clustering tools output a partitioning of the input data into groups, clusters, that share some form of cohesiveness or between-cluster separation requirement[1]. However, in many cases, real data sets, in particular large ones, have on top of such cohesive separated groups, a significant amount of "background" unstructured data. An obvious example of such a scenario is when the input data set is the set of pixels of an image and the goal of the clustering is to detect groups of pixels that correspond to objects in that image. Clustering in such situations is the focus of

---

[1] The assignment to clusters can sometimes be probabilistic, and clusters may be allowed to intersect, but these aspects are orthogonal to the discussion in this paper.

© Springer International Publishing Switzerland 2016
R. Ortner et al. (Eds.): ALT 2016, LNAI 9925, pp. 339–354, 2016.
DOI: 10.1007/978-3-319-46379-7_23

this work. Maybe surprisingly, this topic has received relatively little attention in the clustering research community, and even less so when it comes to theoretical work.

The discussion of finding clustering structure in data sets that also contain subsets that do not conform well to that structure usually falls under the terminology of noise robustness (see e.g., [1,5,10–12]). However, noise robustness, at least in that context, addresses the noisy part of the data as either generated by some specific generative model (like uniform random noise, or Gaussian perturbations) or refers to worst-case adversarially generated noisy data. In this paper we take a different approach. What distinguishes the noise that we consider form the "clean" part of the input data is that it is *structureless*. The exact meaning of such a notion of structurelessness may vary depending on the type of structure the clustering algorithm is aiming to detect in the data. We focus on defining structurelessness as not having significantly large dense subsets. We believe that such a notion is well suited to address "gray background" contrasting with cohesive subsets of the data that are the objects that the clustering aims to detect.

The distinction between structured and unstructured parts of the data requires, of course, a clear notion of relevant structure. For that, we resort to a relatively large body of recent work proposing notions of clusterable data sets. That work was developed mainly to address the gap between the computational hardness of (the optimization problem of) many common clustering objectives and the apparent feasibility of clustering in practical applications. We refer the reader to [6] for a survey of that body of work. Here, we focus on two such notions, one based on the α-center-proximity introduced by [2] and the other, λ-separation, introduced by [7].

Our approach diverges from previous discussions of clusterable inputs in yet another aspect. Much of the theoretical research of clustering algorithms views clustering as an optimization problem. For some predetermined objective function (or clustering cost), the algorithm's task is to find the data partitioning that minimizes that objective. In particular, this approach is shared by all the works surveyed in [6]. We believe that the reality of clustering applications is different. Given a large data set to cluster, there is no way a user may know what is the cost of the optimal clustering of that data, or how close to optimal the algorithm's outcome is. Instead, a user might have a notion of meaningful cluster structure, and will be happy with any outcome that meets such a requirement. Consequently, our algorithms aim to provide meaningful clustering solutions (where "meaningful" is defined in a way inspired by the above mentioned notions of clusterability) without reference to any particular optimization objective function. Our algorithms efficiently compute a hierarchical clustering tree that captures all such meaningful solutions. One should notice that all of those notions of clusterability (those under which it can be show that an objective-minimizing clustering can be found efficiently) assume that there exists an optimal solution that satisfies the meaningfulness condition (such as being perturbation robust, or having significantly smaller distances of points to their own cluster centers than to other centers). Under those assumptions, an algorithm that outputs a tree capturing all meaningful solutions, allows efficient detection of the cost-optimal clustering (in fact, the algorithms of [5] also yield such trees, for clean, noiseless inputs). Consequently, under the assumptions of those previous works, our algorithms yield an efficient procedure for finding such an optimal solution.

1.1   Related Work

The goal of clustering is to partition a set of objects into *dissimilar* subsets of *similar* objects. Based on the definition of similarity, the optimal solution to a clustering task is achieved by optimizing an objective function. Although solving this optimization problem is usually NP-hard, the clustering task is routinely and successfully employed in practice. This gap between theory and practice recommends characterizing the real world data sets by defining mathematical notions of *clusterable* data. As a result, provably efficient clustering algorithms can be found for these so called *nice* data.

In the past few years, there has been a line of work on defining notions of clusterability. The goal of all these methods has been to show that clustering is computationally efficient if the input $\mathcal{X}$ enjoys some nice structure. In [9], a clustering instance is considered to be *stable* if the optimal solution to a given objective function does not change under small multiplicative perturbations of distances between the points. Using this assumption, they give an efficient algorithm to find the max-cut clustering of graphs which are resilient to $O(\sqrt{|\mathcal{X}|})$ perturbations. Using a similar assumption, [1] considered additive perturbations of the underlying metric and designed an efficient algorithm that outputs a clustering with near-optimal cost.

In terms of clusterability conditions, the most relevant previous papers are those addressing clsutering under α-*center proximity* condition (see Definition 5). Assuming that the centers belong to $\mathcal{X}$ (*proper* setting), [2] shows an efficient algorithm that outputs the optimal solution of a given center-based objective assuming that optimal solution satisfies the $(\alpha > 3)$-center proximity. This result was improved to $(\alpha = \sqrt{2} + 1 \approx 2.4)$ when the objective is k-median [5]. In [8] it was shown that unless P=NP such a result cannot be obtained for $(\alpha < 2)$-center proximal inputs.

However, as mentioned above, these results apply only to the noiseless case. Few methods have been suggested for analyzing clusterability in the presence of noise. [5] considers a dataset which has α-center proximity except for an ϵ fraction of the points. They give an efficient algorithm which provides a $1 + O(\epsilon)$-approximation to the cost of the k-median optimal solution when $\alpha > 2 + \sqrt{7} \approx$ 4.6. Note that, while this result applies to adversarial noise as well, it only yields an approximation to the desired solution and the approximation guarantee is heavily influenced by the size of noise.

In a different line of work, [7] studied the problem of robustifying any center-based clustering objective to noise. To achieve this goal, they introduce the notion of *center separation* (look at Definition 7). Informally, an input has center separation when it can be covered by k well-separated set of balls. Given such an input, they propose a paradigm which converts any center-based clustering algorithm into a clustering algorithm which is robust to small amount of noise. Although this framework works for any objective-based clustering algorithm, it requires a strong restriction on the noise and clusterability of the data. For example, when the size of the noise is $\frac{5}{100}|\mathcal{X}|$, their algorithm is able to obtain a robustified version of 2-median, only if $\mathcal{X}$ is covered by k unit balls which are separated with distance 10.

In this work, we consider a natural relaxation of [5,7], with the goal to capture more realistic domains containing arbitrary amount of noise, assuming that noise is *structureless* (in a precise sense defined below). For example, in [5], the size of the noise $|\mathcal{N}| \leq \frac{m(C)}{8}$ (where $m(C)$ is size of the smallest cluster). Our algorithms can handle much larger amount of noise as long as they satisfy the *structureless* condition.

We define a novel notion of "gray background" noise. Informally, we call noise *structureless* if it does not have similar structure to a *nice* cluster at any part of the domain. Under that definition (look at Definition 6), our positive, efficient clustering results, do not depend on any restriction on the size of the noise.

Given a clusterable input $\mathcal{X}$ which contains *structureless* noise, we propose an efficient algorithm that outputs a hierarchical clustering tree of $\mathcal{X}$ that captures all *nice* clusterings of $\mathcal{X}$. Our algorithm perfectly recovers the underlying *nice* clusterings of the input and its performance is independent of number of noisy points in the domain.

We complement our algorithmic results by proving that under more relaxed conditions, either on the level of clusterability of the clean part of the data, or on the unstructuredness requirements on the noise, such results become impossible.

1.2   Outline

The rest of this paper is structured as follows. In Sect. 2, we present our notation and formal definitions. In Sect. 3 we show that the type of noise that we address in this paper is likely to arise under some natural assumptions on the data generating process. In Sect. 4, we present an efficient algorithm that, for any input set $\mathcal{X}$ which contains structureless noise, recovers all the underlying clusterings of non-noise subset of $\mathcal{X}$ that satisfies α-center proximity for $\alpha > 2 + \sqrt{7}$. We complement these results by proving that for $\alpha \leq 2\sqrt{2} + 3$ in the case that we have arbitrary noise and for $\alpha \leq \sqrt{2}+3$ in the case of structureless noise, efficient discovery of all nicely structured subsets is not possible.

In Sect. 5.1, we describe an efficient algorithm that, for any input $\mathcal{X}$, recovers all the underlying clusterings of $\mathcal{X}$ that satisfy λ-center separation for $\lambda \geq 3$. We also prove that it is NP-Hard to improve this to $\lambda \leq 2$. In Sect. 5.2, we consider a similar problem in the presence of either arbitrary or structureless noise. We propose an efficient algorithm that, for any input $\mathcal{X}$ which contains structureless noise, recovers all the underlying clusterings of non-noise subset of $\mathcal{X}$ that satisfy λ-center separation for $\lambda \geq 4$. We will also show that this result is tight for the case of structureless noise. We complement our results by showing that, under arbitrary noise assumption, no similar positive result can be achieved for $\lambda \leq 6$. Note that all our missing proofs can be found in the appendix.

2   Notation and Definition

Let $(\mathbf{M}, d)$ be a metric space. Given a data set $\mathcal{X} \subseteq \mathbf{M}$ and an integer k. A k-clustering of $\mathcal{X}$ denoted by $\mathcal{C}_\mathcal{X}$ is a partition of $\mathcal{X}$ into k disjoints sets. Given points $c_1, \ldots, c_k \in \mathbf{M}$, we define the clustering induced by these points (or *centers*) by assigning each $x \in \mathcal{X}$ to its nearest center. In the *steiner* setting, the centers can be arbitrary points of the metric space $\mathbf{M}$. In the *proper* setting, we restrict our centers to be members of the data set $\mathcal{X}$. In this paper, we will be working in the **proper** setting.

For any set $\mathcal{A} \subseteq \mathcal{X}$ with center $c \in \mathbf{M}$, we define the radius of $\mathcal{A}$ as $r_c(\mathcal{A}) = \max_{x \in \mathcal{A}} d(x, c)$. Throughout the paper, we will use the notation $\mathcal{C}_\mathcal{X}$ to denote the clustering of the set $\mathcal{X}$ and $\mathcal{C}_\mathcal{S}$ to denote the clustering of some $\mathcal{S} \subseteq \mathcal{X}$.

Definition 1 $(r(\mathcal{C}_\mathcal{X})$, $m(\mathcal{C}_\mathcal{X}))$. *Given a clustering* $\mathcal{C}_\mathcal{X} = \{C_1, \ldots, C_k\}$ *induced by centers* $c_1, \ldots, c_k \in \mathbf{M}$, *we define* $m(\mathcal{C}_\mathcal{X}) = \min_i |C_i|$ *and* $r(\mathcal{C}_\mathcal{X}) = \max_i r(C_i)$.

Definition 2 $(\mathcal{C}_\mathcal{X}$ **Restricted to a Set**)**.** *Given* $\mathcal{S} \subseteq \mathcal{X}$ *and a clustering* $\mathcal{C}_\mathcal{X} = \{C_1, \ldots, C_k\}$ *of the set* $\mathcal{X}$. *We define* $\mathcal{C}_\mathcal{X}$ *restricted to the set* $\mathcal{S}$ *as* $\mathcal{C}_{\mathcal{X}|_\mathcal{S}} = \{C_1 \cap \mathcal{S}, \ldots, C_k \cap \mathcal{S}\}$.

Definition 3 $(\mathcal{C}_\mathcal{X}$ **Respects** $\mathcal{C}_\mathcal{S})$**.** *Given* $\mathcal{S} \subseteq \mathcal{X}$, *clusterings* $\mathcal{C}_\mathcal{X} = \{C_1, \ldots, C_k\}$ *and* $\mathcal{C}_\mathcal{S} = \{S_1, \ldots, S_{k'}\}$. *We say that* $\mathcal{C}_\mathcal{X}$ *respects* $\mathcal{C}_\mathcal{S}$ *if* $\mathcal{C}_{\mathcal{X}|_\mathcal{S}} = \mathcal{C}_\mathcal{S}$.

Definition 4 $(\mathcal{T}$ **or** $\mathcal{L}$ **Captures** $\mathcal{C}_\mathcal{S})$**.** *Given a hierarchical clustering tree* $\mathcal{T}$ *of* $\mathcal{X}$ *and a clustering* $\mathcal{C}_\mathcal{S}$ *of* $\mathcal{S} \subseteq \mathcal{X}$. *We say that* $\mathcal{T}$ *captures* $\mathcal{C}_\mathcal{S}$ *if there exists a pruning* $\mathcal{P}$ *which respects* $\mathcal{C}_\mathcal{S}$.

Similarly, given a list of clusterings $\mathcal{L}$ *of* $\mathcal{X}$ *and a clustering* $\mathcal{C}_\mathcal{S}$ *of* $\mathcal{S} \subseteq \mathcal{X}$. *We say that* $\mathcal{L}$ *captures* $\mathcal{C}_\mathcal{S}$ *if there exists a clustering* $\mathcal{C}_\mathcal{X} \in \mathcal{L}$ *which respects* $\mathcal{C}_\mathcal{S}$.

Definition 5 (**α-Center Proximity** [2])**.** *A clustering* $\mathcal{C}_\mathcal{X} = \{C_1, \ldots, C_k\}$ *satisfies* α-*center proximity w.r.t* $\mathcal{X}$ *and* k *if there exist centers* $c_1, \ldots, c_k \in \mathbf{M}$ *such that the following holds. For all* $x \in C_i$ *and* $i \neq j$, $\alpha d(x, c_i) < d(x, c_j)$

Next, we formally define our notion of structureless noise. Roughly, such noise should be scattered sparsely, namely, there should be no significant amount of noise in any small enough ball. Note that such a restriction does not impose any upper bound on the number of noise points.

Definition 6 $((\alpha, \eta)$-**Center Proximity**)**.** *Given* $\mathcal{S} \subseteq \mathcal{X}$, *a clustering* $\mathcal{C}_\mathcal{S} = \{S_1, \ldots, S_k\}$ *has* (α, η)-*center proximity w.r.t* $\mathcal{X}, \mathcal{S}$ *and* k *if there exists centers* $s_1, \ldots, s_k \in \mathbf{M}$ *such that the following holds.*

◇ α-**center proximity**: *For all* $x \in S_i$ *and* $i \neq j$, $\alpha d(x, s_i) < d(x, s_j)$

◇ η-**sparse noise**: *For any ball* B, $r(B) \leq \eta r(\mathcal{C}_S) \implies |B \cap (\mathcal{X} \setminus S)| < \frac{m(\mathcal{C}_S)}{2}$

Definition 7 (λ-Center Separation [7]). *A clustering* $\mathcal{C}_{\mathcal{X}} = \{C_1, \ldots, C_k\}$ *has* λ-*center separation w.r.t* $\mathcal{X}$ *and* k *if there exists centers* $c_1, \ldots, c_k \in \mathbf{M}$ *such that* $\mathcal{C}_{\mathcal{X}}$ *is the clustering induced by these centers and the following holds. For all* $i \neq j$, $d(c_i, c_j) > \lambda r(\mathcal{C}_{\mathcal{X}})$

Definition 8 ((λ, η)-Center Separation). *Given* $S \subseteq \mathcal{X}$, *a clustering* $\mathcal{C}_S$ *has* (λ, η)-*center separation w.r.t* $\mathcal{X}, S$ *and* k *if there exists centers* $s_1, \ldots, s_k \in \mathbf{M}$ *such that* $\mathcal{C}_{\mathcal{X}}$ *is the clustering induced by these centers and the following holds.*

◇ λ-*center separation*: *For all* $i \neq j$, $d(s_i, s_j) > \lambda r(\mathcal{C}_S)$

◇ η-*sparse noise*: *For any ball* B, $r(B) \leq \eta r(\mathcal{C}_S) \implies |B \cap (\mathcal{X} \setminus S)| < \frac{m(\mathcal{C}_S)}{2}$

We denote a ball of radius x at center c by $B(c, x)$. We denote by $P_i(c)$ a collection of i many points sitting on the same location c. If the location is clear from the context, we will use the notation P_i.

3   Justification of Sparse Noise

In this section, we examine our sparseness condition. We will show that if the set of points $\mathcal{N}$ are generated by a non concentrated distribution in a ball in $\mathbf{R}^d$ then with high probability, as long as $\mathcal{N}$ is not too large (so as to " drown" the original data set), it will satisfy the sparse noise condition. The proof is based on the epsilon approximation theorem for classes of finite VC-dimension, applied to the set of balls in $\mathbf{R}^d$. The following, rather natural, definition of non concentrated distribution was introduced in [3].

Definition 9. *A probability distribution over the d-dimensional unit ball is non-concentrated if, for some constant c, the probability density of any point x is at most c times its density under the uniform distribution over that ball.*

Theorem 1 (Noise by Non Concentrated Distribution is Sparse). *Let* $\mathcal{X}$ *be a ball of radius* R *in* $\mathbf{R}^d$ *and* $S \subseteq \mathcal{X}$. *Let* $\mathcal{C}$ *be a clustering of* S *which satisfies* α-*center proximity (or* λ-*center separation). Given parameters* $\epsilon, \delta \in (0, 1)$. *Let* $\mathcal{N} \subseteq \mathcal{X}$ *be picked i.i.d according to a non concentrated probability distribution. If* $|\mathcal{N}| < c \left(\frac{R}{r(C)\eta} \right)^d m(C)$ *then with high probability,* $S \cup \mathcal{N}$ *satisfies* (α, η)-*center proximity (the* (λ, η)-*center separation, respectively).*

Proof. Let $H = \{B \text{ is a ball} : B \subseteq X\}$. Observe that VC-Dim($H$) = $d + 1$. Let $\gamma := \frac{r(C)}{R}$. Since the noise-generating distribution P is c-concentrated, for every ball B, $P(B) \leq c\frac{\text{vol}(B)}{\text{vol}(X)} = c\gamma^d$. Now, the fundamental ϵ-approximation theorem (Theorem 16) establishes the result.   ∎

Note that Theorem 1 shows that the cardinality of the noise set, $|\mathcal{N}|$, can be much bigger than the size of the smallest cluster $m(\mathcal{C})$.

4 Center Proximity

In this section, we study the problem of recovering (α, η)-center proximal clusterings of a set $\mathcal{X}$, in the presence of noise. The goal of our algorithm is to produce an efficient representation (hierarchical clustering tree) of all possible (α, η)-center proximal nice clusterings rather than to output a single clustering or to optimize an objective function. Here is a more precise overview of the results of this section:

- *Positive result under sparse noise* - In Sect. 4.1, we give our main result under sparse noise. If $\alpha \geq 2 + \sqrt{7} \approx 4.6$ and $\eta \geq 1$; for any value of t, Algorithm 1 outputs a tree which captures all clusterings C^* (of a subset of $\mathcal{X}$) which satisfy (α, η)-center proximity and $m(C^*) = t$.
- *Lower bound under sparse noise* - In Sect. 4.2, we show that if $\alpha \leq 2 + \sqrt{3} \approx 3.7$ and $\eta \leq 1$ then there is no tree and no list of 'small' size ($< 2^{k/2}$) which can capture all clusterings C (of a subset of $\mathcal{X}$) which satisfy (α, η)-center proximity even for a fixed value of the size of the smallest cluster ($m(C) = t$).
- *Lower bound with arbitrary noise* - In Sect. 4.3, we show that for a given value of a parameter t, if $\alpha \leq 2\sqrt{2} + 3 \approx 5.8$ and the number of noisy points exceeds $\frac{3}{2}t$ then no tree can capture all clusterings C (of a subset of $\mathcal{X}$) which satisfy α-center proximity even for fixed $m(C) = t$. Identical result holds for 'small' ($< 2^{k/2}$) lists if the number of noisy points exceeds $\frac{3k}{2}t$.

4.1 Positive Result Under Sparse Noise

Given a clustering instance $(\mathcal{X}, d)$ and a parameter t, we introduce an efficient algorithm which outputs a hierarchical clustering tree T of $\mathcal{X}$ with the following property. For every k, for every $S \subseteq \mathcal{X}$ and for every k-clustering C_S which satisfies (α, η)-center proximity (for $\alpha \geq 2 + \sqrt{7}$ and $\eta \geq 1$) and $m(C_S) = t$, T captures C_S. It is important to note that our algorithm only knows $\mathcal{X}$ and has no knowledge of the set S.

Our algorithm has a linkage based structure similar to [5]. However, our method benefits from a novel *sparse distance condition*. We introduce the algorithm in Algorithm 1 and prove its efficiency and correctness in Theorem 3 and Theorem 2 respectively.

Definition 10 (Sparse Distance Condition). *Given a clustering $C = \{C_1, \ldots, C_k\}$ of the set $\mathcal{X}$ and a parameter t. We say that the ball $B \subseteq \mathcal{X}$ satisfies the sparse distance condition w.r.t clustering C when the following holds.*

- $|B| \geq t$.
- *For any $C_i \in C$, if $C_i \cap B \neq \emptyset$, then $C_i \subseteq B$ or $|B \cap C_i| \geq t/2$.*

Intuitively, Algorithm 1 works as follows. It maintains a clustering $C^{(l)}$, which is initialized so that each point is in its own cluster. It then goes over all pairs of points p, q in increasing order of their distance $d(p, q)$. If $B(p, d(p, q))$ satisfies the sparse distance condition w.r.t $C^{(l)}$, then it merges all the clusters which intersect with this ball into a single cluster and updates $\mathcal{C}^{(l)}$. Furthermore, the algorithm builds a tree with the nodes corresponding to the merges performed so far. We will show that for all $\mathcal{S} \subseteq \mathcal{X}$ which are (α, η)-proximal t-min nice and for all clusterings $\mathcal{C}_S$ which have (α, η)-center proximity, Algorithm 1 outputs a tree which captures $\mathcal{C}_S$.

Algorithm 1. Alg. for (α, η)-center proximity with parameter t

Input: $(\mathcal{X}, d)$ and t
Output: A hierarchical clustering tree T of $\mathcal{X}$.

Let $\mathcal{C}^{(l)}$ denote the clustering $\mathcal{X}$ after l merge steps have been performed. Initialize $\mathcal{C}^{(0)}$ so that all points are in their own cluster. That is,
$\mathcal{C}^{(0)} = \{\{x\} : x \in \mathcal{X}\}$.
Go over all pairs of points p, q in increasing order of the distance $d(p, q)$. If $B = B(p, d(p, q))$ satisfies the sparse distance condition then
    Merge all the clusters which intersect with B into a single cluster.

Output clustering tree T. The leaves of T are the points in dataset $\mathcal{X}$. The internal nodes correspond to the merges performed.

Theorem 2. *Given a clustering instance $(\mathcal{X}, d)$ and a parameter t. Algorithm 1 outputs a tree T with the following property. For all k, $\mathcal{S} \subseteq \mathcal{X}$ and for all k-clusterings $\mathcal{C}_S^* = \{S_1^*, \ldots, S_k^*\}$ which satisfy $(2 + \sqrt{7}, 1)$-center proximity the following holds. If $m(\mathcal{C}_S^*) = t$ then T captures $\mathcal{C}_S$.*

Theorem 3. *Given clustering instance $(\mathcal{X}, d)$ and t. Algorithm 1 runs in $poly(|\mathcal{X}|)$.*

Proof. Let $n = |\mathcal{X}|$. Checking if B satisfies the sparse-distance condition takes $O(n)$ time and hence the algorithm runs in $O(n^3)$ time.

4.2   Lower Bound Under Sparse Noise

Theorem 4. *Given the number of clusters k and parameter t. For all $\alpha \leq 2 + \sqrt{3}$ and $\eta \leq 1$ there exists a clustering instance $(\mathcal{X}, d)$ such that any clustering tree T of $\mathcal{X}$ has the following property. There exists $\mathcal{S} \subseteq \mathcal{X}$ and clustering $\mathcal{C}_S$ which satisfies (α, η)-center proximity and $m(\mathcal{C}_S) = t$ but T doesn't capture $\mathcal{C}_S$.*

Theorem 5. *Given the number of clusters k and parameter t. For all $\alpha \leq 2 + \sqrt{3}$, $\eta \leq 1$ there exists $(\mathcal{X}, d)$ such that any list $\mathcal{L}$ (of clusterings of $\mathcal{X}$) has the following property. If $|\mathcal{L}| < 2^{\frac{k}{2}}$ then there exists $\mathcal{S} \subseteq \mathcal{X}$ and clustering $\mathcal{C}_S$ which satisfies (α, η)-center proximity and $m(\mathcal{C}_S) = t$ but $\mathcal{L}$ doesn't capture $\mathcal{C}_S$.*

4.3 Lower Bound Under Arbitrary Noise

Theorem 6. *Given the number of clusters k and a parameter t. For all $\alpha < 2\sqrt{2}+3$ there exists $(\mathcal{X},d)$ such that any clustering tree T of $\mathcal{X}$ has the following property. There exists $\mathcal{S} \subseteq \mathcal{X}$ and there exists clustering $\mathcal{C}_\mathcal{S}$ which satisfies α-center proximity such that $m(\mathcal{C}_\mathcal{S}) = t$ and the following holds. If $|\mathcal{X} \setminus \mathcal{S}| \geq \frac{3t(\mathcal{C}_\mathcal{S})}{2} + 5$, then T doesn't capture $\mathcal{C}_\mathcal{S}$.*

Algorithm 2. Alg. for λ-center separation

Input: $(\mathcal{X},d)$
Output: A hierarchical clustering tree T of $\mathcal{X}$.

Initialize the clustering so that each point is in its own cluster.
Run single-linkage till only a single cluster remains. Output clustering tree T.

Theorem 7. *Given the number of clusters k and parameter t. For all $\alpha \leq 2 + \sqrt{2} + 3$ there exists $(\mathcal{X},d)$ such that any list $\mathcal{L}$ (of clusterings of $\mathcal{X}$) has the following property. There exists $\mathcal{S} \subseteq \mathcal{X}$ and there exists clustering $\mathcal{C}_\mathcal{S}$ which satisfies α-center proximity such that $m(\mathcal{C}_\mathcal{S}) = t$ and the following holds. If $|\mathcal{L}| < 2^{\frac{k}{2}}$ and $|\mathcal{X} \setminus \mathcal{S}| \geq \frac{k}{2}(\frac{3t(\mathcal{C}_\mathcal{S})}{2} + 5)$, then $\mathcal{L}$ doesn't capture $\mathcal{C}_\mathcal{S}$.*

5 Center Separation

5.1 Center Separation Without Noise

In this section, we study the problem of recovering λ-center separated clusterings of a set $\mathcal{X}$, in the absence of noise. We do not want to output a single clustering but to produce an efficient representation (hierarchical clustering tree) of all possible λ-center separated nice clusterings. In Sect. 5.1.1 we give an algorithm that generates a tree of all possible λ-center separated clusterings of $\mathcal{X}$ for $\lambda > 3$. In Sect. 5.1.2, we prove that for $\lambda < 2$, it is NP-Hard to find any such clustering.

5.1.1 Positive Result Under No Noise

Given a clustering instance $(\mathcal{X},d)$, our goal is to output a hierarchical clustering tree T of $\mathcal{X}$ which has the following property. For every k and for every k-clustering $\mathcal{C}_\mathcal{X}$ which satisfies λ-center separation, there exists a pruning $\mathcal{P}$ of the tree which equals $\mathcal{C}_\mathcal{X}$. Our algorithm (Algorithm 2) uses single-linkage to build a hierarchical clustering tree of $\mathcal{X}$. We will show that when $\lambda \geq 3$ our algorithm achieves the above mentioned goal.

Theorem 8. *Given (X,d). For all $\lambda \geq 3$, Algorithm 2 outputs a tree T with the following property. For all k and for all k-clusterings $\mathcal{C}_\mathcal{X}^* = \{C_1^*,\ldots,C_k^*\}$ which satisfy λ-center separation w.r.t X and k, the following holds. For every $1 \leq i \leq k$, there exists a node N_i in the tree T such that $C_i^* = N_i$.*

5.1.2   Lower Bound with No Noise

We will prove that for $\lambda \leq 2$, finding any solution for λ-center separation is NP-Hard. [13] proved that finding any solution for α-center proximity is NP-Hard for $\alpha < 2$. Our reduction is same as the reduction used in Theorem 1 in [13] and hence we omit the proof.

Theorem 9. *Given a clustering instance* $(\mathcal{X}, d)$ *and the number of clusters* k. *For* $\lambda < 2$, *finding a clustering which satisfies* λ-center separation is NP-Hard.

5.2   Center Separation in the Presence of Noise

In this section, we study the problem of recovering (λ, η)-center separated clusterings of a set $\mathcal{X}$, in the presence of noise. Here is a more precise overview of the results of this section:

- *Positive result under sparse noise* - In Sect. 5.2.1, we show that if $\lambda \geq 4$ and $\eta \geq 1$; for any value of parameters r and t, Algorithm 3 outputs a clustering which respects all clusterings $\mathcal{C}^*$ (of a subset of $\mathcal{X}$) which satisfies (λ, η)-center proximity and $m(C^*) = t$ and $r(C^*) = r$.
- *Lower bound under sparse noise* - In Sect. 5.2.2, we show that, if $\lambda < 4$ and $\eta \leq 1$ then there is no tree and no list of 'small' size ($< 2^{k/2}$) which can capture all clusterings $\mathcal{C}$ (of subset of $\mathcal{X}$) which satisfy (λ, η)-center proximity even for fixed values of the size of the smallest cluster ($m(C) = t$) and maximum radius ($r(C) = r$).
- *Lower bound with arbitrary noise* - In Sect. 5.2.3, we show that for a given value of parameters r and t, if $\lambda \leq 6$ and the number of noisy points exceeds $\frac{3}{2}t$ then no tree can capture all clusterings $\mathcal{C}$ (of a subset of $\mathcal{X}$) which satisfy λ-center separation even for fixed $m(\mathcal{C}) = t$ and $r(\mathcal{C}) = r$. Identical result holds for 'small' ($< 2^{k/2}$) lists if the number of noisy points exceeds $\frac{3k}{2}t$.

5.2.1   Positive Result Under Sparse Noise

We are given a clustering instance $(\mathcal{X}, d)$ and parameters r and t. Our goal is to output a clustering $\mathcal{C}_{\mathcal{X}}$ which has the following property. For every k, for every $\mathcal{S} \subseteq \mathcal{X}$ and for every k-clustering $\mathcal{C}_{\mathcal{S}}$ which satisfies (λ, η)-center separation, the clustering $\mathcal{C}_{\mathcal{X}}$ restricted to $\mathcal{S}$ equals $\mathcal{C}_{\mathcal{S}}$.

In the next section, we propose a clustering algorithm (Algorithm 3) and prove (Theorem 10) that our algorithm indeed achieves the above mentioned goal (under certain assumptions on the parameters λ and η). It is important to note that our algorithm only knows $\mathcal{X}$ and has no knowledge of the set $\mathcal{S}$.

Intuitively, Algorithm 3 works as follows. In the first phase, it constructs a list of balls which have radius at most r and contain at least t points. It then constructs a graph as follows. Each ball found in the first phase is represented by a vertex. If two balls have a 'large' intersection then there is an edge between the corresponding vertices in the graph. We then find the connected components in the graph which correspond to the clustering of the original set $\mathcal{X}$.

Algorithm 3. Alg. for (λ, η)-center separation with parameters t and r

Input: $(\mathcal{X}, d), t$ and r
Output: A clustering $\mathcal{C}$ of the set $\mathcal{X}$.

Phase 1
Let $\mathcal{L}$ denote the list of balls found so far. Initialize $\mathcal{L}$ to be the empty set.
$\mathcal{L} = \emptyset$.
Go over all pairs of points $p, q \in \mathcal{X}$ in increasing order of the distance $d(p, q)$.
Let $B := B(p, d(p, q))$. If $|B| \geq t$ and $r(B) \leq r$ then
$\quad \mathcal{L} = \mathcal{L} \cup B$

Output the list of balls $\mathcal{L} = \{B_1, \ldots, B_l\}$ to the second phase of the algorithm.

Phase 2
Construct a graph $G = (V, E)$ as follows. $V = \{v_1, v_2, \ldots, v_l\}$. If $|B_i \cap B_j| \geq t/2$
then construct an edge between v_i and v_j.
Find connected components $(G_1, \ldots, G_k)$ in the graph G.
Merge all the points in the same connected component together to get a
clustering $\mathcal{C} = \{C_1, \ldots, C_k\}$ of the set $\mathcal{X}$.
Assign $x \in \mathcal{X} \setminus \cup_i B_i$ to the closest cluster C_i. That is, $i := \arg\min_{j \in [k]} \min_{y \in C_j} d(x, y)$.
Output $\mathcal{C}$.

Theorem 10. *Given a clustering instance $(\mathcal{X}, d)$ and parameters r and t. For every k, for every $\mathcal{S} \subseteq \mathcal{X}$ and for all k-clusterings $\mathcal{C}_\mathcal{S}^* = \{S_1^*, \ldots, S_k^*\}$ which satisfy $(4, 1)$-center separation such that $m(\mathcal{C}_\mathcal{S}^*) = t$ and $r(\mathcal{C}_\mathcal{S}^*) = r$, the following holds. Algorithm 3 outputs a clustering $\mathcal{C}_\mathcal{X}$ such that $\mathcal{C}_\mathcal{X}|_\mathcal{S} = \mathcal{C}_\mathcal{S}^*$.*

Theorem 11. *Given $(\mathcal{X}, d)$ and parameters r and t. Algorithm 3 runs in $poly(|\mathcal{X}|)$.*

Proof. Let $n = |\mathcal{X}|$. Phase 1 of Algorithm 3 runs in $O(n^2)$ time. Phase 2 gets a list of size l. Constructing G and finding connected components takes $O(l^2)$ time. Hence, the algorithm runs in $O(n^2)$ time.

5.2.2 Lower Bound Under Sparse Noise

Theorem 12. *Given the number of clusters k and parameters r and t. For all $\lambda < 4$ and $\eta \leq 1$, there exists a clustering instance $(\mathcal{X}, d)$ such that any clustering tree $\mathcal{T}$ of $\mathcal{X}$ has the following property. There exists $\mathcal{S} \subseteq \mathcal{X}$ and a k-clustering $\mathcal{C}_\mathcal{S} = \{S_1, \ldots, S_k\}$ which satisfies (λ, η)-center separation such that $m(\mathcal{C}_\mathcal{S}) = t$ and $r(\mathcal{C}_\mathcal{S}) = r$, but $\mathcal{T}$ doesn't capture $\mathcal{C}_\mathcal{S}$.*

Theorem 13. *Given the number of clusters k and parameters r and t. For all $\lambda \leq 4$ and $\eta \leq 1$ there exists a clustering instance $(\mathcal{X}, d)$ such that any list $\mathcal{L}$ (of clusterings of $\mathcal{X}$) has the following property. If $|\mathcal{L}| < 2^{\frac{k}{2}}$ then there exists $\mathcal{S} \subseteq \mathcal{X}$ and clustering $\mathcal{C}_\mathcal{S}$ which satisfies (λ, η)-center separation and $m(\mathcal{C}_\mathcal{S}) = t$ and $r(\mathcal{C}_\mathcal{S}) = r$, but $\mathcal{L}$ doesn't capture $\mathcal{C}_\mathcal{S}$.*

5.2.3   Lower Bound with Arbitrary Noise

Theorem 14. *Given the number of clusters k and parameters r and t. For all $\lambda < 6$, there exists a clustering instance $(\mathcal{X}, d)$ such that any clustering tree $\mathcal{T}$ of $\mathcal{X}$ has the following property. There exists $\mathcal{S} \subseteq \mathcal{X}$ and there exists k-clustering $\mathcal{C}_\mathcal{S}$ which satisfies λ-center separation such that $m(\mathcal{C}_\mathcal{S}) = t$, $r(\mathcal{C}_\mathcal{S}) = r$ and the following holds. If $|\mathcal{X} \setminus \mathcal{S}| \geq \frac{3t}{2} + 5$, then $\mathcal{T}$ doesn't capture $\mathcal{C}_\mathcal{S}$.*

Theorem 15. *Given the number of clusters k and parameters r and t. For all $\lambda \leq 6$ there exists $(\mathcal{X}, d)$ such that any list $\mathcal{L}$ (of clusterings of $\mathcal{X}$) has the following property. There exists $\mathcal{S} \subseteq \mathcal{X}$ and there exists clustering $\mathcal{C}_\mathcal{S}$ which satisfies λ-center separation such that $m(\mathcal{C}_\mathcal{S}) = t$, $r(\mathcal{C}_\mathcal{S}) = r$ and the following holds. If $|\mathcal{L}| < 2^{\frac{k}{2}}$ and $|\mathcal{X} \setminus \mathcal{S}| \geq \frac{k}{2}(\frac{3t(\mathcal{C}_\mathcal{S})}{2} + 5)$, then $\mathcal{L}$ doesn't capture $\mathcal{C}_\mathcal{S}$.*

A   Proofs of Missing Lemmas and Theorems

Proof of Theorem 2 Fix any $\mathcal{S} \subseteq \mathcal{X}$. Let $\mathcal{C}_\mathcal{S}^* = \{S_1^*, \ldots, S_k^*\}$ be a clustering of $\mathcal{S}$ such that $m(\mathcal{C}_\mathcal{S}^*) = t$ and $\mathcal{C}_\mathcal{S}^*$ has (α, η)-center proximity. Denote by $r_i := r(S_i^*)$ and $r = \max r_i$. Define $Y_B^\mathcal{C} := \{C_i \in \mathcal{C} : C_i \subseteq B \text{ or } |B \cap C_i| \geq t/2\}$. Note that whenever a ball B satisfies the sparse-distance condition, all the clusters in $Y_B^{\mathcal{C}^{(l)}}$ are merged together and the clustering $\mathcal{C}^{(l+1)}$ is updated. We will prove the theorem by proving two key facts.

F.1 If the algorithm merges points from a good cluster S_i^* with points from some other good cluster, then at this step the distance being considered $d = d(p, q) > r_i$.

F.2 When the algorithm considers the distance $d = r_i$, it merges all points from S_i^* (and possibly points from $\mathcal{X} \setminus \mathcal{S}$) into a single cluster C_i. Hence, there exists a node in the tree N_i which contains all the points from S_i^* and no points from any other good cluster S_j^*.

Note that the theorem follows from these two facts. Similar reasoning was also used in proof of Lemma 3 in [5]. We now prove both of these facts formally.
Proof of Fact. F.1 Let $\mathcal{C}^{(l)} = \{C_1, \ldots, C_{k'}\}$ be the current clustering of $\mathcal{X}$. Let $l + 1$ be the first merge step which merges points from the good cluster S_i^* with points from some other good cluster. Let $p, q \in \mathcal{X}$ be the pair of points being considered at this step and $B = B(p, d(p, q))$ the ball that satisfies the sparse distance condition at this merge step. Denote by $Y = Y_B^{\mathcal{C}^{(l)}}$. We need to show that $d(p, q) > r_i$. To prove this, we need Claim 1 below.

Claim 1. *Let $p, q \in \mathcal{X}$ and B, Y, S_i^* and $C^{(l)}$ be as defined above. If $d(p, q) \leq r$, then $B \cap S_i^* \neq \emptyset$ and there exists $n \neq i$ such that $B \cap S_n^* \neq \emptyset$.*

$l + 1$ is the first step which merges points from S_i^* with some other good cluster. Hence, $\exists C_i \in Y$ such that $C_i \cap S_i^* \neq \emptyset$ and $\forall n \neq i$, $C_i \cap S_n^* = \emptyset$. Also, $\exists C_j \in Y$ such that $C_j \cap S_j^* \neq \emptyset$ for some S_j^* and $C_j \cap S_i^* = \emptyset$.

$C_i \in Y$. Hence, $C_i \subseteq B$ or $|C_i \cap B| \geq t/2$. The former is trivial. In the latter, for the sake of contradiction, assume that B contains no points from S_i^*. This implies that $B \cap C_i \subseteq B \cap \{\mathcal{X} \setminus \mathcal{S}\}$ and $|B \cap \{\mathcal{X} \setminus \mathcal{S}\}| \geq t/2$. This is a contradiction. The case when $C_j \in Y$ is identical. ∎

Claim 2. *Let the framework be as given in Claim 1. Then, $d(p, q) > r_i$.*

If $d(p, q) > r$, then the claim follows trivially. In the other case, from Claim 1, B contains $p_i \in S_i^*$ and $p_j \in S_j^*$. Let $r_i = d(c_i, q_i)$ for some $q_i \in S_i^*$.

$d(c_i, q_i) < \frac{1}{\alpha} d(q_i, c_j) < \frac{1}{\alpha}[\frac{1}{\alpha} d(p_i, p_j) + \frac{1}{\alpha} d(c_i, q_i) + d(p_i, p_j) + 2d(c_i, q_i)]$ This implies that $(\alpha^2 - 2\alpha - 1)d(q_i, c_i) < (\alpha + 1)d(p_i, p_j)$. For $\alpha \geq 2 + \sqrt{7}$, this implies that $d(c_i, q_i) < d(p_i, p_j)/2$ which implies $d(c_i, q_i) < d(p, q)$. This result was also stated in [5]. ∎

Proof of Fact F.2 Let $\mathcal{C}^{(l)} = \{C_1, \ldots, C_{k'}\}$ be the current clustering of $\mathcal{X}$. Let $l + 1$ be the merge step when $p = s_i$ and $q = q_i$ such that $d(s_i, q_i) = r_i$. We will prove that the ball $B = B(s_i, q_i)$ satisfies the sparse-distance condition.

Claim 3. *Let s_i, q_i, r_i, B and Y be as defined above. Then, B satisfies the sparse distance condition and for all $C \in Y$, for all $j \neq i$, $C \cap S_j^* = \emptyset$.*

$|B| = |S_i^*| \geq t$. Observe that, for all $C \in \mathcal{C}^{(l)}$, $|C| = 1$ or $|C| \geq t$.

- Case 1. $|C| = 1$. If $C \cap B \neq \emptyset \implies C \subseteq B = S_i^*$.
- Case 2. $|C| \geq t$. $C \cap B \neq \emptyset$. Let $h(C)$ denote the height of the cluster in the tree T.
  - Case 2.1. $h(C) = 1$. In this case, there exists a ball B' such that $B' = C$. We know that $r(B') \leq r_i \leq r$. Hence using Claim 2, we get that for all $j \neq i$, $B' \cap S_j^* = \emptyset$. Thus, $|B' \setminus S_i^*| \leq t/2 \implies |B \cap C| = |C| - |C \setminus B| = |C| - |B' \setminus S_i^*| \geq t/2$. Hence, $C \in Y$.
  - Case 2.2. $h(C) > 1$. Then there exists some C' such that $h(C') = 1$ and $C' \subset C$. Now, using set inclusion and the result from the first case, we get that $|B \cap C| \geq |B \cap C'| \geq t/2$. Hence, $C \in Y$. Using Claim 2, we get that for all $j \neq i$, $C \cap S_j^* = \emptyset$. ∎

Proof of Theorem 4 Let $\mathcal{X}, B_1, B_2, B_1', B_2'$ be as shown in Fig. 1. Let $t_1 = \frac{t}{2} + 1$ and $t_2 = \frac{t}{2} - 2$. For $\alpha \leq 2 + \sqrt{3}$, clusterings $\mathcal{C}_\mathcal{S} = \{B_1, B_2, B_3, \ldots, B_k\}$ and $\mathcal{C}_{\mathcal{S}'} = \{B_1', B_2', B_3, \ldots, B_k\}$ satisfy $(\alpha, 1)$-center proximity and $m(\mathcal{C}_\mathcal{S}) = m(\mathcal{C}_{\mathcal{S}}') = t$.

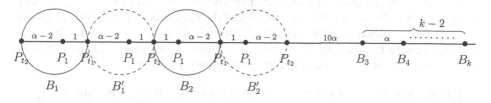

Fig. 1. $\mathcal{X} \subseteq \mathbb{R}$ such that no tree can capture all the (α, η)-proximal clusterings.

Now, a simple proof by contradiction shows that there doesn't exist a tree T and prunings P and P' such that P respects $\mathcal{C}_\mathcal{S}$ and P' respects $\mathcal{C}_{\mathcal{S}'}$. ∎

Proof of Theorem 5 The clustering instance $\mathcal{X}$ is an extension of Fig. 1. Let $G_1 = \{B_1, B_1', B_2, B_2'\}$ be the balls as in Fig. 1. Now, construct $G_2 = \{B_3, B_3', B_4, B_4'\}$ exactly identical to G_1 but far. In this way, we construct $k/2$ copies of G_1. ∎

Proof of Theorem 6 Let $\mathcal{X} \subseteq \mathbb{R}$ be as shown in Fig. 2. Let $t' = \frac{t}{2} - 1$ and let $B_1, B_2, B_3, B_1', B_2', B_3', B_1'', B_2''$ and B_3'' be as shown in Fig. 2. For $\alpha \leq 2\sqrt{2} + 3$, clusterings $\mathcal{C}_\mathcal{S} = \{B_1, B_2, B_3, \ldots, B_k\}$, $\mathcal{C}_{\mathcal{S}'} = \{B_1', B_2', B_3, \ldots, B_k\}$ and $\mathcal{C}_\mathcal{S}'' = \{B_1'', B_2'', B_3, \ldots, B_k\}$ satisfy $(\alpha, 1)$-center proximity. Also, $m(\mathcal{C}_\mathcal{S}) = m(\mathcal{C}_\mathcal{S}') = m(\mathcal{C}_\mathcal{S}'') = t$. Arguing similarly as in Theorem 4 completes the proof. ∎

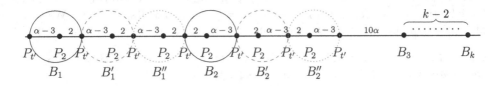

Fig. 2. $\mathcal{X} \subseteq \mathbb{R}$ such that no algorithm can capture all the α-proximal clusterings.

Proofs of Theorems 7, 15 **and** 13 have the exact same ideas as the proof of Theorem 5. To prove the lower bound in the list model, instance constructed in Theorem 5 is a simple extension of the instance in Theorem 4. The instances for the proof of Theorems 7, 15 and 13 are similarly constructed as extensions of their respective tree lower bound instances (Theorems 6, 14 and 12).

Proof of Theorem 8 We will show that $\mathcal{C}_\mathcal{X}^*$ has strong stability ([4]) which will complete the proof (Theorem 8 in [4]). Let $A \subset C_i^*$ and $B \subseteq C_j^*$. Let $p \in A$ and $q \in C_i^* \setminus A$ be points which achieve the minimum distance between A and $C_i^* \setminus A$. If $c_i \in A$ then $d(p, q) \leq d(c_i, q) \leq r$. If $c_i \in C_i^* \setminus A$ then $d(p, q) \leq d(p, c_i) \leq r$. Hence, $d_{min}(A, C_i^* \setminus A) \leq r$. Similarly, we get that $d_{min}(A, B) > r$. ∎

Proof of Theorems 14 **and** 12 are also identical to the proofs of Theorem 6 and 4.

Proof of Theorem 10 Fix $\mathcal{S} \subseteq \mathcal{X}$. Denote by $r_i := r(S_i^*)$. Let $\mathcal{C}_\mathcal{X} = \{C_1, \ldots, C_k\}$ be the clustering outputed by the algorithm. Let $\mathcal{L} = \{B_1, \ldots, B_l\}$ be the list of balls as outputed by Phase 1 of Algorithm 3. Let G be the graph as constructed in Phase 2 of the algorithm. Observe that $B = B(s_i, r_i) = S_i^* \in \mathcal{L}$. WLOG, denote this ball by $B^{(i)}$ and the corresponding vertex in the graph G by $v^{(i)}$. We will prove the theorem by proving two key facts.

F.1 If B_{i1} and B_{i2} intersect S_i^* then the vertices v_{i1} and v_{i2} are connected.

F.2 If B_{i1} intersects S_i^* and B_{j1} intersects S_j^* then v_{i1} and v_{j1} are disconnected in G.

Claim 4. *Let $\mathcal{L}, G, B^{(i)}$ and $v^{(i)}$ be as defined above. Let balls $B_{i1}, B_{i2} \in \mathcal{L}$ be such that $B_{i1} \cap S_i^* \neq \emptyset$ and $B_{i2} \cap S_i^* \neq \emptyset$. Then there exists a path between v_{i1} and v_{i2}.*

Assume that v_{i1} and $v^{(i)}$ are not connected by an edge. Hence, $|B_{i1} \setminus B^{(i)}| \geq t/2$. Since $\lambda > 4$, for all $j \neq i$, $B_{i1} \cap S_j^* = \emptyset$. Thus, $B_{i1} \setminus B^{(i)} \subseteq \mathcal{X} \setminus \mathcal{S}$. which contradicts $|B_{i1} \cap \{\mathcal{X} \setminus \mathcal{S}\}| < t/2$. ∎

Claim 5. *Let the framework be as in Claim 4. Let $B_{i1} \in \mathcal{L}$ be such that $B_{i1} \cap S_i^* \neq \emptyset$ and B_{j1} be such that $B_{j1} \cap S_j^* \neq \emptyset$. Then v_{i1} and v_{j1} are disconnected in G.*

Assume that v_{i1} and v_{j1} are connected. Hence, there exists vertices v_i and v_n such that v_i and v_n are connected by an edge in G and $B_i \cap S_i^* \neq \emptyset$ and $B_n \cap S_n^* \neq \emptyset$ for some $n \neq i$. $|B_i \cap B_n| \geq t/2$. Now, $\lambda \geq 4$, thus $B_i \cap \{\mathcal{S} \setminus S_i^*\} = \emptyset$ and $B_n \cap \{\mathcal{S} \setminus S_n^*\} = \emptyset$. Thus, $B_i \cap B_n \subseteq \mathcal{X} \setminus \mathcal{S}$ which contradicts the sparseness assumption. ∎

Theorem 16 (Vapnik and Chervonenkis [14]). *Let X be a domain set and D a probability distribution over X. Let H be a class of subsets of X of finite VC-dimension d. Let $\epsilon, \delta \in (0,1)$. Let $S \subseteq X$ be picked i.i.d according to D of size m. If $m > \frac{c}{\epsilon^2}(d \log \frac{d}{\epsilon} + \log \frac{1}{\delta})$, then with probability $1 - \delta$ over the choice of S, we have that $\forall h \in H$*

$$\left| \frac{|h \cap S|}{|S|} - P(h) \right| < \epsilon$$

References

1. Ackerman, M., Ben-David, S.: Clusterability: a theoretical study. In: International Conference on Artificial Intelligence and Statistics, pp. 1–8 (2009)
2. Awasthi, P., Blum, A., Sheffet, O.: Center-based clustering under perturbation stability. Inf. Process. Lett. **112**(1), 49–54 (2012)
3. Balcan, M.-F., Blum, A., Fine, S., Mansour, Y.: Distributed learning, communication complexity and privacy. arXiv preprint arXiv:1204.3514 (2012)
4. Balcan, M.-F., Blum, A., Vempala, S.: A discriminative framework for clustering via similarity functions. In: Proceedings of the Fortieth Annual ACM Symposium on Theory of Computing, pp. 671–680. ACM (2008)
5. Balcan, M.F., Liang, Y.: Clustering under perturbation resilience. In: Mehlhorn, K., Pitts, A., Wattenhofer, R., Czumaj, A. (eds.) ICALP 2012, Part I. LNCS, vol. 7391, pp. 63–74. Springer, Heidelberg (2012)
6. Ben-David, S.: Computational feasibility of clustering under clusterability assumptions. arXiv preprint arXiv:1501.00437 (2015)
7. Ben-David, S., Haghtalab, N.: Clustering in the presence of background noise. In: Proceedings of the 31st International Conference on Machine Learning (ICML 2014), pp. 280–288 (2014)
8. Ben-David, S., Reyzin, L.: Data stability in clustering: a closer look. Theor. Comput. Sci. **558**, 51–61 (2014)
9. Bilu, Y., Linial, N.: Are stable instances easy? Comb. Probab. Comput. **21**(05), 643–660 (2012)

10. Cuesta-Albertos, J.A., Gordaliza, A., Matrán, C., et al.: Trimmed k-means: an attempt to robustify quantizers. Ann. Stat. **25**(2), 553–576 (1997)
11. Dave, R.N.: Robust fuzzy clustering algorithms. In: Second IEEE International Conference on Fuzzy Systems, pp. 1281–1286. IEEE (1993)
12. García-Escudero, L.A., Gordaliza, A., Matrán, C., Mayo-Iscar, A.: A general trimming approach to robust cluster analysis. Ann. Stat., 1324–1345 (2008)
13. Reyzin, L.: Data stability in clustering: a closer look. In: Stoltz, G., Vayatis, N., Zeugmann, T., Bshouty, N.H. (eds.) ALT 2012. LNCS, vol. 7568, pp. 184–198. Springer, Heidelberg (2012)
14. Vapnik, V.N., Ya, A.: Chervonenkis: on the uniform convergence of relative frequencies of events to their probabilities. In: Vovk, V., Papadopoulos, H., Gammerman, A. (eds.) Measures of Complexity, pp. 11–30. Springer, Heidelberg (2015)

A Spectral Algorithm with Additive Clustering for the Recovery of Overlapping Communities in Networks

Emilie Kaufmann[1]([✉]), Thomas Bonald[2], and Marc Lelarge[3]

[1] CNRS & CRIStAL, Université Lille, Villeneuve-d'ascq, France
emilie.kaufmann@univ-lille1.fr
[2] Télécom ParisTech, Université Paris-Saclay, Paris, France
[3] Inria-ENS, Paris, France

Abstract. This paper presents a novel *spectral algorithm with additive clustering*, designed to identify overlapping communities in networks. The algorithm is based on geometric properties of the spectrum of the expected adjacency matrix in a random graph model that we call *stochastic blockmodel with overlap* (SBMO). An adaptive version of the algorithm, that does not require the knowledge of the number of hidden communities, is proved to be consistent under the SBMO when the degrees in the graph are (slightly more than) logarithmic. The algorithm is shown to perform well on simulated data and on real-world graphs with known overlapping communities.

1 Introduction

Many datasets (e.g., social networks, gene regulation networks) take the form of graphs whose structure depends on some underlying *communities*. The commonly accepted definition of a community is that nodes tend to be more densely connected within a community than with the rest of the graph. Communities are often hidden in practice and recovering the community structure directly from the graph is a key step in the analysis of these datasets. Spectral algorithms are popular methods for detecting communities [14], that consist in two phases. First, a *spectral embedding* is built, where the n nodes of the graph are projected onto some low dimensional space generated by well-chosen eigenvectors of some matrix related to the graph (e.g., the adjacency matrix or a Laplacian matrix). Then, a *clustering algorithm* (e.g., k-means or k-median) is applied to the n embedded vectors to obtain a partition of the nodes into communities.

It turns out that the structure of many real datasets is better explained by *overlapping* communities. This is particularly true in social networks, in which the neighborhood of any given node is made of several social circles, that naturally overlap [9]. Similarly, in co-authorship networks, authors often belong to several scientific communities and in protein-protein interaction networks, a given protein may belong to several protein complexes [11]. The communities do not form a partition of the graph and new algorithms need to be designed.

© Springer International Publishing Switzerland 2016
R. Ortner et al. (Eds.): ALT 2016, LNAI 9925, pp. 355–370, 2016.
DOI: 10.1007/978-3-319-46379-7_24

This paper presents a novel spectral algorithm, called Spectral Algorithm with Additive Clustering (SAAC). The algorithm consists in a spectral embedding based on the adjacency matrix of the graph, coupled with an additive clustering phase designed to find overlapping communities. The proposed algorithm does not require the knowledge of the number of communities present in the network, and can thus be qualified as adaptive.

SAAC belongs to the family of model-based community detection methods, that are motivated by a random graph model depending on some underlying set of communities. In the non-overlapping case, spectral methods have been shown to perform well under the stochastic block model (SBM), introduced by Holland and Leinhardt [6]. Our algorithm is inspired by the simplest possible extension of the SBM to overlapping communities, we refer to as the stochastic blockmodel with overlaps (SBMO). In the SBMO, each node is associated to a binary membership vector, indicating all the communities to which the node belongs. We show that exploiting an additive structure in the SBMO leads to an efficient method for the identification of overlapping communities. To support this claim, we provide consistency guarantees when the graph is drawn under the SBMO, and we show that SAAC exhibit state-of-the-art performance on real datasets for which ground-truth communities are known.

The paper is structured as follows. In Sect. 2, we cast the problem of detecting overlapping communities into that of estimating a membership matrix in the SBMO model, introduced therein. In Sect. 3, we compare the SBMO with alternative random graph models proposed in the literature, and review the algorithms inspired by these models. In Sect. 4, we exhibit some properties of the spectrum of the adjacency matrix under SBMO, that motivate the new SAAC algorithm, introduced in Sect. 5, where we also formulate theoretical guarantees for an adaptive version of the algorithm. Finally Sect. 6 illustrates the performance of SAAC on both real and simulated data.

Notation. We denote by $||x||$ the Euclidean norm of a vector $x \in \mathbb{R}^d$. For any matrix $M \in \mathbb{R}^{n \times d}$, we let M_i denote its i-th row and $M_{.,j}$ its j-th column. For any $\mathcal{S} \subset \{1, \ldots, d\}$, $|\mathcal{S}|$ denotes its cardinality and $\mathbb{1}_{\mathcal{S}} \in \{0, 1\}^{1 \times d}$ is a row vector such that $(\mathbb{1}_{\mathcal{S}})_{1,i} = \mathbb{1}_{\{i \in \mathcal{S}\}}$. The Frobenius norm of a matrix $M \in \mathbb{R}^{n \times d}$ is

$$||M||_F^2 = \sum_{i=1}^n ||M_i||^2 = \sum_{j=1}^d ||M_{.,j}||^2 = \sum_{1 \le i,j \le n} M_{i,j}^2.$$

The spectral norm of a symmetric matrix $M \in \mathbb{R}^{d \times d}$ with eigenvalues $\lambda_1, \ldots, \lambda_d$ is $||M|| = \max_{i=1..d} |\lambda_i|$. We let $\mathfrak{S}_K$ be the group of permutations of $\{1, \ldots, K\}$ and for $\sigma \in \mathfrak{S}_K$ $P_\sigma \in \mathbb{R}^{K \times K}$ be the permutation matrix associated to σ, defined by $(P_\sigma)_{k,l} = \delta_{\sigma(k),l}$.

2 The Stochastic Blockmodel with Overlaps (SBMO)

2.1 The Model

For any symmetric matrix $A \in [0,1]^{n \times n}$, let $\hat{A}$ be some random symmetric binary matrix whose entries $(\hat{A}_{i,j})_{i \leq j}$ are independent Bernoulli random variables with respective parameters $(A_{i,j})_{i \leq j}$. Then $\hat{A}$ is the adjacency matrix of an undirected random graph with expected adjacency matrix A. In all the paper, we restrict the hat notation to variables that depend on this random graph. For example, the empirical degree of node i and empirical maximal degree observed in the random graph are respectively denoted by $\hat{d}_i = \sum_{j=1}^{n} \hat{A}_{i,j}$ and $\hat{d}_{\max} = \max_i \sum_{j=1}^{n} \hat{A}_{i,j}$, whereas the expected degree of node i and expected maximal degree are respectively denoted by $d_i = \sum_{j=1}^{n} A_{i,j}$ and $d_{\max} = \max_i \sum_{j=1}^{n} A_{i,j}$.

The stochastic block model (SBM) with n nodes and K communities depends on some mapping $k : \{1, \ldots, n\} \to \{1, \ldots, K\}$ that associates nodes to communities and on some symmetric community connectivity matrix $B \in [0,1]^{K \times K}$. In this model, two nodes i and j are connected with probability

$$A_{i,j} = B_{k(i),k(j)} = B_{k(j),k(i)}.$$

Introducing a membership matrix $Z \in \{0,1\}^{n \times K}$ such that $Z_{i,k} = \mathbb{1}_{\{k(i)=k\}}$, the expected adjacency matrix can be written

$$A = ZBZ^T.$$

The stochastic blockmodel with overlap (SBMO) is a slight extension of this model, in which Z is only assumed to be in $\{0,1\}^{n \times K}$ and $Z_i \neq 0$ for all i. Compared to the SBM, the rows of the membership matrix Z are no longer constrained to have only one non-zero entry. Since these n rows give the communities of the respective n nodes of the graph, this means that each node can now belong to several communities.

2.2 Community Detection with Overlap in an Identifiable SBMO

Given some adjacency matrix $\hat{A}$ drawn under the SBMO, our goal is to recover the underlying communities, that is to build an estimate $\hat{Z}$ of the membership matrix Z, up to some permutation of its columns (corresponding to a permutation of the community labels). With $\hat{K}$ the estimate for the number of communities (K is in general unknown), one has $\hat{Z} \in \{0,1\}^{n \times \hat{K}}$. We introduce two performance metrics for this problem. The first is related to the number of nodes that are "well classified", in the sense that there is no error in the estimate of their membership vector. The objective is to minimize the number of misclassified nodes of an estimate $\hat{Z}$ of Z, defined as $\text{MisC}(\hat{Z}, Z) = n$ if $\hat{K} \neq K$ and

$$\text{MisC}(\hat{Z}, Z) = \min_{\sigma \in \mathfrak{S}_K} |\{i : \exists k \leq K \text{ s.t. } \hat{Z}_{i,\sigma(k)} \neq Z_{i,k}\}|$$

otherwise. The second is the fraction of wrong predictions in the membership matrix (again, up to a permutation of the community labels). The estimation error of $\hat{Z}$ is defined as $\mathrm{Error}(\hat{Z}, Z) = 1$ if $\hat{K} \neq K$ and

$$\mathrm{Error}(\hat{Z}, Z) = \frac{1}{nK} \inf_{\sigma \in \mathfrak{S}_K} ||\hat{Z}P_\sigma - Z||_F^2 \leq \frac{\mathrm{MisC}(\hat{Z}, Z)}{n}$$

otherwise.

The communities of a SBMO can only be recovered if the model is *identifiable* in that the equality $Z'B'Z'^T = ZBZ^T$, for some integer K' and matrices $Z' \in \{0,1\}^{n \times K'}$, $B' \in [0,1]^{K' \times K'}$, implies $\mathrm{MisC}(Z', Z) = 0$ (and thus $K' = K$), which means that two SBMO with the same expected adjacency matrices have the same communities, up to a permutation of the community labels. Sufficient conditions for the identifiability of a SBMO with parameter K, Z, B are given in Theorem 1: B should be invertible and each community should contain at least one pure node, that is a node belonging to this community only.

Theorem 1. *We define the following assumptions:*

(SBMO1) B *is invertible;*
(SBMO2) $\forall k \in \{1, \ldots, K\}$ *there exists i such that $Z_{i,k} = \sum_{\ell=1}^K Z_{i,\ell} = 1$.*

Under assumptions (SBMO1) and (SBMO2), the parameters K, B, Z of the SBMO are identifiable (up to a permutation of the community labels).

Our choice for SBMO1-2 is motivated by applications to social networks: homophily will make the matrix B diagonally dominant, hence invertible. A proof of this theorem, as well as a more complete discussion on identifiability, can be found in the extended version of this paper. In the rest of the paper, we assume that the conditions (SBMO1) and (SBMO2) are satisfied.

2.3   Subcommunity Detection

Any SBMO with K overlapping communities may be viewed as a SBM with up to $2^K - 1$ non-overlapping communities, corresponding to groups of nodes sharing exactly the same communities in the SBMO and that we refer to as *subcommunities*. Let K' be the number of subcommunities in the SBMO:

$$K' = |\mathcal{T}|, \quad \text{where} \quad \mathcal{T} = \{z \in \{0,1\}^{1 \times K} : \exists i : Z_i = z\}.$$

The corresponding SBM has K' communities indexed by $z \in \mathcal{T}$, with community connectivity matrix B' given by $B'_{y,z} = yBz^T$. It can be checked that if the initial SBMO satisfies (SBMO1-2) then the corresponding SBM satisfies identifiability conditions for SBM (B' is invertible).

This suggests that community detection in the SBMO reduces to community detection of the corresponding SBM, for which many efficient algorithms are known. However, the knowledge of the subcommunities is not sufficient to recover the initial overlapping communities. It is indeed necessary to map these subcommunities to elements of $\{0,1\}^K \setminus \{0\}$, which is not an easy task: first, the number of communities K is unknown; second, assuming K is known, there are up to $(2^K - 1)!$ such mappings so that a simple approach by enumeration is not feasible in general. Moreover, the performance of clustering algorithms degrades rapidly with the number of communities so that it is preferable to work directly on the K overlapping communities rather than on the K' subcommunities, with K' possibly as large as $2^K - 1$.

As we will see, the SAAC algorithm directly detects the K overlapping communities using the specific geometry of the eigenvectors of the expected adjacency matrix, A. We provide conditions under which these geometric properties hold for the observed adjacency matrix, $\hat{A}$, which guarantees the consistency of the algorithm: the K communities are recovered with probability tending to 1 in the limit of a large number of nodes n.

2.4   Scaling

To study the performance of our algorithm when the number of nodes n grows, we introduce a degree parameter α_n so that the expected adjacency matrix of a graph with n nodes is in fact given by

$$A = \frac{\alpha_n}{n} Z B Z^T,$$

with $B \in [0,1]^{K \times K}$ independent of n and $Z \in \{0,1\}^{n \times K}$. Although Z depends on n, we do not make it explicit in the notation. Observe that the expected degree of each node grows like α_n, since $d_i = \alpha_n \left(\frac{1}{n} Z_i B Z^T \mathbf{1}\right)$, where $\mathbf{1}$ is the vector of ones of dimension n.

We assume that the set of subcommunities $\mathcal{T}$ does not depend on n and that for all $z \in \mathcal{T}$, there exists a constant β_z (independent of n) towards which the proportion of nodes with membership vector z converges:

$$|\{i : Z_i = z\}|/n \to \beta_z > 0. \tag{1}$$

This implies the existence of positive constants L_z and of a matrix $O \in \mathbb{R}^{K \times K}$, such that

$$\forall z \in \mathcal{T}, \quad \frac{1}{n} z B Z^T \mathbf{1} \to L_z, \quad \text{and} \quad \frac{1}{n} Z^T Z \to O. \tag{2}$$

One has $d_i \sim \alpha_n L_z$ for any i such that $Z_i = z$. In the sequel, we assume that the graph is sparse in the sense that $\alpha_n \to \infty$ with $\alpha_n/n \to 0$. Observe also that $O_{k,k}$ is the (limit) proportion of nodes that belong to community k while $O_{k,l}$ is the (limit) proportion of nodes that belong to communities k and l, for $k \neq l$. Hence we refer to O as the *overlap matrix*.

In the following, we will slightly abuse notation by writing $O = \frac{1}{n} Z^T Z$ and $d_i = \alpha_n L_z$ if $Z_i = z$, although these equalities in fact hold only in the limit.

3   Related Work

Models. Several random graph models have been proposed in the literature to model networks with overlapping communities. In these models, each node i is characterized by some community membership vector Z_i that is not always a binary vector, as in the SBMO. In the Mixed-Membership Stochastic Blockmodel (MMSB) [1], introduced as the first model with overlaps, membership vectors are probability vectors drawn from a Dirichlet distribution. In this model, conditionally to Z_i and Z_j, the probability that nodes i and j are connected is $Z_i B Z_j^T$ for some community connectivity matrix B, just like in SBMO. However, the fact that Z_i and Z_j are probability vectors makes the model less interpretable. In particular, the probability that two nodes are connected does not necessarily increase with the number of communities that they have in common, as pointed out by Yang and Leskovec [16], which contradicts a tendency empirically observed in social networks.

A first model that relies on binary membership vectors is the Overlapping Stochastic Block Model (OSBM) [8], in which two nodes i, j are connected with probability $\sigma(Z_i W Z_j^T + Z_i V + Z_j U + w)$, where $W \in \mathbb{R}^{K \times K}$, $U, V \in \mathbb{R}_+^K$, $w \in \mathbb{R}$, and σ is the sigmoid function. Now the probability of connectivity of two nodes increases with the number of communities shared, but the particular form of the probability of connection makes the model hard to analyze. Given a community connectivity matrix B, another natural way to build a random graph model based on binary membership vectors is to assume that two nodes i and j are connected if any pair of communities k, l to which these nodes respectively belong can explain the connection. In other words, i and j are connected with probability $1 - \prod_{k,l=1}^{K}(1 - B_{k,l})^{Z_{i,k} Z_{j,l}}$. Denoting by Q the matrix with entries $Q_{k,l} = -\log(1 - B_{k,l})$, this probability can be written $1 - \exp\left(-Z_i Q Z_j^T\right) \simeq Z_i Q Z_j^T$, where the approximation is valid for sparse networks. In this case, the model is very close to the SBMO, with connectivity matrix Q. The Community-Affiliation Graph Model (AGM) [16] is a particular case of this model in which B is diagonal. The SBMO with a diagonal connectivity matrix can be viewed as a particular instance of an Additive Clustering model [13] and is also related to the 'colored edges' model [4], in which $\hat{A}_{i,j}$ is drawn from a Poisson distribution with mean $\theta_i \theta_j^T$, where $\theta_i \in \mathbb{R}^{1,K}$ is the (non-binary) membership vector of node i. Letting $\theta_i = \sqrt{B_{i,i}} Z_i$ and approximating the Poisson distribution by a Bernoulli distribution, we recover the SBMO.

The Overlapping Continuous Community Assignment Model (OCCAM), proposed by Zhang et al. [19] relies on overlapping communities but also on individual degree parameters, which generalizes the degree-corrected stochastic blockmodel [7]. In the OCCAM, a degree parameter θ_i is associated to each node i. Letting $\Theta = \text{Diag}(\theta_i) \in \mathbb{R}^{n \times n}$, the expected adjacency matrix is $A = \Theta Z B Z^T \Theta$, with a membership matrix $Z \in \mathbb{R}^{n \times K}$. Identifiability of the model is proved assuming that B is positive definite, each row Z_i satisfies $||Z_i|| = 1$, and the degree parameters satisfy $n^{-1} \sum_{i=1}^{n} \theta_i = 1$. The SBMO can be viewed as a particular instance of the OCCAM, for which we provide new identifiability conditions, that allow for binary membership vectors.

Algorithms. Several algorithmic methods have been proposed to identify overlapping community structure in networks [15]. Among the model-based methods, that rely on the assumption that the observed network is drawn under a random graph model, some are approximations of the maximum likelihood or maximum a posteriori estimate of the membership vectors under one of the random graph models discussed above. For example, under the MMSB or the OSBM the membership vectors are assumed to be drawn from a probability (prior) distribution, and variational EM algorithms are proposed to approximate the posterior distributions [1,8]. However, there is no proof of consistency of the proposed algorithms. In the MMSB, a different approach that uses tensor power iteration is proposed in [2] to compute an estimator derived using the moments method, for which the first consistency results are provided.

The first occurrence of a spectral algorithm to find overlapping communities goes back to [18]. The proposed method is an adaptation of spectral clustering with the normalized Laplacian (see e.g., [10]) with a fuzzy clustering algorithm in place of k-means, and its justification is rather heuristic. Another spectral algorithm has been proposed by [19], as an estimation procedure for the (non-binary) membership matrix under the OCCAM. The spectral embedding is a row-normalized version of $\hat{U}\hat{\Lambda}^{1/2} \in \mathbb{R}^{n \times K}$, with $\hat{\Lambda}$ the diagonal matrix containing K leading eigenvalues of $\hat{A}$ and $\hat{U}$ the matrix of associated eigenvectors. The centroids obtained by a k-median clustering algorithm are then used to estimate Z. This algorithm is proved to be consistent under the OCCAM, when moreover degree parameters and membership vectors are drawn according to some distributions. Similar assumptions have appeared before in the proof of consistency of some community detection algorithms in the SBM or DC-SBM [20]. Our consistency results are established for fixed parameters of the model, and hold for relatively sparse graph ($\alpha_n \simeq \log n$), unlike those obtained under the OCCAM.

4   Spectral Analysis of the Adjacency Matrix in the SBMO

Let $\mathcal{Z}$ be the set of membership matrices that contains at least one pure node per community:

$$\mathcal{Z} = \{Z \in \{0,1\}^{n \times K}, \forall k \leq K, \exists i \leq n : Z_i = \mathbb{1}_{\{k\}}\}.$$

From the conditions (SBMO1) and (SBMO2), the expected adjacency matrix $A = ZBZ^T$ is of rank K and Z belongs to $\mathcal{Z}$. Let $U \in \mathbb{R}^{n \times K}$ be a matrix whose columns $u_1, \ldots, u_K \in \mathbb{R}^n$ are normalized orthogonal eigenvectors associated to the K non-zero eigenvalues of A. The structure of U is described in the following proposition. Its first statement follows from the fact that the eigenvectors $u_1, \ldots, u_K$ form a basis of $\mathrm{Im}(A) \subseteq \mathrm{Im}(Z)$. The proof of the second statement, provided below, is also the key ingredient of the proof of Theorem 1.

Proposition 1. *1.* $\exists X \in \mathbb{R}^{K \times K}$ *such that* $U = ZX$.
2. If $U = Z'X'$ *for some* $Z' \in \mathcal{Z}$, $X' \in \mathbb{R}^{K \times K}$, *then there exists* $\sigma \in \mathfrak{S}_K$ *such that* $Z = Z'P_\sigma$.

Proof of Statement 2. Let $Z, Z' \in \mathcal{Z}$ and assume that there exist invertible matrices X, X' such that $U = ZX = Z'X'$. As for all $k = 1, \ldots, K$ there exists some i such that $Z_i = \mathbb{1}_{\{k\}}$, the k-th row of X is a sum of rows in X': $X_k = Z'_i X' = \sum_{l \in \mathcal{S}_k} X'_l$, for some $\mathcal{S}_k \subset \{1, \ldots, K\}$. Similarly, each row of X' is a sum of rows in X, thus for any $k \neq l$, there exist K integers $a_1, \ldots, a_K$:

$$X_k + X_l = \sum_{m=1}^{K} a_m X_m.$$

If $\mathcal{S}_k \cap \mathcal{S}_l \neq \emptyset$, there exists some m such that $a_m \geq 2$. But this is in contradiction with the fact that X is invertible. Hence, $\mathcal{S}_k \cap \mathcal{S}_l = \emptyset$ for all $k \neq l$. The only way for the $\mathcal{S}_k$ to be pairwise disjoint is that there exists a permutation σ such that $X' = P_\sigma X$. Since $ZX = Z'X'$ and X is invertible, this implies $Z = Z'P_\sigma$. $\square$

This decomposition reveals in particular an *additive structure* in U: each row U_i is the sum of rows of pure nodes associated to the communities to which node i belongs. Fixing for each k a pure node i_k in community k, one has indeed

$$\forall i, \ U_i = \sum_{k=1}^{K} U_{i_k} \mathbb{1}_{(Z_{i,k}=1)} \tag{3}$$

Proposition 1 can also be used to establish the following result that relates the eigenvectors of A to those of a $K \times K$ matrix featuring the overlap matrix O introduced in Sect. 2.4. Note that for any $x \in \mathbb{R}^K$, we have $x^T O x = |Zx|^2/n$ so that O has the same rank as Z, equal to K. Hence O is invertible and positive definite, thus the matrix $O^{1/2}$ (resp. its inverse) is well defined.

Proposition 2. *Let* $\mu \neq 0$ *and* $M_0 = O^{1/2}BO^{1/2}$. $u = Zx$ *is an eigenvector of* A *associated to* $\alpha_n \mu$ *if and only if* $O^{1/2}x$ *is an eigenvector of* M_0 *associated to* μ. *In particular, the non-zero eigenvalues of* A *are of the same order as* α_n.

In practice, we observe the adjacency matrix $\hat{A}$, which is a noisy version of A. Our hope is that the K leading eigenvectors of $\hat{A}$ are not too far from the K leading eigenvectors of A, so that, in view of Proposition 1, the solution in Z' of the following optimization problem provides a good estimate of Z:

$$\min_{Z' \in \mathcal{Z}, X' \in \mathbb{R}^{K \times K}} ||\hat{U} - Z'X'||_F,$$

where $\hat{U}$ is the matrix of the K normalized eigenvectors of $\hat{A}$ associated to the K largest eigenvalues. If K is unknown, one should also find an estimate $\hat{K}$, and let $\hat{U} \in \mathbb{R}^{n \times \hat{K}}$ be the matrix of $\hat{K}$ leading eigenvectors.

This hope is supported by the following result on the perturbation of the largest eigenvectors of the adjacency matrix of any random graph, that also provides an adaptive procedure to select the eigenvectors to use in the spectral embedding. $\lambda_{\min}(A)$ is the smallest absolute value of a non-zero eigenvalue of A.

Lemma 1. *Let $\delta \in]0,1[$ and $\eta \in]0,1/2[$. Let $\hat{U}$ be a matrix formed by orthogonal eigenvectors of $\hat{A}$ with an associated eigenvalue λ that satisfy $|\lambda| \geq (2(1+\eta)\,\hat{d}_{\max}\log(4n/\delta))^{1/2}$, and $\hat{K}$ be the number of such eigenvectors. Let U be matrix of $\hat{K}$ largest eigenvectors of A. There exists C_η, $D_\eta > 0$ such that if*

$$d_{\max} \geq C_\eta \log(4n/\delta) \quad and \quad \lambda_{\min}(A)^2/d_{\max} > D_\eta \log(4n/\delta),$$

then with probability larger than $1-\delta$, $\hat{K} = \mathrm{rank}(A)$ and there exists $\hat{P} \in \mathcal{O}_n(\mathbb{R})$:

$$\left\|\hat{U} - U\hat{P}\right\|_F^2 \leq 32\left(1 + \frac{\eta}{\eta+2}\right)\left(\frac{d_{\max}}{\lambda_{\min}(A)^2}\right)\log\left(\frac{4n}{\delta}\right).$$

The proof of Lemma 1 relies on the use of a matrix concentration inequality to bound the spectral norm of $\hat{A} - A$, and on results from linear algebra that relate the eigenvalues and eigenvectors of two matrices that are close in spectral norm (mostly the Davis-Kahan theorem, see, e.g. [12]). A detailed proof can be found in the extended version of this paper.

Under the SBMO, we have $\lambda_{\min}(A) = \Theta(\alpha_n)$ in view of Proposition 2; since $d_{\max} = \Theta(\alpha_n)$, we need $\alpha_n/\log(n) \to +\infty$ to use Lemma 1 to prove that $\hat{U}$ is a good estimate of U. We give in the next section sufficient conditions on the degree parameter α_n to obtain asymptotically exact recovery of the communities.

5   The SAAC Algorithm

The spectral structure of the adjacency matrix suggests that $\hat{Z}$ defined below is a good estimate of the membership matrix Z in the SBMO:

$$(\mathcal{P}): \quad (\hat{Z}, \hat{X}) \in \operatorname*{argmin}_{Z' \in \mathcal{Z}, X' \in \mathbb{R}^{K \times K}} \|\hat{U} - Z'X'\|_F^2, \tag{4}$$

where $\hat{U} \in \mathbb{R}^{n \times K}$ is the matrix of the K normalized leading eigenvectors of $\hat{A}$. In practice, solving $(\mathcal{P})$ is very hard, and the SAAC algorithm, introduced in Sect. 5.1, solves a relaxation of $(\mathcal{P})$ in which Z' is only constrained to have binary entries. In Sect. 5.2, we prove that the estimate $\hat{Z}$ in (4) is consistent.

5.1   Description of the Algorithm

The *spectral algorithm with additive clustering* (SAAC) consists in first computing a matrix $\hat{U} \in \mathbb{R}^{n \times K}$ whose columns are normalized eigenvectors of $\hat{A}$ associated to the K largest eigenvalues (in absolute value), and then computing the solution of the following optimization problem:

$$(\mathcal{P})': \quad (\hat{Z}, \hat{X}) \in \operatorname*{argmin}_{\substack{Z' \in \{0,1\}^{n \times K}: \forall i, Z'_i \neq 0 \\ X' \in \mathbb{R}^{K \times K}}} \|Z'X' - \hat{U}\|_F^2.$$

$(\mathcal{P})'$ is reminiscent of the (NP-hard) k-means problem, in which the same function is minimized under the additional constraint that $\|Z_i\| = 1$ for all i. The name of the algorithm highlights the fact that, rather than finding a clustering of the rows of $\hat{U}$, the goal is to find $\hat{Z}$, containing pure nodes $\hat{i}_1, \ldots, \hat{i}_k$, that reveals the underlying additive structure of $\hat{U}$: for all i, $\hat{U}_i$ is not too far from $\sum_k \hat{U}_{\hat{i}_k} \mathbb{1}_{(\hat{Z}_{i,k}=1)}$, in view of (3).

In practice, just like k-means, we propose to solve $(\mathcal{P})'$ by an alternate minimization over Z' and X'. Let m be some upper bound on the maximum overlap $O_{\max} = \max\{||z||^2, z \in \mathcal{T}\}$, provided to limit the combinatorial complexity of the algorithm. For a fixed X', the optimization in Z' consists in letting the i-th row be $Z'_i = \operatorname{argmin}_{z \in \{0,1\}^{1 \times K}} ||zX' - \hat{U}_i||$, with the extra condition that $1 \leq ||z||^2 \leq m$. Given Z', as long as the matrix $Z'^T Z'$ is invertible, there is a closed form for the optimization in X', given by $X' = (Z'^T Z')^{-1} Z'^T \hat{U}$. If $Z'^T Z'$ is not invertible, Z' does not contain one pure node for each community, which should be the case for a reasonable solution. If this happens, we therefore re-initialize X'.

Alternate minimization is guaranteed to converge, in a finite number of steps, towards a local minimum of the objective. However, the convergence is very sensitive to initialization. We use a k-means++ initialization (see [3]), which is a randomized procedure that picks as initial centroids rows from $\hat{U}$ that should be far from each other. For the first centroid, we choose at random a row in $\hat{U}$ corresponding to a node whose degree is smaller than the median degree in the network. We do so because in the SBMO model, pure nodes tend to have smaller degrees and we expect the algorithm to work well if the initial centroids are chosen not too far from rows in $\hat{U}$ corresponding to pure nodes.

If K is unknown, Theorem 2 suggests an adaptive version of the algorithm: the columns of $\hat{U}$ are normalized eigenvectors of $\hat{A}$ associated to eigenvalues λ such that $|\lambda| \geq (2(1+\eta)\hat{d}_{\max} \log(4n^{1+r}))^{1/2}$, for some positive constants η and r. While heuristics do exist for selecting the number of clusters in spectral clustering (e.g. [14,17]), this thresholding procedure is supported by theory for networks drawn under SBMO. It is reminiscent of the USVT algorithm of [5], that can be used to estimate the expected adjacency matrix in a SBM.

5.2 Consistency of an Adaptive Estimator

We give in Theorem 2 theoretical properties for a slight variant of the estimate $\hat{Z}$ in (4), that is solution of the optimization problem $(\mathcal{P}_\epsilon)$ defined therein, featuring the set of membership matrices for which the proportion of pure nodes in each community is larger than ϵ:

$$\mathcal{Z}_\epsilon(K) = \{Z' \in \mathcal{Z} : \forall k, |\{i : Z'_i = \mathbb{1}_{\{k\}}\}| > \epsilon n\}.$$

Recall the notation introduced in (1) and (2). We assume that ϵ is smaller than $\min_k \beta_{\mathbb{1}_{\{k\}}}$, the smallest proportion of pure nodes, and let $L_{\max} = \max_z L_z$, so that $d_{\max} = \alpha L_{\max}$.

The estimator analyzed is adaptive, for it relies on an estimate $\hat{K}$ of the number of communities, and on $\hat{Z}_\epsilon = \mathcal{Z}_\epsilon(\hat{K})$. We establish its consistency for any fixed matrices B and Z satisfying (SBMO1) and (SBMO2). It is to be noted that while the consistency result for the OCCAM algorithm [19] applies to moderately dense graphs (α_n has to be of order n^α for some $\alpha > 0$), our result handle relatively sparse graphs, in which α_n is of order $(\log(n))^{1+c}$ for some $c > 0$. Theorem 2, whose proof is postponed to Sect. 5.3, features constants defined below, that are related to the overlap matrix O. Note that d_0 in Definition 1 is indeed positive because the matrix $O^{-1/2}$ is invertible.

Definition 1. *Introducing the symmetric $K \times K$ matrix $M_0 := O^{1/2}BO^{1/2}$,*

$$\mu_0 = \min\{|\lambda| : \lambda \in sp(M_0) \setminus \{0\}\}, \quad d_0 := \min_{z \in \{-1,0,1,2\}^{1 \times K}, z \neq 0} \left\| zO^{-1/2} \right\| > 0.$$

Theorem 2. *Let $\eta \in]0, 1/2[$ and $r > 0$. Let $\hat{U}$ be a matrix formed by the orthogonal eigenvectors of $\hat{A}$ with an associated eigenvalue λ satisfying $|\lambda| \geq (2(1 + \eta)\hat{d}_{\max} \log(4n^{1+r}))^{1/2}$. Let $\hat{K}$ be the number of such eigenvectors. Let*

$$(\mathcal{P}_\epsilon): \quad (\hat{Z}, \hat{X}) \in \underset{Z' \in \hat{\mathcal{Z}}_\epsilon, X' \in \mathbb{R}^{\hat{K} \times \hat{K}}}{\arg\min} \|Z'X' - \hat{U}\|_F^2.$$

Assume that $\frac{\alpha_n}{\log n} \to \infty$ and $\epsilon < \min_k \beta_{\mathbb{1}_{\{k\}}}$. There exists two constants $C_1 > 0$ and $C_0(\eta) > 0$ such that if α_n is larger than $(1 + r)(C_0(\eta)/(L_{\max}\mu_0^2))\log(n)$, then, for n large enough, with probability larger than $1 - n^{-r}$, $\hat{K} = K$ and

$$\frac{\mathrm{MisC}(\hat{Z}, Z)}{n} \leq C_1 \frac{K^2 L_{\max}}{d_0^2 \mu_0^2} \frac{\log(4n^{1+r})}{\alpha_n}.$$

5.3   Proof of Theorem 2

Let $U \in \mathbb{R}^{n \times K}$ be a matrix whose columns are K independent normalized eigenvectors of A associated to the non-zero eigenvalues. The proof strongly relies on the following decomposition of U, that is a consequence of Proposition 2.

Lemma 2. *There exists a matrix $V \in \mathcal{O}_K(\mathbb{R})$ of normalized eigenvectors of the matrix $M_0 = O^{1/2}BO^{1/2}$ such that $U = ZX$ with $X = n^{-1/2}O^{-1/2}V$.*

We state below a crucial result characterizing the sensitivity to noise of the decomposition $U = ZX$ of Proposition 2, in terms of the quantity d_0 introduced in Definition 1. Due to space limitation, its proof is provided in the extended version of this paper. It builds on fact that d_0 provides a lower bound on the norm of some particular linear combinations of the rows of X: indeed, one has

$$\forall z \in \{-1, 0, 1, 2\}^{1 \times K} \setminus \{0\}, \quad \|zX\| \geq d_0/\sqrt{n}.$$

Lemma 3. *(Robustness to Noise). Let $Z' \in \mathbb{R}^{n \times K}$, $X' \in \mathbb{R}^{K \times K}$ and $\mathcal{N} \subset \{1, \ldots, n\}$. Assume that*

1. $\forall i \in \mathcal{N}, \|Z_i'X' - U_i\| \leq \frac{d_0}{4K\sqrt{n}}$
2. *there exists $(i_1, \ldots, i_K), (j_1, \ldots, j_K) \in (\mathcal{N}^K): \forall k \in [1, K], Z_{i_k} = Z_{j_k}' = \mathbb{1}_{\{k\}}$*

Then there exists a permutation matrix P_σ such that for all $i \in \mathcal{N}$, $Z_i = (Z'P_\sigma)_i$.

Let $\hat{U}$ the matrix defined in Theorem 2. From Lemma 1 and the fact that $d_{\max} = \alpha_n L_{\max}$ (by definition) and $\lambda_{\min}(A) = \alpha\mu_0$ (by Proposition 2), there exists $C_0(\eta) > 0$ and $\hat{P} \in \mathcal{O}_K(\mathbb{R})$ such that if $\alpha_n \geq (1 + r)(C_0(\eta)/(L_{\max}\mu_0^2))\log(n)$, then with probability larger than $1 - n^{-r}$, $\hat{K} = K$ and

$$||\hat{U} - U\hat{P}||_F^2 \leq 32\left(1 + \frac{\eta}{\eta+2}\right)\frac{L_{\max}}{\mu_0^2}\left(\frac{\log(4n^{1+r})}{\alpha_n}\right). \tag{5}$$

In the sequel, we assume that $\hat{K} = K$ and that this inequality holds.

The estimate $\hat{Z}, \hat{X}$ is then defined by

$$(\hat{Z}, \hat{X}) \in \underset{Z' \in \mathcal{Z}_\epsilon(K), X' \in \mathbb{R}^{K \times K}}{\operatorname{argmin}} ||Z'X' - \hat{U}||_F^2.$$

Let $\hat{X}_1 := \hat{X}\hat{P}^{-1}$ and (Z, X) as in Lemma 2. As $\epsilon < \min_k \beta_{1_{\{k\}}}$, one has $Z \in \mathcal{Z}_\epsilon$. Thus, using notably the above definition of $(\hat{Z}, \hat{X})$,

$$||\hat{Z}\hat{X}_1 - U||_F \leq ||\hat{Z}\hat{X}\hat{P}^{-1} - \hat{U}\hat{P}^{-1}||_F + ||\hat{U}\hat{P}^{-1} - U||_F = ||\hat{Z}\hat{X} - \hat{U}||_F + ||\hat{U} - U\hat{P}||_F$$
$$\leq ||ZX\hat{P} - \hat{U}||_F^2 + ||\hat{U} - U\hat{P}||_F = 2||U\hat{P} - \hat{U}||_F. \tag{6}$$

We now introduce the set of nodes

$$\mathcal{N}_n = \left\{i : ||\hat{Z}_i\hat{X}_1 - U_i|| \leq \frac{d_0}{4K\sqrt{n}}\right\}$$

and show that Assumptions 1 and 2 in Lemma 3 are satisfied for this set and the pair $(\hat{Z}, \hat{X}_1)$, under the condition

$$64K^2/d_0^2||\hat{U} - U\hat{P}||_F^2 \leq \epsilon. \tag{7}$$

By definition of $\mathcal{N}_n$, Assumption 1 holds. We show that, as required by assumption 2., $\mathcal{N}_n$ contains one pure node in each community relatively to Z and $\hat{Z}$.

First, using notably (6), the cardinality of $\mathcal{N}_n^c$ is upper bounded as

$$\frac{|\mathcal{N}_n^c|}{n} = \frac{\sum_{i \in \mathcal{N}_n^c} 1}{n} \leq \frac{16K^2}{d_0^2}\sum_{i=1}^n ||\hat{Z}_i\hat{X}_1 - U_i||^2 \leq \frac{64K^2}{d_0^2}||\hat{U} - U\hat{P}||_F^2.$$

If (7) holds, $|\mathcal{N}_n^c| \leq \epsilon n$. As $\hat{Z} \in \mathcal{Z}_\epsilon(K)$, for all $k \leq K$ the cardinality of the set of nodes i such that $\hat{Z}_i = 1_{\{k\}}$ is strictly larger than ϵn, hence this set cannot be included in $\mathcal{N}_n^c$. Thus, for all k, there exists $j_k \in \mathcal{N}_n$ such that $\hat{Z}_{j_k} = 1_{\{k\}}$. As ϵ is smaller than $\min_k \beta_{1_{\{k\}}}$, the minimal proportion of pure nodes in a community, by a similar argument the set of nodes i such that $Z_i = 1_{\{k\}}$ cannot be included in $\mathcal{N}_n^c$ either. Thus for all k, there exists $i_k \in \mathcal{N}_n$ such that $Z_{i_k} = 1_{\{k\}}$.

Applying Lemma 3, there exists $\sigma \in \mathfrak{S}_K$ such that $\forall i \in \mathcal{N}_n$, $\hat{Z}_{i,\sigma(k)} = Z_{i,k}$: up to a permutation of the community labels, all the communities of nodes in $\mathcal{N}_n$

are recovered. This implies that whenever $\alpha_n \geq (1+r)(C_0(\eta)/(L_{\max}\mu_0^2))\log(n)$, with probability larger than $1 - n^{-r}$,

$$\frac{\text{MisC}(\hat{Z}, Z)}{n} \leq \frac{|\mathcal{N}_n^c|}{n} \leq \frac{64K^2}{d_0^2}||\hat{U} - U\hat{P}||_F^2 \leq \frac{2048K^2L_{\max}}{d_0^2\mu_0^2}\left(1+\frac{\eta}{\eta+2}\right)\frac{\log(4n^{1+r})}{\alpha_n},$$

provided that the final upper bound is smaller that ϵ (which implies that the condition (7) is satisfied), which is the case for n large enough.

6 Experimental Results

6.1 Simulated Data

We compare SAAC to (normalized) spectral clustering using the adjacency matrix, referred to as SC and to the spectral algorithm proposed by Zhang et al. [19] to fit the random graph model called OCCAM. We refer to this algorithm as the OCCAM spectral method. First, we generate networks from SBMO models with $n = 500$ nodes, $K = 5$ communities, $\alpha_n = \log^{1.5}(n)$, $B = \text{Diag}([5, 4, 3, 3, 3])$ and Z drawn at random in such a way that each community has a fraction of pure nodes equal to p/K for some parameter p and the size of the maximum overlap $O_{\max}$ is smaller than 3. Figure 1 (left) shows the error of each method as a function of p, averaged over 100 networks. SAAC significantly outperforms OCCAM, especially when there is a large overlap between communities. As expected, both methods outperform SC, which is designed to handle non-overlapping communities, except when the amount of overlap gets really small. To have a more fair comparison, we also draw networks under a modified version of the model used before, in which the rows of Z are normalized, so that for all i, one has $||Z_i|| = 1$, which is a particular instance of the OCCAM (right). The OCCAM spectral algorithm, designed to fit this model, performs slightly better than the other methods, but the gap between OCCAM and SAAC is very narrow.

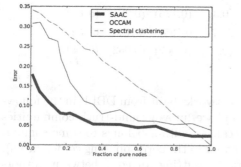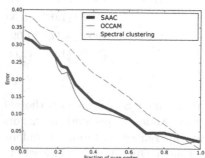

Fig. 1. Comparison of SC, SAAC and the OCCAM spectral algorithm under instances of SBMO (left) and OCCAM (right) random graph models.

6.2   Real Networks

The OCCAM spectral algorithm was shown to outperform other existing algorithms on both simulated data and real data, namely ego networks [9]. Nodes in an ego network are the set of friends of a given central node in a social network, and edges indicate friendship relationships between these nodes. We first apply SAAC on networks from this dataset, that naturally contain overlap. To do so, we use the pre-processing of the networks described in [19], that especially keeps communities if they have at least a fraction of pure nodes equal to 10 % of the network. Additionally, because the focus is on overlapping communities, we keep only networks for which the fraction of nodes that belong to more than one community is larger than 1 %. This leads us to keep only 6 (out of 10) Facebook networks (labeled 0, 414, 686, 698, 1912 and 3437 in the dataset) and 26 (out of 133) Google Plus networks from the original dataset.

Table 1 presents the characteristics of the Facebook networks used, and the performance of SC, OCCAM and SAAC averaged over the 6 networks used (with the standard deviation added). For each algorithm, the estimation error is displayed but also the fraction of false positive (FP) and false negative (FN) entries in $\hat{Z}$ The parameter c corresponds to the average number of communities per node, $c = \sum_{i,k} Z_{i,k}/n$ and $O_{\max}$ is the maximum size of an overlap. OCCAM and SAAC have comparable performance, but there is no significant improvement over spectral clustering. This can be explained by the fact that the amount of overlap (c) is very small in these datasets. The same tendency was observed on the Google Plus networks.

Table 1. Spectral algorithms recovering overlapping friend circles in ego-network.

	n	K	c	$O_{\max}$	FP	FN	Error
SC	190	3.17	1.09	2.17	0.200	0.139	0.120
	(173)	(1.07)	(0.06)	(0.37)	(0.110)	(0.107)	(0.083)
OCCAM	190	3.17	1.09	2.17	0.176	0.113	0.127
	(173)	(1.07)	(0.06)	(0.37)	(0.176)	(0.084)	(0.102)
SAAC	190	3.17	1.09	2.17	0.125	0.101	0.102
	(173)	(1.07)	(0.06)	(0.37)	(0.067)	(0.062)	(0.049)

We then try SAAC on co-authorship networks built from DBLP in the following way. Nodes correspond to authors and we fix as ground-truth communities some conferences (or group of conferences): an author belongs to some community if she/he has published at least one paper in the corresponding conference(s). We then build the network of authors by putting an edge between authors if they have published a paper together in one of the considered conferences. We present results for some conferences with machine learning in their scopes: ICML, NIPS, and two theory-oriented conferences that we group together, ALT and COLT. We compare the three spectral algorithms in terms of estimation error and false positive/false negative rates. Results are presented in Table 2, in which the estimated amount of overlap $\hat{c} = \sum_{i,k} \hat{Z}_{i,k}/n$ is also reported. In this case, SAAC and OCCAM significantly outperform SC, although the error is relatively high. The amount of overlap is under-estimated by both algorithms, but SAAC appears to recover slightly more overlapping nodes. The difficulty of recovering communities in that case may come from the fact that the networks constructed are very sparse.

Table 2. Spectral algorithms recovering overlapping machine learning conferences

$\mathcal{C}_1 = \{ICML\}$, $\mathcal{C}_2 = \{ALT, COLT\}$.

$n = 4374$, $K = 2$, $d_{\mathrm{mean}} = 3.8$

	c	$\hat{c}$	FP	FN	Error
SC	1.09	1.	0.39	0.55	0.46
OCCAM	1.09	1.00	0.2	0.34	0.26
SAAC	1.09	1.03	0.21	0.31	0.25

$\mathcal{C}_1, \mathcal{C}_2$ as in left and $\mathcal{C}_3 = \{NIPS\}$

$n = 9272$, $K = 3$, $d_{\mathrm{mean}} = 4.5$

	c	$\hat{c}$	FP	FN	Error
SC	1.22	1.	0.38	0.39	0.39
OCCAM	1.22	1.02	0.25	0.28	0.27
SAAC	1.22	1.04	0.26	0.28	0.27

7   Conclusion

Most existing algorithms for community detection assume non overlapping communities. Although they may in principle be used to detect all *subcommunities* generated by the various overlaps, this is not sufficient to recover the initial communities due to the combinatorial complexity of the corresponding mapping. We have proposed a spectral algorithm, SAAC, that works directly on the overlapping communities, using the specific geometry of the eigenvectors of the adjacency matrix under the SBMO. We have proved the consistency of this algorithm under the SBMO, provided each community has some positive fraction of pure nodes and the expected node degree is at least logarithmic, and tested its performance on both simulated and real data. This work has raised many interesting issues. First, it would be worth relaxing the assumption that each community has some positive fraction of pure nodes. Next, preliminary experiments on simulated data have shown threshold phenomena in the very sparse regime ($\alpha_n = O(1)$) that should be further explored. Finally, the proof of consistency assumes that the underlying (NP-hard) optimization problem is solved exactly while this is not feasible in practice and heuristics need to be applied, like the proposed alternate optimization routine. Understanding the impact of these heuristics on the performance of the algorithm is an interesting future work.

Acknowledgment. The authors acknowledge the support of the French Agence Nationale de la Recherche (ANR) under reference ANR-11-JS02-005-01 (GAP).

References

1. Airoldi, E., Blei, D., Fienberg, S., Xing, E.: Mixed membership stochastic block-models. J. Mach. Learn. Res. **9**, 1981–2014 (2008)
2. Anandkumar, A., Ge, R., Hsu, D., Kakade, S.: A tensor spectral approach to learning mixed membership community models. JMLR **15**(1), 2239–2312 (2014)
3. Arthur, D., Vassilvitskii, S.: k-means++: the advantage of careful seeding. In: Proceedings of the 18th ACM-SIAM Symposium on Discrete Algorithms (2007)
4. Ball, B., Karrer, B., Newman, M.E.: An efficient and principled way for detecting communities in networks. Phys. Rev. E **84**, 036103 (2011)
5. Chatterjee, S.: Matrix estimation by universal singular value thresholding. Ann. Stat. **43**(1), 177–214 (2015)
6. Holland, P.W., Leinhardt, S.: Stochastic blockmodels: first steps. Soc. Netw. **5**(2), 109–137 (1983)
7. Karrer, B., Newman, M.E.: Stochastic blockmodels and community structure in networks. Phys. Rev. E **83**, 016107 (2011)
8. Latouche, P., Birmelé, E., Ambroise, C.: Overlapping stochastic block models with applications to the French political blogoshpere. Ann. Appl. Stat. **5**(1), 309–336 (2011)
9. Mc Auley, J., Leskovec, J.: Learning to discover social circles in ego networks. In: NIPS, vol. 25, pp. 548–556 (2012)
10. Newman, M.E.: Spectral methods for network community detection and graph partitioning. Phys. Rev. E **88**, 042822 (2013)
11. Palla, G., Derényi, I., Farkas, I., Vicsek, T.: Uncovering the overlapping community structure of complex networks in nature and society. Nature **435**, 814–818 (2005)
12. Rohe, K., Chatterjee, S., Yu, B.: Spectral clustering and the high-dimensional stochastic blockmodel. Ann. Stat. **39**(4), 1978–1915 (2011)
13. Shepard, R.N., Arabie, P.: Additive clustering: representation of similarities as combinations of discrete overlapping properties. Psychol. Rev. **86**(2), 87 (1979)
14. Von Luxburg, U.: A tutorial on spectral clustering. Stat. Comput. **17**, 395–416 (2007)
15. Xie, J., Kelley, S., Szymanski, B.: Overlapping community detection in networks: state of the art and comparative study. ACM Comput. Surv. **45**, 43 (2013)
16. Yang, J., Leskovec, J.: Community-affiliation graph model for overlapping community detection. In: IEEE International Conference on Data Mining (2012)
17. Zelnik-Manor, L., Perona, P.: Self-tuning spectral clustering. In: Advances in Neural Information Processing Systems (2004)
18. Zhang, S., Wang, R.-S., Zhang, X.-S.: Identification of overlapping community structure in complex networks using fuzzy c-means clustering. Phyisca A **374**, 483–490 (2007)
19. Zhang, Y., Levina, E., Zhu, J.: Detecting overlapping communities in networks with spectral methods (2014). arXiv:1412.3432v1
20. Zhao, Y., Levina, E., Zhu, J.: Consistency of community detection in networks under degree-corrected stochastic block models. Ann. Stat. **40**(4), 2266–2292 (2012)

Author Index

Printed in the United States
by Bookmasters

Printed in the United States
By Bookmasters